Genetic Sight: Plant Traits during Postharvest

Guest Editor

Shimeles Tilahun

MDPI

Basel • Beijing • Wuhan • Barcelona • Belgrade • Novi Sad • Cluj • Manchester

Guest Editor
Shimeles Tilahun
Agriculture and Life Sciences
Research Institute
Kangwon National University
Chuncheon
Republic of Korea

Editorial Office
MDPI AG
Grosspeteranlage 5
4052 Basel, Switzerland

This is a reprint of the Special Issue, published open access by the journal *Current Issues in Molecular Biology* (ISSN 1467-3045), freely accessible at: https://www.mdpi.com/journal/cimb/special_issues/plant_traits_genetic.

For citation purposes, cite each article independently as indicated on the article page online and as indicated below:

Lastname, A.A.; Lastname, B.B. Article Title. *Journal Name* **Year**, *Volume Number*, Page Range.

ISBN 978-3-7258-3677-2 (Hbk)
ISBN 978-3-7258-3678-9 (PDF)
https://doi.org/10.3390/books978-3-7258-3678-9

Contents

current issues in molecular biology

MDPI

Editorial

Editorial for the Special Issue "Genetic Sight: Plant Traits during Postharvest"

Shimeles Tilahun [1,2]

1 Agriculture and Life Science Research Institute, Kangwon National University, Chuncheon 24341, Republic of Korea; shimeles@kangwon.ac.kr
2 Department of Horticulture and Plant Sciences, Jimma University, Jimma 378, Ethiopia

Modern breeding alternatives are less costly and sustainable solutions to increase quality, resistance to biotic and abiotic stresses, and to reduce postharvest losses of crops. Omic technologies, namely genomics, transcriptomics, proteomics, metabolomics, and phenomics, provide an efficient way to develop better cultivars. With this regard, review of the established omics technologies for fodder quality improvement through the improvement of the forage nutrition quality, edible quality, and digestibility is included in this special issue [1]. Similarly, studies of genetic relationships among the family, genera or cultivars of crops through molecular markers could assist in developing resistant cultivars to diseases and pests [2], to select super-earliness QTLs for prevention of heat-induced drought stress [3], and assist in seed production through profiling the pattern of protein expression under different crossing periods [4].

As the scope of research has expanded, studies have increasingly focused on the molecular mechanisms regulating a specific trait. This special issue incorporated about the regulation of flower development by CYC-like genes with their different functions and phylogenetic relationships in plant groups [5], function of DUF26 domain-containing genes to regulate the submergence tolerance of wild rice [6], and LbPYLs was suggested as good candidates to enhance Lycium resistance to drought and hot environments [7].

Considering postharvest qualities, candidate genes of plum associated with pulp color, anthocyanin biosynthesis and flavonoid biosynthesis [8], anthocyanin biosynthesis related MYBs in highbush blueberry [9], the potential genetic loci controlling quality traits of melons [10], and genes responsible for softening and ripening in kiwifruit cultivars treated with ethylene [11] were reported.

Generally, as this special issue belongs to the section "Molecular Plant Sciences" and considering the interest of authors to contribute in this special issue, the editorial team tried to observe the broad spectrum, and eleven research articles [2–4,6–13] and two reviews [1,5] were included.

Finally, the Guest Editor would like to thank all authors for their excellent contributions and the Editorial Board for their support.

Funding: This research received no external funding.

Conflicts of Interest: The author declares no conflict of interest.

Citation: Tilahun, S. Editorial for the Special Issue "Genetic Sight: Plant Traits during Postharvest". *Curr. Issues Mol. Biol.* **2023**, *45*, 3515–3516. https://doi.org/10.3390/cimb45040229

Received: 13 April 2023
Accepted: 17 April 2023
Published: 18 April 2023

References

1. Kumar, P.; Singh, J.; Kaur, G.; Adunola, P.M.; Biswas, A.; Bazzer, S.; Kaur, H.; Kaur, I.; Kaur, H.; Sandhu, K.S.; et al. OMICS i Fodder Crops: Applications, Challenges, and Prospects. *Curr. Issues Mol. Biol.* **2022**, *44*, 5440–5473. [CrossRef] [PubMed]
2. Szwarc, J.; Niemann, J.; Kaczmarek, J.; Bocianowski, J.; Weigt, D. Genetic Relationship of Brassicaceae Hybrids with Variou Resistance to Blackleg Is Disclosed by the Use of Molecular Markers. *Curr. Issues Mol. Biol.* **2022**, *44*, 4290–4302. [CrossRef] [PubMed]
3. Sari, H.; Eker, T.; Tosun, H.S.; Mutlu, N.; Celik, I.; Toker, C. Mapping QTLs for Super-Earliness and Agro-Morphological Traits i RILs Population Derived from Interspecific Crosses between *Pisum sativum* × *P. fulvum*. *Curr. Issues Mol. Biol.* **2023**, *45*, 663–67 [CrossRef] [PubMed]
4. Malik, A.; Mor, V.S.; Punia, H.; Duhan, D.S.; Tokas, J.; Bhuker, A.; Alyemeni, M.N.; Shakoor, A. Development and Optimization o Label-Free Quantitative Proteomics under Different Crossing Periods of Bottle Gourd. *Curr. Issues Mol. Biol.* **2023**, *45*, 1349–137 [CrossRef] [PubMed]
5. Chai, Y.; Liu, H.; Chen, W.; Guo, C.; Chen, H.; Cheng, X.; Chen, D.; Luo, C.; Zhou, X.; Huang, C. Advances in Research on the Regulation of Floral Development by CYC-like Genes. *Curr. Issues Mol. Biol.* **2023**, *45*, 2035–2059. [CrossRef] [PubMed]
6. Huang, C.; Wang, D.; Chen, H.; Deng, W.; Chen, D.; Chen, P.; Wang, J. Genome-Wide Identification of DUF26 Domain-Containing Genes in Dongxiang Wild Rice and Analysis of Their Expression Responses under Submergence. *Curr. Issues Mol. Biol.* **2022**, *44*, 3351–3363. [CrossRef] [PubMed]
7. He, W.; Liu, M.; Qin, X.; Liang, A.; Chen, Y.; Yin, Y.; Qin, K.; Mu, Z. Genome-Wide Identification and Expression Analysis of the Aquaporin Gene Family in *Lycium barbarum* during Fruit Ripening and Seedling Response to Heat Stress. *Curr. Issues Mol. Biol.* **2022**, *44*, 5933–5948. [CrossRef] [PubMed]
8. Wang, G.; Weng, W.; Jia, Z.; Zhang, J.; Wang, T.; Xuan, J. Identification of Candidate Genes Associated with Pulp Color by Transcriptomic Analysis of 'Huaxiu' Plum (*Prunus salicina* Lindl.) during Fruit-Ripening. *Curr. Issues Mol. Biol.* **2022**, *44*, 6368–6384. [CrossRef] [PubMed]
9. Zhang, Y.; Huang, D.; Wang, B.; Yang, X.; Wu, H.; Qu, P.; Yan, L.; Li, T.; Cheng, C.; Qiu, D. Characterization of Highbush Blueberry (*Vaccinium corymbosum* L.) Anthocyanin Biosynthesis Related MYBs and Functional Analysis of VcMYB Gene. *Curr. Issues Mol. Biol.* **2023**, *45*, 379–399. [CrossRef] [PubMed]
10. Zhao, H.; Zhang, T.; Meng, X.; Song, J.; Zhang, C.; Gao, P. Genetic Mapping and QTL Analysis of Fruit Traits in Melon (*Cucumis melo* L.). *Curr. Issues Mol. Biol.* **2023**, *45*, 3419–3433. [CrossRef]
11. Choi, H.R.; Baek, M.W.; Jeong, C.S.; Tilahun, S. Comparative Transcriptome Analysis of Softening and Ripening-Related Genes in Kiwifruit Cultivars Treated with Ethylene. *Curr. Issues Mol. Biol.* **2022**, *44*, 2593–2613. [CrossRef] [PubMed]
12. Mardini, M.; Ermolaev, A.; Khrustaleva, L. Hidden Pitfalls of Using Onion Pollen in Molecular Research. *Curr. Issues Mol. Biol.* **2023**, *45*, 1065–1072. [CrossRef] [PubMed]
13. Ma, X.; Yu, Y.; Hu, Z.; Huang, H.; Li, S.; Yin, H. Characterizations of a Class-I BASIC PENTACYSTEINE Gene Reveal Conserved Roles in the Transcriptional Repression of Genes Involved in Seed Development. *Curr. Issues Mol. Biol.* **2022**, *44*, 4059–4069. [CrossRef] [PubMed]

current issues in molecular biology

MDPI

Review

OMICS in Fodder Crops: Applications, Challenges, and Prospects

Pawan Kumar [1,2,†], Jagmohan Singh [3,4,†], Gurleen Kaur [5], Paul Motunrayo Adunola [5], Anju Biswas [6], Sumandeep Bazzer [7], Harpreet Kaur [8], Ishveen Kaur [9], Harpreet Kaur [10], Karansher Singh Sandhu [11], Shailaja Vemula [12], Balwinder Kaur [13], Varsha Singh [14,*] and Te Ming Tseng [14,*]

1 Agrotechnology Division, Council of Scientific and Industrial Research-Institute of Himalayan Bioresource Technology, Palampur 176061, India
2 Department of Genetics and Plant Breeding, CCS Haryana Agricultural University, Hisar 125004, India
3 Division of Plant Pathology, Indian Agricultural Research Institute, New Delhi 110012, India
4 Krishi Vigyan Kendra, Guru Angad Dev Veterinary and Animal Science University, Barnala 148107, India
5 Horticultural Sciences Department, University of Florida, Gainesville, FL 32611, USA
6 Agronomy Department, University of Florida, Gainesville, FL 32611, USA
7 Department of Agronomy, Horticulture, and Plant Science, South Dakota State University, Brookings, WA 57007, USA
8 Department of Plant and Environmental Sciences, New Mexico State University, Las Cruces, NM 88001, USA
9 Department of Biological Sciences, Auburn University, Auburn, AL 36849, USA
10 Department of Agricultural and Environmental Sciences, Tennessee State University, Nashville, TN 37209, USA
11 Department of Crop and Soil Sciences, Washington State University, Pullman, WA 99163, USA
12 Agronomy Department, UF/IFAS Research and Education Center, Belle Glade, FL 33430, USA
13 Department of Entomology, UF/IFAS Research and Education Center, Belle Glade, FL 33430, USA
14 Department of Plant and Soil Sciences, Mississippi State University, Starkville, MS 39759, USA
* Correspondence: vv219@msstate.edu (V.S.); tt1024@msstate.edu (T.M.T.)
† These authors contributed equally to this work.

Citation: Kumar, P.; Singh, J.; Kaur, G.; Adunola, P.M.; Biswas, A.; Bazzer, S.; Kaur, H.; Kaur, I.; Kaur, H.; Sandhu, K.S.; et al. OMICS in Fodder Crops: Applications, Challenges, and Prospects. *Curr. Issues Mol. Biol.* **2022**, *44*, 5440–5473. https://doi.org/10.3390/cimb44110369

Academic Editor: Vijai Bhadauria

Received: 26 September 2022
Accepted: 31 October 2022
Published: 3 November 2022

Abstract: Biomass yield and quality are the primary targets in forage crop improvement programs worldwide. Low-quality fodder reduces the quality of dairy products and affects cattle's health. In multipurpose crops, such as maize, sorghum, cowpea, alfalfa, and oat, a plethora of morphological and biochemical/nutritional quality studies have been conducted. However, the overall growth in fodder quality improvement is not on par with cereals or major food crops. The use of advanced technologies, such as multi-omics, has increased crop improvement programs manyfold. Traits such as stay-green, the number of tillers per plant, total biomass, and tolerance to biotic and/or abiotic stresses can be targeted in fodder crop improvement programs. Omic technologies, namely genomics, transcriptomics, proteomics, metabolomics, and phenomics, provide an efficient way to develop better cultivars. There is an abundance of scope for fodder quality improvement by improving the forage nutrition quality, edible quality, and digestibility. The present review includes a brief description of the established omics technologies for five major fodder crops, i.e., sorghum, cowpea, maize, oats, and alfalfa. Additionally, current improvements and future perspectives have been highlighted.

Keywords: alfalfa; cowpea; genomics; maize; metabolomics; oats; phenomics; proteomics; sorghum; transcriptomics

1. Introduction

Fodder, also known as hay, silage, or forage, is any crop or crop by-product used as feed for livestock, making it an essential source of protein and fat [1]. In crop production, fodder is the end product, the quality of which significantly affects the livestock. The by-products of crops, including cereals, legumes, grasses, and other crops, contribute considerably to fodder. Continuous improvement in animal breeds requires high-quality fodder to feed them.

Biomass yield is the primary target in forage crop improvement programs worldwide. Cultivars or varieties with rapid growth and regrowth are targeted in breeding programs. Therefore, traits such as heading date, flowering time regulation [2–4], and delayed senescence [5] have been targeted for increasing biomass yields. Throughout the world, most fodder (up to 70%) is produced on cultivable land, and the major problem in silage production is the lack of protein in fodder [6]. The increasing population of humans and cattle will create and increase competition in future food and forage production and affect natural resources, particularly land and water [7].

With the increasing challenges in agriculture, we also have great opportunities, especially with booming advances in omic technologies. Advancements in biotechnology and computational sciences have made it feasible to generate omics data for large sets of plants, varieties, or species at a reasonable price [8]. Handling and analyzing large and complex omics data has become possible with the availability of advanced computation and bioinformatics tools. The use of different omics led to the identification of genes, their functions, type of RNA or protein involved, their structure, and pathway involved in the appearance of the final morphological character [9] (Figure 1). Comparative omic analysis of different regions of varying environmental conditions makes it possible to identify genes essential for adaptation [10]. Identified genes can be manipulated or transferred to develop new hybrids or varieties with desirable characteristics.

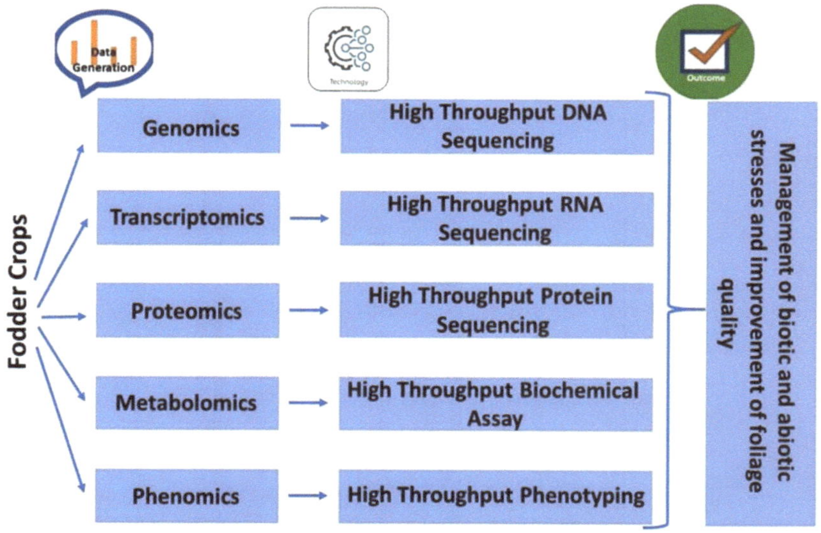

Figure 1. Multiple omics approach for fodder quality improvement in crop plants.

The integration of multi-omics has been successfully performed for yield increment and biotic and abiotic stress tolerance/resistance development in agricultural crops (Figure 2). However, in fodder crops, in the recent past, a handful of studies have been conducted in crop plants using molecular biology approaches of different omic technologies, including genomics, transcriptomics, proteomics, metabolomics, and phenomics [11–16], to achieve higher yields along with quality fodder (Figure 3).

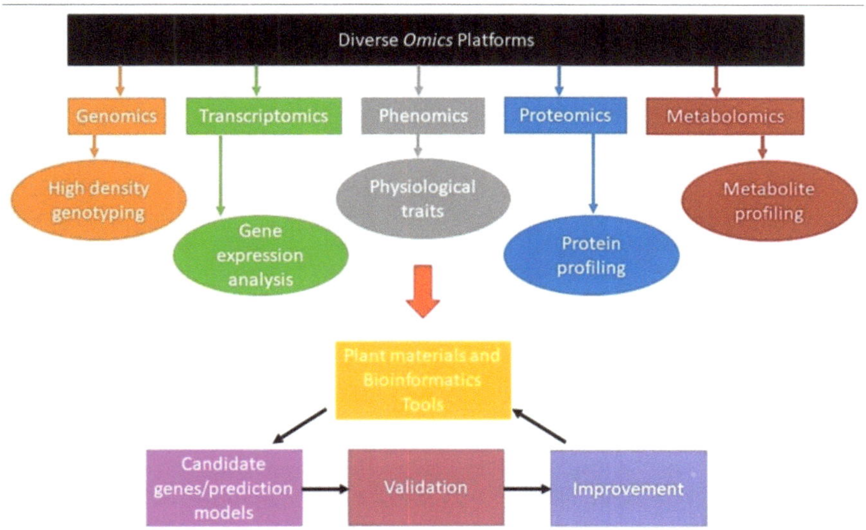

Figure 2. Graphical representation of the application of different omics approaches and their use in crop improvement.

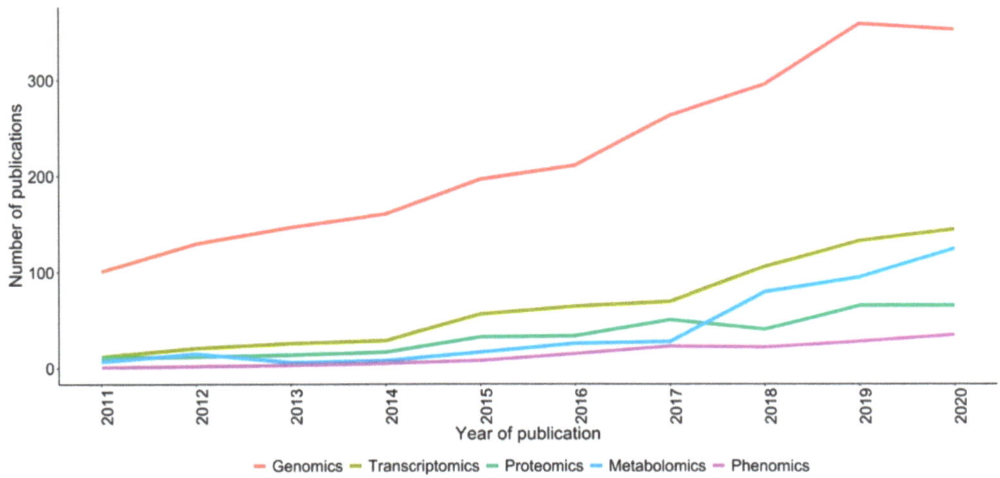

Figure 3. Graphical presentation showing the trend of publications mentioning omic approaches to improve fodder crops in the last decade. The search was made using the associated omics technique and the crop name keyword in the abstract. Source: PubMed, dated 5 August 2022.

Genomics aims to sequence, characterize, and study the genetic compositions, structures, organizations, functions, and networks of an entire plant genome [17]. Genomics approaches are beneficial when dealing with complex traits, as these traits usually have a multi-genic nature and a significant environmental influence [18]. Advances in plant genomics have given new methodologies for plant breeding, which have improved and accelerated the breeding process in many ways (e.g., association mapping, marker-assisted selection, 'breeding by design,' gene pyramiding, genomic selection, etc.) [19–21]. Transcriptomics utilizes high-throughput sequencing platforms to generate transcript data from RNA sequencing, microarray, and serial analysis of gene expression (SAGE) to elucidate non-coding and coding RNA expression profiles to plant biotic and abiotic stresses [22].

Proteomics is the comprehensive study of all proteins in a complex biological system (plants, animals, and humans) at a particular time snapshot [23]. Proteomic analysis is required to estimate the abundance of different proteins, changes observed due to various post-translational modifications, their subsequent function, and localization [24]. It provides a snapshot of different metabolic processes, their consequent interactions, and their effects on other regulatory pathways of biological processes. Hence, a proteomic study is essential to decipher different reactions of pathways at different stresses and times [25].

Metabolomics is an advanced biotechnique that identifies functionally active metabolites, their roles, and the diverse biochemical processes that the metabolites play in plant genotypes and phenotypic expressions [26]. It is commonly applied to explore different aspects of plant breeding, the regulatory mechanisms related to plant growth and development (such as those related to crop productivity and performance), adaptation to biotic and abiotic stresses, nutritional improvement, and the selection of cultivars for agriculture. Metabolomes are simply metabolites (both secondary and primary) having a low molecular weight (usually <1500 Da), including their precursors and intermediates, of the corresponding biosynthetic pathways. Understanding the plant metabolomic processes would be beneficial for improving crop yield and human nutritional aspects in crop-breeding programs [27]. Based on the purpose of the study, metabolomics can be differentiated into two types, targeted and untargeted. Targeted metabolomics deals with the absolute quantification of one or a few metabolites in a set of predefined known substances. Therefore, the targeted approach tends to be highly sensitive and quantitative and can help discover the metabolites associated with specific stress. Targeted metabolomics measures the relative abundances of several hundred to thousands of all detectable metabolites. The untargeted approach, on the contrary, measures mass spectrometric features of unknown metabolites and thus enhances the chances of sensing unintended effects.

All genomics techniques (linkage and association mapping, genome-wide association studies, marker-assisted selection, genome-assisted selection, haplotype-based breeding, etc.) require accurate phenotyping information for selecting a breeding program [27]. Collecting accurate phenotyping information is challenging in large-scale breeding programs, particularly those that screen thousands of genotypes per year at multiple locations. The condition is further worsened for the fodder crops where multi-cuts are performed every season [16]. Recent developments in phenomics have lagged behind genomics and transcriptomics approaches; however, the last decade (2010–2019) witnessed the growth of various phenomics tools for deployment under field and controlled conditions [28]. These phenomic tools have shown the potential to phenotype biotic and abiotic stress and agronomic and physiological traits in fodder crops [29,30]. Several aerial and ground-based platforms carrying various imaging sensors and cameras have been used to accurately, precisely, and rapidly measure traits at multiple stages [31]. More information about these sensors, imagers, platforms, and data analysis methods used for fodder crops is provided in the phenomics section below. The current review discusses the progress in maize, sorghum, oat, alfalfa, and cowpea crops concerning their fodder using various omics, bottlenecks, and future prospects.

1.1. Sorghum

Sorghum (*Sorghum bicolor* L.) is a multipurpose crop used around the world as fodder, feed for poultry, grain for human consumption, and an essential source of biofuel production [32]. Sorghum is resistant to drought and waterlogging and is grown in different soil conditions. Being a water stress-tolerant crop, it can be cultivated in semi-arid areas with low precipitation. Its origin is marked in Africa; being the center of origin, it sustains the highest genetic diversity in sorghum genotypes. The vast diversity of sorghum germplasm is maintained at various centers, such as ~40,000 accessions at the USDA-ARS-National Plant Germplasm System, ~38,000 at ICRISAT, and >16,000 at the National Crop Gene Bank of China [33]. Sorghum ($2n = 20$) is a potential model plant for C4 type plant studies due to its small genome size as compared to other C4 plants such as sugarcane (10 Gbp) and maize (2.5 Gbp) [34]. The first whole-genome sequence of sweet sorghum was generated through the long-read sequencing platform of pacific biosciences [35].

The GWAS studies based on 245 accessions genotyped by 85,585 SNPs and phenotyped under four different environments identified 42 SNPs related to fodder quality traits (crude protein, acid detergent fiber, neutral detergent fiber, cellulose, and hemicellulose) by using mrMLM software [14]. The GWAS panel with 265,487 SNPs is used to detect the inflorescence and plant height traits [36,37], heat and cold stress [38], and disease resistance for anthracnose [39], stalk rot [40], and gray mold disease [41]. The biomass accumulation in the stem is essential in fodder sorghum, which marks the importance of cell wall biosynthesis. The GWAS analysis by the hidden Markov model elucidated 520 genes belonging to 20 gene families associated with cell wall polymers such as pectin, cellulose, and hemicelluloses [42]. The sweet sorghum, with better sugar accumulation in its stem, provides good succulent fodder for animals. The GWAS and QTL mapping identified various QTLs associated with sugar accumulation. The GWAS study reveals the high stem juiciness, juice volume, and sugar yield as compared to grain sorghum cultivars, but with no significant variation in Brix value (sugar concentration) [43]. A QTL responsible for crude protein, acid detergent fiber, and neutral detergent fiber was elucidated [14]. The four QTLs regulating plant height by determining internode length were reported as *dw1, dw2, dw3,* and *dw4* [44] The *dw1* (membrane protein) and *dw2* (protein kinase) are extensively used in sorghum breeding programs [45,46]. The brown mid-rib trait is characterized by low lignin content, and better biomass is of great importance for fodder sorghum. The four alleles, *bmr2*, *bmr6*, *bmr12*, and *bmr19*, governed the brown mid-rib traits, and the bmr12 class reduced lignin and enhanced digestion efficiency [47]. The tb1 gene, a transcription factor influenced by phytochrome B, is involved in sorghum tillering [48]. In another study, the heritability of forage traits such as plant height, leaf length, leaf width, dry weight, number of leaves, green fodder, and dry weight was studied, and the maximum heritability of dry weight was 98.6%, followed by 77.64% for leaf width. Leaf width has a positive correlation and leaf length has a negative correlation with the number of leaves, dry weight, and green fodder [49]. Another important trait of sorghum, i.e., the stay-green feature, was investigated in different sorghum genotypes to determine which genes contribute to this feature. This study identified four main QTLs, Stg1–Stg4, that contribute to the stay-green property in sorghum. Identifying these QTLs is a foundation for further research into stay-green physiology, QTL interaction, and map-based cloning of the genes that drive the stay-green phenotype [50].

The transcriptomic analysis of two drought-challenged sorghum cultivars from plant emergence to post-anthesis, with 400 transcriptomes, revealed the temporal transition in gene expression patterns in leaves and root tissue. The modulation in drought pathways, the difference in transcript photosynthesis, and relative oxygen species indicate the drought tolerance mechanism in sorghum plants to survive under water stress [51]. The dynamics of root architecture concerning apoplastic barriers, such as casparian strips and suberin lamellae, which help survival in barren saline soil, were elucidated by transcriptome analysis of distinct development stages of roots [52]. The transcriptomics studies revealed that sweet sorghum with high sugar content had a better growth rate and showed better biomass product efficiency under abiotic stress such as salinity and water lodging [53]. Another expression study of sweet sorghum (KIT1) and grain sorghum (Razinieh) elucidated that the expression of sugar transporters such as SbSUT1, SbSUT2, and SbSUT6 is higher in the sweet genotypes, and their expression is significantly enhanced when subjected to saline conditions [54]. The transcriptome analysis of grain sorghum, sweet sorghum, and a cross between the two (as line R9188) revealed a specific carbon allocation trend during sugar accumulation in its stem. The line R9188 had starch metabolism and sucrose level, whereas its parent seed sorghum had fully activated sucrose and starch metabolism with high sugar concentration [43].

Proteomic investigation helps identify potential markers among genotypes that can be utilized in breeding programs. Although the sorghum genome sequence was published five years ago, proteomics of sorghum under drought and salt stress is still in its early stages, and there is limited research [55]. Drought tolerance is influenced by stress perception, signal transduction pathways, and changes in gene expression, all of which affect plant physiology and metabolism [56]. Pre- and post-flowering drought stress in sorghum has drastically reduced grain yield. In a study, post-flowering drought stress in sorghum leaf tissue was examined using two-dimensional gel electrophoresis and matrix-assisted laser desorption ionization-time of flight mass spectrometry (MALDI-TOF-MS). This study concludes that the plant increases the activity of key proteins as a defensive mechanism in the sorghum plant to overcome drought stress [57]. Protein expression patterns and physiological analysis of sorghum root to Polyethylene Glycol (PEG)-induced drought stress at the seedling stage. This study achieved protein identification using Coomassie Brilliant Blue-stained 2-DE gels. Out of sixty-five proteins identified, the levels of 43 protein spots increased, and 22 decreased in drought conditions. These proteins are involved in molecular mechanisms such as protein synthesis and carbohydrate and energy metabolism, contributing to drought tolerance in sorghum [10]. In a study, drought-sensitive (ICSB338) and enhanced drought-tolerant (SA1441) sorghum were compared using proteomic techniques. This study showed that drought-tolerant sorghum plants preserve leaf water content through the stomatal shutdown, making them survive under drought-stress conditions compared to drought-sensitive plants [58].

Metabolomic approaches have revealed that drought-stressed plants accumulate a variety of metabolites such as amino acids, organic acids, polyamines, and lipids to protect plant cells against oxidative stresses. In a study by Rajarajan et al. [59], the relative changes of metabolites, such as total carbohydrates, amides, and lipids, were investigated in two sorghum genotypes. The results further revealed a change in these metabolites in the genotypes under drought stress, i.e., DRT1019 had a higher absorbance value than the ICSV95022 genotype. Such alterations in the levels of metabolites in response to drought stress possibly play critical roles in adjusting the cellular metabolism of water-stressed plants. The results of this study provide promising candidate genes for drought tolerance in sorghum that can be used as potential markers for drought tolerance breeding programs. Another recent study on sorghum cultivars highlighted the use of metabolomics to investigate the underlying biochemistry behind changes in seed color [60]. In the past, metabolomics has proven crucial for studying plant × environment interactions and has been applied in the metabolomics-assisted breeding of crops. The metabolomics-based investigation reports on the characterization of multi-parametric metabolic re-programming that underlies the induced defense mechanisms in sorghum plants responding to *Colletotrichum sublineolum* infection. Tugizimana et al. [61] used an LC-MS-untargeted metabolomics approach supported by gene expression analyses, which was aimed at obtaining a comprehensive understanding of the defensive metabolism of sorghum in response to *C. sublineolum* inoculation. Multivariate data analysis identified 72 discriminatory/signatory biomarkers and 23 potential metabolic pathways, with nine being the most significant pathways and collectively defining the metabolic state of the induced resistance in sorghum. The hydrogen cyanide (HCN) in fodder sorghum is an essential trait as it is toxic to animals. The dhurrin [(S)-4-Hydroxymandelnitrile-D-glucopyranoside] in sorghum is metabolized to synthesize HCN. The correlation in transcriptome and metabolome studies undermined the 14 candidate genes of dhurrin metabolism. The p-hydroxybenzaldehyde (pHB) was highest at the early growth stage and reduced with maturity. The candidate genes also showed an expression peak at the seedling stage and decreased at the adult stage [62]. The metabolic profile and expression of dhurrin metabolism genes in sorghum provide the criteria for selecting accession for fodder breeding programs to prevent HCN toxicity in livestock or promote drought tolerance or pathogen resistance.

Furthermore, phenotyping for abiotic stress and biomass accumulation is a tedious and expensive task, but the advent of phenotyping platforms and imaging sensors has opened various new avenues. To maintain stomatal conductance, sorghum can keep its stomata open under different turgor pressures. Canopy temperature depression is a surrogate for stomatal tolerance to select genotypes with better drought tolerance. It has been observed that canopy temperature (CT), water use efficiency, and yield have a high correlation in sorghum, which is possible using several thermal and multispectral imaging techniques [27]. Aerial phenotyping platforms, namely Unmanned Aerial Vehicles (UAVs), satellite imagery, and manned aircraft, have been used to measure biomass, CT, plant height, nitrogen and chlorophyll content, and biotic stresses in sorghum. UAVs are preferred over manned aircraft and satellites due to their high spatial and temporal resolution, low operation and hardware costs, and flexibility in operation. Watanabe et al. [63] used the red, green, and blue (RGB) and near-infrared (NIR) cameras installed on UAVs to measure plant height in sorghum, further deciphering the potential of remote sensing. Moderate to high correlations were observed in manual plant height notes and height data provided by the UAV. The use of robots is still not behind in sorghum, [64] used a low-cost and tracked mobile robot to collect plant height and stem width from each plot using stereo and depth sensors with very low error rates. For bioenergy sorghum, stem characteristics such as standability and juice yield are the key traits; however, their phenotyping is low throughput. Gomez and coworkers [65] developed medical X-ray computed tomography (X-CT) to image morpho-anatomical properties in sorghum to characterize genotypes based on pithiness ratio, stem diameter, and length. The growth angle for nodal roots influences the spatial distribution of mature plants and aids in drought adaptation. Joshi and a group of researchers [66] used the digital images collected in the root chambers using openGelPhoto.tcl software to reveal genetic variation for this trait. Several automated phenotyping platforms are coupled with computational systems to perform data analysis using dimensional reduction, classification algorithms, and a fusion of deep learning models. Most of the studies for sorghum phenotyping are either under controlled conditions or at a small scale; however, in the coming years, we will see the deployment of such tools and methods at a large scale for sorghum breeding.

1.2. *Cowpea*

Cowpea [*Vigna unguiculata* (L.) Walp.] is an herbaceous annual diploid crop ($2n = 22$) belonging to the Fabaceae family with a genome size of ~620 million base pairs [67]. It is mainly grown in the semi-arid tropics in Latin America, Africa, and South Asia [68]. It is a dual-purpose crop as it is an essential contributor to human food and livestock fodder. Being drought-tolerant, cowpea can be grown in semi-arid regions where other legumes do not grow well. It can also grow well in poor soils high in sand content with low organic and phosphorus contents [69,70]. The cowpea is mainly produced in Africa (80%) and some other parts of the world, including Asia, Brazil, and the United States, and all have substantial production [71]. Like other grain legumes, cowpea is a vital food crop in tropical and subtropical countries [72] because of its use mainly as a grain crop, a vegetable, or animal fodder. Although it occupies a smaller proportion of the crop area than cereals, cowpea contributes significantly to household food security in West and Central Africa [73], biotic-abiotic stress [74,75], seed harvests [76], nutrient continent [77], and physiological architecture [78]. Using cowpea as fodder is appealing in mixed crop/livestock practices. Both grain and fodder can be harvested from the same crop. It can also be used as an intercrop with maize, sorghum, sugarcane, cotton, and other crops [79]. It enhances the soil by recycling nutrients by fixing nitrogen with nodulating microbes. Cowpea is mainly grown for grains, it is rich in protein, ranging from 20 to 25% dry weight. Dried cowpea leaves, stems, and pod walls (known as haulm) can provide additional income for farmers as a feed resource. Cowpea fodder contains up to 18.6 g of protein per 100 g of dry weight. With all the benefits, cowpea remains a relevant crop for human and livestock nutrition, food security, and income for subsistence farmers [68]. The improvement of cowpea varieties was initially led by the International Institute of Tropical Agriculture (IITA), Nigeria, which developed and distributed improved cowpea varieties. Initially, the variety improvement at IITA focused on grain cowpea cultivars only [80]. With cowpea's importance as a fodder crop, a systematic breeding program was initiated at the IITA to develop dual-purpose cowpea varieties that focused on combining high yield for both, i.e., grain and fodder, and resistance to biotic and abiotic stresses [68]. As cowpea is a highly self-pollinating crop, improved varieties have been developed using pure line selection, mass selection, pedigree breeding, single-seed descent, and backcrossing methods [81]. Recent studies showed that grain and fodder yield in cowpea are positively correlated, so improving the varieties for both traits is possible. However, there is no relationship between grain yield and fodder quality traits [82].

The genomic efforts for cowpea have been more recent and focused on molecular diversity and genetic linkage mapping [81]. Several marker systems such as allozymes, restriction fragment length polymorphism (RFLP) [83], amplified fragment length polymorphisms (AFLP) [84], DNA amplification fingerprinting (DAF) [85], random amplified polymorphic DNA (RAPD) [86,87], simple sequence repeats (SSRs) [88], cross-species SSRs from Medicago [89], inter-simple sequence repeats [90], sequence-tagged microsatellite sites (STMS) [91], and single nucleotide polymorphism (SNP) markers [91,92] have been used to study origins, domestication, and genetic variation in cowpea. A set of more than 1500 single nucleotide polymorphism (SNP) markers on 442 cowpea landraces identified two major gene pools in cultivated cowpea in Africa [91,92]. Genotyping by sequencing (GBS) was applied to identify cowpea SNPs and estimate genetic diversity, population structure, and phylogenetic relationships. Another diverse set of 768 cultivated cowpea genotypes from 58 countries was studied using GBS SNP markers that revealed three gene pools (America, Africa, and Central West Asia) [91,93]. Lastly, a set of 368 cultivated cowpeas genotyped with 51,128 SNPs revealed six major subpopulations [94]. The genetic research on cowpeas was initiated with the development of the first linkage map for cowpeas using a population of 58 F2 plants from a cross between IT84S-2246-4 and TVNu 1963 [83]. A second cowpea genetic linkage map was developed using 94 F8 recombinant inbred lines (RILs) made from a cross between two cultivated cowpea genotypes, IT84S-2049 and 524B [95]. A third genetic map was made using 94 F8 RILs derived from a cross between a cultivated cowpea line, IT84S-2246-4, and a wild relative (*V. unguiculata* spp. *dekindtiana* var. *pubescens*) TVNu 110-3A [96]. After the development of an Illumina GoldenGate Assay, an SNP consensus map was developed with 928 SNP markers covering a total genetic distance of 680 cM was established based on the genotyping of 741 members of six bi-parental RIL populations derived from the following crosses: 524B × IT84S-2049, CB27 × 24-125B-1, CB46 × IT93K-503-1, Dan Ila × TVu-7778, TVu-14676 × IT84S-2246-4, and Yacine × 58–77 [97]. The resolution of this consensus genetic map was improved by genotyping 579 individuals from additional populations consisting of five RILs (from UCR–US, IITA–Nigeria, ISRA–Senegal, ZAAS–China) and two F4 populations [98]. After the availability of linkage maps, opportunities became available for QTL resolution, map-based cloning, genetic diversity, association mapping, and marker-assisted breeding. Synteny has been reported between cowpea and mung bean [99] based on RFLP-derived separate maps of both crops. Lucas et al. [98] also reported that 941 of 1107 total SNP markers, i.e., 85% mapped in cowpea, show homologs with soybean (*Glycine max*). The markers also showed synteny and collinearity in the soybean genome. Advancements in linkage maps led to the identification of QTLs for several desirable traits in cowpeas, including leaf shape, disease and insect resistance, maturity time, seed-related traits, flowering time, and pod-length variation [81,97,100]. With better comprehension of the genomic basis of variation, genome-wide association studies (GWAS) studies have been highlighted on the subjects of cowpea pod length [100], root architecture [101], cowpea plant improvement traits, as well as the flowering period [94]. These findings are appreciated because cowpea genetic diversity assessment is necessary for strengthening breeding programs to develop high-yielding dual-purpose cultivars with good grain and fodder yields [102]. It has been found that cowpeas have 85% macrosynteny with *Glycine max* and 82% with *Medicago truncatula*, which can help in comparative analysis to identify genomic regions for fodder yield and quality [103]. Paudel and coworkers (104) reported candidate genes for flowering time in cowpeas based on GWAS and showed high H2 estimates (0.72–0.95) for flowering time in cowpeas. Extensive collections of diverse cowpea accessions are conserved in the International Institute of Tropical Agriculture (IITA) ($\sim$15,000 accessions), the United States Department of Agriculture–Genetic Resources Information Network (USDA-GRIN) (7737 accessions), and the University of California, Riverside, CA, United States ($\sim$6000 accessions) [104]. Due to resource limitations in characterizing the whole collection, many conserved accessions in gene banks preclude their direct utilization in a breeding program. Therefore, a mini-core collection consisting of 298 lines from the IITA collection was genotyped based on GBS using 2276 SNP markers to make the germplasm's characterization and utilization more practical [105]. Similarly, another mini-core collection, the University of California-Riverside Minicore (UCR Minicore), consisting of 368 accessions that included landraces and breeding materials from 50 countries, was also developed [94] and genotyped using a publicly available Cowpea iSelect Consortium Array [106]. This array consists of 51,128 assays developed from sequencing 36 diverse accessions and was released to facilitate easy, high-throughput genotyping in cowpeas [106]. A few reports of marker-assisted breeding in cowpea improvement, such as marker-assisted backcross, were used to transfer the Striga resistance gene from the breeding line IT93K-693-2 to three farmers' preferred varieties, IT90K-372-1-2, KVx30-309-6G, and TN5-78 [107]. Three significant QTLs for bacterial blight, one on Vu09 (*qtlblb-1*) and two on Vu04 (*qtlblb-2* and *qtlblb-3*), which were responsible for 30.58%, 10.77%, and 10.63% of phenotypic variations, respectively, have been identified [108]. The QTL on Vu09 was introduced from cultivar V-16 into the bacterial leaf blight susceptible variety C-152 through marker-assisted backcrossing (MABC) [109].

The characterization of different parts of the cowpea plant through transcriptomics has been carried out in studies that express the diverse genes essential for cowpea growth and development. The stress-resilient genes have also been characterized, and their role in the overall improvement of cowpea has also been highlighted [110–113]. There has been minimal use of transcriptomics for cowpeas until now for fodder purposes. Genes for cowpea growth, development, and stress-related genes have been characterized using transcriptomics, which has a role in seed and pod development [110]. Marker-assisted breeding was used to introgress large seed haplotypes into a CB27 background with 22 g/100 seeds using a rare haplotype with large seeds at the Css-1 locus from the cowpea variety IT82E-18 (18.5 g/100 seeds) [114]. During two cycles of backcrossing based on genome-wide SNP markers, foreground and background selections resulted in families with very large seeds (28–35 g/100 seeds). Parasitic weeds, *Striga gesnerioides* (present in the dry savannah areas of West and Central Africa) and *Alectra vogelii* (found in eastern and southern Africa) can cause yield reductions of 73 to 100% [110,115]. One Striga plant can produce up to 90,000 seeds, which can be viable in the soil for 15 to 20 years [116]. There are seven races of Striga identified [117]. Race 1, 2, 3, 4, 5, 6, and 7 were found in Burkina Faso, Mali, Nigeria, and Niger, the Republic of Benin, Cameroon, Senegal, and Zakpota in the Republic of Benin, respectively [118]. Three Striga resistance genes, *Rsg-1*, *Rsg-2*, and *Rsg-3*, were identified in two cowpea lines [119]. Two duplicate genes, *Rav-1* and *Rav-2*, control cowpea resistance to Alectra [120]. Genes conferring resistance to Striga and Alectra were not found to be allelic or linked [121]. Aphid (*Aphis craccivora* Koch) is the first primary insect pest that affects cowpea growth early. It damages the seedlings in drought conditions [122]. A dominant gene (*Rac*) was identified in a germplasm line (TVu-3000) for resistance to aphids [123]. However, this gene became ineffective for aphid resistance. Then a cowpea wild relative, TVNu-1158, was found resistant to aphids in the seedling stage [124]. A set of RILs was developed using this wild relative as a parent, and some of the resistant RILs are now being used as parents in breeding programs to transfer resistance to aphids [125]. Recently, three cultivated cowpea accessions, TVu-6464, TVu-1583, and TVu-15445, with varying resistance levels to *A. craccivora* comparable to the level already found in an existing resistant TVu-801, were reported [126]. All these new sources of resistance could be used in pyramiding to develop new aphid-resistant cowpea varieties. The resistance mechanism in these three accessions was linked to low sucrose levels and high levels of kaempferol and quercetin (aglycones of phenolic compounds) [126]. The development and application of genomic tools for cowpea improvement is limited, and only a few relevant studies have been reported. A little progress has been recorded after identifying molecular markers associated with some desirable traits in the crop, but marker application in variety development is still minimal.

After decades of study on cowpea, many omics datasets are now accessible, which can be used to better understand the genetic relationship between *Vigna unguiculata* ssp. *unguiculata* and other species in the same genus, and their genetic variation. Furthermore, the availability of a genetic map and a genetic transformation protocol increased the opportunity to identify more genes and study their expression [127]. Various salt-tolerant cowpea cultivars have different protein profiles and use other techniques to deal with salt stress. Differential proteome responses in two cowpea cultivars to salt stress have been studied. LC–ESI–MS/MS revealed 22 differentially regulated proteins by salt and recovery. This study speculates that tolerance may be associated with maintaining an optimal enzymatic-protein level required for active energy metabolism, such as photosynthesis in salinity or recovery [128]. A similar study was conducted in cowpea to understand the mechanisms involved in drought-tolerant and drought-sensitive genotypes. One hundred eight differentially expressed proteins were identified using 2D E and MS that may be associated with drought response in both genotypes. Glutamine synthetase, CPN60-2 chaperonin, malate dehydrogenase, heat-shock proteins, and rubisco were identified as drought stress-response peptides expressed differentially in both genotypes. This study concluded that most of the proteins identified were related to photosynthesis, the critical mechanism for plant survival [115]. Protein expression patterns associated with embryogenic cell suspension prepared from friable embryogenic calli (FEC) were investigated in cowpea using two-dimensional gel electrophoresis protein mapping and mass spectrometry analysis. A total of 128 protein spots have been identified, including PR-4 (chitinase) and PR-10 (putative ribonuclease), as significant proteins playing a major role in the differentiation of pro-embryogenic masses into somatic embryos and defense against biotic and abiotic stresses in cowpeas [129].

Metabolomics tools can be deployed in identifying and monitoring physiological responses in plants and the metabolic pathways or linkages arising from the biotic and abiotic stress exerted upon plants. A study on the drought response of three cowpea landraces using leaf physiological and metabolite profiling assessment [130] used gas chromatography time of flight mass spectrometry (GC-TOF-MS) and reported that cowpea landrace A116 genotype performed best with the accumulation of 14 bioactive metabolites including proline, valine, rhamnose, raffinose, isoleucine, fucose, urea, alanine, sucrose, and putrescine. Several metabolites such as galactinol, proline, quercetin, rhamnose, and raffinose involved in drought tolerance have been identified using metabolomics tools in cowpeas [130,131]. Another study on metabolites (polyphenols and carotenoids) in *V. unguiculata* sprouts identified and quantitated 39 hydrophilic compounds using high-performance liquid chromatography (HPLC), electrospray ionization-mass spectrometry (ESI-MS), gas chromatography-mass spectrometry (GC–MS), and gas chromatography [132]. This study provides a new approach for enhancing the carotenoid and phenylpropanoid production of *V. unguiculata*. This metabolite profiling approach has been utilized to understand the molecular response of cowpea plants to abiotic stress and will have an enormous impact on crop yields. Similarly, advances in metabolomics could assist in identifying the various metabolites produced in response to heat stress and thus determine the complex signaling networks contributing to heat stress tolerance in cowpea [133]. Therefore, metabolomics could assist in screening cowpea lines for heat tolerance, but such studies are mainly lacking in cowpeas.

Phenotyping is crucial to work on any trait development. High throughput phenotyping can be a game-changer in plant breeding. Biotic-abiotic stress, seed, plant vigor, nutrition, and flowering time are some of the traits of interest of cowpea to improve for either food or fodder purposes. Despite the importance of phenomics, a minimal amount of work has been conducted on phenomics for this crop [74–78,134]. So, we think phenomic-assisted selection will be prospective for this specific crop.

1.3. *Maize*

Maize (*Zea maize*) is a multi-purpose cereal crop grown for food, feed, and industrial purposes [135–137]. About a billion people worldwide consume maize as a staple food [138]. Globally, it is one of the most important cereal crops, and it is adaptable to grow in different environmental conditions, generating billions of dollars in revenue annually [139–141]. Maize has evolved into one of the most essential model crops for understanding the genetics and genomes of plants [142–144]. The complexity of the diverse maize genome has kept researchers interested in this crop studying cytogenetics and genomics. Aside from rice and wheat, maize is the most studied food crop for the elucidation of cells and tissue molecular composition to reveal the functional machinery of the plant genome. There are approximately 30,000 to 40,000 genes in a maize genome of 2.4 billion base pairs. The variation available throughout the world in the maize crop has allowed plant breeders and geneticists to improve this crop continuously.

Birchler [145] constructed an early genetic map for maize on stained chromosomes based on the cytogenetic position of the relative gene. With the development of molecular [146–149] and sequencing technology [150,151], several attempts have been made for maize genome sequencing. Among all those, B73 RefGen_v4 [152] is the most accurate assembly of maize to date. It was based on PacBio sequencing and high-resolution optical mapping. To separate the genetic architecture of complex traits, linkage mapping and association mapping are the most widely used approaches in maize and other crops. Linkage mapping is a conventional approach and has its limitations. In contrast, association mapping fulfills linkage mapping and covers whole genome sequencing [153]. Whole-genome association studies or genome-wide association studies (GWAS) are becoming a standard tool for rapidly uncovering marker-trait associations in crops such as maize, where the generation of high-density markers is feasible and affordable. With the improvement in quality protein and yield of maize continuously for decades, the digestibility of stock has decreased drastically, ultimately making maize unsuitable for fodder. It is either due to the loss of alleles for digestible cell walls during breeding for stalk standability or genetic drift during breeding for grain yield [154]. Some breeders have attempted to improve maize's fodder quality in the recent past, and genes or alleles have been identified for the same. Mutant genes, for example, the BMR, *wx*, Leafy1 (*Lfy1*), and *floury-2* genes, have been used to develop silage hybrids [155]. The brown-midrib 3 mutant (bm3) in maize (*Z. mays* L.), controlled by the caffeic acid O-methyl transferase (COMT) locus, has a positive influence on maize fodder quality [156].

Zein and a group of researchers [154] evaluated European maize lines for the nucleotide diversity and linkage disequilibrium (LD) pattern across 2.3 kb of the caffeic acid O-methyl transferase (COMT) locus. Andersen and coworkers [157] cloned and sequenced a Phenylalanine Ammonia-Lyase (PAL) genomic sequence from 32 maize inbred lines employed in forage maize breeding programs in Europe. Vinayan and a group of researchers [12] identified candidate genomic regions for fodder quality in testcross progenies of tropical origin using GBS and 55K Infinium chip data sets, rendering GBS the preferred technology to explain the phenotypic variance of complex traits. A total of 196 SPNs associated with acid (75), neutral detergent fiber (41), and in vitro dry matter digestibility (83) were identified by Wang and coworkers [158] in the mature stalk of maize, swaying its forage quality. Zhao and a group of researchers [159,160] studied the genetic influence of cadmium (Cd) and lead (Pb) accumulation in grains and leaves, utilizing GWAS and QTL mapping approaches to develop low-Pb-accumulating maize cultivars.

Vinayan and coworkers [161] identified the genomic regions for fodder traits and conducted a genomic prediction study using 276 elite lines and 1026 DH lines from bi-parental crosses as prediction sets. A set of 955,690 SNPs were generated through GBS v2.7.

Various studies have provided insight into the transcript's composition responsible for a differential gene expressed in specific cells or/and under specific cell treatments, causing phenotypic variation in maize growth, architecture, yield component, environmental response, pest and disease tolerance, and quality traits. A meta-analysis of 187 published articles from 2002 to 2022 revealed a consistent increase in the number of studies on transcriptomics in maize, though this has become significant in the past four years (Figure 4). These studies are primarily focused on the use of transcriptomics to identify tolerance/resistance to biotic and abiotic stresses in maize based on keyword search. A transcriptome study of maize root hair revealed that 3% of all genes were expressed in the root, comparable to the 4% found in Arabidopsis. Most of these genes are functionally related to energy metabolism, suggesting a high energy requirement for rapid cell division and root hair functioning [162]. Liu and Zhang's [163] study identified six genes from transcriptome comparisons and correlation signaling network analysis involved in the regulation of the HY5 module and the MAPK cascade in the presence of blue light, controlling the stomata distribution and development in maize. Transcriptome analysis and qRT-PCR validation experiments of maize roots infected with *Holotrichia parallela* larvae established the expression of 12 differentially expressed genes associated with jasmonic acid mediated signaling and benzoxazinoid biosynthesis pathways responsible for root defense mechanisms against attack in maize [164]. Zhou et al. [165] investigated the mechanism of drought stress tolerance in maize using bulked segregant transcriptome analysis (BSTA). They revealed that alternative splicing, transcription regulation, and hormone metabolism are common mechanisms for maize's response to drought stress. In a similar study, Du et al. [166] found *GRMZM2G055704* as a candidate gene controlling waterlogging tolerance in maize from BSTA, qRT-PCR validation, and QTL association studies. Transcriptome profiling of distinct maize inbred lines led to the discovery of highly expressed four candidate genes on chromosome 2, conferring resistance to Gibberella ear rot disease in maize [167]. From some of these results, it is evident that transcriptomics in maize allows for the large-scale identification of critical regulatory elements for tolerance to biotic and abiotic stresses [168–170], gene function annotation [171–173], and candidate gene identification [25,170,174]. This information will provide breeders with much genetic information for developing improved maize varieties considering the prevailing and anticipated economic, ecological, and environmental challenges to ensure food security.

The field of proteomics has garnered the attention of many scientists to analyze the differences in physiological conditions at proteomic levels under different stresses. For instance, to assess the changes at proteome levels in the case of corn infected by the Asian corn borer (*Ostrinia furnacalis*), Zhang et al. [175] conducted the proteomics of maize leaves and observed the presence of 62 defense-responsive proteins, especially pathogenesis-related protein 1 (PR1) and thioredoxin M-type, chloroplastic precursor, which demonstrated significant impact on the growth of larvae and pupae of the corn borer. Yet, in another study, using Isobaric Tag for Relative and Absolute Quantitation (iTRAQ) sets, Wang et al. [176] conducted comparative proteomic profiling of both susceptible and resistant lines against southern corn rust (*Puccinia polysora*) to find out that one specific remorin protein (ZmREM 1.3) is responsible for impeding resistance in resistant lines, information crucial for future breeding programs. Similarly, differentially expressed proteins have also been observed to protect corn against abiotic stress. For instance, drought stress during the grain filling period drastically reduces the productivity of crops. Hence, in an attempt to tease apart the role of defensive proteins in drought-tolerant varieties, Dong et al. [177] ran the comparative proteomic profiling of both drought-tolerant (ND476) and susceptible (ZX978) lines. They observed that 1655 defense-associated proteins (DAPs) are produced with the downregulation of redundant proteins to help plants save energy and fight stress.

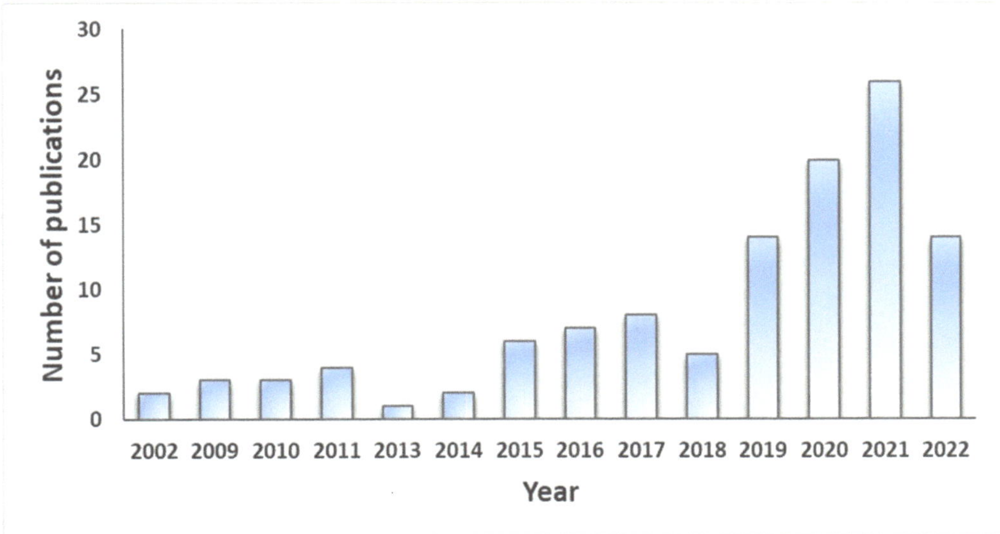

Figure 4. Number of published articles on transcriptomics in maize from 2002 to 2022. A web-search query through Google Scholar was used for a meta-analysis of 187 relevant articles. A keyword search was conducted in the abstract of the transcriptomics studies. Source: Google Scholar, dated 5 May 2022.

The study of metabolomes helps to understand the response of maize plants under different stress conditions such as soil salinity, heat, drought, etc. [136]. A metabolomic study of salt-sensitive (PH4CV) and salt-tolerant (PH6WC) varieties of maize shows the difference in the accumulation of metabolites in roots and different metabolisms in seedlings. Under 100 mM NaCl conditions, glucose metabolism is induced in seedlings of the PH4CV cultivar, whereas significant acid metabolism was induced in seedlings of PH6WC. From the roots, 79 compounds were identified in the salt-sensitive cultivar and 85 were identified in the salt-tolerant cultivar, out of which 30 compounds were common in both cultivars and were associated with the basic metabolism of L-pyroglutamic acid, deoxyadenosine, adenine, cis-9-palmitoleic acid, and galactinol compounds. This knowledge helps us to understand the response of maize seedlings to salt stress [178]. Metabolic pathway analysis also reveals the effect of heat stress on the male sterility of pollen at the most susceptible tetrad stage in maize. The results show a reduction in pyruvate and an increase in sucrose components, whereas various genes associated with auxin production, signaling, and unfolded protein stresses remain unchanged. Genes related to heat stress; metabolic transcriptional regulation pathway altered even though at optimum conditions at which pollen can germinate. This leads to the conclusion that at the tetrad stage of pollen, short heat stress affects the basic metabolic pathways, leading to sterile pollen [179]. A study performed by Ganie et al. [136] in 2015 identified the metabolic pathways under phosphorus stress conditions, which will help develop approaches for increasing phosphorus efficiency. The metabolites were identified using gas chromatography-mass spectroscopy. Under P-limitation, sugar alcohols such as glucitol and mannitol increase, whereas fatty acids such as stigmasterol and cholesterol decrease, which are part of membrane fluidity. In the case of P starvation, plants scavenge the P from these fatty acids and thus disturb the membrane fluidity. It was also shown that the level of serine and glycine increased, which means the rate of photorespiration is also increased [136]. Using the nuclear resonance metabolome technique, the plasticity of leaves in maize plants was studied in response to heat and cold stress. The plastic response of maize plants under heat stress was different than that under cold stress in amino acid derivatives, biomass allocation, and other non-polar metabolites. It was also shown that the metabolic responsiveness in maize lines due to temperature differences was high. In contrast, the functional traits of maize show low plastic responsiveness, which indicates that metabolic and functional plasticity may play different roles in the adaptation of the maize plant to differences in temperature [180]. In another study, metabolic profiles of maize plants through combined effects of different stresses such as salinity, drought, and heat show that the metabolic profile of drought-stressed plants is more like salt-stressed plants than heat-stressed plants. Moreover, it was also found that the metabolic profile of drought-stressed plants, when combined with heat or salt stress, shows a different metabolic profile than either of the individual stresses. This shows the metabolic plasticity in maize to adapt to other environmental conditions. Thus, understanding the metabolic pathways under multi-stress conditions helps further optimize crop breeding for high-yielding plant varieties in changing climatic conditions [135].

The goal of maize forage breeding mainly revolves around yield and its stability, biotic and abiotic stress resistance, wider adaptability with photo and thermo insensitivity, and higher biomass. Due to global dominance, maize is grown under different environmental conditions, which require advanced phenotyping approaches to capture the unexploited variation. With the rise in global temperature, abiotic stresses such as drought are causing a severe threat to maize production. A further hindrance is caused due to high cost, low throughput, and labor-intensive conventional phenotyping. Plant height is usually measured at the end of crop growth and is directly correlated with yield. Adak et al. [181] used the weekly temporal flights and showed that variation captured in plant height at earlier stages is more predominant in predicting yield, and various QTLs are associated with plant height, which was previously assumed to be an oligemic trait. Plant stress phenotyping, especially for biotic and abiotic stress, was recently accomplished using close-range hyperspectral sensors as a promising non-invasive tool. These sensors monitor the physiological and biochemical changes occurring in the plants during different stresses based on the water content, plant organs, photosynthetic apparatus, and internal and external leaf structure. Wu and coworkers [134] performed an extensive study using 368 maize genotypes, where multiple optical images were collected with hyperspectral sensors, color and X-ray computed tomography images over 98 days under drought and controlled conditions. With these vast phenotypic datasets, the authors identified 1529 and 2318 significant QTLs and candidate genes showing drought-tolerant responses. Although various developments have occurred for above-ground phenotyping using proximal and aerial platforms in maize, little is known about the below-ground traits. The complexity of roots can be interfered with by several factors, such as their role in water and nitrogen efficiency, response to biotic/abiotic stresses, and overall plant health.

1.4. Oats

Oat (*Avena sativa* L.) is a nutritionally important cereal crop produced for food, feed, and forage [182,183]. It contains minerals, proteins, fiber, vitamins, lipids, unsaturated fatty acids, as well as other biochemical compounds that play a role in preventing diseases such as colon cancer, type II diabetes, cardiovascular disorders, etc. [184]. Cultivated oat (*Avena sativa* L.) is an allohexaploid ($2n = 6x = 42$; 1C-value = 13.2 (pg) DNA) with three diploid sub-genomes (AA, CC, DD). The oat grain has the highest protein content among cereals, which is approx. 12 to 20%, and lower fat content (<8%) in the groat [185]. Oat is the sixth-ranked cereal crop and has received significant attention for its positive and consistent health benefits and livestock feed [186,187]. It has the benefits of lowering blood cholesterol and reducing the risk of cardiovascular diseases [188] due to the presence of high soluble fiber (β-glucan) and antioxidants such as tocopherol and tocotrienol [189]. The large production is due to the adaptation of this crop to various soil types, where oats perform better than the other small grain cereals, and metabolomics has proven pivotal in studying the adaptive responses of plants to various abiotic and biotic stresses (plant × environment interactions) [190].

Oat genomic research has lagged as compared to the other major crops, such as rice and maize, due to the large size and complexity of its genome, lack of sequence data and sequence redundancy among sub-genomes, numerous chromosomal rearrangements, and chromosome-deficient cytogenetic stocks [191,192]. The genetic research in oat was initiated with the development of the first RFLP map developed by O'Donoughue and co-workers [193] in diploid oat. Different DNA markers such as diversity array technology (DArT) and single nucleotide polymorphism (SNPs) have been used [194] to develop genetic maps using different populations such as 'Kanota × Ogle (K × O)' [193], 'Ogle1040 × TAM O-301 (O × T)' [195], 'Terra × Marion' [196] and 'Ogle × MAM17-5 (O × M)' [197]. The first physical anchored consensus hexaploid oat map based on the previous six populations was developed by combining 985 SNPs and 68 previously published markers and has 21 linkage groups with a total length of 1838 cM [198]. Chaffin and a group of researchers [199] constructed the consensus map for hexaploid oat using cDNA-derived SNPs and GBS from 12 recombinant inbred lines population developed using 19 parents. This consensus map consists of 7202 markers with a total map length of 2843 cM covering 21 consensus chromosomes, which will accelerate the oat genomic research and provide a better understanding of the organization of the oat genome. Recently, in 2021, Pepsico and Corteva Agriscience released the whole genome sequence of hexaploid oat line OT3098, which can be accessed at https://wheat.pw.usda.gov/jb/?data=/ggds/oat-ot3098-pepsico (accessed on 5 May 2022). Advancements in marker technology, along with the development of new maps and integration of maps from multiple populations into a single consensus map, can accelerate oat's genetic and genomic research [200]. These maps have been widely explored in various oat genetic analyses to identify minor quantitative trait loci (QTLs) associated with multiple agronomic traits using experimental bi-parental populations and a diverse panel of oat accessions. In recent years, genome-wide association studies (GWAS) have been reported, primarily focused on biotic stress such as crown rust and quality traits such as β-glucan concentrations [201,202]. Oat crown rust, caused by *Puccinia coronata* f. sp. *avenae*, is a major constraint to oat production worldwide, causing a significant reduction in grain yield, forage, and seed quality [203,204]. Genetic resistance is an effective and economical method of controlling crown rust in oats. Major race-specific resistance genes for crown rust (Pc) have been identified in oat germplasm [205], such as Pc38 on linkage group Mrg02 (chromosome 9D; [206], Pc48 on Mrg20 [206], Pc58a on Mrg02 [207,208], Pc68 on Mrg19 [209], Pc71 on Mrg21 [210], Pc91 on the translocated chromosome 7C-17A [211], and PcKM on Mrg08 [212], but they provided durable resistance only for a short period [213,214]. Alternative strategies such as partial/adult plant resistance can be used to control oat crown rust effectively. Chong [215] identified two genes controlling adult plant crown rust resistance using 157 F7:9 recombinant inbred lines (RILs) developed from an AC Assiniboia × MN841801 cross. Portyanko and co-workers [216] identified four major and three minor QTLs associated with adult plant resistance in a mapping population of MN841801-1 × Noble-2 cross. Acevedo and co-workers [217] also studied crown rust resistance in 150 F6:9 MN841801-1 × Noble-2 RILs and found eight QTLs associated with MN841801-1 alleles. In these studies, adult-plant rust resistance QTLs have been mapped on linkage groups Mrg02, Mrg06, Mrg08, Mrg12, Mrg17, and Mrg20 [216–220]. Association mapping for crown rust resistance has identified QTLs on linkage groups Mrg01, Mrg03, Mrg08, Mrg20, Mrg23, and Mrg28 [221,222] using the Diversity Arrays Technology. Klos and a group of researchers [223] conducted the first multi-environment genome-wide association mapping of crown rust resistance at seedling and adult plant stages using 2972 SNPs genotyped on 631 elite oat lines under both controlled and field conditions. They found 29 SNPs on 12 linkage groups associated with crown rust reaction in at least one experiment. This study identified the QTLs in the genomic regions carrying seedling resistance genes such as *Pc48*, *Pc58a*, *Pc68*, *Pc71*, *Pc91*, and *PcKM*.

Powdery mildew caused by *Blumeria graminis* sp. *avenae* is another primary disease that affects oat yield, and various studies have reported the presence of several genes in wild oat species [224]. Simons and co-workers [225] identified four major genes, *Eg-1*, *Eg-2*,

Eg-3, and *Eg-4*, for powdery mildew. Yu and Herrmann [224] introgressed the powdery mildew resistance gene *Eg-5* from *Avena macrostachya* into cultivated hexaploid oat and mapped this gene on linkage group 22 of the Kanota × Ogle mapping population map [226]. Montilla-Bascon and a group of researchers [222] also conducted association mapping for powdery mildew by using 177 oat accessions genotyped using 31 simple sequence repeats and 1500 DArT markers and found one DArT sequence, oPT-5014, associated with powdery mildew results at the adult plant stage. Oat grain has high nutritional value due to protein, fiber (β-glucan), and oil. Oat grain has <8% oil content in the form of triacylglycerols (TAGs). Quality improvement is an essential objective for oat breeding programs. Kianian et al. [227] conducted the first QTL analysis for oil content using two different populations, the Kanota × Ogle mapping population (K × O) and the Kanota × Marion (K × M) mapping population. They found three QTLs for oil content in each population. In 2000, Kianian et al. [227] used these two populations, K × O and K × M, to study the inheritance of β-glucan and found seven and four QTLs associated with β-glucan in the K × O and K × M populations, respectively. De Koeyer et al. [196] identified six QTLs for oil content in the Terra × Marion mapping population, and Zhu and a group of researchers [197] also reported six QTLs in the Ogle × MAM17-5 (O × M) mapping population. Jackson et al. [189] used the Ogle1040 × TAM O-301 (OT) linkage map and mapping population. They found five, six, and one QTLs associated with α-tocotrienol, α-tocopherol, and total tocopherol concentrations, respectively. Tanhuanpää and coworkers [228] used the Aslak × Matilda (A × M) mapping population and found eight QTLs associated with oil content. Kianian and a group of researchers [227] proposed that acetyl coenzyme A carboxylase (ACCase) might be the potential candidate gene underlying the genomic regions associated with oil content variations in these populations. Eight QTLs were identified by Hizbai and a group of researchers [229] using the Dal × Exeter (D × E) mapping population associated with the oil content. Hizbai et al. [229] also found that the genomic regions associated with variations in different fatty acids (oleic acid, linoleic acid, linolenic acid, etc.) coincided with the oil content QTLs and may have a pleiotropic effect. GWAS for β-glucan was conducted by Asaro and coworkers [202] using 446 elite oat breeding lines genotyped by DArT markers, and 24 DarT markers were found to be associated with β-glucan. The genetic positions of 15 out of these 24 markers colocalized with QTLs reported in previous bi-parental mapping studies [196,201,230]. Based on sequence homology to rice, Newell and a group of researchers [201] found that one of the DArT markers sequences located on rice chromosome 7, which is adjacent to the *β-glucan synthese CslF* gene family, was found in GWAS. More studies are needed to fine-map identified QTLs and develop markers for marker-assisted selection to breed for high protein and fiber content in oats.

Various studies reported the synteny between oats chromosomes with the *Brachypodium distachyon* chromosome, the Oryza sativa chromosome, and wheat chromosomes controlling multiple disease resistance and quality traits. For example, synteny between the 4C oat chromosome and chromosome 4 of *Brachypodium distachyon*, chromosome 5BL of wheat, and chromosome 9 of rice carries the resistance genes for crown rust. Similarly, model grass *Brachypodium distachyon* has accelerated oat genomic research by assisting in rearranging the oat sub-genomes and differentiating the homologous relationships between oat linkage groups [231]. The synteny relationships among the genomes of different cereals accelerate the identification and annotation of genes in oats. In summary, the genomic regions associated with quality traits (protein and oil content) and resistance to biotic stresses (crown rust and powdery mildew) in different mapping populations/accessions will help to elucidate the genes and various metabolic pathways to improve the oat germplasm and transfer of favorable alleles that protect against biotic stress. Advancements in high genotyping technologies (DArT markers and GBS) for high-resolution mapping, new statistical approaches, and implications of genomic selection might be employed to predict the breeding values using marker information to speed the oat breeding process in a short period.

The transcriptome analysis lagged in oats compared to other cereal crops due to the complexity of the genome and the challenges of differentiating the homologs and paralogs. RNA-seq is a powerful tool for transcriptome research, even for crops without reference genomes such as oat. Gutierrez-Gonzalez and a group of researchers [232] conducted the first comprehensive transcriptome study in oat seeds. They successfully studied the expression of genes involved in the biosynthetic pathways of avenanthramides, tocols, and beta-glucans. The resulting *dnOST* (*de novo* Oat Seed Transcriptome) transcript assembly has nearly 75-fold average coverage and will be a valuable tool for further transcriptomic research in oats. Jinqiu et al. [233] studied the transcriptomic expression in oats in response to altitude stress. They found 11,639 differentially expressed genes among low and high altitudes that might provide resistance to altitude stress. Recent advances in high-throughput genome sequencing and transcriptome profiling technologies will enable the quantification of gene expression for yield and quality-related traits in oats.

Therefore, since oats act as a phytoremediation crop, it is plausible to expect that there is some inbuilt adaptive mechanism to help plants combat different stressful conditions. Zhao and coworkers [37] observed the upregulation of 164 proteins with a subsequent downregulation of 241 proteins in shoots under alkali stress. Similarly, many resistance proteins such as GDSL esterase lipase in roots and late embryogenesis abundant (LEA) in shoots are accumulated under high pH and alkali stress. Further, escalating levels of glutathione and ascorbate play an essential role in plant defense against powdery mildew (*Blumeria grammis*) in oats [234].

In one study, it was found that in the Flega variety (susceptible to drought) of oats, there was a dramatic decrease in polar lipid in correlation with the increase of triglycerides (TGA) and free fatty acids (FFA) and in the resistant variety (Patones), there was a slight decrease in polar lipid in the correlation with increased diglycerides in response to drought stress. It was also found that in adverse environmental conditions, plants generally increase the metabolite biosynthesis to replenish the depleted molecule, which was also observed by Xu et al. [182] during the analysis of metabolites in salt-sensitive cultivar BY5 (Baiyan5) and salt-tolerant cultivar BY2 (Baiyan 5) using chromatographic-mass spectroscopy. The results showed the accumulation of more metabolites in BY2 than in By5 under salt stress. This metabolic pathway identification can aid biomarker selection breeding programs for developing salt-tolerant crops. So, the sugars accumulated in salt-tolerant cultivars were sucrose, isomaltose, and sophorose, and in susceptible cultivars were leucrose, trehalose, tagatose, and isomaltose. In salt tolerant genotypes, enrichment of sucrose is believed to be a plant defense strategy against salt stress. Similarly, accumulated amino acids reported were isoleucine, asparagine, serine, and glutamine in the salt-tolerant cultivar and proline, inosine, and asparagine in the susceptible cultivar. The enrichment of these metabolites helps oats improve the neutralization of excess reactive oxygen species and relieve salt stress. The metabolic profiling of oat seeds with Ultra High-Performance Liquid Chromatography-Mass Spectrometry (UHPLC-MS) helps us understand the effect of nitrogen application on seed quality with respect to the accumulation of primary and secondary metabolites. The total organic acid metabolism decreases, whereas the amino acid metabolism increases with nitrogen supplementation [235]. Metabolomic studies of various oat species, subspecies, and individual accessions can also be used as a biochemical fingerprint for identifying oat plants at individual accession levels based on biochemical characteristics. The metabolomic profiles of different oat species (*Avena strigosa* Schreb (diploid), *A. abyssinica* Hochst (tetraploid), and *A. sativa* L. (hexaploid)) give the informative indicators for separation of these species at different ploidy levels. Metabolites such as xylitol, undecylic, glutamic acid, isofucosterol, MAG-118:0, linolenic, methylmalonic, and undecylic acid were identified using an Agilent 6850 gas chromatograph and can be used as informative indicators [184]. A study performed by Pretorious and coworkers where used metabolomics for the identification of signature biochemical compounds that can be used for developing the differential metabolic profiles of various oat cultivars (Dunnart, Magnifico, Pallinup, SWK001, and Overberg). The metabolites were analyzed with ultra-high-performance liquid chromatography (UHPLC) coupled with a mass spectrometer analytical platform. Biomarker compounds among the respective cultivars were profiled into different classes, including amino acids, fatty acids, carboxylic acids, phenolic compounds like hydroxybenzoic and hydroxycinnamic acids, and flavonoids. Therefore, metabolomics provides us with great insight into understanding the biochemistry and physiology of crops, which will be helpful in overcoming the limitations of marker-assisted breeding programs for crop improvement [190].

Phenomics-assisted selection is becoming popular in oats. Some studies have been conducted on the different traits of interest till today using multispectral and hyperspectral cameras. Some crucial characteristics of oats are biomass, nutrients, seed/kernel, biotic-abiotic stresses, etc. [29]. So, we can predict the better future of oat breeding for fodder purposes using phenomics-assisted selection.

1.5. *Alfalfa*

Alfalfa (*Medicago sativa* L.) is the most essential perennial forage crop globally and is grown on more than 30 million hectares worldwide [236]. It is the fourth major cash crop in the United States, accounting for $9.7 billion in production (USDA-NASS, 2021) after corn, soybeans, and wheat (www.naaic.org, accessed on 12 June 2022) and in the world because of its highly nutritious forage and broad adaptability [237]. Cultivated alfalfa (*Medicago sativa* L.) is a perennial, autotetraploid ($2n = 4x = 32$), allogamous, and heterozygous species with a basic chromosome number of eight and a genome size of 800–1000 Mb [238]. Primary breeding goals in alfalfa include increasing yield, enhancing nutritive quality, and improving tolerance to biotic and abiotic factors that challenge alfalfa production. Alfalfa exhibits severe inbreeding depression, precluding the development of inbred lines [239].

Most of the early efforts in alfalfa breeding were focused on improving resistance to biotic stresses and forage nutritional quality and not on yield. Yields in alfalfa have been stagnant for the last few decades. Major QTL related to yield and morphological traits, including fall dormancy and winter-hardiness, persistence, viability, self-fertility, and resistance to various biotic and abiotic stresses, have been mapped in alfalfa, primarily in tetraploid populations [240]. A good review has already been published in alfalfa for progress in breeding using traditional breeding until 2018 [240,241]. Initially, the genetic dissection of essential traits, including forage quality and yield in tetraploid alfalfa, was difficult due to the complex segregation of alleles in tetrasomic inheritance and difficulties in analyzing polyploid linkage relationships [242]. Thus, early genetic linkage mapping [242–244] was conducted in diploid ($2n = 2x = 16$) species of alfalfa using molecular markers such as restriction fragment length polymorphism (RFLP), random amplified polymorphic DNA (RAPD), and SSR markers. Later, genetic mapping was extended to tetraploid alfalfa in backcross populations using single-dose allele analysis with RFLP and SSR markers, where the presence or absence of each marker allele was scored independently of the other alleles at the same locus [244–247]. Such alleles follow disomic inheritance, which could be mapped using diploid linkage mapping software. With the development of TetraploidMap [248] software, a linkage mapping tool and its updates thereafter for autotetraploid species, genetic mapping in tetraploid alfalfa was greatly enhanced. As DNA marker technology advanced, a saturated genetic linkage map of autotetraploid *M. sativa* was generated by Li et al. [249] using genotyping-by-sequencing (GBS) methods. That map contained 3561 SNP markers on 64 linkage groups across both parents, with an average density of one marker per 1.5 and 1.0 cM for the maternal and paternal haplotype maps, respectively. Research from multiple projects has demonstrated that the alfalfa linkage groups were found to be highly syntenic with the diploid *M. truncatula* model *Medicago* genome sequence reference [246,249–252].

In the last decade, several association mapping studies have been conducted using diploid or tetraploid alfalfa breeding populations to identify marker-trait associations for biomass yield and stem composition [253–255], fiber-related traits and digestibility [158], and CP concentrations [256] using SSR markers. Recent advances in next-generation sequencing have provided a new strategy to generate cost-effective, high-density, genome-wide single nucleotide polymorphism (SNP) sets [257]. Moreover, it is difficult and costly, to date, to achieve 60x coverage in genome sequencing or GBS approaches to account for tetraploid allele dosage and distinguish between three heterozygous classes to conduct GWAS and GS. Earlier studies combined the three heterozygous genotype classes to conduct a genomic study [237] or use a single-dose allele in a bi-parental population [258]. Zhang and coworkers [259] mapped quantitative trait loci (QTL) for flowering time in autotetraploid alfalfa (*Medicago sativa* L.) using SNP markers. With the development of new software such as GWASpoly [260], it is now possible to account for three different heterozygous classes using biallelic SNP markers and conduct genomic analyses using different polyploid gene action models. Now that the autotetraploid alfalfa genome has been sequenced [261], this resource can be used along with *M. truncatula* as reference genomes for GBS analysis in alfalfa.

The application of genomic predictions to alfalfa biomass yield and forage quality breeding has been initiated recently [13,237,262,263]. Li et al. [262] used clonal ramets from 185 to 190 individuals from two selection cycles of a tetraploid alfalfa breeding population to conduct genomic predictions in a three-location study and reported prediction accuracies of 0.43 to 0.66 within each location. They also attempted to validate the models across two locations and reported 0.21 to 0.61 prediction accuracies. Jia et al. [264] used 322 individual genotypes from 75 diverse alfalfa populations to test three Bayesian genomic prediction models for 25 agronomic traits in alfalfa. These included 15 forage quality traits and ten other traits such as dry matter, plant regrowth, fall dormancy, leaf to stem ratio, etc. In their study, prediction accuracies ranged between 0.0021 and 0.6485, and no significant differences were observed between BayesA, BayesB, and BayesC models. Medina et al. [263] also conducted genomic prediction among 304 clonally propagated alfalfa plants for biomass yield under salt stress. They reported that prediction accuracies ranged from 0.087 to 0.457 in different harvests. Predicting the breeding value of candidate parent genotypes for synthetic variety development of outbred species, such as alfalfa, can also be pursued by genotyping a set of parent plants and phenotyping their half-sib progenies [236,237,240] genotyped parents and phenotyped half-sib progenies to test seven different parametric and machine learning genomic selection models for predicting biomass yield in two different alfalfa populations. The support vector machine (SVM) models performed best, with the highest prediction accuracies. Similarly, Biazzi et al. [13] used half-sib progenies of 152 genotypes for phenotyping forage quality traits in alfalfa and achieved prediction accuracies ranging from 0.1 to 0.4 for different traits for five other GS models. In this later study, rrBLUP, BayesA and B, and Bayesian LASSO tended to outperform the SVM model with a linear kernel. Efforts are being made to fully sequence cultivated alfalfa at the diploid level (CADL) and to develop the pangenome to explore core genes in alfalfa. Through comparing cultivar sequence information and using GWAS for traits such as nutritional quality and yield, forage crops can be improved.

The Affymetrix GeneChip® Medicago genome array was released in 2005 and included over 52,000 Medicago probe sets designed from 32,167 *M. truncatula* ESTs, 18,733 gene predictions from *M. truncatula* genome sequences, and 1896 cDNAs from M. sativa. This GeneChip® was used for cross-species transcriptomic studies in alfalfa [265–267]. Lignin content in plants is a biochemically well characterized trait and directly affects the forage quality as a livestock feed. Therefore, cell wall composition, which mainly consists of cellulose and lignin, in stems was chosen for transcript profiling in earlier studies in alfalfa [267–269]. Later, several studies contributed to the transcriptome sequencing of alfalfa targeting different genes for salt tolerance [270–272], drought tolerance [273], freezing tolerance [274], and bacterial stem blight disease [275] traits using NGS technologies or

RNA-Seq. In 2011, the first publicly available *Medicago sativa* gene index (MSGI 1.0) was developed by Yang et al. [268], using the elongating stem and post-elongation stem internodes from two alfalfa genotypes with differing cell wall composition in stems. The MGSI 1.0 has 124,025 unique sequences, including 22,729 tentative consensus sequences (TCs), 22,315 singletons, and 78,981 pseudo-singletons. A high-resolution melting [HRM] technology has also been implemented to obtain 54,216 unique sequences in a population generated using two diverse genotypes differing in water stress sensitivity [273]. Li and coworkers [269] conducted transcript profiling using Illumina short reads from stem tissues at different growth stages with 27 diverse alfalfa germplasm, including commercial and wild genotypes, and identified 604,164 SNPs and InDels in the alfalfa population. Large scale transcriptomic analysis was conducted by Liu et al. [276] using 15 types of different tissues, including germinating seeds, young leaves, young stems, mature stems, mature pods to callus cells from one alfalfa cultivar. This extensive RNA-Seq analysis generated 40,433 unigenes in addition to 1649 EST-SSRs markers. The MSGI 1.2 was assembled by adding transcripts from multiple plant tissues such as roots, nodules, leaves, flowers, and stem internodes at multiple growth stages, representing 112,626 unique genes using two diverse subspecies of *M. sativa* [277]. Shu and coworkers [274] reported that C-repeat binding factors (CBF) genes play an important role in freezing tolerance in alfalfa. Later in 2018, Luo and coworkers [278] used the PacBio Iso-Seq sequencing technology to generate the full-length transcripts from alfalfa roots under salt stress conditions in a single cultivar to identify 5011 and 4546 differentially expressed (DEG) genes with NaCl and mannitol after 24 h of treatment. Recently, Illumina short read transcripts and Iso-Seq long read transcripts were used for generating a pan-transcriptome to identify 1,124,275 unique isoforms and 91,378 genes in response to drought and salt stress using three diverse alfalfa germplasms for these two traits [279]. Jiang and coworkers [280] combined genomic and transcriptomic analysis in an F1 population developed using two parental genotypes with varying leaf sizes. They identified seven candidate genes associated with leaf development in five major QTL regions using association studies and 2443 leaf-specific genes and 3770 differentially expressed genes using RNA-Seq analysis. With the advancement in transcriptomic technologies, studies can be conducted to identify genes for various diseases using different genotypes from various plant tissue replicates at multiple stages, which is currently lacking in alfalfa.

Recent advances in metabolomics and proteomics technologies greatly expedite the identification and characterization of natural products and their associated metabolites. Proteomics and metabolomics can remarkably examine the balance between carbon and nitrogen metabolism under stress conditions in alfalfa during interactions with nitrogen-fixing bacteria [281,282]. Metabolomic analysis of alfalfa (*Medicago sativa* L.) root-symbiotic rhizobia responses under alkali stress. Water stress limits nitrogen fixation in nodules by reducing nitrogenase activity [282] and Rubisco availability in leaves [282]. Metabolomics has been utilized to elucidate the internal causes of nutrient change at different developmental stages in alfalfa plants [283]. Furthermore, a study found that a water deficit significantly enhanced the alfalfa's freezing tolerance. This was correlated with increased soluble sugar, amino acid, and lipid and lipid-like molecule content using metabolomic analyses. This study improves our understanding of the relationships between metabolites and freezing tolerance following cold acclimation and freezing temperatures, suggesting that metabolites play essential roles in enhancing the freezing tolerance of alfalfa.

Saline-alkali stress is the chief abiotic stress in alfalfa, significantly affecting the yield and quality of the crop. Several attempts have been made to understand the adaptive mechanisms against these stresses in resistant crops. One such study elucidated the induction of 226 differentially abundant proteins (DAP) under salt stress, leading to an elevation in glutathione, an antioxidant and oxidation-reduction pathway to adapt the plant to salt stress [10]. In another similar study, proteomic analysis between drought tolerant variety of alfalfa Longzhong and drought-sensitive variety named Gannong No.3 revealed the accumulation of 142 DAPs involved in stress, defense, transmembrane transport, and cytoskeleton metabolism to increase the osmotic adjustment capacity of the Longzhang variety [284]. Comparative proteomic analysis has helped tremendously in deciphering different types of proteins responsible for adaptive mechanisms in alfalfa.

Moreover, metabolomics can be utilized to evaluate the metabolic response of the root-nodule symbiosis in alfalfa under alkali stress [281]. Physiological analysis and metabolic profiling using GC-TOF/MS comparative analysis employed to identify metabolites and pathways that change after *Rhizobium* inoculation revealed that RI plants accumulated more antioxidants (SOD, POD, GSH), osmolytes (sugar, glycols, proline), organic acids (succinic acid, fumaric acid, and alpha-ketoglutaric acid), and metabolites that are involved in nitrogen fixation. This study revealed a distinct metabolic profile is induced in nodulized plants with putative alkali tolerance compared to non-nodulated alfalfa plants. The effects of *Bacillus subtilis* inoculation on the growth and Cd uptake of alfalfa were evaluated using metabolomics analyses [285]. The results indicated that inoculation significantly decreased the amount of plant malondialdehyde (MDA) and improved the activities of plant antioxidant enzymes and soil nutrient cycling-involved enzymes, thereby promoting biomass by 29.4%. Moreover, a study by [286] found the metabolic mechanisms underlying the response of alfalfa reproductive organs to boron deficiency and surplus, which could provide new strategies for improving seed yield and quality. Boron deficiency leads to the excessive accumulation of sugars in flowers and phenolic compounds in seeds, causing abscission of reproductive organs and then reduced yield and quality of seeds. Boron surplus caused a severe reduction in the metabolites associated with amino acid and carbohydrate metabolism, causing flowers to fall before seed set, thus reducing seed yield.

Phenotyping for biomass yield requires significant resources [287], which are time and labor-intensive. Some critical traits for alfalfa breeding are biomass, plant height, persistence, biotic-abiotic stress, root architecture, etc. Biomass [141,288–290] is the most important trait for alfalfa breeding. Remote sensing techniques have enabled efficient and non-destructive estimation of biomass in alfalfa [289], such as screening large breeding populations [291]. Breeding alfalfa for biomass involves repeated and numerous phenotyping efforts, which is laborious and costly [292]. Biswas and coworkers [292] found that phenomic-assisted selection can reduce up to 70% of manual data collection but still predict the biomass yield correctly, help select high-yielding alfalfa cultivars, and significantly contribute to breeding. According to Cazenave et al. [291], high throughput phenotyping (HTP) can identify minor differences in alfalfa yield when screening diverse germplasm. More recently, HTP enhanced the efficacy of the selection process for biomass in small plots in alfalfa breeding populations [293] and provided a good prediction of biomass in bigger plots [289]. Considering all that research, phenomics can lessen the challenge of biomass measurement in larger populations for alfalfa breeding programs aimed at improving biomass yield. Therefore, applying HTP or phenomics can simplify the phenotyping process for alfalfa biomass. To our knowledge, few studies [141,290,291,294,295] have been conducted on phenomics-assisted research for other traits such as root architecture [295], abiotic stress [291], nutritional value [296], seed identification or separation from other seeds [141,290,296]. However, it is becoming popular in other crops for reducing time and labor. Genomic selection (GS) is becoming a popular perennial crop breeding technique, requiring phenotyping data to make the prediction model predict yield or other traits. We found only one study [297] that used phenomic-assisted selection with GS to make the prediction model. Though some other research has been conducted on phenomic-assisted selection in alfalfa, still, there is a vast prospect of utilizing phenomic-assisted selection or high throughput phenotyping in alfalfa breeding.

2. Conclusions and Future Prospect

There is no argument for the fact that the field of omics has grown by leaps and bounds, providing fast, robust, cutting-edge technology that enables fast and accurate detection of different problems in complex biological samples. However, many challenges still need to be conquered on the journey to produce smart, resilient, high-grain quality crops. Work has been conducted on the grain quality. However, the fodder traits still needed to be targeted in the crop improvement programs using techniques including genomics, transcriptomics, proteomics, metabolomics, and phenomics. Advances in genomic tools for breeding and developing improved cultivars for several crops have been made. However, there is scope for developing better performing fodder crop varieties that will be characterized by traits like stay-green, a higher number of tillers per plant, higher total biomass, good nutritional quality, and tolerance to biotic/abiotic stresses.

The functional study is faster in crops such as Arabidopsis and rice, which have well-developed transformation systems. However, a high-efficiency and stable transgenic system for fodder crops is required to make the functional study less time-consuming. Transcriptomics techniques such as RNA sequencing and NGS have proved to be excellent tools for identifying genes/QTLs and constructing gene maps, which have expedited the process of crop improvement. There is still a considerable void in analyzing whole proteome levels of non-model species and agriculturally essential crops. Research is in its infancy stage in exploring the full potential of emerging technologies such as peptidomics, phosphoproteomics, and redox proteomics to dive deep into protein and molecular interactions. Phenomics brings genetics and physiology together and gives the possibility of studying the under-explored fields of plant science and revealing the molecular basis of several unmanageable plant activities. Breeding cultivars tolerant to stressed environments can be speeded by collecting, integrating, and utilizing phenomic data. However, technological advances for collecting, handling, and processing extensive data must be employed. Along with technological improvement, more robust and advanced bioinformatics tools need to be developed to interpret the large amounts of information gathered by all the above-mentioned omics. The advancement in omic technologies in terms of being eco-friendly, low-cost, and time-efficient will improve fodder crops.

Funding: This research received no external funding.

Institutional Review Board Statement: Not applicable.

Informed Consent Statement: Not applicable.

Data Availability Statement: Not applicable.

Conflicts of Interest: The authors declare no conflict of interest.

References

1. Skibbe, A. Some fodder plants and feeding stuffs-their culture and chemical composition. *J. Dep. Agric.* **1922**, *4*, 338–349.
2. Fè, D.; Cericola, F.; Byrne, S.; Lenk, I.; Ashraf, B.H.; Pedersen, M.G.; Roulund, N.; Asp, T.; Janss, L.; Jensen, C.S.; et al. Genomic dissection and prediction of heading date in perennial ryegrass. *BMC Genom.* **2015**, *16*, 921. [CrossRef] [PubMed]
3. Skøt, L.; Sanderson, R.; Thomas, A.; Skøt, K.; Thorogood, D.; Latypova, G.; Asp, T.; Armstead, I. Allelic variation in the perennial ryegrass FLOWERING LOCUS T gene is associated with changes in flowering time across a range of populations. *Plant Physiol.* **2011**, *155*, 1013–1022. [CrossRef] [PubMed]
4. Shinozuka, H.; Cogan, N.O.I.; Spangenberg, G.C.; Forster, J.W. Quantitative Trait Locus (QTL) meta-analysis and comparative genomics for candidate gene prediction in perennial ryegrass (*Lolium perenne* L.). *BMC Genet.* **2012**, *13*, 101. [CrossRef]
5. Li, Z.; Yagi, K.; Sakai, H.; Kobayashi, K. Influence of elevated CO_2 and nitrogen nutrition on rice plant growth, soil microbial biomass, dissolved organic carbon and dissolved CH4. *Plant Soil* **2004**, *258*, 81–90. [CrossRef]
6. Lukyanova, M.; Kovshov, V.; Zalilova, Z.; Lukyanov, V.; Araslanbaev, I. A systemic comparative economic approach efficiency of fodder production. *J. Innov. Entrep.* **2021**, *10*, 48. [CrossRef]
7. Kumar, S.; Bhat, V. Application of omics technologies in forage crop improvement. In *OMICS Applications in Crop Science*; CRC Press: Boca Raton, FL, USA, 2013; pp. 523–548. ISBN 978-1-4665-8525-6.
8. Li, Q.; Yan, J. Sustainable agriculture in the era of omics: Knowledge-driven crop breeding. *Genome Biol.* **2020**, *21*, 154. [CrossRef]

9. Groen, S.C.; Ćalić, I.; Joly-Lopez, Z.; Platts, A.E.; Choi, J.Y.; Natividad, M.; Dorph, K.; Mauck, W.M.; Bracken, B.; Cabral, C.L.U.; et al. The strength and pattern of natural selection on gene expression in rice. *Nature* **2020**, *578*, 572–576. [CrossRef]
10. Li, H.; Li, Y.; Ke, Q.; Kwak, S.-S.; Zhang, S.; Deng, X. Physiological and Differential Proteomic Analyses of Imitation Drought Stress Response in Sorghum bicolor Root at the Seedling Stage. *Int. J. Mol. Sci.* **2020**, *21*, 9174. [CrossRef]
11. Somegowda, V.K.; Rayaprolu, L.; Rathore, A.; Deshpande, S.P.; Gupta, R. Genome-Wide Association Studies (GWAS) for Traits Related to Fodder Quality and Biofuel in Sorghum: Progress and Prospects. *Protein Pept. Lett.* **2021**, *28*, 843–854. [CrossRef]
12. Vinayan, M.T.; Babu, R.; Jyothsna, T.; Zaidi, P.H.; Blümmel, M. A note on potential candidate genomic regions with implications for maize stover fodder quality. *Field Crops Res.* **2013**, *153*, 102–106. [CrossRef]
13. Biazzi, E.; Nazzicari, N.; Pecetti, L.; Brummer, E.C.; Palmonari, A.; Tava, A.; Annicchiarico, P. Genome-Wide Association Mapping and Genomic Selection for Alfalfa (*Medicago sativa*) Forage Quality Traits. *PLoS ONE* **2017**, *12*, e0169234. [CrossRef]
14. Li, J.; Tang, W.; Zhang, Y.-W.; Chen, K.-N.; Wang, C.; Liu, Y.; Zhan, Q.; Wang, C.; Wang, S.-B.; Xie, S.-Q.; et al. Genome-Wide Association Studies for Five Forage Quality-Related Traits in Sorghum (*Sorghum bicolor* L.). *Front. Plant Sci.* **2018**, *9*, 1146. [CrossRef]
15. Al-Qurainy, F.; Alshameri, A.; Gaafar, A.-R.; Khan, S.; Nadeem, M.; Alameri, A.A.; Tarroum, M.; Ashraf, M. Comprehensive Stress-Based De Novo Transcriptome Assembly and Annotation of Guar (*Cyamopsis tetragonoloba* (L.) Taub.): An Important Industrial and Forage Crop. *Int. J. Genomics* **2019**, *2019*, 7295859. [CrossRef]
16. Araus, J.L.; Kefauver, S.C.; Zaman-Allah, M.; Olsen, M.S.; Cairns, J.E. Translating High-Throughput Phenotyping into Genetic Gain. *Trends Plant Sci.* **2018**, *23*, 451–466. [CrossRef]
17. Abdurakhmonov, I.Y. *Plant Genomics*; InTech: London, UK, 2016; ISBN 978-953-51-2455-9.
18. Pérez-de-Castro, A.M.; Vilanova, S.; Cañizares, J.; Pascual, L.; Blanca, J.M.; Díez, M.J.; Prohens, J.; Picó, B. Application of genomic tools in plant breeding. *Curr. Genom.* **2012**, *13*, 179–195. [CrossRef]
19. Lorenz, A.J.; Chao, S.; Asoro, F.G.; Heffner, E.L.; Hayashi, T.; Iwata, H.; Smith, K.P.; Sorrells, M.E.; Jannink, J.-L. Genomic Selection in Plant Breeding. In *Advances in Agronomy*; Elsevier: Amsterdam, The Netherlands, 2011; Volume 110, pp. 77–123. ISBN 9780123855312.
20. Peleman, J.D.; van der Voort, J.R. Breeding by design. *Trends Plant Sci.* **2003**, *8*, 330–334. [CrossRef]
21. Collard, B.C.Y.; Mackill, D.J. Marker-assisted selection: An approach for precision plant breeding in the twenty-first century. *Philos. Trans. R. Soc. Lond. B Biol. Sci.* **2008**, *363*, 557–572. [CrossRef]
22. Cramer, G.R.; Urano, K.; Delrot, S.; Pezzotti, M.; Shinozaki, K. Effects of abiotic stress on plants: A systems biology perspective. *BMC Plant Biol.* **2011**, *11*, 163. [CrossRef]
23. Suhre, K.; McCarthy, M.I.; Schwenk, J.M. Genetics meets proteomics: Perspectives for large population-based studies. *Nat. Rev. Genet.* **2021**, *22*, 19–37. [CrossRef]
24. Mustafa, G.; Komatsu, S. Plant proteomic research for improvement of food crops under stresses: A review. *Mol. Omics* **2021**, *17*, 860–880. [CrossRef] [PubMed]
25. Guo, J.; Li, C.; Zhang, X.; Li, Y.; Zhang, D.; Shi, Y.; Song, Y.; Li, Y.; Yang, D.; Wang, T. Transcriptome and GWAS analyses reveal candidate gene for seminal root length of maize seedlings under drought stress. *Plant Sci.* **2020**, *292*, 110380. [CrossRef] [PubMed]
26. Führs, H.; Götze, S.; Specht, A.; Erban, A.; Gallien, S.; Heintz, D.; Van Dorsselaer, A.; Kopka, J.; Braun, H.-P.; Horst, W.J. Characterization of leaf apoplastic peroxidases and metabolites in Vigna unguiculata in response to toxic manganese supply and silicon. *J. Exp. Bot.* **2009**, *60*, 1663–1678. [CrossRef] [PubMed]
27. Kaur, B.; Sandhu, K.S.; Kamal, R.; Kaur, K.; Singh, J.; Röder, M.S.; Muqaddasi, Q.H. Omics for the improvement of abiotic, biotic, and agronomic traits in major cereal crops: Applications, challenges, and prospects. *Plants* **2021**, *10*, 1989. [CrossRef] [PubMed]
28. Falk, K.G.; Jubery, T.Z.; O'Rourke, J.A.; Singh, A.; Sarkar, S.; Ganapathysubramanian, B.; Singh, A.K. Soybean Root System Architecture Trait Study through Genotypic, Phenotypic, and Shape-Based Clusters. *Plant Phenomics* **2020**, *2020*, 1925495. [CrossRef]
29. Gill, T.; Gill, S.K.; Saini, D.K.; Chopra, Y.; de Koff, J.P.; Sandhu, K.S. A comprehensive review of high throughput phenotyping and machine learning for plant stress phenotyping. *Phenomics* **2022**, *2*, 156–183. [CrossRef]
30. Arya, S.; Sandhu, K.S.; Singh, J.; Kumar, S. Deep learning: As the new frontier in high-throughput plant phenotyping. *Euphytica* **2022**, *218*, 47. [CrossRef]
31. Sankaran, S.; Quirós, J.J.; Miklas, P.N. Unmanned aerial system and satellite-based high resolution imagery for high-throughput phenotyping in dry bean. *Comput. Electron. Agric.* **2019**, *165*, 104965. [CrossRef]
32. Calviño, M.; Messing, J. Sweet sorghum as a model system for bioenergy crops. *Curr. Opin. Biotechnol.* **2012**, *23*, 323–329. [CrossRef]
33. Silva, T.N.; Thomas, J.B.; Dahlberg, J.; Rhee, S.Y.; Mortimer, J.C. Progress and challenges in sorghum biotechnology, a multipurpose feedstock for the bioeconomy. *J. Exp. Bot.* **2022**, *73*, 646–664. [CrossRef]
34. Paterson, A.H.; Bowers, J.E.; Bruggmann, R.; Dubchak, I.; Grimwood, J.; Gundlach, H.; Haberer, G.; Hellsten, U.; Mitros, T.; Poliakov, A.; et al. The Sorghum bicolor genome and the diversification of grasses. *Nature* **2009**, *457*, 551–556. [CrossRef]
35. Cooper, E.A.; Brenton, Z.W.; Flinn, B.S.; Jenkins, J.; Shu, S.; Flowers, D.; Luo, F.; Wang, Y.; Xia, P.; Barry, K.; et al. A new reference genome for Sorghum bicolor reveals high levels of sequence similarity between sweet and grain genotypes: Implications for the genetics of sugar metabolism. *BMC Genom.* **2019**, *20*, 420. [CrossRef]

36. Zhang, D.; Kong, W.; Robertson, J.; Goff, V.H.; Epps, E.; Kerr, A.; Mills, G.; Cromwell, J.; Lugin, Y.; Phillips, C.; et al. Genetic analysis of inflorescence and plant height components in sorghum (Panicoidae) and comparative genetics with rice (*Oryzoidae*). *BMC Plant Biol.* **2015**, *15*, 107. [CrossRef]
37. Zhao, Z.; Liu, J.; Jia, R.; Bao, S.; Haixia; Chen, X. Physiological and TMT-based proteomic analysis of oat early seedlings in response to alkali stress. *J. Proteom.* **2019**, *193*, 10–26. [CrossRef]
38. Ortiz, D.; Ferruzzi, M.G. Identification and Quantification of Carotenoids and Tocochromanols in Sorghum Grain by High-Performance Liquid Chromatography. *Methods Mol. Biol.* **2019**, *1931*, 141–151. [CrossRef]
39. Cuevas, H.E.; Prom, L.K.; Cooper, E.A.; Knoll, J.E.; Ni, X. Genome-wide association mapping of anthracnose (Colletotrichum sublineolum) resistance in the US sorghum association panel. *Plant Genome* **2018**, *11*, 170099. [CrossRef]
40. Adeyanju, A.; Little, C.; Yu, J.; Tesso, T. Genome-Wide Association Study on Resistance to Stalk Rot Diseases in Grain Sorghum. *G3: Genes Genomes Genet.* **2015**, *5*, 1165–1175. [CrossRef]
41. Nida, H.; Girma, G.; Mekonen, M.; Tirfessa, A.; Seyoum, A.; Bejiga, T.; Birhanu, C.; Dessalegn, K.; Senbetay, T.; Ayana, G.; et al. Genome-wide association analysis reveals seed protein loci as determinants of variations in grain mold resistance in sorghum. *Theor. Appl. Genet.* **2021**, *134*, 1167–1184. [CrossRef]
42. Rai, K.M.; Thu, S.W.; Balasubramanian, V.K.; Cobos, C.J.; Disasa, T.; Mendu, V. Identification, Characterization, and Expression Analysis of Cell Wall Related Genes in *Sorghum bicolor* (L.) Moench, a Food, Fodder, and Biofuel Crop. Front. *Plant Sci.* **2016**, *7*, 1287. [CrossRef]
43. Li, Y.; Mehta, R.; Messing, J. A new high-throughput assay for determining soluble sugar in sorghum internode-extracted juice. *Planta* **2018**, *248*, 785–793. [CrossRef]
44. Quinby, J.R.; Karper, R.E. Inheritance of height in sorghum. *Agron. J.* **1954**, *46*, 211–216. [CrossRef]
45. Hilley, J.; Truong, S.; Olson, S.; Morishige, D.; Mullet, J. Identification of dw1, a regulator of sorghum stem internode length. *PLoS ONE* **2016**, *11*, e0151271. [CrossRef] [PubMed]
46. Hilley, J.L.; Weers, B.D.; Truong, S.K.; McCormick, R.F.; Mattison, A.J.; McKinley, B.A.; Morishige, D.T.; Mullet, J.E. Sorghum dw2 encodes a protein kinase regulator of stem internode length. *Sci. Rep.* **2017**, *7*, 4616. [CrossRef] [PubMed]
47. Da Silva, M.J.; Carneiro, P.C.S.; de Souza Carneiro, J.E.; Damasceno, C.M.B.; Parrella, N.N.L.D.; Pastina, M.M.; Simeone, M.L.F.; Schaffert, R.E.; da Costa Parrella, R.A. Evaluation of the potential of lines and hybrids of biomass sorghum. *Ind. Crops Prod.* **2018**, *125*, 379–385. [CrossRef]
48. Kebrom, T.H.; Burson, B.L.; Finlayson, S.A. Phytochrome B represses Teosinte Branched1 expression and induces sorghum axillary bud outgrowth in response to light signals. *Plant Physiol.* **2006**, *140*, 1109–1117. [CrossRef] [PubMed]
49. Toor, M.D.; Adnan, M.; Javed, M.S.; Habibah, U.; Arshad, A.; Din, M.M.; Ahmad, R. Foliar application of Zn: Best way to mitigate drought stress in plants; A review. *Int. J. Appl. Res.* **2020**, *6*, 16–20.
50. Harris, K.; Subudhi, P.K.; Borrell, A.; Jordan, D.; Rosenow, D.; Nguyen, H.; Klein, P.; Klein, R.; Mullet, J. Sorghum stay-green QTL individually reduce post-flowering drought-induced leaf senescence. *J. Exp. Bot.* **2007**, *58*, 327–338. [CrossRef]
51. Varoquaux, N.; Cole, B.; Gao, C.; Pierroz, G.; Baker, C.R.; Patel, D.; Madera, M.; Jeffers, T.; Hollingsworth, J.; Sievert, J.; et al. Transcriptomic analysis of field-droughted sorghum from seedling to maturity reveals biotic and metabolic responses. *Proc. Natl. Acad. Sci. USA* **2019**, *116*, 27124–27132. [CrossRef]
52. Wei, X.; Yang, Z.; Han, G.; Zhao, X.; Yin, S.; Yuan, F.; Wang, B. The developmental dynamics of the sweet sorghum root transcriptome elucidate the differentiation of apoplastic barriers. *Plant Signal. Behav.* **2020**, *15*, 1724465. [CrossRef]
53. Ngara, R.; Goche, T.; Swanevelder, D.Z.H.; Chivasa, S. Sorghum's Whole-Plant Transcriptome and Proteome Responses to Drought Stress: A Review. *Life* **2021**, *11*, 704. [CrossRef]
54. Kanbar, A.; Shakeri, E.; Alhajturki, D.; Riemann, M.; Bunzel, M.; Morgano, M.T.; Stapf, D.; Nick, P. Sweet versus grain sorghum: Differential sugar transport and accumulation are linked with vascular bundle architecture. *Ind. Crops Prod.* **2021**, *167*, 113550. [CrossRef]
55. Ngara, R.; Ndimba, B.K. Understanding the complex nature of salinity and drought-stress response in cereals using proteomics technologies. *Proteomics* **2014**, *14*, 611–621. [CrossRef]
56. Bray, E.A. Plant responses to water deficit. *Trends Plant Sci.* **1997**, *2*, 48–54. [CrossRef]
57. Woldesemayat, A.A.; Modise, D.M.; Ndimba, B.K. Identification of proteins in response to terminal drought stress in sorghum (*Sorghum bicolor* (L.) Moench) using two-dimensional gel-electrophoresis and MALDI-TOF-TOF MS/MS. *Ind. J. Plant Physiol.* **2018**, *23*, 24–39. [CrossRef]
58. Goche, T.; Shargie, N.G.; Cummins, I.; Brown, A.P.; Chivasa, S.; Ngara, R. Comparative physiological and root proteome analyses of two sorghum varieties responding to water limitation. *Sci. Rep.* **2020**, *10*, 11835. [CrossRef]
59. Rajarajan, K.; Ganesamurthy, K.; Raveendran, M.; Jeyakumar, P.; Yuvaraja, A.; Sampath, P.; Prathima, P.T.; Senthilraja, C. Differential responses of sorghum genotypes to drought stress revealed by physio-chemical and transcriptional analysis. *Mol. Biol. Rep.* **2021**, *48*, 2453–2462. [CrossRef]
60. Zhou, Y.; Wang, Z.; Li, Y.; Li, Z.; Liu, H.; Zhou, W. Metabolite Profiling of Sorghum Seeds of Different Colors from Different Sweet Sorghum Cultivars Using a Widely Targeted Metabolomics Approach. *Int. J. Genomics* **2020**, *2020*, 6247429. [CrossRef]

61. Tugizimana, F.; Djami-Tchatchou, A.T.; Steenkamp, P.A.; Piater, L.A.; Dubery, I.A. Metabolomic Analysis of Defense-Related Reprogramming in Sorghum bicolor in Response to Colletotrichum sublineolum Infection Reveals a Functional Metabolic Web of Phenylpropanoid and Flavonoid Pathways. *Front. Plant Sci.* **2018**, *9*, 1840. [CrossRef]
62. Choi, S.C.; Chung, Y.S.; Lee, Y.G.; Kang, Y.; Park, Y.J.; Park, S.U.; Kim, C. Prediction of dhurrin metabolism by transcriptome and metabolome analyses in sorghum. *Plants* **2020**, *9*, 1390. [CrossRef]
63. Watanabe, K.; Guo, W.; Arai, K.; Takanashi, H.; Kajiya-Kanegae, H.; Kobayashi, M.; Yano, K.; Tokunaga, T.; Fujiwara, T.; Tsutsumi, N.; et al. High-Throughput Phenotyping of Sorghum Plant Height Using an Unmanned Aerial Vehicle and Its Application to Genomic Prediction Modeling. *Front. Plant Sci.* **2017**, *8*, 421. [CrossRef]
64. Young, S.N.; Kayacan, E.; Peschel, J.M. Design and field evaluation of a ground robot for high-throughput phenotyping of energy sorghum. *Precision Agric.* **2019**, *20*, 697–722. [CrossRef]
65. Gomez, F.E.; Carvalho, G.; Shi, F.; Muliana, A.H.; Rooney, W.L. High throughput phenotyping of morpho-anatomical stem properties using X-ray computed tomography in sorghum. *Plant Methods* **2018**, *14*, 59. [CrossRef] [PubMed]
66. Joshi, D.C.; Singh, V.; Hunt, C.; Mace, E.; van Oosterom, E.; Sulman, R.; Jordan, D.; Hammer, G. Development of a phenotyping platform for high throughput screening of nodal root angle in sorghum. *Plant Methods* **2017**, *13*, 56. [CrossRef] [PubMed]
67. Boukar, O.; Belko, N.; Chamarthi, S.; Togola, A.; Batieno, J.; Owusu, E.; Haruna, M.; Diallo, S.; Umar, M.L.; Olufajo, O.; et al. Cowpea (Vigna unguiculata): Genetics, genomics and breeding. *Plant Breed.* **2019**, *138*, 415–424. [CrossRef]
68. Singh, B.B.; Ajeigbe, H.A.; Tarawali, S.A.; Fernandez-Rivera, S.; Abubakar, M. Improving the production and utilization of cowpea as food and fodder. *Field Crops Res.* **2003**, *84*, 169–177. [CrossRef]
69. Kolawole, G.O.; Tian, G.; Singh, B.B. Differential response of cowpea lines to aluminum and phosphorus application. *J. Plant Nutr.* **2000**, *23*, 731–740. [CrossRef]
70. Sanginga, N.; Lyasse, O.; Singh, B.B. Phosphorus use efficiency and nitrogen balance of cowpea breeding lines in a low P soil of the derived savanna zone in West Africa. *Plant Soil* **2000**, *220*, 119–128. [CrossRef]
71. Quinn, J.; Myers, R. *Cowpea: A Versatile Legume for Hot, Dry Conditions*; Jefferson Institute: Columbia, MO, USA, 1999.
72. Chinma, C.E.; Emelife, I.G.; Alemede, I.C. Physicochemical and functional properties of some nigerian cowpea varieties. *Pak. J. Nutr.* **2007**, *7*, 186–190. [CrossRef]
73. Langyintuo, A.S.; Ntoukam, G.; Murdock, L.; Lowenberg-DeBoer, J.; Miller, D.J. Consumer preferences for cowpea in Cameroon and Ghana. *Agric. Econ.* **2004**, *30*, 203–213. [CrossRef]
74. Rego, C.H.Q.; França-Silva, F.; Gomes-Junior, F.G.; de Moraes, M.H.D.; Medeiros, A.D.; Silva, C.B.D. Using Multispectral Imaging for Detecting Seed-Borne Fungi in Cowpea. *Agriculture* **2020**, *10*, 361. [CrossRef]
75. Zhou, Z.; Zang, Y.; Shen, B.; Zhou, X.; Luo, X. Detection of cowpea weevil (Callosobruchus maculatus (F.)) in soybean with hyperspectral spectrometry and a backpropagation neural network. In Proceedings of the 2010 Sixth International Conference on Natural Computation, Yantai, China, 10–12 August 2010; pp. 1223–1227.
76. ElMasry, G.; Mandour, N.; Ejeez, Y.; Demilly, D.; Al-Rejaie, S.; Verdier, J.; Belin, E.; Rousseau, D. Multichannel imaging for monitoring chemical composition and germination capacity of cowpea (*Vigna unguiculata*) seeds during development and maturation. *Crop J.* **2021**, *10*, 1399–1411. [CrossRef]
77. Amaral, J.B.C.; Lopes, F.B.; de Magalhães, A.C.M.; Kujawa, S.; Taniguchi, C.A.K.; dos Teixeira, A.S.; de Lacerda, C.F.; Queiroz, T.R.G.; de Andrade, E.M.; da Araújo, I.C.S.; et al. Quantifying nutrient content in the leaves of cowpea using remote sensing. *Appl. Sci.* **2022**, *12*, 458. [CrossRef]
78. Burridge, J.; Jochua, C.N.; Bucksch, A.; Lynch, J.P. Legume shovelomics: High—Throughput phenotyping of common bean (*Phaseolus vulgaris* L.) and cowpea (*Vigna unguiculata* subsp, unguiculata) root architecture in the field. *Field Crops Res.* **2016**, *192*, 21–32. [CrossRef]
79. Singh, P. Current status of feed and forage in management of livestock in India. In *Managing Agriculture for a Better Tomorrow: The Indian Experience*; MD Publishing: New Delhi, India, 1998; p. 85.
80. Singh, B.B.; Chambliss, O.; Sharma, B. *Recent Advances in Cowpea Breeding. Recent Advances in Cowpea Breeding*; IITA: Montpellier, France, 1997.
81. Boukar, O.; Fatokun, C.A.; Huynh, B.-L.; Roberts, P.A.; Close, T.J. Genomic tools in cowpea breeding programs: Status and perspectives. *Front. Plant Sci.* **2016**, *7*, 757. [CrossRef]
82. Samireddypalle, A.; Boukar, O.; Grings, E.; Fatokun, C.A.; Kodukula, P.; Devulapalli, R.; Okike, I.; Blümmel, M. Cowpea and groundnut haulms fodder trading and its lessons for multidimensional cowpea improvement for mixed crop livestock systems in west africa. *Front. Plant Sci.* **2017**, *8*, 30. [CrossRef]
83. Fatokun, C.A. A linkage map for cowpea (*Vigna unguiculata* (L.) Walp) based on DNA markers (2n = 22). A Compilation of Linkage and Restriction Maps of Genetically Studied Organisms. *Genetic Maps* **1993**, *2*, 6256–6258.
84. Fatokun, C.A.; Young, N.; Myers, G. *Molecular Markers and Genome Mapping in Cowpea*; IITA: Montpellier, France, 1997.
85. Spencer, M.; Ndiaye, M.A.; Gueye, M.; Diouf, D.; Ndiaye, M.; Gresshoff, P.M. DNA-based relatedness of cowpea [*Vigna unguiculata* (L.) Walp.] genotypes using DNA amplification fingerprinting. *Physiol. Mol. Biol. Plants* **2000**, *6*, 81–88.
86. Mignouna, H.D.; Ellis, N.T.; Knox, M.R.; Asiedu, R.; Ng, Q.N. Analysis of genetic diversity in Guinea yams (*Dioscorea* spp.) using AFLP fingerprinting. *Trop. Agric.* **1998**, *75*, 224–229.

87. Zannouou, A.; Kossou, D.K.; Ahanchede, A.; Zoundjihékpon, J.; Agbicodo, E.; Struik, P.C.; Sanni, A. Genetic variability of cultivated cowpea in Benin assessed by random amplified polymorphic DNA. *Afr. J. Biotechnol.* **2008**, *7*, 24.
88. Ogunkanmi, L.A.; Ogundipe, O.T.; Ng, N.Q.; Fatokun, C.A. Genetic diversity in wild relatives of cowpea (Vigna unguiculata) as revealed by simple sequence repeats (SSR) markers. *J. Food Agric. Env.* **2008**, *6*, 263–268.
89. Mahamadou, S.; Jeremy, T.O.; Bhavani, S.G.; Michael, P.T. Genetic diversity of cowpea (Vigna unguiculata L. Walp) cultivars in Burkina Faso resistant to Striga gesnerioides. *Afr. J. Biotechnol.* **2010**, *9*, 8146–8153. [CrossRef]
90. Ghalmi, N.; Malice, M.; Jacquemin, J.-M.; Ounane, S.-M.; Mekliche, L.; Baudoin, J.-P. Morphological and molecular diversity within Algerian cowpea (*Vigna unguiculata* (L.) Walp.) landraces. *Genet. Resour. Crop Evol.* **2010**, *57*, 371–386. [CrossRef]
91. Choumane, W.; Winter, P.; Weigand, F.; Kahl, G. Conservation and variability of sequence-tagged microsatellite sites (STMSs) from chickpea (*Cicer aerietinum* L.) within the genus Cicer. *Theor. Appl. Genet.* **2000**, *101*, 269–278. [CrossRef]
92. Huynh, B.-L.; Close, T.J.; Roberts, P.A.; Hu, Z.; Wanamaker, S.; Lucas, M.R.; Chiulele, R.; Cissé, N.; David, A.; Hearne, S.; et al. Gene pools and the genetic architecture of domesticated cowpea. *Plant Genome* **2013**, *6*, 122. [CrossRef]
93. Xiong, H.; Shi, A.; Mou, B.; Qin, J.; Motes, D.; Lu, W.; Ma, J.; Weng, Y.; Yang, W.; Wu, D. Genetic Diversity and Population Structure of Cowpea (Vigna unguiculata L. Walp). *PLoS ONE* **2016**, *11*, e0160941. [CrossRef]
94. Muñoz-Amatriaín, M.; Lo, S.; Herniter, I.A.; Boukar, O.; Fatokun, C.; Carvalho, M.; Castro, I.; Guo, Y.; Huynh, B.; Roberts, P.A.; et al. The UCR Minicore: A resource for cowpea research and breeding. *Legume Sci.* **2021**, *3*, e95. [CrossRef]
95. Menéndez, C.M.; Hall, A.E.; Gepts, P. A genetic linkage map of cowpea (*Vigna unguiculata*) developed from a cross between two inbred, domesticated lines. *Theor. Appl. Genet.* **1997**, *95*, 1210–1217. [CrossRef]
96. Ubi, B.E.; Mignouna, H.; Thottappilly, G. Construction of a Genetic Linkage Map and QTL Analysis Using a Recombinant Inbred Population Derived from an Intersubspecific Cross of Cowpea (*Vigna unguiculata* (L.) Walp.). *Breed. Sci.* **2000**, *50*, 161–172. [CrossRef]
97. Muchero, W.; Diop, N.N.; Bhat, P.R.; Fenton, R.D.; Wanamaker, S.; Pottorff, M.; Hearne, S.; Cisse, N.; Fatokun, C.; Ehlers, J.D.; et al. A consensus genetic map of cowpea [*Vigna unguiculata* (L) Walp.] and synteny based on EST-derived SNPs. *Proc. Natl. Acad. Sci. USA* **2009**, *106*, 18159–18164. [CrossRef]
98. Lucas, M.R.; Diop, N.-N.; Wanamaker, S.; Ehlers, J.D.; Roberts, P.A.; Close, T.J. Cowpea–Soybean Synteny Clarified through an Improved Genetic Map. *Plant Genome J.* **2011**, *4*, 218. [CrossRef]
99. Menancio-Hautea, D.; Fatokun, C.A.; Kumar, L.; Danesh, D.; Young, N.D. Comparative genome analysis of mungbean (*Vigna radiata* L. Wilczek) and cowpea (*V. unguiculata* L. Walpers) using RFLP mapping data. *Theor. Appl. Genet.* **1993**, *86*, 797–810. [CrossRef]
100. Xu, P.; Wu, X.; Muñoz-Amatriaín, M.; Wang, B.; Wu, X.; Hu, Y.; Huynh, B.-L.; Close, T.J.; Roberts, P.A.; Zhou, W.; et al. Genomic regions, cellular components and gene regulatory basis underlying pod length variations in cowpea (*V. unguiculata* L. Walp). *Plant Biotechnol. J.* **2017**, *15*, 547–557. [CrossRef]
101. Burridge, J.D.; Schneider, H.M.; Huynh, B.-L.; Roberts, P.A.; Bucksch, A.; Lynch, J.P. Genome-wide association mapping and agronomic impact of cowpea root architecture. *Theor. Appl. Genet.* **2017**, *130*, 419–431. [CrossRef]
102. Omomowo, O.I.; Babalola, O.O. Constraints and prospects of improving cowpea productivity to ensure food, nutritional security and environmental sustainability. *Front. Plant Sci.* **2021**, *12*, 751731. [CrossRef]
103. Kulkarni, K.P.; Tayade, R.; Asekova, S.; Song, J.T.; Shannon, J.G.; Lee, J.-D. Harnessing the potential of forage legumes, alfalfa, soybean, and cowpea for sustainable agriculture and global food security. *Front. Plant Sci.* **2018**, *9*, 1314. [CrossRef]
104. Paudel, D.; Dareus, R.; Rosenwald, J.; Muñoz-Amatriaín, M.; Rios, E.F. Genome-Wide Association Study Reveals Candidate Genes for Flowering Time in Cowpea (*Vigna unguiculata* [L.] Walp.). *Front. Genet.* **2021**, *12*, 667038. [CrossRef]
105. Fatokun, C.; Girma, G.; Abberton, M.; Gedil, M.; Unachukwu, N.; Oyatomi, O.; Yusuf, M.; Rabbi, I.; Boukar, O. Genetic diversity and population structure of a mini-core subset from the world cowpea (*Vigna unguiculata* (L.) Walp.) germplasm collection. *Sci. Rep.* **2018**, *8*, 16035. [CrossRef]
106. Muñoz-Amatriaín, M.; Mirebrahim, H.; Xu, P.; Wanamaker, S.I.; Luo, M.; Alhakami, H.; Alpert, M.; Atokple, I.; Batieno, B.J.; Boukar, O.; et al. Genome resources for climate-resilient cowpea, an essential crop for food security. *Plant J.* **2017**, *89*, 1042–1054. [CrossRef]
107. Salifou, M.; Tignegre, J.B.L.S.; Tongoona, P.; Offei, S.; Ofori, K.; Danquah, E. Differential responses of 15 cowpea genotypes to three Striga hot spots in Niger. *Int. J. Biol. Chem. Sci.* **2017**, *11*, 1413. [CrossRef]
108. Dinesh, H.B.; Lohithaswa, H.C.; Viswanatha, K.P.; Singh, P.; Rao, A.M. Identification and marker-assisted introgression of QTL conferring resistance to bacterial leaf blight in cowpea (*Vigna unguiculata* (L.) Walp.). *Plant Breed.* **2016**, *135*, 506–512. [CrossRef]
109. Boukar, O.; Abberton, M.; Oyatomi, O.; Togola, A.; Tripathi, L.; Fatokun, C. Introgression Breeding in Cowpea [*Vigna unguiculata* (L.) Walp.]. *Front. Plant Sci.* **2020**, *11*, 567425. [CrossRef] [PubMed]
110. Yao, S.; Jiang, C.; Huang, Z.; Torres-Jerez, I.; Chang, J.; Zhang, H.; Udvardi, M.; Liu, R.; Verdier, J. The Vigna unguiculata Gene Expression Atlas (VuGEA) from de novo assembly and quantification of RNA-seq data provides insights into seed maturation mechanisms. *Plant J.* **2016**, *88*, 318–327. [CrossRef] [PubMed]

111. Amorim, L.L.B.; Ferreira-Neto, J.R.C.; Bezerra-Neto, J.P.; Pandolfi, V.; de Araújo, F.T.; da Silva Matos, M.K.; Santos, M.G.; Kido, E.A.; Benko-Iseppon, A.M. Cowpea and abiotic stresses: Identification of reference genes for transcriptional profiling by qPCR. *Plant Methods* **2018**, *14*, 88. [CrossRef] [PubMed]
112. Spriggs, A.; Henderson, S.T.; Hand, M.L.; Johnson, S.D.; Taylor, J.M.; Koltunow, A. Assembled genomic and tissue-specific transcriptomic data resources for two genetically distinct lines of Cowpea (*Vigna unguiculata* (L.) Walp) [version 2; peer review: 3 approved]. *Gates Open Res.* **2018**, *2*, 7. [CrossRef]
113. Chen, H.; Chen, H.; Hu, L.; Wang, L.; Wang, S.; Wang, M.L.; Cheng, X. Genetic diversity and a population structure analysis of accessions in the Chinese cowpea [*Vigna unguiculata* (L.) Walp.] germplasm collection. *Crop J.* **2017**, *5*, 363–372. [CrossRef]
114. Lucas, M.R.; Huynh, B.-L.; Roberts, P.A.; Close, T.J. Introgression of a rare haplotype from Southeastern Africa to breed California blackeyes with larger seeds. *Front. Plant Sci.* **2015**, *6*, 126. [CrossRef]
115. Lima, E.N.; dos Silva, M.L.S.; de Abreu, C.E.B.; Mesquita, R.O.; Lobo, M.D.P.; de Monteiro-Moreira, A.C.O.; Gomes-Filho, E.; de Bertini, C.H.C.M. Research Article Differential proteomics in contrasting cowpea genotypes submitted to different water regimes. *Genet. Mol. Res.* **2019**, *18*. [CrossRef]
116. Parker, C. Protection of crops against parasitic weeds. *Crop Prot.* **1991**, *10*, 6–22. [CrossRef]
117. Lane, J.A.; Moore, T.H.; Child, D.V.; Bailey, J.A. Variation in virulence of Striga gesnerioides on cowpea: New sources of crop resistance. *Adv. Cowpea Res.* **1997**, *84*, 225–230.
118. Botanga, C.J.; Timko, M.P. Phenetic relationships among different races of Striga gesnerioides (Willd.) Vatke from West Africa. *Genome* **2006**, *49*, 1351–1365. [CrossRef]
119. Fery, R.L.; Singh, B.B. *Cowpea Genetics: A Review of the Recent Literature*; International Institute of Tropical Agriculture: Ibadan, Nigeria, 1997; pp. 13–29.
120. Singh, B.B.; Emechebe, A.M.; Atokple, I.D.K. Inheritance of alectra resistance in cowpea genotype B 301. *Crop Sci.* **1993**, *33*, 70–72. [CrossRef]
121. Atokple, I.D.K.; Singh, B.B.; Emechebe, A.M. Genetics of resistance to striga and alectra in cowpea. *J. Hered.* **1995**, *86*, 45–49. [CrossRef]
122. Singh, S.E.; Jackai, L.E. Insect pests of cowpeas in Africa: Their life cycle, economic importance and potential for control. *Cowpea Res. Prod. Util.* **1985**, *2*, 217–231.
123. Bata, H.D.; Singh, B.B.; Singh, S.R.; Ladeinde, T.A.O. Inheritance of resistance to aphid in cowpea1. *Crop Sci.* **1987**, *27*, 892–894. [CrossRef]
124. Souleymane, A.; Aken'Ova, M.E.; Fatokun, C.A.; Alabi, O.Y. Screening for resistance to cowpea aphid (Aphis craccivora Koch) in wild and cultivated cowpea (Vigna unguiculata L. Walp.) accessions. *Int. J. Sci. Environ. Technol.* **2013**, *2*, 611–621.
125. Lo, S.; Muñoz-Amatriaín, M.; Boukar, O.; Herniter, I.; Cisse, N.; Guo, Y.-N.; Roberts, P.A.; Xu, S.; Fatokun, C.; Close, T.J. Identification of QTL controlling domestication-related traits in cowpea (Vigna unguiculata L. Walp). *Sci. Rep.* **2018**, *8*, 6261. [CrossRef]
126. Togola, A.; Boukar, O.; Servent, A.; Chamarthi, S.; Tamò, M.; Fatokun, C. Identification of sources of resistance in cowpea mini core accessions to Aphis craccivora Koch (Homoptera: Aphididae) and their biochemical characterization. *Euphytica.* **2020**, *216*, 88. [CrossRef]
127. Diouf, D. Recent advances in cowpea [*Vigna unguiculata* (L.) Walp.]"omics" research for genetic improvement. *Afr. J. Biotechnol.* **2011**, *10*, 2803–2810.
128. Carneiro, R.G.S.; Oliveira, D.C.; Isaias, R.M.S. Developmental anatomy and immunocytochemistry reveal the neo-ontogenesis of the leaf tissues of Psidium myrtoides (Myrtaceae) towards the globoid galls of Nothotrioza myrtoidis (Triozidae). *Plant Cell Rep.* **2014**, *33*, 2093–2106. [CrossRef]
129. Nogueira, F.C.S.; Gonçalves, E.F.; Jereissati, E.S.; Santos, M.; Costa, J.H.; Oliveira-Neto, O.B.; Soares, A.A.; Domont, G.B.; Campos, F.A.P. Proteome analysis of embryogenic cell suspensions of cowpea (Vigna unguiculata). *Plant Cell Rep.* **2007**, *26*, 1333–1343. [CrossRef]
130. Gomes, A.M.F.; Rodrigues, A.P.; António, C.; Rodrigues, A.M.; Leitão, A.E.; Batista-Santos, P.; Nhantumbo, N.; Massinga, R.; Ribeiro-Barros, A.I.; Ramalho, J.C. Drought response of cowpea (*Vigna unguiculata* (L.) Walp.) landraces at leaf physiological and metabolite profile levels. *Environ. Exp. Bot.* **2020**, *175*, 104060. [CrossRef]
131. Goufo, P.; Moutinho-Pereira, J.M.; Jorge, T.F.; Correia, C.M.; Oliveira, M.R.; Rosa, E.A.S.; António, C.; Trindade, H. Cowpea (*Vigna unguiculata* L. Walp.) Metabolomics: Osmoprotection as a Physiological Strategy for Drought Stress Resistance and Improved Yield. *Front. Plant Sci.* **2017**, *8*, 586. [CrossRef] [PubMed]
132. Yeo, H.J.; Park, C.H.; Lee, K.B.; Kim, J.K.; Park, J.S.; Lee, J.-W.; Park, S.U. Metabolic Analysis of Vigna unguiculata Sprouts Exposed to Different Light-Emitting Diodes. *Nat. Prod. Commun.* **2018**, *13*, 1934578X1801301. [CrossRef]
133. Ramalingam, A.; Kudapa, H.; Pazhamala, L.T.; Weckwerth, W.; Varshney, R.K. Proteomics and metabolomics: Two emerging areas for legume improvement. *Front. Plant Sci.* **2015**, *6*, 1116. [CrossRef] [PubMed]
134. Wu, X.; Sun, T.; Xu, W.; Sun, Y.; Wang, B.; Wang, Y.; Li, Y.; Wang, J.; Wu, X.; Lu, Z.; et al. Unraveling the Genetic Architecture of Two Complex, Stomata-Related Drought-Responsive Traits by High-Throughput Physiological Phenotyping and GWAS in Cowpea (Vigna unguiculata L. Walp). *Front. Genet.* **2021**, *12*, 743758. [CrossRef]

135. Sun, C.X.; Gao, X.X.; Li, M.Q.; Fu, J.Q.; Zhang, Y.L. Plastic responses in the metabolome and functional traits of maize plants to temperature variations. *Plant Biol.* **2016**, *18*, 249–261. [CrossRef]
136. Ganie, A.H.; Ahmad, A.; Pandey, R.; Aref, I.M.; Yousuf, P.Y.; Ahmad, S.; Iqbal, M. Metabolite Profiling of Low-P Tolerant and Low-P Sensitive Maize Genotypes under Phosphorus Starvation and Restoration Conditions. *PLoS ONE* **2015**, *10*, e0129520. [CrossRef]
137. Yadav, O.P.; Hossain, F.; Karjagi, C.G.; Kumar, B.; Zaidi, P.H.; Jat, S.L.; Chawla, J.S.; Kaul, J.; Hooda, K.S.; Kumar, P.; et al. Genetic improvement of maize in india: Retrospect and prospects. *Agric. Res.* **2015**, *4*, 325–338. [CrossRef]
138. Shiferaw, B.; Prasanna, B.M.; Hellin, J.; Bänziger, M. Crops that feed the world 6. Past successes and future challenges to the role played by maize in global food security. *Food Sec.* **2011**, *3*, 307–327. [CrossRef]
139. Gaffney, I.; Sallach, J.B.; Wilson, J.; Bergström, E.; Thomas-Oates, J. Metabolomic Approaches to Studying the Response to Drought Stress in Corn (*Zea mays*) Cobs. *Metabolites* **2021**, *11*, 438. [CrossRef]
140. Adunola, M.P.; Fayeun, L.S.; Fadara, A.B. Impact of climate change on armyworm infestation on maize in Nigeria: A review. *J. Plant Breed. Crop Sci.* **2021**, *13*, 158–167. [CrossRef]
141. Wang, X.; Zhang, H.; Song, R.; He, X.; Mao, P.; Jia, S. Non-Destructive Identification of Naturally Aged Alfalfa Seeds via Multispectral Imaging Analysis. *Sensors* **2021**, *21*, 5804. [CrossRef]
142. De Yassitepe, J.E.C.T.; da Silva, V.C.H.; Hernandes-Lopes, J.; Dante, R.A.; Gerhardt, I.R.; Fernandes, F.R.; da Silva, P.A.; Vieira, L.R.; Bonatti, V.; Arruda, P. Maize transformation: From plant material to the release of genetically modified and edited varieties. *Front. Plant Sci.* **2021**, *12*, 766702. [CrossRef]
143. Liu, J.; Fernie, A.R.; Yan, J. The Past, Present, and Future of Maize Improvement: Domestication, Genomics, and Functional Genomic Routes toward Crop Enhancement. *Plant Commun.* **2020**, *1*, 100010. [CrossRef]
144. Strable, J.; Scanlon, M.J. Maize (Zea mays): A model organism for basic and applied research in plant biology. *Cold Spring Harb. Protoc.* **2009**, *2009*, pdb.emo132. [CrossRef]
145. Birchler, J.A. The cytogenetic localization of the alcohol dehydrogenase-1 locus in maize. *Genetics* **1980**, *94*, 687–700. [CrossRef]
146. Gardiner, J.M.; Coe, E.H.; Melia-Hancock, S.; Hoisington, D.A.; Chao, S. Development of a core RFLP map in maize using an immortalized F2 population. *Genetics* **1993**, *134*, 917–930. [CrossRef]
147. Weber, D.; Helentjaris, T. Mapping RFLP loci in maize using B-A translocations. *Genetics* **1989**, *121*, 583–590. [CrossRef]
148. Burr, B.; Burr, F.A.; Thompson, K.H.; Albertson, M.C.; Stuber, C.W. Gene mapping with recombinant inbreds in maize. *Genetics* **1988**, *118*, 519–526. [CrossRef]
149. Helentjaris, T.; Slocum, M.; Wright, S.; Schaefer, A.; Nienhuis, J. Construction of genetic linkage maps in maize and tomato using restriction fragment length polymorphisms. *Theor. Appl. Genet.* **1986**, *72*, 761–769. [CrossRef]
150. Dong, Q.; Roy, L.; Freeling, M.; Walbot, V.; Brendel, V. ZmDB, an integrated database for maize genome research. *Nucleic Acids Res.* **2003**, *31*, 244–247. [CrossRef] [PubMed]
151. Sanger, F.; Nicklen, S.; Coulson, A.R. DNA sequencing with chain-terminating inhibitors. *Proc. Natl. Acad. Sci. USA* **1977**, *74*, 5463–5467. [CrossRef] [PubMed]
152. Schnable, P.S.; Ware, D.; Fulton, R.S.; Stein, J.C.; Wei, F.; Pasternak, S.; Liang, C.; Zhang, J.; Fulton, L.; Graves, T.A.; et al. The B73 maize genome: Complexity, diversity, and dynamics. *Science* **2009**, *326*, 1112–1115. [CrossRef] [PubMed]
153. Chakradhar, T.; Hindu, V.; Reddy, P.S. Genomic-based-breeding tools for tropical maize improvement. *Genetica* **2017**, *145*, 525–539. [CrossRef] [PubMed]
154. Zein, I.; Wenzel, G.; Andersen, J.R.; Lübberstedt, T. Low Level of Linkage Disequilibrium at the COMT (Caffeic Acid O-methyl Transferase) Locus in European Maize (*Zea mays* L.). Genet. Resour. *Crop Evol.* **2007**, *54*, 139–148. [CrossRef]
155. Pinter, J.; Glenn, F.; Pen, S.; Pok, I.; Hegyi, Z.; Zsubori, Z.T.; Hadi, G.; Marton, C.L. Utilizing Leafy genes as resources in quality silage maize breeding. *Maydica* **2012**, *56*, 243–250.
156. Barrière, Y.; Argillier, O. Brown-midrib genes of maize: A review. *Agronomie* **1993**, *13*, 865–876. [CrossRef]
157. Andersen, J.R.; Zein, I.; Wenzel, G.; Krützfeldt, B.; Eder, J.; Ouzunova, M.; Lübberstedt, T. High levels of linkage disequilibrium and associations with forage quality at a phenylalanine ammonia-lyase locus in European maize (*Zea mays* L.) inbreds. *Theor. Appl. Genet.* **2007**, *114*, 307–319. [CrossRef]
158. Wang, H.; Li, K.; Hu, X.; Liu, Z.; Wu, Y.; Huang, C. Genome-wide association analysis of forage quality in maize mature stalk. *BMC Plant Biol.* **2016**, *16*, 227. [CrossRef]
159. Zhao, X.; Luo, L.; Cao, Y.; Liu, Y.; Li, Y.; Wu, W.; Lan, Y.; Jiang, Y.; Gao, S.; Zhang, Z.; et al. Genome-wide association analysis and QTL mapping reveal the genetic control of cadmium accumulation in maize leaf. *BMC Genom.* **2018**, *19*, 91. [CrossRef]
160. Zhao, L.; Xie, J.; Zhang, H.; Wang, Z.; Jiang, H.; Gao, S. Enzymatic activity and chlorophyll fluorescence imaging of maize seedlings (*Zea mays* L.) after exposure to low doses of chlorsulfuron and cadmium. *J. Integr. Agric.* **2018**, *17*, 826–836. [CrossRef]
161. Vinayan, M.T.; Seetharam, K.; Babu, R.; Zaidi, P.H.; Blummel, M.; Nair, S.K. Genome wide association study and genomic prediction for stover quality traits in tropical maize (*Zea mays* L.). *Sci. Rep.* **2021**, *11*, 686. [CrossRef]
162. Hey, S.; Baldauf, J.; Opitz, N.; Lithio, A.; Pasha, A.; Provart, N.; Nettleton, D.; Hochholdinger, F. Complexity and specificity of the maize (*Zea mays* L.) root hair transcriptome. *J. Exp. Bot.* **2017**, *68*, 2175–2185. [CrossRef]
163. Liu, T.; Zhang, X. Transcriptome and metabolomic analyses reveal regulatory networks controlling maize stomatal development in response to blue light. *Int. J. Mol. Sci.* **2021**, *22*, 5393. [CrossRef]

164. Pan, Y.; Zhao, S.-W.; Tang, X.-L.; Wang, S.; Wang, X.; Zhang, X.-X.; Zhou, J.-J.; Xi, J.-H. Transcriptome analysis of maize reveals potential key genes involved in the response to belowground herbivore Holotrichia parallela larvae feeding. *Genome* **2020**, *63*, 1–12. [CrossRef]
165. Zhou, K.; Zeng, X.; Zhang, B.; Aslam, M.; Xin, H.; Liu, W.; Zou, H. Bulk segregant transcriptome analysis based differential expression of drought response genes in maize. *Pak. J. Agric. Sci.* **2020**, *57*, 909–923.
166. Du, H.; Zhu, J.; Su, H.; Huang, M.; Wang, H.; Ding, S.; Zhang, B.; Luo, A.; Wei, S.; Tian, X.; et al. Bulked segregant RNA-seq reveals differential expression and SNPs of candidate genes associated with waterlogging tolerance in maize. *Front. Plant Sci.* **2017**, *8*, 1022. [CrossRef]
167. Kebede, A.Z.; Johnston, A.; Schneiderman, D.; Bosnich, W.; Harris, L.J. Transcriptome profiling of two maize inbreds with distinct responses to Gibberella ear rot disease to identify candidate resistance genes. *BMC Genom.* **2018**, *19*, 131. [CrossRef]
168. Yue, R.; Lu, C.; Qi, J.; Han, X.; Yan, S.; Guo, S.; Liu, L.; Fu, X.; Chen, N.; Yin, H.; et al. Transcriptome analysis of cadmium-treated roots in maize (*Zea mays* L.). *Front. Plant Sci.* **2016**, *7*, 1298. [CrossRef]
169. Zhang, X.; Zhang, R.; Li, L.; Yang, Y.; Ding, Y.; Guan, H.; Wang, X.; Zhang, A.; Wen, H. Negligible transcriptome and metabolome alterations in RNAi insecticidal maize against Monolepta hieroglyphica. *Plant Cell Rep.* **2020**, *39*, 1539–1547. [CrossRef]
170. He, W.; Zhu, Y.; Leng, Y.; Yang, L.; Zhang, B.; Yang, J.; Zhang, X.; Lan, H.; Tang, H.; Chen, J.; et al. Transcriptomic analysis reveals candidate genes responding maize gray leaf spot caused by Cercospora zeina. *Plants* **2021**, *10*, 2257. [CrossRef] [PubMed]
171. Sun, G.; Yu, H.; Wang, P.; Guerrero, M.L.; Mural, R.V.; Mizero, O.N.; Grzybowski, M.; Song, B.; van Dijk, K.; Schachtman, D.P.; et al. A role for heritable transcriptomic variation in maize adaptation to temperate environments. *BioRxiv* **2022**, *39*. [CrossRef]
172. Dukowic-Schulze, S.; Sundararajan, A.; Mudge, J.; Ramaraj, T.; Farmer, A.D.; Wang, M.; Sun, Q.; Pillardy, J.; Kianian, S.; Retzel, E.F.; et al. The transcriptome landscape of early maize meiosis. *BMC Plant Biol.* **2014**, *14*, 118. [CrossRef] [PubMed]
173. Teoh, K.T.; Requesens, D.V.; Devaiah, S.P.; Johnson, D.; Huang, X.; Howard, J.A.; Hood, E.E. Transcriptome analysis of embryo maturation in maize. *BMC Plant Biol.* **2013**, *13*, 19. [CrossRef] [PubMed]
174. Wu, X.; Wang, B.; Xie, F.; Zhang, L.; Gong, J.; Zhu, W.; Li, X.; Feng, F.; Huang, J. QTL mapping and transcriptome analysis identify candidate genes regulating pericarp thickness in sweet corn. *BMC Plant Biol.* **2020**, *20*, 117. [CrossRef]
175. Zhang, Y.T.; Zhang, Y.L.; Chen, S.X.; Yin, G.H.; Yang, Z.Z.; Lee, S.; Liu, C.G.; Zhao, D.D.; Ma, Y.K.; Song, F.Q.; et al. Proteomics of methyl jasmonate induced defense response in maize leaves against Asian corn borer. *BMC Genom.* **2015**, *16*, 224. [CrossRef]
176. Wang, S.; Chen, Z.; Tian, L.; Ding, Y.; Zhang, J.; Zhou, J.; Liu, P.; Chen, Y.; Wu, L. Comparative proteomics combined with analyses of transgenic plants reveal ZmREM1.3 mediates maize resistance to southern corn rust. *Plant Biotechnol. J.* **2019**, *17*, 2153–2168. [CrossRef]
177. Dong, A.; Yang, Y.; Liu, S.; Zenda, T.; Liu, X.; Wang, Y.; Li, J.; Duan, H. Comparative proteomics analysis of two maize hybrids revealed drought-stress tolerance mechanisms. *Biotechnol. Biotechnol. Equip.* **2020**, *34*, 763–780. [CrossRef]
178. Yue, J.Y.; Wang, L.H.; Dou, X.T.; Wang, Y.J.; Wang, H.Z. Comparative metabolomic profiling in the roots of salt-tolerant and salt-intolerant maize cultivars treated with NaCl stress. *Biol. Plant.* **2020**, *64*, 569–577. [CrossRef]
179. Begcy, K.; Nosenko, T.; Zhou, L.-Z.; Fragner, L.; Weckwerth, W.; Dresselhaus, T. Male sterility in maize after transient heat stress during the tetrad stage of pollen development. *Plant Physiol.* **2019**, *181*, 683–700. [CrossRef]
180. Sun, C.X.; Li, M.Q.; Gao, X.X.; Liu, L.N.; Wu, X.F.; Zhou, J.H. Metabolic response of maize plants to multi-factorial abiotic stresses. *Plant Biol.* **2016**, *18* (Suppl. 1), 120–129. [CrossRef]
181. Adak, A.; Murray, S.; Anderson, S.L. Phenomic data-driven prediction through field-based high throughput phenotyping, and integration with genomic data in maize. In Proceedings of the Plant and Animal Genome XXIX Conference, San Diego, CA, USA, 8–12 January 2022.
182. Xu, Z.; Chen, X.; Lu, X.; Zhao, B.; Yang, Y.; Liu, J. Integrative analysis of transcriptome and metabolome reveal mechanism of tolerance to salt stress in oat (*Avena sativa* L.). *Plant Physiol. Biochem.* **2021**, *160*, 315–328. [CrossRef]
183. Andon, M.B.; Anderson, J.W. State of the Art Reviews: The Oatmeal-Cholesterol Connection: 10 Years Later. *Am. J. Lifestyle Med.* **2008**, *2*, 51–57. [CrossRef]
184. Loskutov, I.; Shelenga, T.; Blinova, E.; Gnutikov, A.; Konarev, A. Metabolomic profiling in evaluation of cultivated oat species with different ploidy level. *BIO Web Conf.* **2021**, *36*, 01026. [CrossRef]
185. Peterson, D.M. Composition and nutritional characteristics of oat grain and products. In *Oat Science and Technology*; Marshall, H.G., Sorrells, M.E., Eds.; Agronomy Monographs; American Society of Agronomy, Crop Science Society of America: Madison, WI, USA, 1992; pp. 265–292. ISBN 9780891181101.
186. Newell, M.A.; Kim, H.J.; Asoro, F.G.; Lauter, A.M.; White, P.J.; Scott, M.P.; Jannink, J.-L. Microenzymatic evaluation of oat (*Avena sativa* L.) β-Glucan for high-throughput phenotyping. *Cereal Chem. J.* **2014**, *91*, 183–188. [CrossRef]
187. Stevens, E.J.; Armstrong, K.W.; Bezar, H.J.; Griffin, W.B. *Fodder Oats: An Overview*; Food and Agriculture Organization of the United Nations: Rome, Italy, 2004; pp. 1–9.
188. Maki, K.C.; Galant, R.; Samuel, P.; Tesser, J.; Witchger, M.S.; Ribaya-Mercado, J.D.; Blumberg, J.B.; Geohas, J. Effects of consuming foods containing oat beta-glucan on blood pressure, carbohydrate metabolism and biomarkers of oxidative stress in men and women with elevated blood pressure. *Eur. J. Clin. Nutr.* **2007**, *61*, 786–795. [CrossRef]
189. Jackson, E.W.; Wise, M.; Bonman, J.M.; Obert, D.E.; Hu, G.; Peterson, D.M. QTLs affecting α-tocotrienol, α-tocopherol, and total tocopherol concentrations detected in the Ogle/TAM O-301 oat mapping population. *Crop Sci.* **2008**, *48*, 2141–2152. [CrossRef]

190. Pretorius, C.J.; Tugizimana, F.; Steenkamp, P.A.; Piater, L.A.; Dubery, I.A. Metabolomics for biomarker discovery: Key signatory metabolic profiles for the identification and discrimination of oat cultivars. *Metabolites* **2021**, *11*, 165. [CrossRef]
191. Ladizinsky, G. Chromosome rearrangements in the hexaploid oats. *Heredity* **1970**, *25*, 457–461. [CrossRef]
192. Kianian, S.F.; Wu, B.C.; Fox, S.L.; Rines, H.W.; Phillips, R.L. Aneuploid marker assignment in hexaploid oat with the C genome as a reference for determining remnant homoeology. *Genome* **1997**, *40*, 386–396. [CrossRef]
193. O'Donoughue, L.S.; Wang, Z.; Röder, M.; Kneen, B.; Leggett, M.; Sorrells, M.E.; Tanksley, S.D. An RFLP-based linkage map of oats based on a cross between two diploid taxa (*Avena atlantica* × *A. hirtula*). *Genome* **1992**, *35*, 765–771. [CrossRef]
194. Tinker, N.A.; Kilian, A.; Wight, C.P.; Heller-Uszynska, K.; Wenzl, P.; Rines, H.W.; Bjørnstad, A.; Howarth, C.J.; Jannink, J.-L.; Anderson, J.M.; et al. New DArT markers for oat provide enhanced map coverage and global germplasm characterization. *BMC Genom.* **2009**, *10*, 39. [CrossRef] [PubMed]
195. Portyanko, V.A.; Hoffman, D.L.; Lee, M.; Holland, J.B. A linkage map of hexaploid oat based on grass anchor DNA clones and its relationship to other oat maps. *Genome* **2001**, *44*, 249–265. [CrossRef] [PubMed]
196. De Koeyer, D.L.; Tinker, N.A.; Wight, C.P.; Deyl, J.; Burrows, V.D.; O'Donoughue, L.S.; Lybaert, A.; Molnar, S.J.; Armstrong, K.C.; Fedak, G.; et al. A molecular linkage map with associated QTLs from a hulless x covered spring oat population. *Theor. Appl. Genet.* **2004**, *108*, 1285–1298. [CrossRef] [PubMed]
197. Zhu, S.; Rossnagel, B.G.; Kaeppler, H.F. Genetic analysis of quantitative trait loci for groat protein and oil content in oat. *Crop Sci.* **2004**, *44*, 254–260. [CrossRef]
198. Oliver, R.E.; Tinker, N.A.; Lazo, G.R.; Chao, S.; Jellen, E.N.; Carson, M.L.; Rines, H.W.; Obert, D.E.; Lutz, J.D.; Shackelford, I.; et al. SNP discovery and chromosome anchoring provide the first physically-anchored hexaploid oat map and reveal synteny with model species. *PLoS ONE* **2013**, *8*, e58068. [CrossRef]
199. Chaffin, A.S.; Huang, Y.-F.; Smith, S.; Bekele, W.A.; Babiker, E.; Gnanesh, B.N.; Foresman, B.J.; Blanchard, S.G.; Jay, J.J.; Reid, R.W.; et al. A consensus map in cultivated hexaploid oat reveals conserved grass synteny with substantial subgenome rearrangement. *Plant Genome* **2016**, *9*. [CrossRef]
200. Huang, Y.-F.; Poland, J.A.; Wight, C.P.; Jackson, E.W.; Tinker, N.A. Using genotyping-by-sequencing (GBS) for genomic discovery in cultivated oat. *PLoS ONE* **2014**, *9*, e102448. [CrossRef]
201. Newell, M.A.; Asoro, F.G.; Scott, M.P.; White, P.J.; Beavis, W.D.; Jannink, J.-L. Genome-wide association study for oat (*Avena sativa* L.) beta-glucan concentration using germplasm of worldwide origin. *Theor. Appl. Genet.* **2012**, *125*, 1687–1696. [CrossRef]
202. Asoro, F.G.; Newell, M.A.; Scott, M.P.; Beavis, W.D.; Jannink, J. Genome-wide association study for beta-glucan concentration in elite north american oat. *Crop Sci.* **2013**, *53*, 542–553. [CrossRef]
203. Long, J.; Holland, J.B.; Munkvold, G.P.; Jannink, J.-L. Responses to selection for partial resistance to crown rust in oat. *Crop Sci.* **2006**, *46*, 1260–1265. [CrossRef]
204. Simons, M.D. Crown Rust. In *Diseases, Distribution, Epidemiology, and Control*; Elsevier: Cambridge, MA, USA, 1985; pp. 131–172. ISBN 9780121484026.
205. Chong, J.; Brown, P.D. Genetics of resistance to Puccinia coronata f. sp. avenae in two Avena Sativa Accessions. *Can. J. Plant Pathol.* **1996**, *18*, 286–292. [CrossRef]
206. Wight, C.P.; O'Donoughue, L.S.; Chong, J.; Tinker, N.A.; Molnar, S.J. Discovery, localization, and sequence characterization of molecular markers for the crown rust resistance genes Pc38, Pc39, and Pc48 in cultivated oat (*Avena sativa* L.). *Mol. Breed.* **2005**, *14*, 349–361. [CrossRef]
207. Hoffman, D.L.; Chong, J.; Jackson, E.W.; Obert, D.E. Characterization and mapping of a crown rust resistance gene complex (pc58) in TAM O-301. *Crop Sci.* **2006**, *46*, 2630–2635. [CrossRef]
208. Jackson, E.W.; Obert, D.E.; Menz, M.; Hu, G.; Avant, J.B.; Chong, J.; Bonman, J.M. Characterization and mapping of oat crown rust resistance genes using three assessment methods. *Phytopathology* **2007**, *97*, 1063–1070. [CrossRef]
209. Kulcheski, F.R.; Graichen, F.A.S.; Martinelli, J.A.; Locatelli, A.B.; Federizzi, L.C.; Delatorre, C.A. Molecular mapping of Pc68, a crown rust resistance gene in Avena sativa. *Euphytica* **2010**, *175*, 423–432. [CrossRef]
210. Bush, A.L.; Wise, R.P. High-resolution mapping adjacent to the Pc71 crown-rust resistance locus in hexaploid oat. *Mol. Breed.* **1998**, *4*, 13–21. [CrossRef]
211. Gnanesh, B.N.; Mitchell Fetch, J.; Menzies, J.G.; Beattie, A.D.; Eckstein, P.E.; McCartney, C.A. Chromosome location and allele-specific PCR markers for marker-assisted selection of the oat crown rust resistance gene Pc91. *Mol. Breed.* **2013**, *32*, 679–686. [CrossRef]
212. Gnanesh, B.N.; McCartney, C.A.; Eckstein, P.E.; Mitchell Fetch, J.W.; Menzies, J.G.; Beattie, A.D. Genetic analysis and molecular mapping of a seedling crown rust resistance gene in oat. *Theor. Appl. Genet.* **2015**, *128*, 247–258. [CrossRef]
213. Carson, M.L. Virulence in Oat Crown Rust (Puccinia coronata f. sp. avenae) in the United States from 2006 through 2009. *Plant Dis.* **2011**, *95*, 1528–1534. [CrossRef]
214. Chong, J.; Leonard, K.J.; Salmeron, J.J. A North American System of Nomenclature for Puccinia coronata f. sp. avenae. *Plant Dis.* **2000**, *84*, 580–585. [CrossRef] [PubMed]
215. Chong, J. Inheritance of resistance to two Puccinia coronata isolates in a partial resistant oat line MN841801. *Acta Phytopathol. Entomol. Hung.* **2000**, *35*, 37–40.

216. Portyanko, V.A.; Chen, G.; Rines, H.W.; Phillips, R.L.; Leonard, K.J.; Ochocki, G.E.; Stuthman, D.D. Quantitative trait loci for partial resistance to crown rust, Puccinia coronata, in cultivated oat, *Avena sativa* L. *Theor. Appl. Genet.* **2005**, *111*, 313–324. [CrossRef] [PubMed]
217. Acevedo, M.; Jackson, E.W.; Chong, J.; Rines, H.W.; Harrison, S.; Bonman, J.M. Identification and validation of quantitative trait loci for partial resistance to crown rust in oat. *Phytopathology* **2010**, *100*, 511–521. [CrossRef]
218. Babiker, E.M.; Gordon, T.C.; Jackson, E.W.; Chao, S.; Harrison, S.A.; Carson, M.L.; Obert, D.E.; Bonman, J.M. Quantitative Trait Loci from Two Genotypes of Oat (Avena sativa) Conditioning Resistance to Puccinia coronata. *Phytopathology* **2015**, *105*, 239–245. [CrossRef]
219. Lin, Y.; Gnanesh, B.N.; Chong, J.; Chen, G.; Beattie, A.D.; Mitchell Fetch, J.W.; Kutcher, H.R.; Eckstein, P.E.; Menzies, J.G.; Jackson, E.W.; et al. A major quantitative trait locus conferring adult plant partial resistance to crown rust in oat. *BMC Plant Biol.* **2014**, *14*, 250. [CrossRef]
220. Zhu, S.; Kaeppler, H.F. Identification of quantitative trait loci for resistance to crown rust in oat line MAM17-5. *Crop Sci.* **2003**, *43*, 358–366. [CrossRef]
221. Winkler, L.R.; Michael Bonman, J.; Chao, S.; Admassu Yimer, B.; Bockelman, H.; Esvelt Klos, K. Population Structure and genotype-phenotype associations in a collection of oat landraces and historic cultivars. *Front. Plant Sci.* **2016**, *7*, 1077. [CrossRef]
222. Montilla-Bascón, G.; Rispail, N.; Sánchez-Martín, J.; Rubiales, D.; Mur, L.A.J.; Langdon, T.; Howarth, C.J.; Prats, E. Genome-wide association study for crown rust (*Puccinia coronata* f. sp. avenae) and powdery mildew (*Blumeria graminis* f. sp. avenae) resistance in an oat (*Avena sativa*) collection of commercial varieties and landraces. *Front. Plant Sci.* **2015**, *6*, 103. [CrossRef]
223. Klos, K.E.; Yimer, B.A.; Babiker, E.M.; Beattie, A.D.; Bonman, J.M.; Carson, M.L.; Chong, J.; Harrison, S.A.; Ibrahim, A.M.H.; Kolb, F.L.; et al. Genome-Wide Association Mapping of Crown Rust Resistance in Oat Elite Germplasm. *Plant Genome* **2017**, *10*, 103. [CrossRef]
224. Yu, J.; Herrmann, M. Inheritance and mapping of a powdery mildew resistance gene introgressed from Avena macrostachya in cultivated oat. *Theor. Appl. Genet.* **2006**, *113*, 429–437. [CrossRef]
225. Simons, M.D. *Oats: A Standardized System of Nomenclature for Genes and Chromosomes and Catalog of Genes Governing Characteristers*; U.S. Dept. of Agriculture, Science and Education Administration: Washington, DC, USA, 1978.
226. Wight, C.P.; Tinker, N.A.; Kianian, S.F.; Sorrells, M.E.; O'Donoughue, L.S.; Hoffman, D.L.; Groh, S.; Scoles, G.J.; Li, C.D.; Webster, F.H.; et al. A molecular marker map in "Kanota" x "Ogle" hexaploid oat (*Avena* spp.) enhanced by additional markers and a robust framework. *Genome* **2003**, *46*, 28–47. [CrossRef]
227. Kianian, S.F.; Egli, M.A.; Phillips, R.L.; Rines, H.W.; Somers, D.A.; Gengenbach, B.G.; Webster, F.H.; Livingston, S.M.; Groh, S.; O'Donoughue, L.S.; et al. Association of a major groat oil content QTL and an acetyl-CoA carboxylase gene in oat. *Theor. Appl. Genet.* **1999**, *98*, 884–894. [CrossRef]
228. Tanhuanpää, P.; Manninen, O.; Kiviharju, E. QTLs for important breeding characteristics in the doubled haploid oat progeny. *Genome* **2010**, *53*, 482–493. [CrossRef]
229. Hizbai, B.T.; Gardner, K.M.; Wight, C.P.; Dhanda, R.K.; Molnar, S.J.; Johnson, D.; Frégeau-Reid, J.; Yan, W.; Rossnagel, B.G.; Holland, J.B.; et al. Quantitative trait loci affecting oil content, oil composition, and other agronomically important traits in oat. *Plant Genome* **2012**, *5*, 104. [CrossRef]
230. Kianian, S.F.; Phillips, R.L.; Rines, H.W.; Fulcher, R.G.; Webster, F.H.; Stuthman, D.D. Quantitative trait loci influencing β-glucan content in oat (Avena sativa, 2n=6x=42). *Theor. Appl. Genet.* **2000**, *101*, 1039–1048. [CrossRef]
231. Gutierrez-Gonzalez, J.J.; Garvin, D.F. Reference Genome-Directed Resolution of Homologous and Homeologous Relationships within and between Different Oat Linkage Maps. *Plant Genome J.* **2011**, *4*, 178. [CrossRef]
232. Gutierrez-Gonzalez, J.J.; Tu, Z.J.; Garvin, D.F. Analysis and annotation of the hexaploid oat seed transcriptome. *BMC Genom.* **2013**, *14*, 471. [CrossRef]
233. Jinqiu, Y.; Bing, L.; Tingting, S.; Jinglei, H.; Zelai, K.; Lu, L.; Wenhua, H.; Tao, H.; Xinyu, H.; Zengqing, L.; et al. Integrated Physiological and Transcriptomic Analyses Responses to Altitude Stress in Oat (*Avena sativa* L.). *Front. Genet.* **2021**, *12*, 638683. [CrossRef]
234. Zechmann, B. Subcellular Roles of Glutathione in Mediating Plant Defense during Biotic Stress. *Plants* **2020**, *9*, 1067. [CrossRef]
235. Allwood, J.W.; Xu, Y.; Martinez-Martin, P.; Palau, R.; Cowan, A.; Goodacre, R.; Marshall, A.; Stewart, D.; Howarth, C. Rapid UHPLC-MS metabolite profiling and phenotypic assays reveal genotypic impacts of nitrogen supplementation in oats. *Metabolomics* **2019**, *15*, 42. [CrossRef] [PubMed]
236. Annicchiarico, P. Alfalfa forage yield and leaf/stem ratio: Narrow-sense heritability, genetic correlation, and parent selection procedures. *Euphytica* **2015**, *205*, 409–420. [CrossRef]
237. Annicchiarico, P.; Barrett, B.; Brummer, E.C.; Julier, B.; Marshall, A.H. Achievements and challenges in improving temperate perennial forage legumes. *CRC Crit. Rev. Plant Sci.* **2015**, *34*, 327–380. [CrossRef]
238. Blondon, F.; Marie, D.; Brown, S.; Kondorosi, A. Genome size and base composition in Medicago sativa and M. truncatula species. *Genome* **1994**, *37*, 264–270. [CrossRef] [PubMed]
239. Parajuli, A.; Yu, L.-X.; Peel, M.; See, D.; Wagner, S.; Norberg, S.; Zhang, Z. Self-incompatibility, Inbreeding Depression, and Potential to Develop Inbred Lines in Alfalfa. In *The Alfalfa Genome*; Yu, L.-X., Kole, C., Eds.; Compendium of Plant Genomes; Springer International Publishing: Cham, Switzerland, 2021; pp. 255–269. ISBN 978-3-030-74465-6.

240. Li, X.; Brummer, E.C. Applied genetics and genomics in alfalfa breeding. *Agronomy* **2012**, *2*, 40–61. [CrossRef]
241. Hawkins, C.; Yu, L.-X. Recent progress in alfalfa (*Medicago sativa* L.) genomics and genomic selection. *Crop J.* **2018**, *6*, 565–575. [CrossRef]
242. Brummer, E.C.; Bouton, J.H.; Kochert, G. Development of an RFLP map in diploid alfalfa. *Theor. Appl. Genet.* **1993**, *86*, 329–332. [CrossRef]
243. Echt, C.S.; Kidwell, K.K.; Knapp, S.J.; Osborn, T.C.; McCoy, T.J. Linkage mapping in diploid alfalfa (Medicago sativa). *Genome* **1994**, *37*, 61–71. [CrossRef]
244. Diwan, N.; Bouton, J.H.; Kochert, G.; Cregan, P.B. Mapping of simple sequence repeat (SSR) DNA markers in diploid and tetraploid alfalfa. *Theor. Appl. Genet.* **2000**, *101*, 165–172. [CrossRef]
245. Brouwer, D.J.; Duke, S.H.; Osborn, T.C. Mapping Genetic Factors Associated with Winter Hardiness, Fall Growth, and Freezing Injury in Autotetraploid Alfalfa. *Crop Sci.* **2000**, *40*, 1387. [CrossRef]
246. Julier, B.; Flajoulot, S.; Barre, P.; Cardinet, G.; Santoni, S.; Huguet, T.; Huyghe, C. Construction of two genetic linkage maps in cultivated tetraploid alfalfa (*Medicago sativa*) using microsatellite and AFLP markers. *BMC Plant Biol.* **2003**, *3*, 9. [CrossRef]
247. Sledge, M.K.; Ray, I.M.; Jiang, G. An expressed sequence tag SSR map of tetraploid alfalfa (*Medicago sativa* L.). *Theor. Appl. Genet.* **2005**, *111*, 980–992. [CrossRef]
248. Hackett, C.A.; Luo, Z.W. TetraploidMap: Construction of a linkage map in autotetraploid species. *J. Hered.* **2003**, *94*, 358–359. [CrossRef]
249. Li, X.; Wei, Y.; Acharya, A.; Jiang, Q.; Kang, J.; Brummer, E.C. A saturated genetic linkage map of autotetraploid alfalfa (*Medicago sativa* L.) developed using genotyping-by-sequencing is highly syntenous with the Medicago truncatula genome. *G3: Genes Genomes Genet.* **2014**, *4*, 1971–1979. [CrossRef]
250. Choi, H.-K.; Kim, D.; Uhm, T.; Limpens, E.; Lim, H.; Mun, J.-H.; Kalo, P.; Penmetsa, R.V.; Seres, A.; Kulikova, O.; et al. A sequence-based genetic map of Medicago truncatula and comparison of marker colinearity with M. sativa. *Genetics* **2004**, *166*, 1463–1502. [CrossRef]
251. Li, X.; Wei, Y.; Moore, K.J.; Michaud, R.; Viands, D.R.; Hansen, J.L.; Acharya, A.; Brummer, E.C. Association mapping of biomass yield and stem composition in a tetraploid alfalfa breeding population. *Plant Genome* **2011**, *4*, 122. [CrossRef]
252. Ray, I.M.; Han, Y.E.L.; Meenach, C.D.; Santantonio, N.; Sledge, M.K.; Pierce, C.A.; Sterling, T.M.; Kersey, R.K.; Bhandari, H.S.; Monteros, M.J. Identification of quantitative trait loci for alfalfa forage biomass productivity during drought stress. *Crop Sci.* **2015**, *55*, 2012–2033. [CrossRef]
253. Li, X.; Wang, X.; Wei, Y.; Brummer, E.C. Prevalence of segregation distortion in diploid alfalfa and its implications for genetics and breeding applications. *Theor. Appl. Genet.* **2011**, *123*, 667–679. [CrossRef]
254. Sakiroglu, M.; Sherman-Broyles, S.; Story, A.; Moore, K.J.; Doyle, J.J.; Charles Brummer, E. Patterns of linkage disequilibrium and association mapping in diploid alfalfa (*M. sativa* L.). *Theor. Appl. Genet.* **2012**, *125*, 577–590. [CrossRef]
255. Kang, J.; Zhang, T.; Wang, M.; Zhang, Y.; Yang, Q. Research progress in the quantitative trait loci (QTL) and genomic selection of alfalfa. *Acta Prataculturae Sin.* **2014**, *23*, 304–312.
256. Jia, C.; Wu, X.; Chen, M.; Wang, Y.; Liu, X.; Gong, P.; Xu, Q.; Wang, X.; Gao, H.; Wang, Z. Identification of genetic loci associated with crude protein and mineral concentrations in alfalfa (Medicago sativa) using association mapping. *BMC Plant Biol.* **2017**, *17*, 97. [CrossRef]
257. Elshire, R.J.; Glaubitz, J.C.; Sun, Q.; Pol, J.A.; Kawamoto, K.; Buckler, E.S. A robust, simple genotyping-by-sequencing (GBS) approach for high diversity species. *PLoS ONE* **2011**, *6*, e19379. [CrossRef]
258. Adhikari, L.; Lindstrom, O.M.; Markham, J.; Missaoui, A.M. Dissecting key adaptation traits in the polyploid perennial Medicago Sativa using GBS-SNP mapping. *Front. Plant Sci.* **2018**, *9*, 934. [CrossRef]
259. Zhang, F.; Kang, J.; Long, R.; Yu, L.-X.; Sun, Y.; Wang, Z.; Zhao, Z.; Zhang, T.; Yang, Q. Construction of high-density genetic linkage map and mapping quantitative trait loci (QTL) for flowering time in autotetraploid alfalfa (*Medicago sativa* L.) using genotyping by sequencing. *Plant Genome* **2020**, *13*, e20045. [CrossRef]
260. Rosyara, U.R.; De Jong, W.S.; Douches, D.S.; Endelman, J.B. Software for genome-wide association studies in autopolyploids and its application to potato. *Plant Genome* **2016**, *9*, 123. [CrossRef]
261. Shen, C.; Du, H.; Chen, Z.; Lu, H.; Zhu, F.; Chen, H.; Meng, X.; Liu, Q.; Liu, P.; Zheng, L.; et al. The chromosome-level genome sequence of the autotetraploid alfalfa and resequencing of core germplasms provide genomic resources for alfalfa research. *Mol. Plant* **2020**, *13*, 1250–1261. [CrossRef]
262. Li, X.; Wei, Y.; Acharya, A.; Hansen, J.L.; Crawford, J.L.; Viands, D.R.; Michaud, R.; Claessens, A.; Brummer, E.C. Genomic prediction of biomass yield in two selection cycles of a tetraploid alfalfa breeding population. *Plant Genome* **2015**, *8*, 134. [CrossRef]
263. Medina, C.A.; Hawkins, C.; Liu, X.-P.; Peel, M.; Yu, L.-X. Genome-wide association and prediction of traits related to salt tolerance in autotetraploid alfalfa (*Medicago sativa* L.). *Int. J. Mol. Sci.* **2020**, *21*, 3361. [CrossRef]
264. Jia, C.; Zhao, F.; Wang, X.; Han, J.; Zhao, H.; Liu, G.; Wang, Z. Genomic Prediction for 25 Agronomic and Quality Traits in Alfalfa (Medicago sativa). *Front. Plant Sci.* **2018**, *9*, 1220. [CrossRef]
265. Yang, S.S.; Xu, W.W.; Tesfaye, M.; Lamb, J.F.S.; Jung, H.G.; Samac, D.A.; Vance, C.P.; Gronwald, J.W. Single-feature polymorphism discovery in the transcriptome of tetraploid alfalfa. *Plant Genome* **2009**, *2*, 155. [CrossRef]
266. Tesfaye, M.; Silverstein, K.A.T.; Bucciarelli, B.; Samac, D.A.; Vance, C.P. The AffymetrixMedicago GeneChiparray is applicable for transcript analysis of alfalfa (*Medicago sativa*). *Functional Plant Biol.* **2006**, *33*, 783. [CrossRef] [PubMed]

267. Yang, S.S.; Xu, W.W.; Tesfaye, M.; Lamb, J.F.S.; Jung, H.-J.G.; VandenBosch, K.A.; Vance, C.P.; Gronwald, J.W. Transcript profiling of two alfalfa genotypes with contrasting cell wall composition in stems using a cross-species platform: Optimizing analysis by masking biased probes. *BMC Genom.* **2010**, *11*, 323. [CrossRef] [PubMed]
268. Yang, S.S.; Tu, Z.J.; Cheung, F.; Xu, W.W.; Lamb, J.F.S.; Jung, H.-J.G.; Vance, C.P.; Gronwald, J.W. Using RNA-Seq for gene identification, polymorphism detection and transcript profiling in two alfalfa genotypes with divergent cell wall composition in stems. *BMC Genom.* **2011**, *12*, 199. [CrossRef] [PubMed]
269. Li, X.; Acharya, A.; Farmer, A.D.; Crow, J.A.; Bharti, A.K.; Kramer, R.S.; Wei, Y.; Han, Y.; Gou, J.; May, G.D.; et al. Prevalence of single nucleotide polymorphism among 27 diverse alfalfa genotypes as assessed by transcriptome sequencing. *BMC Genom.* **2012**, *13*, 568. [CrossRef] [PubMed]
270. Postnikova, O.A.; Shao, J.; Nemchinov, L.G. Analysis of the alfalfa root transcriptome in response to salinity stress. *Plant Cell Physiol.* **2013**, *54*, 1041–1055. [CrossRef]
271. Dong, W.; Liu, X.; Li, D.; Gao, T.; Song, Y. Transcriptional profiling reveals that a MYB transcription factor MsMYB4 contributes to the salinity stress response of alfalfa. *PLoS ONE* **2018**, *13*, e0204033. [CrossRef]
272. Lei, Y.; Xu, Y.; Hettenhausen, C.; Lu, C.; Shen, G.; Zhang, C.; Li, J.; Song, J.; Lin, H.; Wu, J. Comparative analysis of alfalfa (*Medicago sativa* L.) leaf transcriptomes reveals genotype-specific salt tolerance mechanisms. *BMC Plant Biol.* **2018**, *18*, 35. [CrossRef]
273. Han, Y.; Kang, Y.; Torres-Jerez, I.; Cheung, F.; Town, C.D.; Zhao, P.X.; Udvardi, M.K.; Monteros, M.J. Genome-wide SNP discovery in tetraploid alfalfa using 454 sequencing and high resolution melting analysis. *BMC Genom.* **2011**, *12*, 350. [CrossRef]
274. Shu, Y.; Li, W.; Zhao, J.; Zhang, S.; Xu, H.; Liu, Y.; Guo, C. Transcriptome sequencing analysis of alfalfa reveals CBF genes potentially playing important roles in response to freezing stress. *Genet. Mol. Biol.* **2017**, *40*, 824–833. [CrossRef]
275. Nemchinov, L.G.; Shao, J.; Lee, M.N.; Postnikova, O.A.; Samac, D.A. Resistant and susceptible responses in alfalfa (Medicago sativa) to bacterial stem blight caused by Pseudomonas syringae pv. *syringae*. *PLoS ONE* **2017**, *12*, e0189781. [CrossRef]
276. Liu, Z.; Chen, T.; Ma, L.; Zhao, Z.; Zhao, P.X.; Nan, Z.; Wang, Y. Global transcriptome sequencing using the Illumina platform and the development of EST-SSR markers in autotetraploid alfalfa. *PLoS ONE* **2013**, *8*, e83549. [CrossRef]
277. O'Rourke, J.A.; Fu, F.; Bucciarelli, B.; Yang, S.S.; Samac, D.A.; Lamb, J.F.S.; Monteros, M.J.; Graham, M.A.; Gronwald, J.W.; Krom, N.; et al. The Medicago sativa gene index 1.2: A web-accessible gene expression atlas for investigating expression differences between Medicago sativa subspecies. *BMC Genom.* **2015**, *16*, 502. [CrossRef]
278. Luo, D.; Zhou, Q.; Wu, Y.; Chai, X.; Liu, W.; Wang, Y.; Yang, Q.; Wang, Z.; Liu, Z. Full-length transcript sequencing and comparative transcriptomic analysis to evaluate the contribution of osmotic and ionic stress components towards salinity tolerance in the roots of cultivated alfalfa (*Medicago sativa* L.). *BMC Plant Biol.* **2019**, *19*, 32. [CrossRef]
279. Medina, C.A.; Samac, D.A.; Yu, L.-X. Pan-transcriptome identifying master genes and regulation network in response to drought and salt stresses in Alfalfa (*Medicago sativa* L.). *Sci. Rep.* **2021**, *11*, 17203. [CrossRef]
280. Jiang, X.; Yang, X.; Zhang, F.; Yang, T.; Yang, C.; He, F.; Gao, T.; Wang, C.; Yang, Q.; Wang, Z.; et al. Combining QTL mapping and RNA-Seq unravels candidate genes for alfalfa (*Medicago sativa* L.) leaf development. *BMC Plant Biol.* **2022**, *22*, 485. [CrossRef]
281. Song, T.; Xu, H.; Sun, N.; Jiang, L.; Tian, P.; Yong, Y.; Yang, W.; Cai, H.; Cui, G. Metabolomic analysis of alfalfa (*Medicago sativa* L.) root-symbiotic rhizobia responses under alkali stress. *Front. Plant Sci.* **2017**, *8*, 1208. [CrossRef]
282. Aranjuelo, I.; Molero, G.; Erice, G.; Avice, J.C.; Nogués, S. Plant physiology and proteomics reveals the leaf response to drought in alfalfa (*Medicago sativa* L.). *J. Exp. Bot.* **2011**, *62*, 111–123. [CrossRef]
283. Fan, W.; Ge, G.; Liu, Y.; Wang, W.; Liu, L.; Jia, Y. Proteomics integrated with metabolomics: Analysis of the internal causes of nutrient changes in alfalfa at different growth stages. *BMC Plant Biol.* **2018**, *18*, 78. [CrossRef]
284. Zhang, C.; Shi, S. Physiological and proteomic responses of contrasting alfalfa (*Medicago sativa* L.) varieties to peg-induced osmotic stress. *Front. Plant Sci.* **2018**, *9*, 242. [CrossRef]
285. Li, Q.; Xing, Y.; Fu, X.; Ji, L.; Li, T.; Wang, J.; Chen, G.; Qi, Z.; Zhang, Q. Biochemical mechanisms of rhizospheric Bacillus subtilis-facilitated phytoextraction by alfalfa under cadmium stress—Microbial diversity and metabolomics analyses. *Ecotoxicol. Environ. Saf.* **2021**, *212*, 112016. [CrossRef]
286. Chen, L.; Xia, F.; Wang, M.; Wang, W.; Mao, P. Metabolomic analyses of alfalfa (Medicago sativa L. cv. 'Aohan') reproductive organs under boron deficiency and surplus conditions. *Ecotoxicol. Environ. Saf.* **2020**, *202*, 111011. [CrossRef] [PubMed]
287. Annicchiarico, P.; Nazzicari, N.; Brummer, E.C. Alfalfa genomic selection: Challenges, strategies, transnational cooperation. In *Breeding in a World of Scarcity*; Roldán-Ruiz, I., Baert, J., Reheul, D., Eds.; Springer International Publishing: Cham, Switzerland, 2016; pp. 145–149. ISBN 978-3-319-28930-4.
288. Chandel, A.K.; Khot, L.R.; Yu, L.-X. Alfalfa (*Medicago sativa* L.) crop vigor and yield characterization using high-resolution aerial multispectral and thermal infrared imaging technique. *Comput. Electron. Agric.* **2021**, *182*, 105999. [CrossRef]
289. Feng, L.; Zhang, Z.; Ma, Y.; Du, Q.; Williams, P.; Drewry, J.; Luck, B. Alfalfa Yield Prediction Using UAV-Based Hyperspectral Imagery and Ensemble Learning. *Remote Sens.* **2020**, *12*, 2028. [CrossRef]
290. Yang, W.; Feng, H.; Zhang, X.; Zhang, J.; Doonan, J.H.; Batchelor, W.D.; Xiong, L.; Yan, J. Crop phenomics and high-throughput phenotyping: Past decades, current challenges, and future perspectives. *Mol. Plant* **2020**, *13*, 187–214. [CrossRef] [PubMed]
291. Cazenave, A.; Shah, K.; Trammell, T.; Komp, M.; Hoffman, J.; Motes, C.M.; Monteros, M.J. High-throughput approaches for phenotyping alfalfa germplasm under abiotic stress in the field. *Plant Phenome J.* **2019**, *2*, 1–13. [CrossRef]

292. Biswas, A.; Andrade, M.H.M.L.; Acharya, J.P.; de Souza, C.L.; Lopez, Y.; de Assis, G.; Shirbhate, S.; Singh, A.; Munoz, P.; Rios, E.F. Phenomics-assisted selection for herbage accumulation in alfalfa (*Medicago sativa* L.). *Front. Plant Sci.* **2021**, *12*, 756768. [CrossRef]
293. Tang, Z.; Parajuli, A.; Chen, C.J.; Hu, Y.; Revolinski, S.; Medina, C.A.; Lin, S.; Zhang, Z.; Yu, L.-X. Validation of UAV-based alfalfa biomass predictability using photogrammetry with fully automatic plot segmentation. *Sci. Rep.* **2021**, *11*, 3336. [CrossRef]
294. Hu, X.; Yang, L.; Zhang, Z.; Wang, Y. Differentiation of alfalfa and sweet clover seeds via multispectral imaging. *Seed Sci. Technol.* **2020**, *48*, 83–99. [CrossRef]
295. Bucciarelli, B.; Xu, Z.; Ao, S.; Cao, Y.; Monteros, M.J.; Topp, C.N.; Samac, D.A. Phenotyping seedlings for selection of root system architecture in alfalfa (*Medicago sativa* L.). *Plant Methods* **2021**, *17*, 125. [CrossRef]
296. Feng, L.; Zhang, Z.; Ma, Y.; Sun, Y.; Du, Q.; Williams, P.; Drewry, J.; Luck, B. Multitask Learning of Alfalfa Nutritive Value From UAV-Based Hyperspectral Images. *IEEE Geosci. Remote Sens. Lett.* **2022**, *19*, 1–5. [CrossRef]
297. Santantonio, N. *Evaluating Approaches to High-Throughput Phenotyping and Genotyping for Genomic Selection in Alfalfa*; U.S. Alfalfa Farmer Research Initiative: Manhattan, KS, USA, 2021; pp. 1–21.

current issues in molecular biology

Article

Genetic Relationship of Brassicaceae Hybrids with Various Resistance to Blackleg Is Disclosed by the Use of Molecular Markers

Justyna Szwarc [1,*], Janetta Niemann [1], Joanna Kaczmarek [2], Jan Bocianowski [3] and Dorota Weigt [1]

[1] Department of Genetics and Plant Breeding, Poznań University of Life Sciences, Dojazd 11, 60-632 Poznan, Poland
[2] Institute of Plant Genetics of the Polish Academy of Sciences, Strzeszyńska 34, 60-479 Poznań, Poland
[3] Department of Mathematical and Statistical Methods, Poznań University of Life Sciences, Wojska Polskiego 28, 60-637 Poznań, Poland
* Correspondence: justyna.szwarc@up.poznan.pl

Abstract: *Brassica napus* is an important oil source. Its narrow gene pool can be widened by interspecific hybridization with the Brassicaceae species. One of the agronomically important traits, that can be transferred through the hybridization, is the resistance to blackleg, a dangerous disease mainly caused by *Leptosphaeria maculans*. Hybrid individuals can be analyzed with various molecular markers, including Simple Sequence Repeats (SSR). We investigated the genetic similarity of 32 Brassicaceae hybrids and 19 parental components using SSR markers to reveal their genetic relationship. Furthermore, we compared the field resistance to blackleg of the interspecific progenies. The tested set of 15 SSR markers proved to be useful in revealing the genetic distances in the Brassicaceae hybrids and species. However, genetic similarity of the studied hybrids could not be correlated with the level of field resistance to *L. maculans*. Moreover, our studies confirmed the usefulness of the Brassicaceae hybrids in terms of blackleg management.

Keywords: Brassicaceae; interspecific hybrids; SSR markers; *Leptosphaeria maculans*; genetic similarity

Citation: Szwarc, J.; Niemann, J.; Kaczmarek, J.; Bocianowski, J.; Weigt, D. Genetic Relationship of Brassicaceae Hybrids with Various Resistance to Blackleg Is Disclosed by the Use of Molecular Markers. *Curr. Issues Mol. Biol.* **2022**, *44*, 4290–4302. https://doi.org/10.3390/cimb44090295

Academic Editor: Julius Liobikas

Received: 1 August 2022
Accepted: 15 September 2022
Published: 17 September 2022

1. Introduction

Brassicaceae is a family of high agroeconomic importance comprising fodder, oilseed plants, vegetables, ornamental species, as well as plants of medical and scientific importance. Furthermore, ecological, morphological, and genetic diversity of this family makes it a perfect model for relationship and evolution studies [1]. The genus *Brassica* contains three diploid species, i.e., *B. rapa* (AA genome), *B. nigra* (BB genome), and *B. oleracea* (CC genome), and allotetraploid species obtained as a result of natural interspecific crosses, namely *B. napus* (AACC), *B. juncea* (AABB), and *B. carinata* (BBCC). Another representative of the Brassicaceae family is *Sinapis alba*, a yellow mustard plant closely related to *Brassica*, well known for possessing many potentially useful traits [2].

Brassica napus (rapeseed) is one of the most important oil crops, accounting for over 12% of worldwide oil production (USDA). Due to a relatively short history of cultivation and use of conventional breeding methods, rapeseed displays limited genetic diversity [3,4]; therefore, it seems crucial to expand the *B. napus* gene pool. One of the most effective approaches to solve this problem is interspecific hybridization [5]. Crossing the rapeseed with different species may help to enrich the *B. napus* germplasm and to enable the transfer of genome fragments carrying desirable traits, which could further improve the cultivar's characteristics. The sexual incompatibility and differences in the genome sizes of parental components may result in hybridization failure [6]. Barriers of interspecific hybridization can be overcome by implementing in vitro techniques, including ovary, ovule, and embryo rescue [7]. The Department of Genetics and Plant Breeding of Poznań University of Life

Sciences has great experience in creating interspecific Brassicaceae hybrids, which are profoundly analyzed in terms of chromosomal constitution, morphology, as well as insect and pathogen resistance. Recently developed hybrids showed a significant variability of blackleg resistance in field conditions. Blackleg, mainly caused by *L. maculans*, is a fungal disease which can cause significant yield losses [8]. The reliance on commercial cultivars with a single resistance source increases the selective pressure on pathogens and accelerates its evolution. Management of blackleg disease includes proper agronomic practices (such as crop rotation and tillage), the use of fungicides, weed control, the use of certified seeds and the use of resistant cultivars [9]. The breeding of resistant cultivars is environmentally friendly and is a reliable method of controlling blackleg disease [10]. It relies on the existence of naturally resistant genotypes, which can be used as a donor of certain genes conferring blackleg resistance.

Various molecular marker systems such as RFLP, SSR, and RAPD can be used to determine the genetic distance of the Brassicaceae species [11]. Simple Sequence Repeats (SSR) or microsatellites are defined as tandem repeats of short nucleotide motifs, usually consisting of 1–6 base pairs [12]. They occur frequently in eucaryotic organisms, and the variation in repeat numbers results in a high degree of polymorphism. The random distribution of SSR loci in plant genomes allows for genetic differentiation within and between species [13]. Moreover, it defines the high utility of SSR markers for cruciferous plants, as the microsatellite loci among members of the Brassicaceae family show high variation in length, which subsequently permits the differentiation of species [14,15]. The SSR markers had been previously used in numerous Brassicaceae studies, including unraveling the genetic variation and species diversity [14,16,17], species and cultivar differentiation [13,18], and estimation of genetic distances [19].

We are aiming to gain insight into the genetic relationship between hybrids with different parental components, which are diverse in terms of resistance to blackleg. Therefore, the objectives of this research are to determine the genetic similarity of hybrid and parental genotypes from the Brassicaceae family and to evaluate the usefulness of the chosen SSR markers for genetic diversity analysis.

2. Materials and Methods

A total of 32 various Brassicaceae hybrids of F_9 and F_{10} generation and 19 parental genotypes were used as research material (Table 1). Interspecific hybrids of the F_1 generation were developed at the Department of Genetics and Plant Breeding (Poznań University of Life Sciences), with the use of in vitro cultures. Next, selected combinations were self-pollinated multiple times in order to obtain stable hybrid lines.

Table 1. List of Brassicaceae genotypes used in this study and groups for the analysis of molecular variance (AMOVA).

No of Genotype	Combination/Species	Group
1	*B. napus* cv. Jet Neuf × *B. carinata* PI 649091	1
2	*B. napus* cv. Lisek × *B. carinata* Dodola	1
3	*B. napus* cv. Jet Neuf × *B. carinata*—PI 649094	1
4	*B. napus* cv. Jet Neuf × *B. carinata*—PI 649096	1
5	*B. carinata* 1	1
6	*B. carinata* 2	1
7	*B. carinata* 3	1
8	*B. carinata* 4	1
9	*B. carinata* cv. Dodola	1
10	*B. carinata* PI 596534	1
11	*B. napus* cv. Górczański × *B. rapa* ssp. *chinensis*	2
12	*B. napus* cv. Zhongshuang9 × *B. rapa* ssp. *chinensis* 08 007574	2
13	*B. rapa* ssp. *chinensis* (COBORU)	2
14	*B. rapa* ssp. *chinensis* PI430485 98CI	2

Table 1. *Cont.*

No of Genotype	Combination/Species	Group
15	*B. napus* cv. Lisek × *B. rapa* Pak Choi 08, 007574	2
16	*B. napus* cv. Lisek × *B. rapa* Pak Choi 08, 007569	2
17	*B. napus* cv. Górczański × *B. rapa* Pak Choi 08, 007574	2
18	*B. fruticulosa* PI 649097	3
19	*B. napus* cv. Californium × *B. fruticulosa*—PI649097	3
20	*B. napus* cv. Lisek × *B. fruticulosa*—PI649097	3
21	*B.napus* cv. Anderson	1, 2, 3, 4, 5, 6, 7
22	*B. napus* cv. Monolit	1, 2, 3, 4, 5, 6, 7
23	*B. napus* cv. Skrzeszowicki	1, 2, 3, 4, 5, 6, 7
24	*B. napus* cv. Lisek	1, 2, 3, 4, 5, 6, 7
25	*B. napus* cv. Californium × *B. oleracea* var. *alboglabra*	4
26	*B. napus* cv. Jet Neuf × *B. rapa* ssp. *pekinensis* 08 007569	5
27	*B. napus* cv. Jet Neuf × *B. rapa* ssp. *pekinensis* 08 007574	5
28	*B. napus* cv. Górczański × *B. rapa* ssp. *pekinensis* 08.007574	5
29	*B. napus* cv. Górczański × *B. rapa* ssp. *pekinensis* 08.007569	5
30	*B. napus* cv. Californium × *B. rapa* ssp. *pekinensis* 08 007574	5
31	*B. napus* cv. Californium × *B. rapa* ssp. *pekinensis* 08 007574-1	5
32	*B. napus* cv. Californium × *B. rapa* ssp. *pekinensis* 08 007574-2	5
33	*B. napus* cv. Californium × *B. rapa* ssp. *pekinensis* 08 007574-3	5
34	*B. napus* cv. Zhongshuang9 × *B. rapa* ssp. *pekinensis* 08 006169	5
35	*B. napus* MS8 line × *B. rapa* ssp. *pekinensis* 08 006169	5
36	*B. napus* MS8 line × *B. rapa* ssp. *pekinensis* 08 006169	5
37	*B. napus* MS8 line × *B. rapa* ssp. *pekinensis* 08 006169	5
38	*B. rapa* ssp. *pekinensis* 08, 007569	5
39	*B. rapa* ssp. *pekinensis* 08, 007574	5
40	*B. rapa* ssp. *pekinensis* (COBORU)	5
41	*B. napus* cv. Lisek × *B. oleracea* var. *alboglabra*	4
42	*B. napus* cv. Jet Neuf × *S. alba* cv. Bamberka	6
43	*B. napus* cv. Lisek × *B. fruticulosa*—PI649099	3
44	*B. napus* cv. Lisek × *S. alba* cv. Bamberka	6
45	*B. napus* cv. Lisek × *B. tournefortii*	7
46	*B. napus* cv. Jet Neuf × *B. oleracea* var. *alboglabra*	4
47	*B. napus* cv. Californium × *S. alba* cv. Bamberka	6
48	*B. rapa* ssp. *pekinensis* 08 006169	5
49	*B. oleracea* var. *alboglabra*	4
50	*S. alba* cv. Bamberka	6
51	*B. napus* cv. Zhongshuang9 × *B. rapa* ssp. *pekinensis* 08 006169 2	5

2.1. Molecular Analysis

15 SSR markers were selected to characterize the genetic background of the research material. The markers were chosen according to the literature data [20]. This set of microsatellites was developed from *B. rapa* using the ISSR-suppression-PCR method. Preliminary screening was performed in order to assess their usefulness in the present study. Genomic DNA was extracted from young seedling leaves of the studied individuals using the Genomic Mini AX Plant kit (A&A Biotechnology, Gdańsk, Poland) according to the manufacturer's protocol. PCR was performed in a total volume of 12.5 µL (6.25 µL OptiTaq Master Mix (EURx, Gdańsk, Poland), 2 × 0.5 µL primers, 4.25 µL H_2O, and 1 µL DNA template) under the following conditions: initial denaturation at 94 °C for 5 min, thirty-five cycles of amplification (denaturing at 94 °C for 45 s, annealing at primer specific temperature for 45 s, extension at 72 °C for 1.5 min), followed by a final extension step at 72 °C for 7 min. Primer sequences and annealing temperatures are presented in Table 2.

Table 2. Primer sequences and annealing temperatures of SSR markers used in the study.

SSR Marker	Primer Sequences	Annealing Temperature
mstg001	F: CAT GAG TTT TCA TAA ATA AAA R: TAT GCA ACT TGT CTT TGA TAT	41 °C
mstg004	F: CAT ATA TAG CAT GAG TGG TGC R: CTT AAA GGG CAC TCT TTC ATG	47 °C
mstg008	F: TCT CTT TGA AAT CTC AAC CCA R: AGA TGG CAT GTT AAA CTG AAC	47 °C
mstg012	F: TGA TAC ATA GAC TTG GTG GTG R: CGG CAT TAT CTT GAA CAC GTT	48 °C
mstg013	F: AGA TTT GGC TTA CAC GAC GAC R: ATA TAC CAG GTA CCG TCA CTC	50 °C
mstg016	F: CGT TAC ATT CGG GTA TCA CTA R: TCA TCG AAA GCC TTG TAA CTG	48 °C
mstg025	F: AGA GGC AGT TAC GTT CAC GTC R: CAT CGC ACT CGT GTC TCT TTC	52 °C
mstg027	F: CTC TTT TGG TCA GCT TCC TCA R: TTG TTA GTT AGA TCC TCG TAG	48 °C
mstg028	F: GCC AAG AAG ACG AAG ATT CTC R: AGG TTC TCG ATT TAG GAA CCG	49 °C
mstg033	F: ATG TAA GCA TCT TTG ATC TGC R: CTT GAT CTT CCT GAT GTA CTC	46 °C
mstg034	F: CGA CTG GTA ATA TTC TGA TAC R: CAT GAA AGA CTC TCA AAT CCC	46 °C
mstg038	F: GAA TGG TGG TTC TTG TGT GTC R: CAA AGC GAA GCT CTT GAA TTG	49 °C
mstg039	F: TAC TCG CTC TTG TTG AAG CTG R: GAC AAT CTT GGA GTC ATC TCG	50 °C
mstg042	F: GAT ATT CGA TCC GCT TCG ACA R: CGA ATA TCT CAT CCA CTT TGT	49 °C
mstg052	F: AGT AAC ATG TTT TCT TTT GTG R: CAT CAG ATG CTC AAG GAA CTT	46 °C
mstg055	F: ACA CGC GCC TAT GCA GAA TAC R: CTT AGC GAT TAC GGT GAA GCC	52 °C

Electrophoresis was performed on agarose gel stained with Midori Green Advance (Nippon Genetics, Düren, Deutchland), 5 µL per 100 mL of TBE buffer. All image data obtained from the electrophoresis gels were examined in the same way: for each marker, the presence or absence of a band of particular size was scored as '1' or '0', respectively. Next, a binary data matrix was created which was further analyzed with Peak Scanner Software v1.0 (Applied Biosystems, Waltham, MA, USA).

2.2. Statistical Analysis

The polymorphic information content (PIC) was calculated for each marker using the formula:

$$PIC_i = 1 - \sum_{j=1}^{k} p_{ij}{}^2,$$

where p_{ij} denotes frequency of the jth allele for i-th marker among a total of k alleles [21,22].

Genetic similarity (GS) was estimated for each pair of genotypes on the basis of Nei and Li [23]:

$$GS = \frac{2N_{AB}}{N_A + N_B},$$

where N_{AB} denotes the number of bands in genotypes **A** and **B**, N_A and N_B denote the number of bands in **A** and **B**, respectively. The similarity matrix was used to construct a dendrogram using the unweighted pair group method with arithmetic mean (UPGMA) to determine genetic relationships among the genotypes studied. The principal component analysis (PCA) was calculated on the basis of the similarity matrix. All the analyses were conducted using the GenStat 18.2 edition (VSN International Ltd., Hemel Hempstead, UK) statistical software package. The analysis of molecular variance (AMOVA) was made using GenAlEx 6.5 [24]. AMOVA estimated and partitioned the total molecular variance between and within the groups of genotypes and tested the partitioned variance components [25]. The population genetic structure coefficient (F_{ST}) was calculated using the formula:

$$F_{ST} = \frac{H_T - H_S}{H_T},$$

where H_T denotes the probability that two alleles drawn at random from the entire group differ in state and H_S denotes the probability that two alleles drawn at random from a subgroup differ in state. Groups for AMOVA, presented in the Table 1, were created by organizing the genotypes on the basis of their parental components, e.g., *B. napus* × *S. alba* hybrids were grouped together with *S. alba*. Four *B. napus* cultivars were added to each group.

2.3. Resistance to Blackleg

All hybrid combinations have been studied in terms of resistance to phoma leaf spotting/blackleg in field conditions. The assessment was carried out in testing fields at the Poznań University of Life Sciences experimental station Dłoń, located in Wielkopolska Voivodeship. The soil and weather conditions were typical for this region of Poland, and no fungicides or pesticides were used on the testing field. The agricultural practices were optimal for the local ecological conditions. The experiment was set up in a completely randomized block design with five replications; the size of a single plot was 10 m^2 with a 0.30 m row distance and a sowing density of 60 seeds per square meter. The assessment was performed in two terms, i.e., in November, BBCH 19 (term I), and July, BBCH 70-89 (term II). Phoma leaf spotting (term I) was evaluated according to the scale from 0 to 4, where 0 was no visible disease symptoms and 4 was numerous (over 10) leaf spots per plant [26]. The blackleg symptoms (term II) were assessed according to a scale from 0 to 9, where 0 was no visible symptoms and 9 was a plant totally damaged by the disease [26]. Obtained scale values were subsequently transformed into percentage values. For every genotype, 10 randomly chosen individuals were examined, and for each genotype, the average values from 10 replications were calculated.

3. Results

3.1. Genetic Similarity Assessment

The set of 15 primer pairs allowed for the detection of 2 monomorphic and 98 polymorphic alleles (Table 3, Figure 1). The average number of polymorphic alleles per marker was 6.533, ranging from 2 to 15. Monomorphic alleles were observed only for two markers: mstg028 and mstg042. The SSR markers used in this study generated highly informative loci with the PIC values ranging from 0.594 for mstg016 to 0.989 for mstg039, with the mean 0.848 (Table 3).

Table 3. Quantity of detected alleles and PIC values for SSR markers.

SSR Marker	Quantity of Polymorphic Alleles	Quantity of Monomorphic Alleles	Percentage of Polymorphic Alleles (%)	PIC (Polymorphism Information Content)
mstg004	2	0	100	0.962
mstg008	8	0	100	0.969
mstg012	7	0	100	0.771
mstg016	8	0	100	0.594
mstg025	4	0	100	0.838
mstg028	7	1	87.5	0.769
mstg033	3	0	100	0.988
mstg038	9	0	100	0.841
mstg039	15	0	100	0.989
mstg042	2	1	66.7	0.913
mstg052	7	0	100	0.893
mstg055	9	0	100	0.776
mstg001	4	0	100	0.908
mstg034	5	0	100	0.686
mstg027	8	0	100	0.822
Mean	6.533	0.133	96.947	0.848

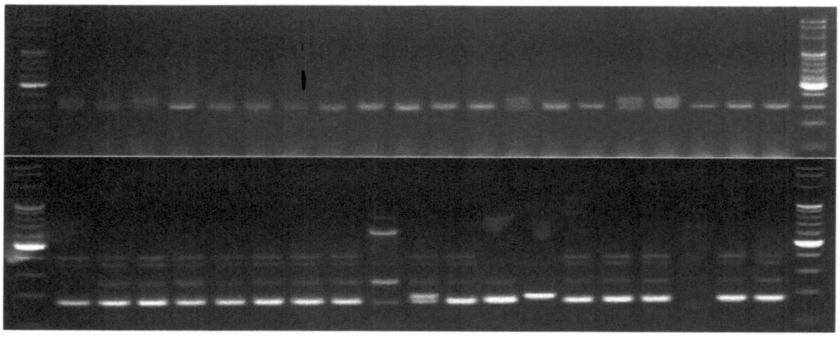

Figure 1. Example of electrophorograms with visible PCR products. Results for genotypes 21–40, marker mstg004 (**above**) and mstg008 (**below**).

The data were computed to estimate genetic similarity between the studied rapeseed genotypes based on Nei and Li's coefficients. The highest genetic similarity (equal to 0.97) was found between genotypes *B. napus* cv. Zhongshuang9 × *B. rapa* ssp. *pekinensis* 08 006169 (34) and *B. napus* cv. Zhongshuang9 × *B. rapa* ssp. *pekinensis* 08 006169 (51), whereas the lowest genetic similarity (0.22) was found for *B. carinata* (7) and *B. fruticulosa* PI 649097 (18). The mean value of genetic similarity was 0.63. The SSR marker data were used to group cultivars by the UPGMA method. The relationships between genotypes are presented in the form of a dendrogram (Figure 2), in which nine clusters were clearly distinguished. Cluster I comprised only one individual, genotype 13 (*B. rapa* ssp. *chinensis* (COBORU)), which had less than a 0.5 similarity with other genotypes; Cluster II comprised only one individual, genotype 18 (*B. fruticulosa* PI 649097); Cluster III comprised only one individual, genotype 43 (*B. napus* cv. Lisek × *B. fruticulosa*—PI649099); Cluster IV comprised genotypes 42, 44, 47, and 50 (*B. napus* cv. Jet Neuf × *S. alba* cv. Bamberka, *B. napus* cv. Lisek × *S. alba* cv. Bamberka, *B. napus* cv. Californium × *S. alba* cv. Bamberka, and *S. alba* cv. Bamberka); Cluster V, 14, 38, 39, 40, and 48 (*B. rapa* ssp. *chinensis* PI430485 98CI, *B. rapa* ssp. *pekinensis* 08, 007569, *B. rapa* ssp. *pekinensis* 08, 007574, *B. rapa* ssp. *pekinensis* (COBORU), and *B. rapa* ssp. *pekinensis* 08 006169); Cluster VI comprised only one individual, genotype 49 (*B. oleracea* var. *alboglabra*); Cluster VII comprised genotypes 1, 2, 3, 4, 5, 6, 7, 8, 9, and 10 (*B. napus* cv. Jet Neuf × *B. carinata* PI 649091, *B. napus* cv. Lisek × *B. carinata* Dodola, *B. napus* cv. Jet Neuf × *B. carinata*—PI 649094, *B. napus* cv. Jet Neuf × *B. carinata*—PI 649096, *B.*

carinata 1, *B. carinata* 2, *B. carinata* 3, *B. carinata* 4, *B. carinata* cv. Dodola, and *B. carinata* PI 596534); Cluster VIII comprised two genotypes, 19 and 20 (*B. napus* cv. Californium × *B. fruticulosa*—PI649097 and *B. napus* cv. Lisek × *B. fruticulosa*—PI649097), while the ninth cluster contained the remaining 26 genotypes (Figure 2).

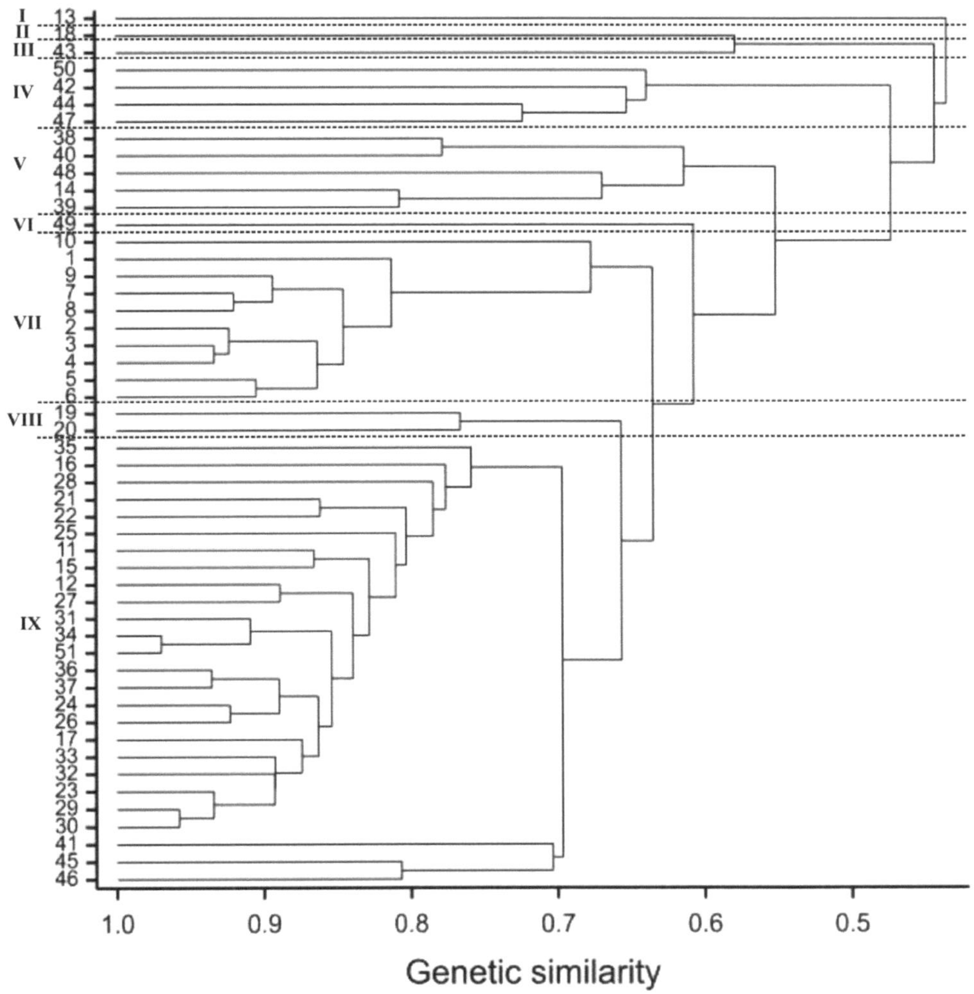

Figure 2. Dendrogram obtained from SSR data showing the genetic relationship of studied genotypes (numbers according to Table 1). Genotypes were grouped hierarchically using the UPGMA method. The scale at the bottom of the dendrogram indicates the level of similarity between individual plants.

The significant differentiation (F_{ST} = 0.059; p = 0.011) between the genotypes among the groups presented in Table 1 was further supported by the AMOVA results. The intra- and inter-genotype variabilities were found to be significant, with 6% of the genetic variance contributed by the differentiation between the groups, whereas 94% was partitioned within the groups. The largest variability was observed in the first group (mean squares within the group was equal to 9.582), while the smallest was in group number 7 (4.160) (Table 4).

Table 4. Values of differentiation F_{ST} (below diagonal) and probability based on non-parametric permutational testing procedures with 999 permutations (above diagonal) between groups of genotypes.

Group	1	2	3	4	5	6	7
1	0.000	0.045	0.002	0.016	0.005	0.002	0.055
2	0.041 *	0.000	0.153	0.072	0.421	0.398	0.433
3	0.154 **	0.028	0.000	0.001	0.012	0.052	0.181
4	0.077 *	0.052	0.191 ***	0.000	0.083	0.009	0.060
5	0.066 **	0.000	0.092 *	0.050	0.000	0.289	0.393
6	0.135 **	0.000	0.103	0.160 *	0.017	0.000	0.384
7	0.077	0.000	0.046	0.099	0.000	0.000	0.000
Mean squares within group	9.582	8.132	4.281	8.797	8.853	4.250	4.160

* $p < 0.05$, ** $p < 0.01$, *** $p < 0.001$

Statistical significant differences were observed between the following pairs of groups of genotypes: 1–2, 1–3, 1–4, 1–5, 1–6, 3–4, 3–5, and 4–6 (Table 4).

The PCA for 51 genotypes based on the distance matrix was presented in Figure 3. The first two PCs explained a total of 31.54% SSR marker variation (16.69% and 14.85%, respectively).

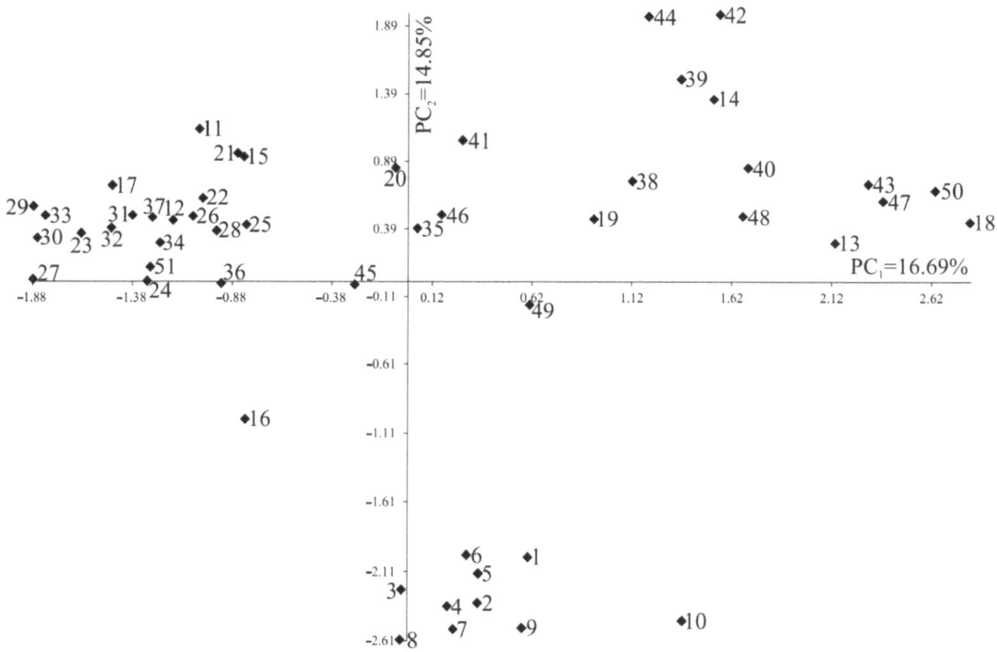

Figure 3. Principal component analysis of 51 genotypes based on 100 detected PCR products, numbers 1–51 according to Table 1.

3.2. Field Resistance to Blackleg

The performed analysis allowed to distinguish the genotypes with the highest resistance level to blackleg (Table 5). Sixteen hybrid combinations belonged to the statistically best group (group f) in both terms, which indicates their ability to maintain stable and low susceptibility to pathogen infestation. These include hybrids with *B. carinata*, *B. fruticulosa*, and *S. alba* as a parental component. The lowest level of blackleg resistance was observed for *B. napus* cv. Górczański × *B. rapa* ssp. *chinensis* in both terms (23.33% and 25% infestation), although those genotypes are still considered as moderately resistant. Examples of lesions observed on hybrid combinations are presented in Figure 4.

Table 5. Results of blackleg field resistance assessment for hybrid plants. The level of infestation is expressed as a percentage.

No of Genotype	Combination	Infestation Level—Term I	Infestation Level—Term II
1	*B. napus* cv. Jet Neuf × *B. carinata* PI 649091	0 f *	3 ef
2	*B. napus* cv. Lisek × *B. carinata* Dodola	0 f	3 ef
3	*B. napus* cv. Jet Neuf × *B. carinata*—PI 649094	0 f	4 ef
4	*B. napus* cv. Jet Neuf × *B. carinata*—PI 649096	0 f	3 ef
11	*B. napus* cv. Górczański × *B. rapa* ssp. *chinensis*	23.33 a	25 a
12	*B. napus* cv. Zhongshuang9 × *B. rapa* ssp. *chinensis* 08 007574	15 b	22 ab
15	*B. napus* cv. Lisek × *B. rapa* Pak Choi 08, 007574	8 bcde	8 def
16	*B. napus* cv. Lisek × *B. rapa* Pak Choi 08, 007569	8 bcde	9 cdef
17	*B. napus* cv. Górczański × *B. rapa* Pak Choi 08, 007574	7 cdef	8 def
19	*B. napus* cv. Californium × *B. fruticulosa*—PI649097	0 f	4 ef
20	*B. napus* cv. Lisek × *B. fruticulosa*—PI649097	0 f	5 ef
25	*B. napus* cv. Californium × *B. oleracea* var. *alboglabra*	9.33 bcde	2.08 f
26	*B. napus* cv. Jet Neuf × *B. rapa* ssp. *pekinensis* 08 007569	8 bcde	8 def
27	*B. napus* cv. Jet Neuf × *B. rapa* ssp. *pekinensis* 08 007574	5 def	6 ef
28	*B. napus* cv. Górczański × *B. rapa* ssp. *pekinensis* 08.007574	12.33 bc	15 bcd
29	*B. napus* cv. Górczański × *B. rapa* ssp. *pekinensis* 08.007569	11 bcd	6 ef
30	*B. napus* cv. Californium × *B. rapa* ssp. *pekinensis* 08 007574	5 def	15 bcd
31	*B. napus* cv. Californium × *B. rapa* ssp. *pekinensis* 08 007574-1	4 def	16 bc
32	*B. napus* cv. Californium × *B. rapa* ssp. *pekinensis* 08 007574-2	5.25 def	13.33 cd
33	*B. napus* cv. Californium × *B. rapa* ssp. *pekinensis* 08 007574-3	6 cdef	14 cd
34	*B. napus* cv. Zhongshuang9 × *B. rapa* ssp. *pekinensis* 08 006169	3.33 ef	9 cdef
35	*B. napus* MS8 line × *B. rapa* ssp. *pekinensis* 08 006169 1	4 def	6 ef
36	*B. napus* MS8 line × *B. rapa* ssp. *pekinensis* 08 006169 2	6 cdef	6 ef
37	*B. napus* MS8 line × *B. rapa* ssp. *pekinensis* 08 006169 3	6 cdef	6 ef
41	*B. napus* cv. Lisek × *B. oleracea* var. *alboglabra*	10 bcde	10 cde
42	*B. napus* cv. Jet Neuf × *S. alba* cv. Bamberka	0 f	3 ef
43	*B. napus* cv. Lisek × *B. fruticulosa*—PI649099	0 f	5 ef
44	*B. napus* cv. Lisek × *S. alba* cv. Bamberka	4 def	4 ef
45	*B. napus* cv. Lisek × *B. tournefortii*	8 bcde	6 ef
46	*B. napus* cv. Jet Neuf × *B. oleracea* var. *alboglabra*	10.33 bcde	10 cde
47	*B. napus* cv. Californium × *S. alba* cv. Bamberka	0 f	3 ef
51	*B. napus* cv. Zhongshuang9 × *B. rapa* ssp. *pekinensis* 08 006169 2	6 cdef	15 bcd

* Values with different letters in columns are significantly different.

Figure 4. Examples of leaf damage on hybrid genotypes caused by *L. maculans*.

4. Discussion

The assessment of diversity between species is important for the management of germplasm resources and for the curation of genetic databases. As the phenotypic assessments partially relay on environmental conditions, they do not allow for a clear discrimination of related species. Thus, in this study, genotypic analysis using SSR markers was performed for the unbiased determination of genetic diversity.

Molecular DNA markers are important tools for genetic similarity studies. SSR markers are especially valuable, as they enable multi-allelic detection and can be applied using various laboratory systems [27]. The markers selected for this study derived only from *B. rapa* (AA, 2n = 20) and were developed using the ISSR-suppression-PCR method by Tamura et al., [20]; however, the applicability of these markers for a wider group of *Brassica* species has been suggested by the aforementioned authors. The Brassicaceae family consists of approximately 3000 species [28] with diverse genomic composition, e.g., the U triangle (A, B, and C genome), *S. alba* (S genome), and *B. fruticulosa* (F genome), although conserved regions of gene content and gene order are present among the family [29]. This attribute, combined with the before mentioned unique features of the microsatellite loci that are widely spread among the Brassicaceae, allows to detect sequences originating from one species in the genomes of its relatives. We managed to confirm that the selected SSR markers can be used for genetic similarity studies in the Brassicaceae family, as the markers enabled the detection of allelic variation.

Polymorphism Information Content (PIC) is an indicator that allows to evaluate the discriminatory ability of molecular markers and to study the genetic diversity [30]. The PIC value can vary from 0 to 1, and markers with a PIC value exceeding 0.7 are considered highly informative [31]. Therefore, it can be concluded that twelve out of fifteen tested markers are particularly effective in detecting the polymorphism in the studied population.

The UPGMA allowed for the distinction of nine groups, based on genetic similarity. Generally, the applied method permitted the assessment of the genetic distance of the studied hybrids and their parents, but not all of the results are in line with the predictions. For example, *B. rapa* ssp. *chinensis* (COBORU) shows weak connection to their progeny or other genotypes form the same species. Furthermore, the distinctiveness of this genotype was confirmed with the PCA method. The weaker-than-expected association between species can be explained by a different origin (geographical distribution) or outbreeding [32].

The PCA analysis was conducted to confirm the complicated structure of the studied individuals, and the results confirmed a close relationship for *B. rapa* and *B. carinata* and their hybrid progeny. The rest of the genotypes were generally more scattered around the diagram. However, attention should be drawn to the short distance revealed for two pairs of genotypes: *B. napus* cv. Lisek × *B. fruticulosa* PI649099 and *B. napus* cv. Californium × *S. alba* cv. Bamberka, and *B. napus* MS8 line × *B. rapa* ssp. *pekinensis* 08 006169 and *B. napus* cv. Jet Neuf × *B. oleracea* var. *alboglabra*. These hybrids' male parental components present entirely different genomic structures, however their genetic similarity can be explained by the unequal inheritance of the *B. napus* genome during hybridization. It should also be emphasized that the markers used in this study derived from *B. rapa,* which possess A genome, which might have an impact on the obtained PCR products.

The genetic similarity of the studied genotypes varied from 0.22 to 0.97. The extensive range of the similarity coefficient values show that the Brassicaceae germplasm collection reflects a diverse and varied population. These results are in line with the findings of Kumari et al. [33], as well as other researchers [34], who studied the genetic diversity in nine genotypes of *Brassica* and their wild relatives.

The level of field resistance to blackleg varied between the studied genotypes. We managed to select sixteen combinations with the lowest pathogen infestation, which might be especially valuable in future studies focusing on finding a durable resistance to *L. maculans* and incorporating their germplasm into the *B. napus* gene pool. All individuals that had *B. carinata*, *B. fruticulosa*, and *S. alba* as one of the interspecific cross components showed the lowest infestation level. This indicates that particular attention should be paid to these parental species, as they may hold valuable resistance genes that could help to control the disease. This is especially important considering the previously reported resistance breakdowns [35]. The aforementioned species have been previously characterized as potentially significant resistance gene sources [36–38], which is in line with our findings.

Hybrid individuals with the lowest blackleg infestation could be found in five out of nine groups distinguished with UPGMA and were spread evenly on the PCA diagram. This indicates that the genetic similarity of the studied hybrid genotypes is not correlated with their level of field resistance. On the other hand, it might be simply explained by the fact that applied molecular markers are not linked to the regions of the genome containing the resistance genes.

In conclusion, the tested SSR markers proved to be useful in revealing the genetic distances in Brassicaceae hybrids and species. The ability to properly characterize and organize the genetic resources is key to the effective conservation of accessions. More precise and quick determination of the relationship of genotypes and the amount of variation within or among accessions in a collection can be accomplished by using molecular diagnostic techniques. Other than successfully maintaining the collections, genetic markers are invaluable for crop improvement and plant breeding programs. Moreover, our studies confirmed the usefulness of the Brassicaceae hybrids in terms of blackleg management and the importance of searching new sources of *L. maculans* resistance outside the *B. napus* gene pool.

Author Contributions: Conceptualization, J.N. and J.S.; methodology, J.S., J.N., J.K., J.B., and D.W.; software, J.B.; validation, J.S., J.N., and J.K.; formal analysis, J.N., J.S., D.W., and J.K.; investigation, J.N. and J.K.; resources, J.N.; data curation, J.S., J.N., and J.B.; writing—original draft preparation, J.S. and J.N.; writing—review and editing, J.S., J.N., and J.B.; visualization, J.S. and J.B.; supervision, J.N., J.S., and J.K.; project administration, J.N.; funding acquisition, J.N. All authors have read and agreed to the published version of the manuscript.

Funding: This research was funded by the Polish Ministry of Agriculture and Rural Development, project number 27.

Institutional Review Board Statement: Not applicable.

Informed Consent Statement: Not applicable.

Data Availability Statement: Not applicable.

Conflicts of Interest: The authors declare no conflict of interest.

References

1. Hasanuzzaman, M. *The Plant Family Brassicaceae*; Springer Nature Singapore Pte Ltd.: Singapore, 2020; ISBN 978-981-15-6344-7.
2. Kumari, P.; Bisht, D.S.; Bhat, S.R. Stable, fertile somatic hybrids between *Sinapis alba* and *Brassica juncea* show resistance to *Alternaria brassicae* and heat stress. *Plant Cell Tissue Organ Cult.* **2018**, *133*, 77–86. [CrossRef]
3. Prakash, S.; Xiao Ming, W.; Bhat, S.R. History, evolution, and domestication of *Brassica* crops. *Plant Breed. Rev.* **2012**, *35*, 19–84.
4. Fu, Y.-B.; Gugel, R.K. Genetic diversity of Canadian elite summer rape (*Brassica napus* L.) cultivars from the pre- to post-canola quality era. *Can. J. Plant Sci.* **2010**, *90*, 23–33. [CrossRef]
5. Hu, D.; Jing, J.; Snowdon, R.J.; Mason, A.S.; Shen, J.; Meng, J.; Zou, J. Exploring the gene pool of *Brassica napus* by genomics-based approaches. *Plant Biotechnol. J.* **2021**, *19*, 1693–1712. [CrossRef]
6. Kamiński, P.; Marasek-Ciolakowska, A.; Podwyszyńska, M.; Starzycki, M.; Starzycka-Korbas, E.; Nowak, K. Development and Characteristics of Interspecific Hybrids between *Brassica oleracea* L. and *B. napus* L. *Agronomy* **2020**, *10*, 1339. [CrossRef]
7. Sharma, B.B.; Kalia, P.; Singh, D.; Sharma, T.R. Introgression of Black Rot Resistance from *Brassica* carinata to Cauliflower (*Brassica oleracea botrytis* Group) through Embryo Rescue. *Front. Plant Sci.* **2017**, *8*, 1255. [CrossRef]
8. Hwang, S.-F.; Strelkov, S.E.; Peng, G.; Ahmed, H.; Zhou, Q.; Turnbull, G. Blackleg (*Leptosphaeria maculans*) Severity and Yield Loss in Canola in Alberta, Canada. *Plants* **2016**, *5*, 31. [CrossRef]
9. Kutcher, H.; Fernando, D.; Turkington, T.; Mclaren, D. Best Management Practices for Blackleg Disease of Canola. *Prairie Soils Crops* **2011**, *4*, 122–134.
10. Fernando, W.G.D.; Chen, Y.; Ghanbarnia, K. Breeding for Blackleg Resistance: The Biology and Epidemiology. In *Advances in Botanical Research*; Rapeseed Breeding; Academic Press: Cambridge, MA, USA, 2007; Volume 45, pp. 271–311.

11. Hasan, M.; Seyis, F.; Badani, A.; Pons-Kühnemann, J.; Friedt, W.; Lühs, W.; Snowdon, R. Analysis of Genetic Diversity in the *Brassica napus* L. Gene Pool Using SSR Markers. *Genet. Resour. Crop Evol.* **2006**, *53*, 793–802. [CrossRef]
12. Gupta, P.K.; Balyan, H.S.; Sharma, P.C.; Ramesh, B. Microsatellites in plants: A new class of molecular markers. *Curr. Sci.* **1996**, *70*, 45–54.
13. Yu, Q.; Wang, Q.; Wu, G.; Ma, Y.; He, X.; Wang, X.; Xie, P.; Hu, L.; Liu, J. Genetic differentiation and delimitation of Pugionium dolabratum and Pugionium cornutum (*Brassicaceae*). *Plant Syst. Evol.* **2013**, *299*, 1355–1365. [CrossRef]
14. Thakur, A.K.; Singh, K.H.; Singh, L.; Nanjundan, J.; Khan, Y.J.; Singh, D. SSR marker variations in *Brassica* species provide insight into the origin and evolution of *Brassica* amphidiploids. *Hereditas* **2017**, *155*, 6. [CrossRef]
15. Matuszczak, M. Markery molekularne w badaniach rzepaku (*Brassica napus* L.). I. Przegląd stosowanych technik. *Rośliny Oleiste-Oilseed Crops* **2013**, *34*, 129–150.
16. Abbas, S.J.; Farhatullah; Marwat, K.B.; Khan, I.A.; Munir, I. Molecular analysis of genetic diversity in *Brassica* species. *Pak. J. Bot. Pak.* **2009**, *41*, 167–176.
17. Wang, J.; Kaur, S.; Cogan, N.O.I.; Dobrowolski, M.P.; Salisbury, P.A.; Burton, W.A.; Baillie, R.; Hand, M.; Hopkins, C.; Forster, J.W.; et al. Assessment of genetic diversity in Australian canola (*Brassica napus* L.) cultivars using SSR markers. *Crop Pasture Sci.* **2009**, *60*, 1193–1201. [CrossRef]
18. Zhai, L.; Xu, L.; Wang, Y.; Cheng, H.; Chen, Y.; Gong, Y.; Liu, L. Novel and useful genic-SSR markers from de novo transcriptome sequencing of radish (*Raphanus sativus* L.). *Mol. Breed.* **2014**, *33*, 611–624. [CrossRef]
19. Plieske, J.; Struss, D. Microsatellite markers for genome analysis in *Brassica*. I. development in *Brassica napus* and abundance in *Brassicaceae* species. *Theor. Appl. Genet.* **2001**, *102*, 689–694. [CrossRef]
20. Tamura, K.; Nishioka, M.; Hayashi, M.; Zhang, Z.; Lian, C.; Hougetsu, T.; Harada, K. Development of Microsatellite Markers by ISSR-suppression-PCR Method in *Brassica rapa*. *Breed. Sci.* **2005**, *55*, 247–252. [CrossRef]
21. Anderson, J.A.; Churchill, G.A.; Autrique, J.E.; Tanksley, S.D.; Sorrells, M.E. Optimizing parental selection for genetic linkage maps. *Genome* **1993**, *36*, 181–186. [CrossRef]
22. Wolko, Ł.; Bocianowski, J.; Antkowiak, W.; Słomski, R. Genetic diversity and population structure of wild pear (*Pyrus pyraster* (L.) Burgsd.) in Poland. *Open Life Sci.* **2014**, *10*, 19–29. [CrossRef]
23. Nei, M.; Li, W.H. Mathematical model for studying genetic variation in terms of restriction endonucleases. *Proc. Natl. Acad. Sci. USA* **1979**, *76*, 5269–5273. [CrossRef]
24. Peakall, R.; Smouse, P.E. genalex 6: Genetic analysis in Excel. Population genetic software for teaching and research. *Mol. Ecol. Notes* **2006**, *6*, 288–295. [CrossRef]
25. Excoffier, L.; Smouse, P.E.; Quattro, J.M. Analysis of molecular variance inferred from metric distances among DNA haplotypes: Application to human mitochondrial DNA restriction data. *Genetics* **1992**, *131*, 479–491. [CrossRef]
26. Brachaczek, A.; Kaczmarek, J.; Jedryczka, M. Warm and Wet Autumns Favour Yield Losses of Oilseed Rape Caused by Phoma Stem Canker. *Agronomy* **2021**, *11*, 1171. [CrossRef]
27. Shirasawa, K.; Oyama, M.; Hirakawa, H.; Sato, S.; Tabata, S.; Fujioka, T.; Kimizuka-Takagi, C.; Sasamoto, S.; Watanabe, A.; Kato, M.; et al. An EST-SSR Linkage Map of Raphanus sativus and Comparative Genomics of the *Brassicaceae*. *DNA Res.* **2011**, *18*, 221–232. [CrossRef]
28. Flannery, M.; Mitchell, F.; Coyne, S.; Kavanagh, T.; Burke, J.; Salamin, N.; Dowding, P.; Hodkinson, T. Plastid genome characterisation in *Brassica* and *Brassicaceae* using a new set of nine SSRs. *Theor. Appl. Genet. Theor. Angew. Genet.* **2006**, *113*, 1221–1231. [CrossRef]
29. Suwabe, K.; Iketani, H.; Nunome, T.; Ohyama, A.; Hirai, M.; Fukuoka, H. Characteristics of Microsatellites in *Brassica rapa* Genome and their Potential Utilization for Comparative Genomics in Cruciferae. *Breed. Sci.* **2004**, *54*, 85–90. [CrossRef]
30. Serrote, C.M.L.; Reiniger, L.R.S.; Silva, K.B.; dos Santos Rabaiolli, S.M.; Stefanel, C.M. Determining the Polymorphism Information Content of a molecular marker. *Gene* **2020**, *726*, 144175. [CrossRef]
31. Berg, P.; Singer, M. *Dealing with Genes: The Language of Heredity*; University Science Books: Mill Valley, CA, USA, 1992; ISBN 978-0-935702-69-9.
32. Das, S.; Rajagopal, J.; Bhatia, S.; Srivastava, P.S.; Lakshmikumaran, M. Assessment of genetic variation within *Brassica* campestris cultivars using amplified fragment length polymorphism and random amplification of polymorphic DNA markers. *J. Biosci.* **1999**, *24*, 433–440. [CrossRef]
33. Kumari, P.; Rathore, R.K.S.; Yadav, R.; Singh, K.P.; Kumar, R. Utility of SSR and ISSR markers for assessment of genetic diversity in *Brassicas* and their related genera. *Progress. Agric.* **2009**, *9*, 71–78.
34. Westman, A.L.; Kresovich, S. Simple sequence repeat (SSR)-based marker variation in *Brassica* nigra genebank accessions and weed populations. *Euphytica* **1999**, *109*, 85–92. [CrossRef]
35. Van de Wouw, A.P.; Marcroft, S.J.; Ware, A.; Lindbeck, K.; Khangura, R.; Howlett, B.J. Breakdown of resistance to the fungal disease, blackleg, is averted in commercial canola (*Brassica napus*) crops in Australia. *Field Crops Res.* **2014**, *166*, 144–151. [CrossRef]
36. Rimmer, S.R.; van den Berg, C.G.J. Resistance of oilseed *Brassica spp.* to blackleg caused by Leptosphaeria maculans. *Can. J. Plant Pathol.* **1992**, *14*, 56–66. [CrossRef]

37. Marcroft, S.J.; Wratten, N.; Purwantara, A.; Salisbury, P.A.; Potter, T.D.; Barbetti, M.J.; Khangura, R.; Howlett, B.J. Reaction of a range of *Brassica* species under Australian conditions to the fungus, *Leptosphaeria maculans*, the causal agent of blackleg. *Aust. J. Exp. Agric.* **2002**, *42*, 587–594. [CrossRef]
38. Niemann, J.; Kaczmarek, J.; Wojciechowski, A.; Jędryczka, M. Resistance to stem canker (*Leptosphaeria spp.*) in interspecific *Brassica* hybrids. *Prog. Plant Prot.* **2016**, *56*, 245–250.

current issues in molecular biology

MDPI

Article

Mapping QTLs for Super-Earliness and Agro-Morphological Traits in RILs Population Derived from Interspecific Crosses between *Pisum sativum* × *P. fulvum*

Hatice Sari [1,2,*], Tuba Eker [1], Hilal Sule Tosun [3], Nedim Mutlu [4], Ibrahim Celik [5] and Cengiz Toker [1,*]

1 Faculty of Agriculture, Department of Field Crops, Akdeniz University, Antalya 07070, Turkey
2 Department of Crop and Soil Science, Washington State University, Pullman, WA 99164, USA
3 Faculty of Agriculture, Department of Plant Protection, Akdeniz University, Antalya 07070, Turkey
4 Faculty of Agriculture, Department of Ag-Biotech, Akdeniz University, Antalya 07070, Turkey
5 Department of Agricultural and Livestock Production, Pamukkale University, Denizli 20700, Turkey
* Correspondence: haticesari@akdeniz.edu.tr (H.S.); toker@akdeniz.edu.tr (C.T.)

Abstract: Earliness in crop plants has a vital role in prevention of heat-induced drought stress and in combating global warming, which is predicted to exacerbate in the near future. Furthermore, earliness may expand production into northern areas or higher altitudes, having relatively shorter growing season and may also expand arable lands to meet global food demands. The primary objective of the present study was to investigate quantitative trait loci (QTLs) for super-earliness and important agro-morphological traits in a recombinant inbred line (RIL) population derived from an interspecific cross. A population of 114 RILs developed through single-seed descent from an interspecific cross involving *Pisum sativum* L. and *P. fulvum* Sibth. et Sm. was evaluated to identify QTLs for super-earliness and important agro-morphological traits. A genetic map was constructed with 44 SSRs markers representing seven chromosomes with a total length of 262.6 cM. Of the 14 QTLs identified, two were for super-earliness on LG2, one for plant height on LG3, six for number of pods per plant on LG2, LG4, LG5 and LG6, one for number of seeds per pod on LG6, one for pod length on LG4 and three for harvest index on LG3, LG5, and LG6. AA205 and AA372-1 flanking markers for super-earliness QTLs were suggested for marker-assisted selection (MAS) in pea breeding programs due to high heritability of the trait. This is the first study to map QTLs originating from *P. sativum* and *P. fulvum* recently identified species with super-earliness character and the markers (AA205 and AA372-1) linked to QTLs were valuable molecular tools for pea breeding.

Keywords: *Pisum sativum*; *Pisum fulvum*; earliness; agronomic traits; morphological traits; SSRs; QTLs; recombinant inbred lines (RILs)

Citation: Sari, H.; Eker, T.; Tosun, H.S.; Mutlu, N.; Celik, I.; Toker, C. Mapping QTLs for Super-Earliness and Agro-Morphological Traits in RILs Population Derived from Interspecific Crosses between *Pisum sativum* × *P. fulvum*. *Curr. Issues Mol. Biol.* **2023**, *45*, 663–676. https://doi.org/10.3390/cimb45010044

Academic Editor: Vijai Bhadauria

Received: 21 December 2022
Revised: 30 December 2022
Accepted: 2 January 2023
Published: 11 January 2023

1. Introduction

The *Pisum* L. genus is classified in the Fabaceae (Legumes) family, Fabaoideae (Papilionoideae) sub-family and Fabeae Rchb. tribe. The genus *Pisum* consists of three cultivated species including *P. sativum* L. (garden pea), *P. arvense* (L.) Poir. (field pea) and *P. abyssinicum* A.Br. (Dekoko or Abyssinian/Ethiopian pea) and there are seven taxa in the genus. These taxa were classified as follows: *P. sativum* L subsp. *sativum* var. *sativum*, *P. sativum* subsp. *sativum* var. *arvense* and *P. sativum* subsp. *abyssinicum* as cultivated species, while *P. elatius* (M.Bieb.) Asch. and Graebn. complex contains three varieties including *P. sativum* subsp. *elatius* var. *elatius*, *P. sativum* subsp. *elatius* var. *pumilio* Meikle and *P. sativum* subsp. *elatius* var. *brevipedinculatum* Davis and Meikle [1]. *P. fulvum* Sibth. et Sm. with a small distribution in Middle East and Turkey is the most distinct relative of the garden pea [2,3].

Pea has a central place in the history of genetics as an experimental plant since Mendel studied the famous laws of heredity [4]. Garden pea is among the most important food legumes, fodders and vegetable crops. It is grown in 99 countries worldwide. World annual production quantity of garden pea was reported to be 14.6 million tons for dry pea and 19.9 million tons for vegetable pea in 2020 [5]. It is used for various purposes, including food (leaves, green pods, unripe fresh seeds and dry mature seeds) and feed (direct grazing and silage). Garden pea is quite rich in protein (21–33%), starch (37–49%), soluble sugars (5%), fiber (2–9%), minerals and vitamins [2,6,7].

Garden pea, like faba bean (*Vicia faba* L.), is a cool season food legume and more susceptible to droughts than chickpea (*Cicer arietinum* L.), lentil (*Lens culinaris* Medik.) and grass pea (*Lathyrus sativus* L.) [8]. Like other crop plants, earliness provides many advantages in garden pea cultivation, such as prevention from drought-induced heat stress. Three resistance mechanisms have been reported for heat and drought: (i) escape, (ii) avoidance and (iii) tolerance. Escape is provided by early phenology including earliness [9,10]. Drought-induced heat stress has already increased due to climate change and is predicted to worsen due to a rise in temperature up to 1.5–4 °C in the near future [11]. Breeding for heat-tolerant garden pea has crucial importance with early flowered and matured cultivars [12]. Success in the selection for heat-tolerant garden pea depends on accuracy of selection with few efforts and short times by marker-assisted selection (MAS) [13,14]. New genetic sources for earliness, yield and yield-related traits should therefore be studied and mapped in garden pea.

Pea has a quite large genome of about 4.45 Gb [15]. Thanks to high-resolution genetic maps, it is possible to identify genes or QTLs controlling important agro-morphological and desirable traits. QTL mapping studies were performed in pea for many characteristics using maps constructed with molecular markers [16–30]. Dirlewanger et al. [16] identified a QTL for earliness on chromosome 6 (LG2) in the F_2 population of pea. Prioul et al. [19] discovered three QTLs for days to flowering on LGs 2, 3 and 6 in RILs, derived from intraspecific crosses. Timmerman-Vaughan et al. [20] identified four QTLs for days to flowering on LGs 1, 2b, 3 and 5b in A26 × Rovar population and three QTLs on LGs 1, 3 and 5b in A88 × Rovar population. Fondevilla et al. [23] detected four QTLs for earliness on LG2, LG3, and LG6 in RILs, derived from the cross between *P. sativum* subsp. *syriacum* and *P. sativum*. Furthermore, four QTLs were determined for days to flowering on LGs 3, 4 and 5 in interspecific crosses between *P. sativum* and *P. fulvum* by Jha et al. [28]. Huang et al. [30] identified three QTLs for days to flowering on LGs 2, 3 and 6b in intraspecific crosses.

There are also important QTL studies on pea seed quality. QTLs were determined in Va and Vb of the linkage map for seed protein content [31]. In addition to important minerals such as Ca, Fe, K, Mg, Mn, Mo, and P [32], QTLs of starch, fiber, and phytate contents [33] in pea were also defined. Various QTL mapping studies have been performed for agro-morphological traits of pea. The majority of QTLs for plant height were determined on LG3 [18,19,28,33–37]. Furthermore, several QTLs were identified for morphological traits, like number of pods per plant, number of seeds per pod, pod size, biological yield, seed yield and harvest index on all seven linkage groups [31,38–41]. However, because quantitative traits are polygenic and influenced by environmental conditions, it may be scattered in linkage groups. The QTL studies using plant populations derived from interspecific crosses in pea are very limited [28,42].

Promising interspecific crosses of pea were previously reported to have a potential for further improvement of super-earliness [43]. Thus, the objectives of this study were to: (i) generate a genetic linkage map of a pea recombinant inbred line (RIL) population derived from a cross between *P. sativum* (♀) and *P. fulvum* (♂) × (ii) discover novel QTLs associated with super-earliness and important agro-morphological traits.

2. Materials and Methods

2.1. Plant Material

A total of 114 F_4 recombinant inbred lines (RILs) derived from the *P. sativum* × *P. fulvum* interspecific crosses were used as the plant material of the present study. The parents were selected based on contrasting or significant differences in morphological and phenological characteristics. The female parent, ACP 20 (*P. sativum*), is a large, wrinkled and cream color-seeded and early-flowered genotype, while the male parent, AWP 600 (*P. fulvum*), is a small, smooth and black color-seeded and late-flowered genotype [43].

ACP 20 is a landrace from Antalya, Turkey, whereas AWP 600 originated in Turkey and was obtained from USDA GRIN in the United States. Each recombinant inbred line was advanced as five seeds after the F_2 population. That is, from F_3 to F_4, each line was advanced as a family consisting of five individuals. Both parents and RILs were evaluated in the years 2019 and 2020 under glasshouse conditions.

2.2. Phenotyping

Phenotyping for QTL was recorded on 12 characteristics, namely as flower color (FC), days to flowering (DF; days), days to pod setting (DP; days), plant height (PH; cm), first pod height (FH; cm), internode length (IN; cm), number of pods per plant (PP), number of seeds per pod (SP), pod length (PL; cm), biological yield per plant (BY; g), seed yield per plant (SY; g) and harvest index per plant (HI; %). Phenotyping was recorded in parents and F_4 lines derived from interspecific crosses *P. sativum* × *P. fulvum*. Phenotyping was evaluated by averaging the five families grown on F_4 lines for each characteristic. DF was recorded as the number of days after germination until the first flowering. DP was recorded as the number of days after germination until the first pod setting. PH and FH were recorded in cm as the height of a plant from the ground to the top of the plant and as the height from ground to the first pod, respectively. Internode length (IN) was measured at the distance between two stipules with a ruler. PP and SP were recorded as the total number of pods per plant and seeds per pod, respectively. PL was recorded in cm as the length of a pod. BY was recorded in grams (g) as the total weight of a plant after harvest, while SY was recorded in g as the weight of seeds per plant after harvest. HI was calculated in percentage (%), as the ratio of SY to BY multiplied by 100. For PL, three randomly selected pods of each plant were used and SP of the same pods was recorded.

2.3. Genotyping

DNA isolation was carried out according to the CTAB method developed by Doyle and Doyle [44] using young leaves. In order to create the genetic map and determine the QTLs, a total of 70 SSR markers, 10 from each linkage group (LG), were selected from the SSR markers mapped by Loridon et al. [21]. Forty-five SSRs markers showing polymorphism in both the female and the male parent were used in this study. Some information about polymorphic markers is presented in Table 1.

The polymerase chain reaction (PCR) mix was prepared in a total volume of 15.47 μL, 1.5 μL dNTP, 1.25 μL MgCl2, 1.5 μL PCR buffer, 0.1 μL Taq DNA polymerase, 7.62 μL ultrapure water, 1 μL F primer, 1 μL R primer and 1.5 μL DNA for each sample. In PCR amplification, initial denaturation at 95 °C for 3 min, then at 94 °C for 50 s, at 45–55 °C for 40 s, at 72 °C for 50 s, final extension after 35 cycles at 72 °C for 5 min were completed. After the PCR amplification was completed, 3% agarose gel electrophoresis was used to visually examine and score the separation of the formed bands based on the size differences of the obtained PCR products. A 1 kB plus marker (Thermo Scientific GeneRuler 1kB plus DNA Ladder) was used to determine product sizes (molecular weights). The electrophoresis tank, which is connected to the power source, was run at 75 volts for about 100 min and the bands were separated from each other.

Of the F_4 RILs, those with the same band size as the female parent were scored as "A", those with the same band size as the male parent as "B" and those with both parent bands were scored as "H". Lines that did not show bands were scored as "-".

Table 1. List of SSRs primers used for this study with primer sequences.

Primers	Linkage Groups	Forward (5′-3′)	Reverse (3′-5′)
AD147	LG1	AGCCCAAGTTTCTTCTGAATCC	AAATTCGCAGAGCGTTTGTTAC
AA474	LG1	GCCCACACAAGTGGTTCTATAAAT	ATTAGTCGTTTTTCTGAAACATCAAAG
AA67	LG1	CCCATGTGAAATTCTCTTGAAGA	GCATTTCACTTGATGAAATTTCG
AC75	LG1	CGCTCACCAAATGTAGATGATAA	TCATGCATCAATGAAAGTGATAAA
D21	LG1	TATTCTCCTCCAAAATTTCCTT	GTCAAAATTAGCCAAATTCCTC
AB28	LG1	CCTGAGTCATCACATAGGAGAT	GCAGAAGTATTTGACTTGATGGAA
AA205	LG2	TACGCAATCATAGAGTTTGGAA	AATCAAGTCAATGAAACAAGCA
AB149	LG2	ACAAAGGATGATGAAAGACCCG	TCATTACTCAAAGAATGCACCCAC
AB33	LG2	CATTGAATTTGTGGGAGAAAGG	TGTGGATGTTGCAATTTCGT
AA504	LG2	TGAGTGCAGTTGCAATTTCG	TCAGATGAAGAGCATGTGGG
AD148	LG2	GAAACATCATTGTGTCTTCTTG	TTCCATCACTTGATTGATAAAC
AA372.1	LG2	GAGTGACCAAAGTTTTGTGAA	CCTTGAACCCATTTTTAAGAGT
D23	LG2	ATGGTTGTCCCAGGATAGATAA	GAAAACATTGGAGAGTGGAGTA
AA153	LG2	TTTGATAGTCCGACTTTTCCAT	GTGACAAAAGAATTCAAAACGC
AD73	LG3	CAGCTGGATTCAATCATTGGTG	ATGAGTAATCCGACGATGCCTT
AD270	LG3	CTCATCTGATGCGTTGGATTAG	AGGTTGGATTTGTTGTTTGTTG
AA5	LG3	TGCCAATCCTGAGGTATTAACACC	CATTTTTGCAGTTGCAATTTCGT
AA122	LG4	GGGTCTGCATAAGTAGAAGCCA	AAGGTGTTTCCCCTAGACATCA
AD61	LG4	CTCATTCAATGATGATAATCCTA	ATGAGGTACTTGTGTGAGATAAA
A9	LG4	GTGCAGAAGCATTTGTTCAGAT	CCCACATATATTTGGTTGGTCA
AA315	LG4	AGTGGGAAGTAAAAGGTGTAG	TTTCACTAGATGATATTTCGTT
AA92	LG4	AAGGTCTGAAGCTGAACCTGAAGG	GCAGCCCACAGAAGTGCTTCAA
AB45	LG4	ATTACACCAACAATCTCCCACT	TGTAGAAGCATTTGGGTAGTTG
AB23	LG5	TCAGCCTTTATCCTCCGAACTA	GAACCCTTGTGCAGAAGCATTA
AC58	LG5	TCCGCAATTTGGTAACACTG	CGTCCATTTCTTTTATGCTGAG
AA399	LG5	CCATTGGTATATGAAAGATCGCT	TCCCAATTAATATGGCTAGGCT
PSGAPA-1	LG5	GACATTGTTGCCAATAACTGG	GGTTCTGTTCTCAATACAAG
AA163.2	LG5	TAGTTTCCAATTCAATCGACCA	AGTGTATTGTAAATGCACAAGGG
AD55.2	LG5	AACACATTAACTAAGTCCACAC	AAACCTATCACTTTAGAAACCT
AA99	LG5	AACAATAACATGGCAAAGATT	ACCTTGCGATATAATTGATG
AD51	LG6	ATGAAGTAGGCATAGCGAAGAT	GATTAAATAAAGTTCGATGGCG
AB20	LG6	TTGCATCCCACACAAGTGGT	ACCTCCAGGTTCTGCCTTATCT
AC76a	LG6	CCCAATCCAATAAATAAAGAAA	AATGGTTGTTATGCCATTTT
AD159	LG6	AGCTTGGAACCACAAGATTAGT	GTGAATGATAATTCTCACCCTC
AA285	LG6	TCGCCTAATCTAGATGAGAATA	CTTAACATTTTAGGTCTTGGAG
AC74	LG6	CCTTAGTGTTCTTCAACTC	ACAGAACCAAGTTATCAATA
AD146	LG7	TGCTCAAGTCAATATATGAAGA	CAAGCAAATAGTTGTTTTGTTA
AA19	LG7	GCAGTTGTTGTACCCTAAAATT	TGTATTAGATGAAATTTTGTTTCTC
AA339	LG7	GTGTAGAAGTATTTTACTTGATG	CATCTATTGAAGGAAAATTAT
AD237	LG7	AGATCATTTGGTGTCATCAGTG	TGTTTAATACAACGTGCTCCTC
AA206	LG7	CTGAGAACTCAACGCTCAGACG	CGAGGGTCGAGTTCTGAGATTT
AA90	LG7	CCCTTACCATATTTCGTTTCT	TGCGACTCCATTCTAGTATTG

2.4. Genetic Mapping

The genetic linkage map was created using the Join-Map 4.1 software [45]. Markers were assigned to linkage groups (LGs) with a LOD greater than 3 using the Kosambi map method of Join-Map 4.1. The linkage groups identified in this study were aligned to seven pea chromosomes based on common markers in the pea genetic map previously reported [21].

2.5. QTL Analyses

The QGene software was used for QTL analysis of days to flowering, days to pod setting, plant height, first pod height, internodes, number of pods per plant, number of seeds per pod, pod length, biological yield, seed yield and harvest index. Using the Composite Interval Mapping (CIM) method, the QTL was determined for each quantitative trait with LOD > 3. MapChart 2.0 [46] software was used to mark the determined QTLs on the created genetic map.

Gene action was calculated by dividing the absolute value of the estimated dominance effect (|d|) by the absolute value of the estimated additive effect (|a|) [47]. |d|/|a| = 0–0.20 additive (A); partial dominance (PD) between 0.21–0.80; dominance (D) between 0.81–1.20; and >1.20 have been classified as over-dominance (OD).

3. Results

3.1. Phenotypic Characteristics

Days to flowering was recorded as 53 days for *P. sativum* and 117 days for *P. fulvum*, while the earliest lines flowered 24 days after germination in the F_4 population. Days to pod setting was 62 days for *P. sativum* and 128 days for *P. fulvum*. On the other hand, the earliest lines formed pods in 30 days in the F_4 population (Table 2). Plant height for *P. sativum* and *P. fulvum* was 90 and 41 cm, respectively, whereas it varied between 15 to 266.7 cm for the F_4 population. The number of pods per plant was 11 for *P. sativum* and 23 for *P. fulvum*, ranging between 2 to 197 in the F_4 population. Pod length was 10 cm for *P. sativum* and 4 cm for *P. fulvum*, while it was between 3 to 9 cm in F_4 lines. The number of seeds per pod in *P. sativum* and *P. fulvum* was seven and three, respectively, and it was between one to seven in F_4 lines. The harvest index was 47% for *P. sativum* and 15% for *P. fulvum*, whereas it ranged from 1% to 58% in F_4 lines (Table 2). Distributions of parents and lines for each characteristic are presented in Figure 1.

Table 2. Minimum (Min) and maximum (Max) values, means ($\overline{X}$) and ± standard errors ($S_{\overline{X}}$) for phenological and morphological traits in parents and F_4 population originated from interspecific crosses *P. sativum* × *P. fulvum*.

Traits	ACP 20		AWP 600		F_4	
	Min-Max	$\overline{X} \pm S_{\overline{X}}$	Min-Max	$\overline{X} \pm S_{\overline{X}}$	Min-Max	$\overline{X} \pm S_{\overline{X}}$
Days to flowering	50–56	53.1 ± 0.7	115–119	116.7 ± 0.7	24–103	54.3 ± 1.8
Days to pod setting	59–63	61.7 ± 0.4	127–130	128.3 ± 0.6	30–114	63.1 ± 1.9
Plant height (cm)	62–117	89.8 ± 8.1	39–43	41 ± 0.7	15–266.7	98.8 ± 6.1
First pod height (cm)	17–40	27.8 ± 3.9	5–6	5.3 ± 0.2	2–75.4	14.9 ± 1.4
No of Internodes	8–9	8.8 ± 0.2	2–3	2.5 ± 0.2	1.6–11.7	5.5 ± 0.2
No of pods per plant	4–21	11.3 ± 2.5	21–24	22.8 ± 0.5	2–197	44.4 ± 4.8
No of seeds per pod	5–8	6.7 ± 0.4	2–3	2.7 ± 0.2	1–7	3.1 ± 0.1
Pod length (cm)	9–11	10.3 ± 0.3	4–4.5	4.2 ± 0.1	2.8–9	5 ± 0.1
Biological yield (g/plant)	16.4–265.6	96.4 ± 38.7	26.7–28.7	27.5 ± 0.3	2–371.1	91.5 ± 8.6
Seed yield (g/plant)	9.6–102.5	40.9 ± 14.8	3.7–4.3	4 ± 0.1	0.4–94.9	20.7 ± 2.2
Harvest index	35.2–58.3	47 ± 3.7	13.9–15.6	14.6 ± 0.3	0.6–58.4	27.3 ± 1.4

3.2. Genetic Mapping

A population of 114 F_4 RILs obtained from *P. sativum* × *P. fulvum* interspecific crosses was screened with 70 codominant SSR markers. Of these 70 SSRs, 45 of them showed polymorphism between parents (Figure 2). After each marker was scored on the population, the markers with a LOD greater than three were selected and a genetic linkage map was generated, resulting seven linkage groups. The 44 SSR markers were mapped on LGs. The total length of the map is 262.6 cM, with an average marker resolution of 5.9 cM (Table 3). The number, names and resolution of the SSR markers for each LG are given in Table 3.

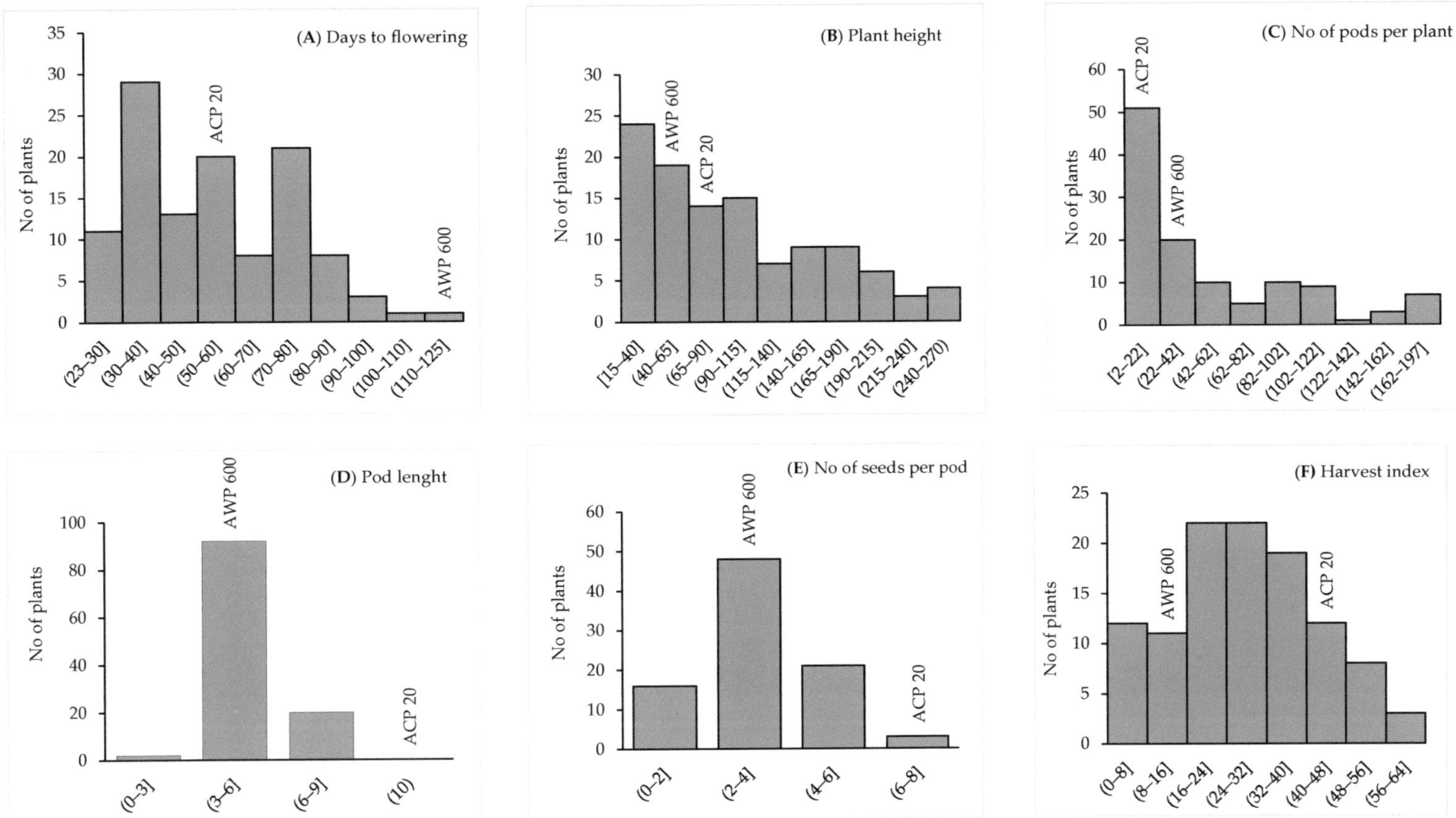

Figure 1. Frequency distribution of (**A**) days to flowering, (**B**) plant height, (**C**) no of pods per plant, (**D**) pod length, (**E**) no of seeds per pod, and (**F**) harvest index in the ACP 20 (*P. sativum*), AWP 600 (*P. fulvum*) and F_4 population derived from interspecific crosses between *P. sativum* × *P. fulvum*.

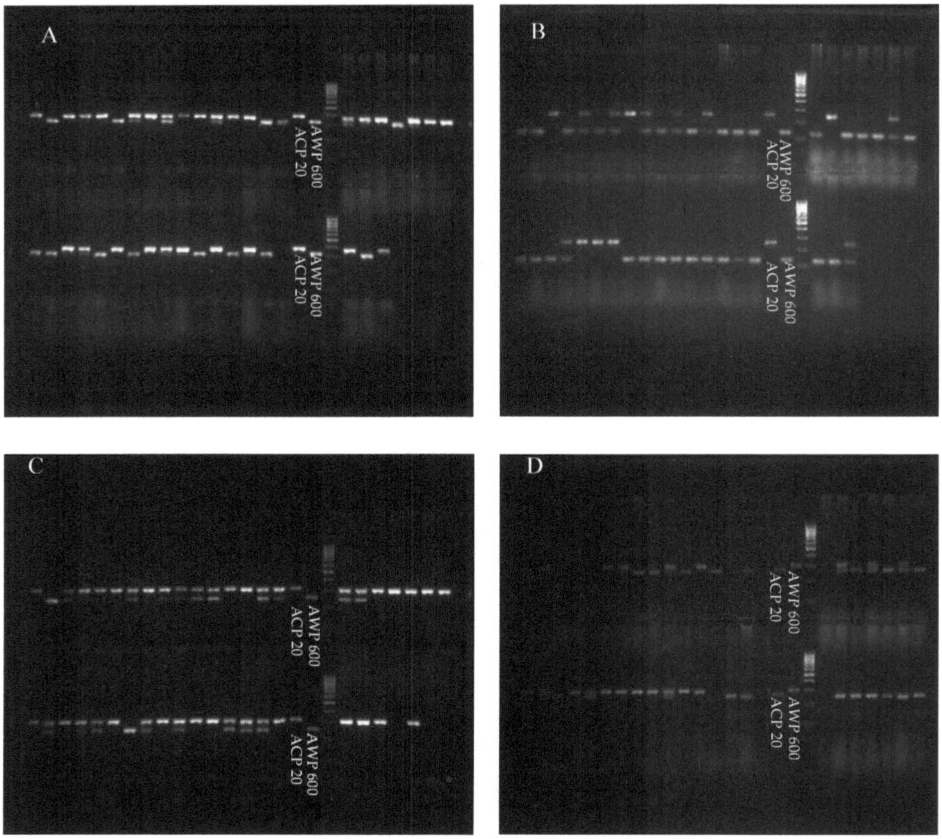

Figure 2. Agarose gel images of markers (**A**) AA67, (**B**) AD148, (**C**) AD270, and (**D**) AA285 between parents and F_4 lines derived from interspecific crosses involving *P. sativum* × *P. fulvum*.

Table 3. Linkage group (LG), number of SSR markers, length (cM) and average marker distance (cM) of the linkage groups developed with 44 SSRs using F_4 lines involving *P. sativum* × *P. fulvum*.

LGs	Map Length (cM)	No. of Markers	Resolution of Markers (cM)
LG1	27.6 cM	6	4.6 cM
LG2	46.7 cM	8	5.8 cM
LG3	25.3 cM	3	8.4 cM
LG4	38.4 cM	6	6.4 cM
LG5	39.6 cM	7	5.6 cM
LG6	44.7 cM	7	6.3 cM
LG7	40.3 cM	7	5.7 cM
Total	262.6 cM	44	5.9 cM

3.3. QTLs Analyses

A total of 14 QTLs on five different linkage groups were determined for earliness and important agro-morphological traits. Two QTLs were determined on LG2 for flowering time. The first QTL FLO2.1 had a LOD value of 3.6. The AA205 marker, which explained 14% of the phenotypic variance, was the closest to the QTL (Table 4). The second QTL FLO2.2 explained 14% of the phenotypic variance with a LOD of 3.2. The marker AA372.1 was the closest to the QTL (Figure 3). FLO2.1 and FLO2.2 showed a dominance/additive (d/a) ratio of 1.06 and 0.71, indicating a dominant and partially dominant gene action, respectively (Table 4). The QTLs for flowering time explained 28% of the total phenotypic variance.

Table 4. QTLs detected for days to flowering, plant height, no. of pods per plant, no. of seeds per pod, pod length and harvest index under glasshouse conditions.

Characters	QTL	Linkage Group (LG)	Map Position (cM)	$R^{2.3}$ *	Flanking Markers	Genetic Effect		Gene Action ***
						Add **	Dom	|d|/|a|
Days to flowering	*FLO2.1*	LG2	0–2 cM	0.13	AA205	6.5	−6.9	1.06 D
	FLO2.2	LG2	30–32 cM	0.13	AA372.1	6.5	−4.6	0.71 PD
Plant height	*PH3.1*	LG3	0–12 cM	0.16	AD73; AD270	−28.6	47.2	1.65 OD
Pods per plant	*NP2.1*	LG2	0–2 cM	0.16	AA205	10.4	−17.9	1.72 OD
	NP2.2	LG2	26–30 cM	0.13	AD148	10.6	−24.3	2.29 OD
	NP2.3	LG2	46–46 cM	0.13	AA153	11.0	−3.3	0.30 PD
	NP4.1	LG4	38–38 cM	0.14	AB45	12.7	11.2	0.88 D
	NP5.1	LG5	38–38 cM	0.14	AA99	10.0	−23.1	2.31 OD
	NP6.1	LG6	2–12 cM	0.17	AB20; AD51-2	9.2	463.5	50.38 OD
Seeds per pod	*NS6.1*	LG6	16–16 cM	0.13	AC76a	1.8	19.8	11.0 OD
Pod length	*PL4.1*	LG4	36–36 cM	0.12	AA92	0.42	0.47	1.1 D
Harvest index	*HX3.1*	LG3	0–8 cM	0.13	AD73	8.0	−3.2	0.39 PD
	HX5.1	LG5	38–38 cM	0.12	AA99	3.6	−15.8	4.38 OD
	HX6.1	LG6	8–8 cM	0.12	AD51-2	4.0	−300.2	75.0 OD

* R^2 is the percentage of phenotypic variation individually explained by each QTL; ** A negative sign reflects that the QTL alleles which increased were contributed by the wild parent, whereas a positive value means that alleles were donated by the cultivated parent; *** Gene action shows dominance (D), partial dominance (PD) or over-dominance (OD).

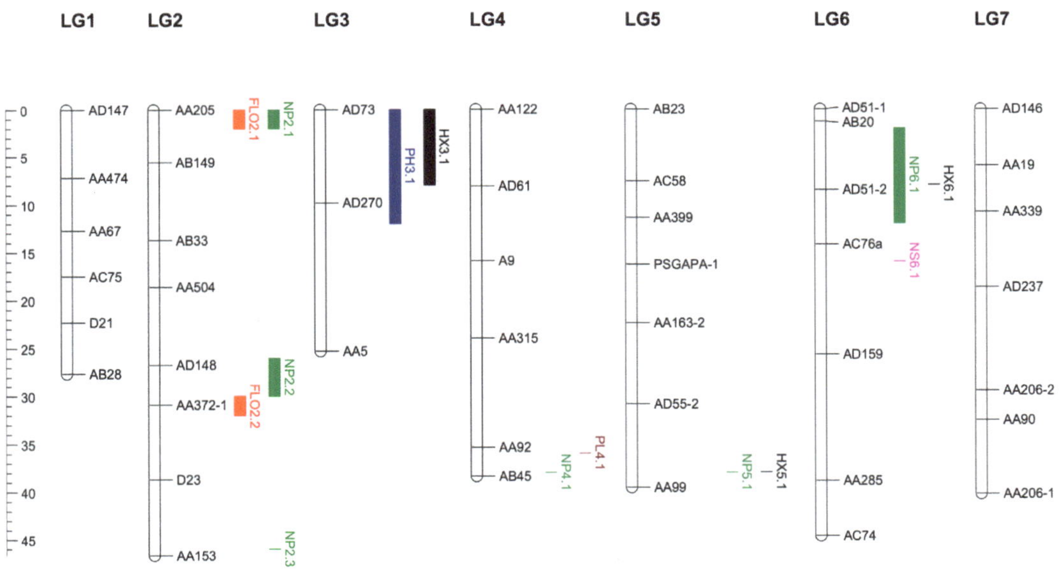

Figure 3. QTLs mapping for FLO: days to flowering, PH: plant height, NP: number of pods per plant, NS: number of seeds per pod, PL: pod length, HX: harvest index in 114 RILs developed from interspecific crosses between *P. sativum* and *P. fulvum*.

One QTL for plant height was identified on LG3 with a LOD value of 4.35. Two flanking markers were determined for PH3.1 QTL (Figure 3). The QTL flanked by markers AD73 and AD270 explained 16% of phenotypic variation. PH3.1 showed d/a ratio of 1.65, indicating an over-dominance gene action (Table 4).

Six QTLs were determined for the number of pods per plant, one of the most important yield components. Three QTLs were mapped on LG2, one on LG4, one on LG5 and one on LG6. The NP2.1, NP2.2 and NP2.3 on LG2 were mapped with LOD values of 4.3, 3.5 and 3.6, respectively (Figure 3, Table 4). The NP2.1 and NP2.2 explained 16% and 13% of the phenotypic variation, respectively.

The NP2.3, NP4.1 and NP5.1 explained 13%, 14% and 14% of the variation (Table 4, Figure 3). The NP6.1 QTL associated with the number of pods per plant explained 17% of the phenotypic variation (Table 4). The six QTL determined for the number of pods per plant within the scope of this study explained a total of 87% of the variation. NP2.1, NP2.2, NP5.1 and NP 6.1 had over dominance gene action. The HX3.1 QTL associated with the harvest index explained 13% of the phenotypic variation (Table 4). The second QTL (HX5.1) explained 12% of the phenotypic variation for the same trait. The third QTL (HX6.1) on LG6 explained 12% of the phenotypic variation. The three QTLs for harvest index explained 37% of the total phenotypic variation. The PL4.1 QTL associated with the pod length explained 12% of the phenotypic variation (Figure 3, Table 4). The NS6.1 associated with the number of seeds per pod explained 13% of the phenotypic variation (Table 4).

4. Discussion

A total of 114 F_4 RILs derived from *P. sativum* × *P. fulvum* interspecific crosses were used for phenotyping and genotyping. The 70 SSR markers were selected from the genetic map created by Loridon et al. [21]. The 45 SSRs showed parental polymorphisms. The linkage map with seven linkage groups was created using 44 SSRs with a LOD greater than 3.0 using the Kosambi function. Each LG represents a pea chromosome and the total map length was 262.6 cM with an average marker resolution of 5.9 cM (Table 3). The pea genetic map with the highest number of SSR markers was reported by Loridon et al. [21] with 239 polymorphic markers. In this study, a genetic linkage map was created by using of the RIL population developed from interspecific crossing. Common markers used in both studies are indicative of cross-population transferability.

Days to flowering, days to pod setting, plant height, first pod height, internode, number of pods per plant, number of seeds per pod, pod length, biological yield, seed yield and harvest index were evaluated to determine QTLs in this study. The evaluated characteristics are important targets for pea breeding. Of these characteristics, 14 QTLs were determined for a total of six traits: days to flowering, plant height, number of pods per plant, number of seeds per pod, pod length and harvest index. In the QTL analyses, composite interval mapping (CIM) was used. The F_4 RIL population of 114 individuals derived from *P. sativum* × *P. fulvum* interspecific crosses was used to determine the QTLs. In a study comparing RIL populations for QTL detection, it was concluded that the F_4 RIL population may be as effective as the $F_{6\text{-}7}$ populations [48].

Flowering time is one of the main determinants of adaptation to different ecological and geographical regions. Early-flowering genotypes in pea play an important role in minimizing bottlenecks such as abiotic and biotic stresses. There are growing global concerns about the impact of climate change on food production, livelihoods and food security [49,50]. Global warming is thought to harm agricultural production and is one of the most serious threats to food supply. The second threat is the increasing world population, estimated at 8 billion by 2030, which will require a 60% increase in current food production [51,52]. The majority of the world's population lives in cities, and considering the reasons for migration from rural areas to cities, it is inevitable that the consumption rate will create even more food deficits [50]. According to the data of the International Panel on Climate Change (IPCC), global warming will exceed 1.5 °C by 2030, causing permanent loss of the most sensitive ecosystems. It is thought to cause a crisis for societies in underdeveloped and developing countries. Super-early individuals from the previous study could escape high temperature stress, while late-maturing individuals were exposed to heat stress during the flowering and pod setting periods [43]. The earliest lines in the F_4 population flowered 24 days after germination, while *P. sativum* flowered in 53 days and *P. fulvum* flowered in 117 days (Table 2). *P. sativum* required 62 days to reach pod setting, while *P. fulvum* required 128 days. In the F_4 population, the earliest lines developed pods in 30 days (Figure 1). In a previous study, the earliest days to flowering in the F_2 and F_3 generations of the same population were 17 and 13 days under short-days, respectively [43].

More than 20 loci related to flowering time and flowering development had been identified in pea and the interactions of these loci determined flowering time. Late-flowering (*Lf*) [53], high-response (*Hr*), sterile nodes (*Sn*), early (*E*), photoperiod (*Ppd*) [54] and die Neutralis (*Dne*) loci are the most important ones [55–57]. The *Ppd* and *Lf* loci were mapped on LG2 [54,56], while the *Hr* and *Dne* loci were mapped on LG3 [58,59]. In this study, two QTLs associated with the markers AA205 and AA372.1 were found for flowering time, which are in the same linkage group (on LG2) as the *Ppd* and *Lf* loci. Guindon et al. [40] reported that seed diameter and seed weight characteristics were associated with the AA205 marker in peas. Three genomic regions controlling flowering time were identified on LG2, LG3, and LG6 by Prioul et al. [19]. QTL *flo1* was mapped on LG2, the same linkage group as the QTL found in this study, contributing most of the variation [19]. QTL determined on LG2 was associated with the AB33 marker. The marker flanking the AB33 marker was AA372.1 and it was linked with the *FLO2.2* determined in this study (Table 3). In addition, QTL *flo2* was mapped on LG3 and QTL *mpIII-3* was reported to be in the same region with the pea blight resistance QTL. Resistance alleles in the blight resistance-related QTLs had been associated with alleles that delay flowering time [19]. Burstin et al. [60] mapped one QTL in 49 cM of LGV where the *Det* gene is located for flowering time. Foucher et al. [61] reported that the *Det* gene played a role in the regulation of flowering time. Fondevilla et al. [62] determined two QTLs on LG3 for earliness in pea. In addition, it was reported that the QTLs were close to the AB64 and AA175 markers. QTL was mapped for earliness in pea on LG2 by Dirlewanger et al. [16]. Jha et al. [28] identified four QTLs for flowering time at LGs 3, 4 and 5. In a recent study, three QTLs, two on LG1 and one on LG2, were mapped for flowering time in F_2 and F_3 populations obtained from DDR14 and Explorer intraspecific crosses [40]. Fondevilla et al. [23] defined QTLs for flowering time on LG6 and LG3. QTL on LG3 determined by Fondevilla et al. [23] was reported to be related to earliness in the study by Timmerman-Vaughan et al. [20]. Although QTLs determined for flowering time in the previous studies were close to AA205 and AA372.1 markers, it was not directly related. QTL studies on flowering time in peas are limited and two more new QTLs were found on LG2 with this study.

Major and minor QTLs have been identified for plant height in peas in previous studies. Tar'an et al. [18] determined three main QTLs with a total variation of 64.6% and Hamon et al. [34] identified three minor QTLs on LG3. Three QTLs were determined for plant height in LG2, LG3, and LG7 by Prioul et al. [19] and it was emphasized that the QTL on LG3 explained 63% of the variation. Gali et al. [36] identified a major QTL for plant height on LG3, explaining 33–65% of the phenotypic variance in the three RIL populations. Also, Ferrari et al. [35] mapped QTL for plant height on LG3. Gali et al. [33] identified four loci on LG3 associated with plant height using the GWAS (genome-wide association) method. Guindon et al. [40] found QTL on LG2 for plant height. In one study, two QTLs were found, one QTL on LG3 and one QTL on LG5 [37]. In Jha et al. [28], in which quantitative loci of blight disease in pea were studied, five QTLs associated with plant height were identified. Three of the QTLs were positioned on LG3, LG4 and LG7. Although QTLs were found in different linkage groups related to plant height in previous studies, the majority of QTLs that explain the phenotypic variation were identified on LG3, as reported in this study.

Previous studies using RIL populations have identified multiple QTLs associated with the number of pods per plant in more than one linkage group. For the number of pods, a total of five QTLs were determined on LG1, LG2, LG3, LG5 and LG6 [38]. Guindon et al. [40] detected a QTL in the LG1 for the same trait. Sadras et al. [41] determined the QTL in LG2 for the number of pods per m^2. QTL was determined for the number of seeds per pod on LG2 in an RIL population [31]. Two QTLs were identified for the number of seeds per pod on LG1 by Guindon et al. [40]. Timmerman-Vaughan et al. [39] determined a total of seven QTLs on LG1, LG2, LG3, LG4 and LG7 linkage groups for the number of seeds per m^2. Sadras et al. [41] found QTL on LG2 and LG3 for the number of seeds per pod. In addition, it was reported that the QTLs of flowering time and yield components were mostly on LG2 [41]. Four QTLs were identified, explaining 40% of the total phenotypic variation for number of seeds per plant by Timmerman-Vaughan et al. [39]. Two of these QTLs were mapped on LG3, one on LG1 and one on LG2 [39]. A QTL (*PL4.1*) associated with the pod length was detected in LG4 with a LOD value of 3.08 (Table 1). It explained 12% of the phenotypic variation. The closest marker to the PL4.1 QTL was AA92 and its position on the map was 36 cM (Figure 3). One QTL for pod size was mapped on LG2 [40].

Another QTL mapped in this study was the harvest index. The *HX3.1* QTL explained 13%, and *HX5.1* and *HX6.1* QTL each explained 12% of the phenotypic variation. The three QTLs explained a total of 37% of the phenotypic variation (Table 4). In a similar study, four QTLs were identified that explained a total of 40% of phenotypic variation. Two of these were mapped on LG3, one on LG1 and one on LG2 [39]. Yield components can be included in many linkage groups due to the characteristics that are easily affected by the environment and the populations.

QTL studies were carried out in pea for various characteristics. In addition to physiological traits, yield and yield components, QTL studies were also conducted against biotic and abiotic stress factors. In a study investigating the relationship between lodging resistance and plant height in pea, lodging resistance was mapped on LG3 [18], and in pea blight disease resistance on LG2, LG3, LG5 and LG6 [19,28], seed color, grain weight, grain yield, biological yield, protein content, broomrape resistance and powdery mildew resistance QTLs were reported in all seven linkage groups for many characteristics [22,63–66]. Studying QTLs mapped in the same linkage group in future studies may strengthen the functionality of the markers used.

The super-earliness character has a high heritability in this interspecific population [43] and thus MAS can be employed to transfer it into cultivated types.

In conclusion, a 262.6 cM long genetic map was constructed with 44 SSRs markers. A total of 14 QTLs were mapped, two QTLs for super-earliness on LG2, one for plant height on LG3, six QTLs for number of pods per plant on LG2, LG4, LG5 and LG6, one for number of seeds per pod on LG6, one for pod length on LG4 and three for harvest index on LG3, LG5, and LG6. The SSR markers AA205 and AA372-1 flanking super-earliness QTLs can potentially contribute significantly to future marker-assisted pea breeding programs.

Author Contributions: Conceptualization, H.S. and C.T.; methodology, H.S., N.M. and I.C.; software, H.S. and I.C.; validation, T.E. and H.S.T.; formal analysis, H.S., C.T. and N.M.; investigation, H.S., T.E. and H.S.T.; writing—original draft preparation, H.S. and C.T.; writing—review and editing, H.S., C.T., N.M. and I.C.; visualization, H.S. and T.E.; supervision, C.T.; project administration, C.T. All authors have read and agreed to the published version of the manuscript.

Funding: This research was funded by Akdeniz University, grant number FDK-2020-5257.

Institutional Review Board Statement: Not applicable.

Informed Consent Statement: Not applicable.

Data Availability Statement: All data are available in this article.

Conflicts of Interest: The authors declare no conflict of interest.

References

1. Davis, P.H. *Flora of Turkey and the East Aegean Islands.*; Davis, P.H., Ed.; University Press: Edinburgh, Scotland, 1970; Volume 3.
2. Ladizinsky, G.; Abbo, S. *The Search for Wild Relatives of Cool Season Legumes*; Springer: Berlin/Heidelberg, Germany, 2015.

3. Smýkal, P.; Coyne, C.J.; Ambrose, M.J.; Maxted, N.; Schaefer, H.; Blair, M.W.; Berger, J.; Greene, S.L.; Nelson, M.N.; Besharat, N. Legume Crops Phylogeny and Genetic Diversity for Science and Breeding. *Crit. Rev. Plant Sci.* **2015**, *34*, 43–104. [CrossRef]
4. Mendel, G. Experiments in Plant Hybridization (1865). In *Verhandlungen Naturforschenden Ver. Brünn Available Online*; 1996; Available online: http://old.esp.org/foundations/genetics/classical/gm-65-a.pdf (accessed on 20 December 2022).
5. FAOSTAT. Food and Agriculture Organisation Statistics Database. Available online: https://search.library.wisc.edu/database/UWI12320 (accessed on 20 December 2022).
6. Dahl, W.J.; Foster, L.M.; Tyler, R.T. Review of the Health Benefits of Peas (*Pisum sativum* L.). *Br. J. Nutr.* **2012**, *108*, S3–S10. [CrossRef] [PubMed]
7. Smýkal, P.; Aubert, G.; Burstin, J.; Coyne, C.J.; Ellis, N.T.; Flavell, A.J.; Ford, R.; Hýbl, M.; Macas, J.; Neumann, P. Pea (*Pisum sativum* L.) in the Genomic Era. *Agronomy* **2012**, *2*, 74–115. [CrossRef]
8. Toker, C.; Yadav, S.S. Legumes Cultivars for Stress Environments. In *Climate change and management of cool season grain legume crops*; Springer: Berlin/Heidelberg, Germany, 2010; pp. 351–376.
9. Blum, A. *Plant Breeding for Stress Environments*; CRC Press: Boca Raton, FL, USA, 2018.
10. Toker, C.; Lluch, C.; Tejera, N.A.; Serraj, R.; Siddique, K.H.M. 23 Abiotic Stresses. *Chickpea Breed. Manag.* **2007**, 474. Available online: https://www.researchgate.net/profile/Cengiz-Toker/publication/285964620_Conventional_breeding_methods/links/5b19399a45851587f2987c0c/Conventional-breeding-methods.pdf#page=500 (accessed on 20 December 2022).
11. Masson-Delmotte, V.; Zhai, P.; Pörtner, H.-O.; Roberts, D.C.; Skea, J.; Shukla, P.R.; Pirani, A.; Moufouma-Okia, W.; Péan, C.; Pidcock, R. *Global Warming of 1.5 °C: Summary for Policy Makers*; IPCC: Geneva, Switzerland, 2018.
12. Jiang, Y.; Lindsay, D.L.; Davis, A.R.; Wang, Z.; MacLean, D.E.; Warkentin, T.D.; Bueckert, R.A. Impact of Heat Stress on Pod-Based Yield Components in Field Pea (*Pisum sativum* L.). *J. Agron. Crop Sci.* **2020**, *206*, 76–89. [CrossRef]
13. Parihar, A.K.; Dixit, G.P.; Bohra, A.; Sen Gupta, D.; Singh, A.K.; Kumar, N.; Singh, D.; Singh, N.P. Genetic Advancement in Dry Pea (*Pisum sativum* L.): Retrospect and Prospect. In *Accelerated Plant Breeding, Volume 3*; Springer: Berlin/Heidelberg, Germany, 2020; pp. 283–341.
14. Tafesse, E.G.; Gali, K.K.; Lachagari, V.R.; Bueckert, R.; Warkentin, T.D. Genome-Wide Association Mapping for Heat Stress Responsive Traits in Field Pea. *Int. J. Mol. Sci.* **2020**, *21*, 2043. [CrossRef]
15. Kreplak, J.; Madoui, M.-A.; Cápal, P.; Novák, P.; Labadie, K.; Aubert, G.; Bayer, P.E.; Gali, K.K.; Syme, R.A.; Main, D.; et al. A Reference Genome for Pea Provides Insight into Legume Genome Evolution. *Nat. Genet.* **2019**, *51*, 1411–1422. [CrossRef] [PubMed]
16. Dirlewanger, E.; Isaac, P.G.; Ranade, S.; Belajouza, M.; Cousin, R.; De Vienne, D. Restriction Fragment Length Polymorphism Analysis of Loci Associated with Disease Resistance Genes and Developmental Traits in *Pisum sativum* L. *Theor. Appl. Genet.* **1994**, *88*, 17–27. [CrossRef] [PubMed]
17. Irzykowska, L.; Wolko, B.; Swiecicki, W.K. Interval Mapping of QTLs Controlling Some Morphological Traits in Pea. *Cell. Mol. Biol. Lett.* **2002**, *7*, 417–422.
18. Tar'an, B.; Warkentin, T.; Somers, D.J.; Miranda, D.; Vandenberg, A.; Blade, S.; Woods, S.; Bing, D.; Xue, A.; DeKoeyer, D. Quantitative Trait Loci for Lodging Resistance, Plant Height and Partial Resistance to Mycosphaerella Blight in Field Pea (*Pisum sativum* L.). *Theor. Appl. Genet.* **2003**, *107*, 1482–1491. [CrossRef]
19. Prioul, S.; Frankewitz, A.; Deniot, G.; Morin, G.; Baranger, A. Mapping of Quantitative Trait Loci for Partial Resistance to Mycosphaerella Pinodes in Pea (*Pisum sativum* L.), at the Seedling and Adult Plant Stages. *Theor. Appl. Genet.* **2004**, *108*, 1322–1334. [CrossRef] [PubMed]
20. Timmerman-Vaughan, G.M.; Frew, T.J.; Butler, R.; Murray, S.; Gilpin, M.; Falloon, K.; Johnston, P.; Lakeman, M.B.; Russell, A.; Khan, T. Validation of Quantitative Trait Loci for Ascochyta Blight Resistance in Pea (*Pisum sativum* L.), Using Populations from Two Crosses. *Theor. Appl. Genet.* **2004**, *109*, 1620–1631. [CrossRef] [PubMed]
21. Loridon, K.; McPhee, K.; Morin, J.; Dubreuil, P.; Pilet-Nayel, M.-L.; Aubert, G.; Rameau, C.; Baranger, A.; Coyne, C.; Lejeune-Henaut, I. Microsatellite Marker Polymorphism and Mapping in Pea (*Pisum sativum* L.). *Theor. Appl. Genet.* **2005**, *111*, 1022–1031. [CrossRef]
22. Pilet-Nayel, M.-L.; Muehlbauer, F.J.; McGee, R.J.; Kraft, J.M.; Baranger, A.; Coyne, C.J. Consistent Quantitative Trait Loci in Pea for Partial Resistance to Aphanomyces Euteiches Isolates from the United States and France. *Phytopathology* **2005**, *95*, 1287–1293. [CrossRef] [PubMed]
23. Fondevilla, S.; Satovic, Z.; Rubiales, D.; Moreno, M.T.; Torres, A.M. Mapping of Quantitative Trait Loci for Resistance to Mycosphaerella Pinodes in *Pisum sativum* subsp. *syriacum*. *Mol. Breed.* **2008**, *21*, 439–454. [CrossRef]
24. Dumont, E.; Fontaine, V.; Vuylsteker, C.; Sellier, H.; Bodèle, S.; Voedts, N.; Devaux, R.; Frise, M.; Avia, K.; Hilbert, J.-L. Association of Sugar Content QTL and PQL with Physiological Traits Relevant to Frost Damage Resistance in Pea under Field and Controlled Conditions. *Theor. Appl. Genet.* **2009**, *118*, 1561–1571. [CrossRef]
25. Fondevilla, S.; Fernández-Aparicio, M.; Satovic, Z.; Emeran, A.A.; Torres, A.M.; Moreno, M.T.; Rubiales, D. Identification of Quantitative Trait Loci for Specific Mechanisms of Resistance to Orobanche Crenata Forsk. in Pea (*Pisum sativum* L.). *Mol. Breed.* **2010**, *25*, 259–272. [CrossRef]
26. Ubayasena, L.; Bett, K.; Tar'an, B.; Vijayan, P.; Warkentin, T. Genetic Control and QTL Analysis of Cotyledon Bleaching Resistance in Green Field Pea (*Pisum sativum* L.). *Genome* **2010**, *53*, 346–359. [CrossRef]

27. Feng, J.; Hwang, R.; Chang, K.F.; Conner, R.L.; Hwang, S.F.; Strelkov, S.E.; Gossen, B.D.; McLaren, D.L.; Xue, A.G. Identification of Microsatellite Markers Linked to Quantitative Trait Loci Controlling Resistance to Fusarium Root Rot in Field Pea. *Can. J. Plant Sci.* **2011**, *91*, 199–204. [CrossRef]
28. Jha, A.B.; Tar'an, B.; Stonehouse, R.; Warkentin, T.D. Identification of QTLs Associated with Improved Resistance to Ascochyta Blight in an Interspecific Pea Recombinant Inbred Line Population. *Crop Sci.* **2016**, *56*, 2926–2939. [CrossRef]
29. Timmerman-Vaughan, G.M.; Moya, L.; Frew, T.J.; Murray, S.R.; Crowhurst, R. Ascochyta Blight Disease of Pea (*Pisum sativum* L.): Defence-Related Candidate Genes Associated with QTL Regions and Identification of Epistatic QTL. *Theor. Appl. Genet.* **2016**, *129*, 879–896. [CrossRef] [PubMed]
30. Huang, S.; Gali, K.K.; Tar'an, B.; Warkentin, T.D.; Bueckert, R.A. Pea Phenology: Crop Potential in a Warming Environment. *Crop Sci.* **2017**, *57*, 1540–1551. [CrossRef]
31. Krajewski, P.; Bocianowski, J.; Gaw\lowska, M.; Kaczmarek, Z.; Pniewski, T.; Święcicki, W.; Wolko, B. QTL for Yield Components and Protein Content: A Multienvironment Study of Two Pea (*Pisum sativum* L.) Populations. *Euphytica* **2012**, *183*, 323–336. [CrossRef]
32. Ma, Y.; Coyne, C.J.; Grusak, M.A.; Mazourek, M.; Cheng, P.; Main, D.; McGee, R.J. Genome-Wide SNP Identification, Linkage Map Construction and QTL Mapping for Seed Mineral Concentrations and Contents in Pea (*Pisum sativum* L.). *BMC Plant Biol.* **2017**, *17*, 43. [CrossRef] [PubMed]
33. Gali, K.K.; Sackville, A.; Tafesse, E.G.; Lachagari, V.R.; McPhee, K.; Hybl, M.; Mikić, A.; Smýkal, P.; McGee, R.; Burstin, J. Genome-Wide Association Mapping for Agronomic and Seed Quality Traits of Field Pea (*Pisum sativum* L.). *Front. Plant Sci.* **2019**, *10*, 1538. [CrossRef]
34. Hamon, C.; Coyne, C.J.; McGee, R.J.; Lesné, A.; Esnault, R.; Mangin, P.; Hervé, M.; Le Goff, I.; Deniot, G.; Roux-Duparque, M. QTL Meta-Analysis Provides a Comprehensive View of Loci Controlling Partial Resistance to Aphanomyces Euteichesin Four Sources of Resistance in Pea. *BMC Plant Biol.* **2013**, *13*, 1–19. [CrossRef]
35. Ferrari, B.; Romani, M.; Aubert, G.; Boucherot, K.; Burstin, J.; Pecetti, L.; Huart, M.; Klein, A.; Annicchiarico, P. Association of SNP Markers with Agronomic and Quality Traits of Field Pea in Italy. *Czech J. Genet. Plant Breed.* **2016**, *52*, 83–93. [CrossRef]
36. Gali, K.K.; Liu, Y.; Sindhu, A.; Diapari, M.; Shunmugam, A.S.; Arganosa, G.; Daba, K.; Caron, C.; Lachagari, R.V.; Tar'an, B. Construction of High-Density Linkage Maps for Mapping Quantitative Trait Loci for Multiple Traits in Field Pea (*Pisum sativum* L.). *BMC Plant Biol.* **2018**, *18*, 1–25. [CrossRef] [PubMed]
37. Gawlowska, M.; Knopkiewicz, M.; Święcicki, W.; Boros, L.; Wawer, A. Quantitative Trait Loci for Stem Strength Properties and Lodging in Two Pea Biparental Mapping Populations. *Crop Sci.* **2021**, *61*, 1682–1697. [CrossRef]
38. Irzykowska, L.; Wolko, B. Interval Mapping of QTLs Controlling Yield-Related Traits and Seed Protein Content in *Pisum sativum*. *J. Appl. Genet.* **2004**, *45*, 297–306.
39. Timmerman-Vaughan, G.M.; Mills, A.; Whitfield, C.; Frew, T.; Butler, R.; Murray, S.; Michael, L.; John, M.; Adrian, R.; Derek, W. Linkage Mapping of QTL for Seed Yield, Yield Components, and Developmental Traits in Pea. *Crop Sci.* **2005**, *45*, 1336–1344. [CrossRef]
40. Guindon, M.F.; Martin, E.; Cravero, V.; Gali, K.K.; Warkentin, T.D.; Cointry, E. Linkage Map Development by GBS, SSR, and SRAP Techniques and Yield-Related QTLs in Pea. *Mol. Breed.* **2019**, *39*, 1–16. [CrossRef]
41. Sadras, V.O.; Lake, L.; Kaur, S.; Rosewarne, G. Phenotypic and Genetic Analysis of Pod Wall Ratio, Phenology and Yield Components in Field Pea. *Field Crop Res.* **2019**, *241*, 107551. [CrossRef]
42. Aryamanesh, N.; Zeng, Y.; Byrne, O.; Hardie, D.C.; Al-Subhi, A.M.; Khan, T.; Siddique, K.H.M.; Yan, G. Identification of Genome Regions Controlling Cotyledon, Pod Wall/Seed Coat and Pod Wall Resistance to Pea Weevil through QTL Mapping. *Theor. Appl. Genet.* **2014**, *127*, 489–497. [CrossRef] [PubMed]
43. Sari, H.; Sari, D.; Eker, T.; Toker, C. *De Novo* Super-Early Progeny in Interspecific Crosses *Pisum sativum* L. × *P. fulvum* Sibth. et Sm. *Sci. Rep.* **2021**, *11*, 19706. [CrossRef] [PubMed]
44. Doyle, J.J. Isolation of Plant DNA from Fresh Tissue. *Focus* **1990**, *12*, 13–15.
45. Van Ooijen, J.W. Multipoint Maximum Likelihood Mapping in a Full-Sib Family of an Outbreeding Species. *Genet. Res.* **2011**, *93*, 343–349. [CrossRef]
46. Voorrips, R.E. MapChart: Software for the Graphical Presentation of Linkage Maps and QTLs. *J. Hered.* **2002**, *93*, 77–78. [CrossRef]
47. Stuber, C.W.; Edwards, M.D.; Wendel, J.F. Molecular Marker-Facilitated Investigations of Quantitative Trait Loci in Maize. II. Factors Influencing Yield and Its Component Traits1. *Crop Sci.* **1987**, *27*, 639–648. [CrossRef]
48. Takuno, S.; Terauchi, R.; Innan, H. The Power of QTL Mapping with RILs. *PLoS ONE* **2012**, *7*, e46545. [CrossRef]
49. Onyekachi, O.G.; Bonifeca, O.O.; Gemlack, N.F.; Nicholas, N. *Abiotic and Biotic Stress in Plants*; BoD—Books on Demand: Norderstedt, Germany, 2019; ISBN 978-1-78923-811-2.
50. Schiermeier, Q. Eat Less Meat: UN Climate-Change Report Calls for Change to Human Diet. *Nature* **2019**, *572*, 291–293. [CrossRef] [PubMed]
51. FAO. *The Future of Food and Agriculture–Trends and Challenges. Annual Report*; Food Agriculture Organization: Rome, Italy, 2017; pp. 1–180.
52. FAO; IFAD; UNICEF; WFP; WHO. *The State of Food Security and Nutrition in the World 2018: Building Climate Resilience for Food Security and Nutrition*; Food & Agriculture Organization: Rome, Italy, 2018; ISBN 978-92-5-130571-3.

53. White, O.E. Studies of Inheritance in Pisum. II. The Present State of Knowledge of Heredity and Variation in Peas. *Proc. Am. Philos. Soc.* **1917**, *56*, 487–588.
54. Murfet, I.C. Flowering in Pisum. Three Distinct Phenotypic Classes Determined by the Interaction of a Dominant Early and a Dominant Late Gene. *Heredity* **1971**, *26*, 243–257. [CrossRef]
55. King, W.M.; Murfet, I.C. Flowering in Pisum: A Sixth Locus, Dne. *Ann. Bot.* **1985**, *56*, 835–846. [CrossRef]
56. Arumingtyas, E.L.; Murfet, I.C. Flowering in Pisum: A Further Gene Controlling Response to Photoperiod. *J. Hered.* **1994**, *85*, 12–17. [CrossRef]
57. Weller, J.L.; Ortega, R. Genetic Control of Flowering Time in Legumes. *Front. Plant Sci.* **2015**, *6*, 207. [CrossRef]
58. Lejeune-Hénaut, I.; Hanocq, E.; Béthencourt, L.; Fontaine, V.; Delbreil, B.; Morin, J.; Petit, A.; Devaux, R.; Boilleau, M.; Stempniak, J.-J.; et al. The Flowering Locus Hr Colocalizes with a Major QTL Affecting Winter Frost Tolerance in *Pisum sativum* L. *Theor. Appl. Genet.* **2008**, *116*, 1105–1116. [CrossRef] [PubMed]
59. Liew, L.C.; Hecht, V.; Laurie, R.E.; Knowles, C.L.; Vander Schoor, J.K.; Macknight, R.C.; Weller, J.L. *DIE NEUTRALIS* and *LATE BLOOMER 1* Contribute to Regulation of the Pea Circadian Clock. *Plant Cell* **2009**, *21*, 3198–3211. [CrossRef]
60. Burstin, J.; Marget, P.; Huart, M.; Moessner, A.; Mangin, B.; Duchene, C.; Desprez, B.; Munier-Jolain, N.; Duc, G. Developmental Genes Have Pleiotropic Effects on Plant Morphology and Source Capacity, Eventually Impacting on Seed Protein Content and Productivity in Pea. *Plant Physiol.* **2007**, *144*, 768–781. [CrossRef]
61. Foucher, F.; Morin, J.; Courtiade, J.; Cadioux, S.; Ellis, N.; Banfield, M.J.; Rameau, C. *DETERMINATE* and *LATE FLOWERING* Are Two *TERMINAL FLOWER1/CENTRORADIALIS* Homologs That Control Two Distinct Phases of Flowering Initiation and Development in Pea. *Plant Cell* **2003**, *15*, 2742–2754. [CrossRef]
62. Fondevilla, S.; Almeida, N.F.; Satovic, Z.; Rubiales, D.; Vaz Patto, M.C.; Cubero, J.I.; Torres, A.M. Identification of Common Genomic Regions Controlling Resistance to *Mycosphaerella pinodes*, Earliness and Architectural Traits in Different Pea Genetic Backgrounds. *Euphytica* **2011**, *182*, 43–52. [CrossRef]
63. Timmerman-Vaughan, G.M.; McCallum, J.A.; Frew, T.J.; Weeden, N.F.; Russell, A.C. Linkage Mapping of Quantitative Trait Loci Controlling Seed Weight in Pea (*Pisum sativum* L.). *Theor. Appl. Genet.* **1996**, *93*, 431–439. [CrossRef] [PubMed]
64. Tar'an, B.; Warkentin, T.; Somers, D.J.; Miranda, D.; Vandenberg, A.; Blade, S.; Bing, D. Identification of Quantitative Trait Loci for Grain Yield, Seed Protein Concentration and Maturity in Field Pea (*Pisum sativum* L.). *Euphytica* **2004**, *136*, 297–306. [CrossRef]
65. Valderrama, M.R.; Román, B.; Satovic, Z.; Rubiales, D.; Cubero, J.I.; Torres, A.M. Locating Quantitative Trait Loci Associated with Orobanche Crenata Resistance in Pea. *Weed Res.* **2004**, *44*, 323–328. [CrossRef]
66. Fondevilla, S.; Torres, A.M.; Moreno, M.T.; Rubiales, D. Identification of a New Gene for Resistance to Powdery Mildew in *Pisum fulvum*, a Wild Relative of Pea. *Breed. Sci.* **2007**, *57*, 181–184. [CrossRef]

current issues in molecular biology

Article

Development and Optimization of Label-Free Quantitative Proteomics under Different Crossing Periods of Bottle Gourd

Anurag Malik [1], Virender Singh Mor [1,*,†], Himani Punia [2,*,†], D. S. Duhan [3], Jayanti Tokas [2], Axay Bhuker [1], Mohammed Nasser Alyemeni [4] and Awais Shakoor [5]

1 Department of Seed Science & Technology, College of Agriculture, CCS Haryana Agricultural University, Hisar 125 004, India
2 Department of Biochemistry, College of Basic Sciences & Humanities, CCS Haryana Agricultural University, Hisar 125 004, India
3 Department of Vegetable Science, College of Agriculture, CCS Haryana Agricultural University, Hisar 125 004, India
4 Botany and Microbiology Department, College of Science, King Saud University, Riyadh 11451, Saudi Arabia
5 Department of Environment and Soil Sciences, University of Lleida, 25198 Lleida, Spain
* Correspondence: virendermor@hau.ac.in (V.S.M.); puniahimani@hau.ac.in (H.P.)
† This author contributes equally to this work.

Abstract: Bottle gourd, a common vegetable in the human diet, has been valued for its medicinal and energetic properties. In this experiment, the time-resolved analysis of the changes in the proteins' electrophoretic patterning of the seed development at different crossing periods was studied in bottle gourd using label-free quantitative proteomics. Hybrid HBGH-35 had the highest observed protein levels at the 4th week of the crossing period (F_4) compared to the parental lines, viz. G-2 (M) and Pusa Naveen (F). The crossing period is significantly correlated with grain filling and reserve accumulation. The observed protein expression profile after storage was related to seed maturation and grain filling in bottle gourds. A total of 2517 proteins were identified in differentially treated bottle gourd fruits, and 372 proteins were differentially expressed between different crossing periods. Proteins related to carbohydrate and energy metabolism, anthocyanin biosynthesis, cell stress response, and fruit firmness were characterized and quantified. Some proteins were involved in the development, while others were engaged in desiccation and the early grain-filling stage. F4 was distinguished by an increase in the accumulation of low molecular weight proteins and enzymes such as amylase, a serine protease, and trypsin inhibitors. The seed vigor also followed similar patterns of differential expression of seed storage proteins. Our findings defined a new window during seed production, which showed that at F_4, maximum photosynthetic assimilates accumulated, resulting in an enhanced source–sink relationship and improved seed production. Our study attempts to observe the protein expression profiling pattern under different crossing periods using label-free quantitative proteomics in bottle gourd. It will facilitate future detailed investigation of the protein associated with quality traits and the agronomic importance of bottle gourd through selective breeding programs.

Keywords: bottle gourd; crossing periods; label-free quantitation; plant proteomics; seed vigor; SDS-PAGE

Citation: Malik, A.; Mor, V.S.; Punia, H.; Duhan, D.S.; Tokas, J.; Bhuker, A.; Alyemeni, M.N.; Shakoor, A. Development and Optimization of Label-Free Quantitative Proteomics under Different Crossing Periods of Bottle Gourd. *Curr. Issues Mol. Biol.* **2023**, *45*, 1349–1372. https://doi.org/10.3390/cimb45020088

Academic Editor: Vijai Bhadauria

Received: 22 November 2022
Revised: 4 January 2023
Accepted: 17 January 2023
Published: 6 February 2023

1. Introduction

Bottle gourd (*Lagenaria siceraria* (Molina) Standl) is an edible, useful, and medicinal vegetable crop that belongs to the Cucurbitaceae family. It is a cross-pollinated vegetable, and the amount of pollination ranges from 94% to 99%. The degree of cross-pollination also depends on several factors: flowering time, temperature, wind velocity and direction, planting design, insect population, and genotypic nature, ultimately determining the kernel quality [1]. In the cucurbit's vegetable crop, signs of the reproductive phase emerge approximately six to seven weeks after the planting when the flowering starts. Seed quality is a significant aspect of agricultural production and food security, especially during the growing uncertainty caused by climate change and abiotic factors [2]. It has a significant impact on output and resource efficiency when evaluating crop sustainability [3]. Hybrids hold the plant's whole genetic combination, making them a distribution platform for agricultural biotechnology and crop enhancement [4]. Manufacturers demand high-quality seeds to maximize their profits for crop cultivation, so their benefits would not be compromised during their field trials. Farmers constantly need high-quality sources to ensure efficient and effective plantations; thus, companies must sell high-quality seeds to maintain their competitive positions in markets [5].

The practical and inexpensive processing of seeds is an essential prerequisite for the successful hybrid production of seeds [6]. The efficiency of hybrid seed production depends on multiple factors, such as selecting an appropriate agro-climate location, a suitable season, improved floral synchronization for enhanced seed setting through the right crossing period and supplementary pollination techniques, etc. [1]. Owing to the agronomical, economic, and environmental significance and understanding of the molecular and genetic dynamics governing seed development processes, recent genome-wide studies have focused on programs considering quality and yield [5,7]. Another important agronomical trait is seed size, which directly influences grain yield and is determined by storage reserves at physiological maturity [8].

Higher value-added seeds could be produced by knowing the seed storage proteins, which account for 40–70% of storage proteins, enzymes, housekeeping proteins, and other proteins [9]. Quality is an essential aspect of crop production, and it demands specific quality and functionality during development [10]. During germination, enzymatic hydrolysis of storage proteins occurs, where proteases convert them into soluble peptides to produce free amino acids that are transported to the growing embryonic axis to maintain growth and a source of energy [11]. Several complex and interrelated processes are involved in seed development, which is highly malleable [8]. The development of the embryo is accompanied by desiccation tolerance, the final phase of seed maturation [12]. Under adverse conditions, the quality of the seed deteriorates [13]. Seed deterioration is a biological process determined by a range of physical, biochemical, and physiological changes that commence at physiological maturity and subsequently progress, which negatively affect metabolic activities, loss of seed viability, reduced seed quality, and culminate in seed death [14]. Therefore, seed vigor could be described as the energy level of viable seeds in sustainable agriculture, and the complicated interplay determines the interaction between genetic and environmental factors. Therefore, the justifications for the variability in this performance are cumbersome and remain unexplained.

Label-free proteomics is a method based on the spectral count of mass peaks and is convenient and reliable in proteomics studies [15,16]. Improvements in normalization methods have greatly improved the accuracy of label-free proteomics that rely on mass spectrometry stability [17–19]. Moreover, the unlabeled method reduces experimental error and is more accurate compared with the labeling method, and label-free quantitative proteomic approaches recognize the deepest proteome coverage, which is advantageous for the initial exploration of interesting proteins produced in bottle gourd during different crossing periods [20–22].

The proteomics study consolidates the information needed to visualize, discover, and compare the proteins and mechanisms linked with crop physiology and has also become the most commonly used approach for identifying proteins present in fruits and vegetables' biological systems [23,24]. Analysis of electrophoretic patterns via proteomics better indicates cellular and metabolic activities in bottle gourd [2,7,25–27]. Very little information is available on the protein extraction of vegetables, chiefly bottle gourd. To our knowledge, this is the first time-resolved study at the proteome level under different crossing periods for seed production in bottle gourd. Quality seed plays an essential role in enhancing agricultural productivity as well as the national economy. To achieve agricultural development goals, the supply of viable and vigorous seeds during planting time is crucial, as a healthy seed is a prerequisite for exploiting additional inputs' full potential. Therefore, the present study aims to understand and elucidate the proteome and transcriptome functional interactions, particularly those with unique features, evaluate the best crossing period for quality seed development, and subsequently regulate the deposition of storage reserves during seed maturation.

2. Materials and Methods

2.1. Plant Materials and Experimental Design

Seeds of the bottle gourd parental lines, G-2 Line (male) and Pusa Naveen (female), were procured from the Department of Vegetable Science to produce Hybrid HBGH-35. The experiment was conducted at Seed Science and Technology (29.1416° N, 75.7112° E, with an average elevation of 215 m (705 ft) above mean sea level) during the Kharif season of 2017–2018. The soil type of the experimental field is sandy loam. The concentrations of organic matter percent, total nitrogen, available phosphate, and rapidly available phosphate in the uppermost 20–30 cm of soil were 0.49%, 182 kg/ha, 18 kg/ha, and 285 kg/ha, respectively. The climate is semi-arid, with freezing winters and hot, dry, desiccating winds during the summer. All the laboratory analyses were performed at the Seed Biotechnology Laboratory and the Centre of Bio-nanotechnology (Central Laboratory) of CCS Haryana Agricultural University, Hisar, India.

2.2. Experimental Manipulations

The seed of parental lines was sown in the field during 2019–2020 and 2020–2021. Being an indeterminate crop, flowering in bottle gourds continues for around two months. Sowing took place on 17 July, and flowering began in both parents in the first week of September. The crossing period is divided into five meteorological weeks: F1 (4–10 September), F2 (11–17 September), F3 (18–24 September), F4 (25 September to 1 October), and F5 (2–8 October). The emasculation and pollination work were performed in both parental lines with the help of needles, scalpers, and forceps. The pollen was collected from male flowers and dusted on the stigmatic surface of emasculated female flowers for hybrid seed production. Manual emasculation and dusting were continued throughout the crossing period. Subsequently, all the floral buds emerging beyond 15 days after the crossing period were pinched off to ensure better growth and production of crossed fruits. At the start of the crossing period in hybrid and parental line seed production, the male flower was chosen in the parental line and bagged with paper bags the previous evening, pinched off the next morning, and carefully kept in a glass container.

The leaves and seeds were harvested at different crossing periods: fruit set during the first week of the crossing period (F1), fruit set during the second week of the crossing period (F2), fruit set during the third week of the crossing period (F3), fruit set during the fourth week of the crossing period (F4), and fruit set during the fifth week of the crossing period (F5), and kept for hybrid and parental line seed production. Tender leaves present at the axillary buds were chosen for this study. The weather data during different agro-meteorological crossing weeks are shown in Tables 1 and 2. The leaves were excised from the vines (growing points) from each crossing period and immediately flash-frozen in liquid nitrogen and stored at −80 °C. Every experiment was carried out in three biological replicates and five technical triplicates. For protein analysis, seeds were manually extracted from each crossing period. After extraction, seeds were dried in the shade to lower the moisture content to 8% and were frozen in liquid nitrogen, and stored in sealed containers at −80 °C.

2.3. Seed Vigor

Seedling vigor indexes I and II were calculated as per the formula described [28].

Seedling vigor index-I = germination (%) × seedling length (cm).

Seedling vigor index-II = germination (%) × seedling dry weight (mg).

Table 1. Weather data during the field experiment.

Month	Maximum Temperature °C	Minimum Temperature °C	RH (%) M	RH (%) E	BRI SUN HRS	PAN Evaporation (mm)	RAIN Fall (mm)	Average WS KM/H
July	34.4	27.0	90.7	70.9	6.5	4.6	2.6	7.3
August	34.7	26.3	89.7	69.3	6.3	4.2	3.1	5.6
September	34.9	23.5	87.2	49.5	6.8	4.2	1.9	2.9
October	35.0	17.2	84.8	28.0	6.6	3.6	0.0	1.9
November	27.2	10.8	90.1	39.8	3.4	2.8	0.0	2.0
December	23.8	7.3	89.7	34.5	3.7	1.4	0.0	0.7
Mean	31.7	18.7	88.7	48.7	5.5	3.5	1.3	3.4

BRI = bright sunshine hours; PAN = evapotranspiration.

Table 2. Temperature conditions under different crossing periods of bottle gourd during the field experiment.

	DOC	Max. T (°C)	Min. T (°C)		DOC	Max. T (°C)	Min. T (°C)
F_1	4-September-2017	35.8	25.6	F_2	11-September-2017	37.2	26.5
	5-September-2017	34.9	24		12-September-2017	34.9	24.5
	6-September-2017	34.6	25.3		13-September-2017	35.9	25.6
	7-September-2017	35.9	22.5		14-September-2017	36.4	26
	8-September-2017	33.4	24.5		15-September-2017	35.4	23.9
	9-September-2017	34.2	25		16-September-2017	35.8	23.9
	10-September-2017	34.4	25.9		17-September-2017	32.0	24.2
	Mean	34.7	24.7		Mean	35.4	24.9
F_3	18-September-2017	35.6	20	F_4	25-September-2017	34.4	22.7
	19-September-2017	36.6	19.9		26-September-2017	36.4	23.5
	20-September-2017	36.8	19		27-September-2017	36	22.9
	21-September-2017	36.8	22.9		28-September-2017	36.9	22.9
	22-September-2017	36.4	24.3		29-September-2017	36.6	20.5
	23-September-2017	33.4	24		30-September-2017	37	19.8
	24-September-2017	30.6	20.2		1-October-2017	37.4	18.5
	Mean	35.2	21.5		Mean	36.4	21.5
F_5	3-October-2017	37	18.8				
	4-October-2017	36.4	19.5				
	5-October-2017	36.9	19				
	6-October-2017	36.2	19.5				
	7-October-2017	36.4	18.2				
	8-October-2017	35.6	18				
	Mean	36.5	19.0				

2.4. Optimization of the Methodology with Protein Extraction Methods

2.4.1. Extraction from the Lysis Buffer

The freshly collected bottle gourd leaves under different crossing periods (F1, F2, F3, and F4) were grounded in liquid nitrogen (−196 °C) by using a pre-chilled pestle and mortar for the extraction of proteins. To this fine powder, lysis buffer (2% CHAPS, 30 mM Tris 2 M, thiourea, and 7 M urea) was added for precipitation, and sonication was performed in a sonicator (UP200S, Hielscher) for about 10 s for up to four cycles. The precipitated solution was kept on a rotator for 2 h for mixing, and after that, it was centrifuged at 15,000× *g* for 20 min. The supernatant obtained was stored at −80 °C for further use.

2.4.2. Sucrose Extraction

Proteins were extracted using sucrose [29], with some modifications. Using a homogenizer, 5 g of bottle gourd leaf and seed were homogenized in liquid nitrogen using 20 mL of extraction buffer (50 mM sodium borate, 50 mM ascorbic acid, 1% mercaptoethanol, and 1 mM PMSF) at 4 °C for 1 min (IKAT-18, Staufen, Germany). Following that, the material was centrifuged at 35,000× g for 30 min at 4 °C (Thermo Scientific, Loughborough, UK). Then, 50 mL of cold 0.1 M ammonium acetate in methanol was added to the clear supernatant. Protein precipitation was achieved by incubating the samples overnight at −20 °C.

Using a homogenizer (IKAT 18, Staufen, Germany), 5 g of bottle gourd leaf and seed were homogenized in liquid nitrogen with 20 mL of extraction buffer (50 mM sodium borate, 50 mM ascorbic acid, 1% mercaptoethanol, and 1 mM PMSF) at 4 °C for 1 min. The sample was then centrifuged (Thermo Scientific, Loughborough, UK) at 35,000× g for 30 min at 4 °C. To this clear supernatant, 50 mL of cold 0.1 M ammonium acetate in methanol was added. The samples were incubated overnight at −20 °C for protein precipitation.

2.5. TCA Extraction Methods

2.5.1. 10% TCA with 0.07% β-ME and 1 mM PMSF

Fresh leaves and seeds were homogenized in liquid nitrogen and precipitated in 10% trichloroacetic acid (TCA) diluted in cold acetone containing 0.07% mercaptoethanol and 1 mM of phenylmethylsulfonyl fluoride. The suspension was centrifuged at 15,000× g for 30 min at 4 °C, and the pellet was air-dried.

2.5.2. 10% TCA with 0.07% β-ME

The seed and leaf samples were homogenized in liquid nitrogen. After crushing, 10% trichloroacetic acid dissolved in cold acetone containing 0.07% mercaptoethanol was added. The suspension was sonicated for 1 min with a 10 s interval and kept at −20 °C for 1 h. It was centrifuged at 20,000× g for 20 min at 4 °C, and the protein pellet was washed.

2.5.3. Acetone-Phenol Extraction

The tissue was re-suspended in 10 mL of cold acetone, vortexed for 5 min, and centrifuged at 10,000× g for 5 min at 4 °C. The polyphenol-free sample was then air-dried before protein extraction, and this procedure was repeated three times. Then, 1 g of tissue was mixed with 2.5 mL of chilled Tris-buffered phenol, pH 8.8 (Fisher Scientific, Loughborough, UK), and 1 mL of chilled extraction buffer (0.2 M Tris-base pH 8.5, 10 mM EDTA, 0.4% mercaptoethanol, and 1 M sucrose), then vortexed on ice for 30 min and centrifuged for 10 min at 5000× g. The aqueous phase was again vortexed for 1–2 min after adding extraction buffer and Tris-buffered phenol. The material was then centrifuged for 10 min at 4 °C at 5000× g. The phenol phase was mixed with the initial phenol extraction phase before being centrifuged for 5 min at 10,000× g. The clear aqueous phase was decanted into a new tube. The phenolic phase was vortexed well after adding 25 mL of cool 0.1 M ammonium acetate in cold methanol. The material was then incubated overnight at −20 °C. The precipitate was collected by centrifugation at 20,000× g for 20 min at 4 °C.

2.5.4. Sodium Dodecyl Sulfate (SDS) Extraction

In this method, proteins were extracted using SDS according to the procedure proposed by Toledo et al. [29], with slight modifications. Here, 500 µL of extraction buffer containing 0.0625 M Tris-HCl, pH 6.25, 2% sodium dodecyl sulfate (SDS), 10% glycerol, 5% ME, and 0.001% bromophenol blue was used to extract the sample. The samples were vortexed and left at room temperature for an overnight period. Then, 4% CHAPS was added into the protein sample and then boiled for 5 min, and the precipitate was collected by centrifugation at 18,000× g for 15 min at 15 °C.

2.6. Precipitation Methods

As was stated above, the soluble proteins in the supernatant were precipitated using different methods.

2.6.1. Precipitation with Acetone Containing β-ME and PMSF

The protein pellet was washed three times with cold acetone containing 0.07% mercaptoethanol and 1 mM of PMSF, then centrifuged at 10,000× *g* for 30 min and incubated at −20 °C for 1 h. Again, the pellet was centrifuged at 25,000× *g* for 25 min and then dissolved in lysis buffer.

2.6.2. Precipitation with Acetone Containing β-ME

The protein pellet was washed twice in acetone containing 0.07% ME and 0.07% (*w*/*v*) DTT for 15 min at 20,000× *g*. The mixture was stored at −20 °C for 1 h and centrifuged at 20,000× *g* for 25 min. Finally, the pellet obtained was resuspended in a lysis buffer for further use.

2.6.3. Protein Precipitation with Ammonium Acetate

The protein pellet was washed once with 0.1 M ammonium acetate in cold methanol, twice with cold 80% acetone, and once with cold 70% methanol. The final resulting pellet was air-dried.

2.6.4. Clean-Up

The supernatant containing soluble protein samples was precipitated, followed by their clean-up. All the interfering compounds, such as detergents, lipids, nucleic acids, and other salts, were removed using a 2D-Clean-up kit (Bio-Rad Laboratories, Hercules, CA, USA). To the supernatant, precipitation agents were added for precipitation, and it was then centrifuged at 15,000× *g* for 5 min. The pellet was again washed with washing reagents (1 and 2), followed by centrifugation at 12,000× *g* for 10 min. The clear pellet was incubated for 1 h at −20 °C. After the incubation, it was again centrifuged. The pellet was air-dried at room temperature and resuspended in lysis buffer with gentle shaking (to avoid bubbling).

2.7. Quantification of Proteins

The total soluble protein in the re-suspended pellet was quantified by a 2D Quant kit (Thermo Fisher Scientific Inc., Waltham, MA, USA) per the manufacturer's instructions and measured in an ELISA Plate Reader (M200 pro-NanoQuant, TECAN) using bovine serum albumin (BSA) as a standard.

2.8. Electrophoresis

The proteins present in the leaf and seed extracts were analyzed by sodium dodecyl sulfate-polyacrylamide gel electrophoresis (SDS-PAGE). The stacking gel (4%) and resolving gel (12%) were used to separate the proteins in the MiniVE gel electrophoresis apparatus (Bio-Rad Laboratories, Hercules, CA, USA). Then, 30 µg of sample, as quantified by the BSA standard, was loaded in each well. Coomassie Brilliant Blue (CBB) G 250 (Bio-Rad Laboratories, Hercules, CA, USA) was used to stain the gels, followed by dissolving and de-staining in acetic acid and methanol. The Gel Pro Analyzer version 3.3 was used to scan the gel slab. The standard molecular weight marker determined the polypeptides' molecular weight (Thermo Fisher Scientific Inc., Waltham, MA, USA).

2.9. In-Gel Digestion of the SDS-PAGE Separated Proteins

For in-gel digestion, 30 µg of protein was loaded into each well, excised using a scalpel as indicated, and then added to Eppendorf tubes (1.5 mL). Excised pieces were subjected to de-staining using 50 mM of DTT, diluted in 50 mM of ABC, and incubated at 56 °C for 2 h. After the reduction step, the DTT solution was removed and 100 mM of IAA in 50 mM of ABC was added for the alkylation of cysteine residues. Then, the IAA solution was removed from the gel pieces and washed three times with 50 mM of ABC. The gel pieces were then air-dried three times for 5 min at room temperature. Following this, 50 mM of ABC buffer, containing 12.5 ng/µL of sequencing-grade modified trypsin (freshly prepared), was added. Then, 40 mM of ABC was added to cover the gel pieces and incubated overnight at 37 °C.

2.10. Peptide Extraction

After trypsin digestion, 10% v/v formic acid and the supernatant were incubated for 10 min at 37 °C. Peptides were extracted from the digested pieces by sonicating for 10 min with ACN/H_2O (50:50, v/v; 50 µL) + 0.1% v/v TFA. The pooled supernatant was dried by spinning tubes in a speed vac and stored at −80 °C. For use in mass pac, the lyophilized sample was reconstituted with 1% trifluoroacetic acid (TFA) (after this step, desalting was performed).

2.11. Peptide Desalting

The dried samples were reconstituted with 0.1% formic acid. Zip-tip C18 (Millipore Corporation, Bedford, MA, USA) was activated with 100% ACN. The samples were loaded onto a zip-tip, and the peptide was bound to C18 material. A zip-tip was loaded with 20 µL of 0.1% formic acid and 60% ACN, and the elute was collected in a new Eppendorf tube and stored at −80 °C until LC-MS/MS quantification.

2.12. Nano LC-MS/MS

The samples were then analyzed using an LC system (EASY-nLC 1200; Thermo Fisher Scientific) coupled to an MS (Orbitrap Fusion ETD MS; Thermo Fisher Scientific), equipped with a LC pre-column (75 µm × 2 cm, Nanoviper C18, 3 µm) at a flow rate of 300 nL/min. The peptide ions were detected using the MS (Orbitrap LC-MS, Thermo Fisher Scientific) with the installed Xcalibur software (version 2.0.7; Thermo Fisher Scientific). The MS was used to acquire full-scan mass spectra ranging from 375 to 1800 m/z with a resolution of 12,000. The acquired MS spectra were used for protein identification.

2.13. Differential Analysis of MS Data

Label-free quantification was also performed with Proteome Discoverer (PD) 2.2 (Version 2.2.0.388; Thermo Scientific), and the differential analysis of the relative abundance of proteins between samples was calculated. To determine the functional role of the proteins identified in the MS analysis, functional categorization of identified proteins was performed using Gene Ontology (GO) software, PANTHER (Protein Analysis Through Evolutionary Relationships). Pathway mapping of identified proteins was performed using the Kyoto Encyclopedia of Genes and Genomes (KEGG) databases.

2.14. Data Analysis

The LFQ intensities derived from all the evaluated PD samples were considered for statistical analysis. Each distinctive band was visually labeled with a number. The presence and absence were scored as 1 or 0, respectively. The data were statistically analyzed by the Gel Doc 2000 Bio-Rad system.

3. Results

3.1. Agro-Metrological Conditions and Productivity

Currently, bottle gourd is considered neglected and is an underutilized species globally. It has immense potential, but its possible widespread adoption is restricted due to a lack of evidence regarding its morphology, physiology, and nutritional benefits. The adaptation and distribution of bottle gourd are bi-hemispherical, and therefore, it grows well in both tropical and temperate regions. The weather data during different agro-meteorological crossing weeks are shown in Tables 1 and 2. During the Kharif (summer) season, the optimum temperature favoring the seed set at the F4 crossing period was 35 °C (Table 2). In contrast, relative humidity, bright sunshine hours (BRI), evaporation rate, and maximum precipitation rate were around 87%, 6.8, 4.2, and 1.9 mm, respectively, favoring the maximum seed production with full vigor potential (Table 1).

3.2. Seed Vigor Analysis

The results revealed that at the fourth week of the crossing period (F4), the accumulation of seed reserves was the maximum in Hybrid HBGH-35, which thus had full seed vigor potential, followed by the female parental line (Pusa Naveen), while the minimum was in the male parental line (G2-line) (Figure 1a,b). The temperature plays a crucial role in seed setting and fruit development (Tables 1 and 2). At a temperature of 34 °C and a relative humidity of 87%, the F4 crossing period had maximum seed potential. The interaction plot between the bottle gourd parental lines and the hybrid revealed that both the hybrid HBGH-35 and F4 crossing periods had the highest seed vigor I and II (Figure 2a,b).

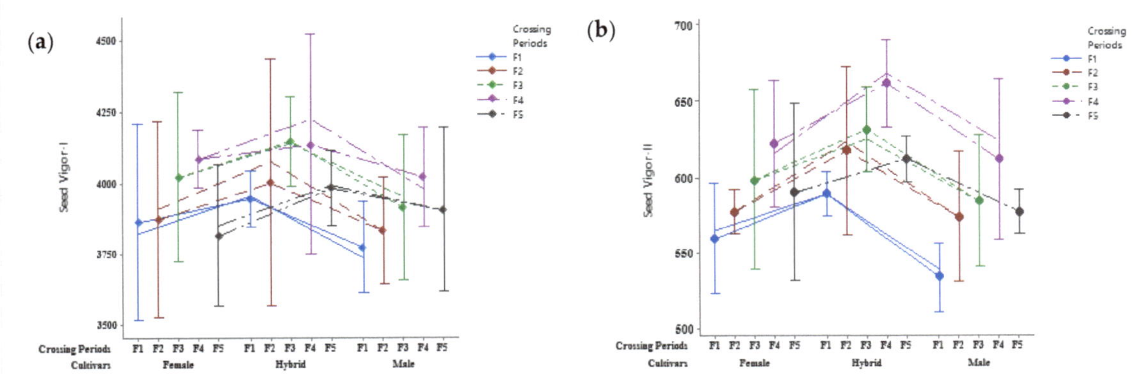

Figure 1. The seed vigor potential of bottle gourd hybrid HBGH-35 and two parental lines, female Pusa Naveen and male G2-line. (**a**) Seed vigor I and (**b**) seed vigor II.

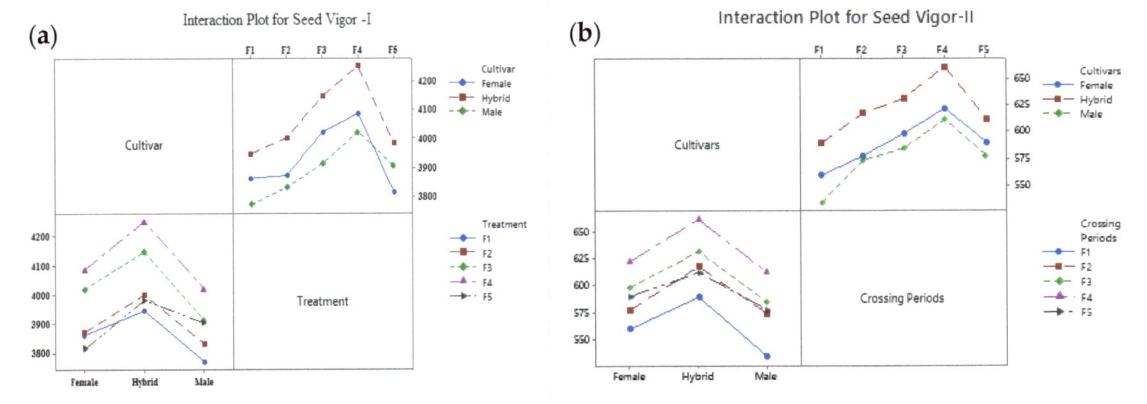

Figure 2. As an interaction plot, the seed vigor potential of bottle gourd hybrid HBGH-35 and two parental lines, female Pusa Naveen and male G2-line, were plotted. (**a**) Seed vigor I and (**b**) seed vigor II.

3.3. Optimization of the Methodology

Preliminary research was conducted to establish the appropriate method to extract protein from bottle gourd. The protein extracted from the SDS reagent (Figure 3a, lane 1) and TCA containing ME, as well as TCA containing mercaptoethanol and 1 mM of PMSF, did not yield reproducible results (Figure 3a, lanes 3 and 4). The lysis buffer protein extraction method exhibited substantial products with sharp band intensities and band strengths, followed by their clean-up (Figure 3a, lane 6). The protein extracted with sucrose

(Figure 3a, lane 5) which precipitated with a 2D clean-up kit also displayed a sharp band. Still, precipitation with cold acetone did not show clear bands, and the proteins were not effectively extracted with the acetone-phenol (Figure 3a, lane 2) method as no clear bands were detected. Therefore, proteins extracted with the lysis buffer method exhibited reproducible results in bottle gourd leaves. The proteins extracted with the lysis buffer method showed precise band intensities and band strengths (Figure 3b, lane 1), followed by their clean-up. The sucrose and acetone-phenol extractions (Figure 3b, lanes 2 and 5), even after precipitation with acetone and a clean-up kit, did not effectively extract the proteins. When precipitated with ammonium acetate, the extractions with 10% TCA and 0.07% ME + PMSF and TCA with ME (Figure 3b, lanes 3 and 4) showed clear bands in seeds but not in leaves. Due to its reducing nature, SDS extraction (Figure 3b, lane 6) displayed bands but with low intensities. After protein extraction, the supernatant was exposed to different protein precipitation methods, where cold acetone and ammonium acetate-washed proteins showed the smeared bands. With the protein being hydrophobic, acetone reduced the protein's solubility and amplified its non-polar environment; thus, the proteins coagulated and did not disperse uniformly due to disturbance of the charge-to-mass ratio. Protein precipitated with 2D-clean-up dissolved the pellets entirely and uniformly, and the protein concentration was also enhanced when quantified.

3.4. Polypeptide Composition of Extractable Proteins

In this study, one bottle gourd hybrid HBGH-35 and two parenteral lines, G2 line (M) and Pusa Naveen (F), which differ in their quality and yield traits, were evaluated based on physiological characteristics (Figure 3c–f). SDS-PAGE separated the extractable bottle gourd proteins (leaves and seeds) into multiple components. The electrophoretic protein analysis exhibited a total of 102 protein bands in both parental lines and the hybrid. SDS-PAGE gels resolved 30 µg of protein sample from three biological replicates of the hybrid and two parenteral lines, and electrophoresed proteins were visualized after CBB staining. The banding pattern was consistent across all bottle gourd leaf proteins, resulting in 29 protein bands at various crossing times (see Figure 3c,d for positions of these polypeptides). Some bands were unique and specific markers for genotypes, which permitted their gel electropherograms' identification and characterization. Six prominent peptide bands were identified with average molecular weights in the range of 39.1–52.4 kDa and 55.0–62.1 kDa (Figure 3c,d). These bands represent the major seed storage proteins and photosynthetic apparatus, accounting for 40–60% of the extracted bottle gourd protein. The 40 kDa polypeptide in hybrid HBGH-35 (Figure 3c) represents the photosynthetic apparatus with higher band strength and band intensity at F4 in comparison to the female line (Figure 3d, lanes 6–10) and the male line (Figure 3d, lanes 1–5). The relative protein expression was more prominent at F4 in all the cultivars, indicating the maximum accumulation of photosynthetic assimilates and storage reserves than at F1, F2, F3, and F5. The significant differences detected in band numbers 4, 37, 38, 42, 49, 62, and 63 differed in appearance and band intensity depending on the storage proteins (Figure 3c,d). Thus, the results indicated that the higher band strengths and apparent band intensities were higher in hybrid HBGH-35, followed by the Pusa Naveen (F) (Figure 3c,d, lanes 6–10) line at the F4 crossing period.

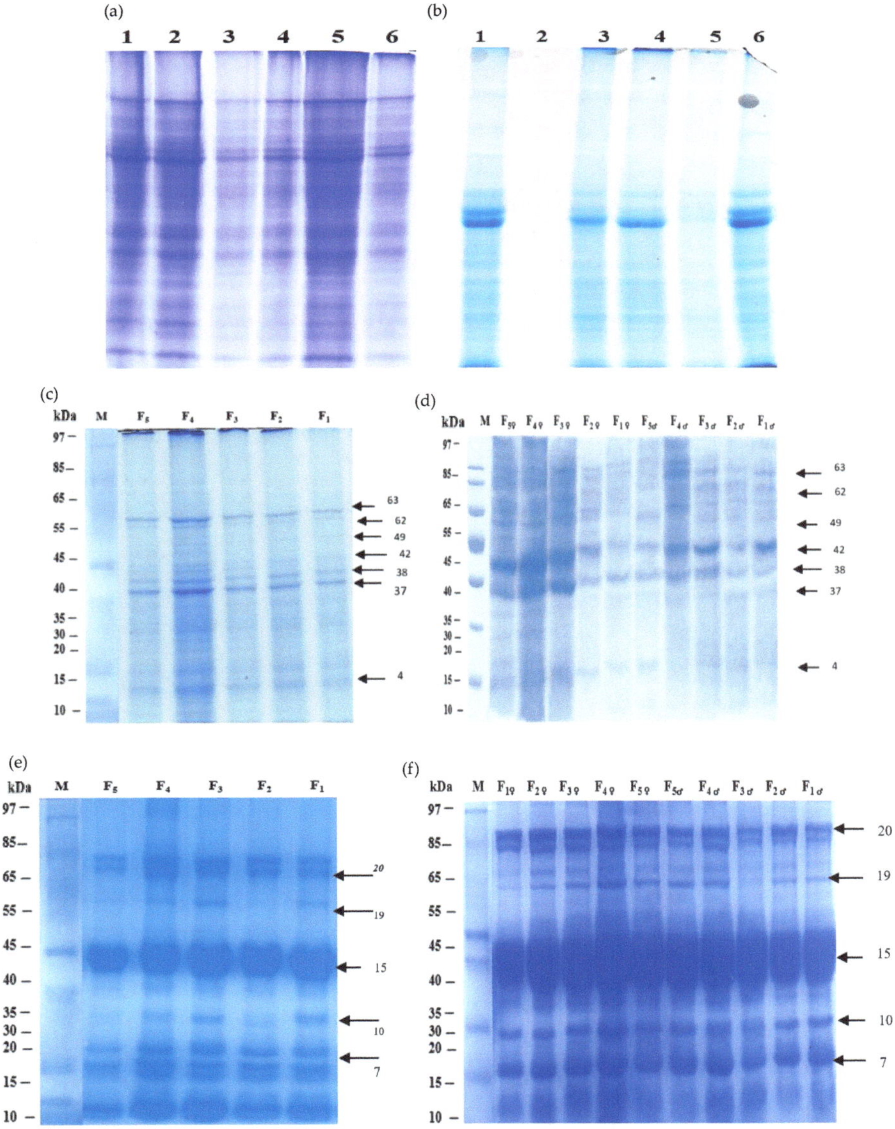

Figure 3. (**a**) Gel showing different protein extraction methods in bottle gourd leaves. Lane 1, SDS; Lane 2, acetone-phenol; Lane 3, 10% TCA containing β-ME; Lane 4, TCA with β-ME and 1 mM PMSF; Lane 5, sucrose; Lane 6: lysis buffer. (**b**) Lane 1, lysis buffer; Lane 2, sucrose; Lane 3, 10% TCA containing β-ME and 1 mM PMSF; Lane 4, TCA with β-ME; Lane 5, acetone-phenol; Lane 6, SDS. (**c**) SDS-PAGE of extractable proteins in bottle gourd leaves. Hybrid HBGH-35: Lane 1, F1; Lane 2, F2; Lane 3, F3; Lane 4, F4; Lane 5, F5. (**d**) Lane 1, F1 in male; Lane 2, F2 in male; Lane 3, F3 in male; Lane 4, F4 in male; Lane 5, F5 in male; Lane 6, F1 in the female; Lane 7, F2 in the female; Lane 8, F3 in the female; Lane 9, F4 in the female; Lane 10, F5 in the female. (**e**) SDS-PAGE analysis of extractable proteins in bottle gourd seeds. Hybrid HBGH-35: Lane 1, F1; Lane 2, F2; Lane 3, F3; Lane 4, F4; Lane 5, F5. (**f**) Lane 1, F1 male; Lane 2, F2 male; Lane 3, F3 male; Lane 4, F4 male; Lane 5, F5 male; Lane 6, F1 female; Lane 7, F2 female; Lane 8, F3 female; Lane 9, F4 female; Lane 10, F5 female; M: marker; kDa: kilodaltons (molecular weight standards).

The proteins extracted from bottle gourd seeds had a different protein pattern than the proteins extracted from the leaves (Figure 3e,f). The gel analysis demonstrated that the protein subunits had a molecular weight ranging from 12.64 to 93.04 kDa. The band intensity and number varied under different crossing periods, hybrid lines, and parental lines. Data showed that hybrid HBG-35 (Figure 3e) was resolved into 20 bands, while the male (Figure 3f, lanes 6–10) and female lines (Figure 3f, lanes 1–5) were resolved into 19 bands, respectively, under different crossing periods. The major protein bands identified in the samples were calculated with a molecular weight of 45.6, 62.8, 85.0, and 87.3 kDa. These bands represent the major seed storage proteins, particularly albumins, prolamins, and gliadins. The 45 kDa polypeptide expressed a significant portion of extractable proteins in bottle gourd seeds. The considerable differences in band numbers 7, 10, 15, 19, and 20 differed in band appearance and intensity. The relative protein expression was more prominent at the F4 crossing period, chiefly in the hybrid (Figure 3e), followed by the female line (Figure 3f, lanes 1–5), indicating the maximum accumulation of photosynthetic assimilates and storage reserves in comparison to F1, F2, F3, and F5. The percentage of each polypeptide varied significantly between bottle gourd lines and crossing periods (F1 to F5). As far as polymorphism is concerned, there were five monomorphic, seven polymorphic, and four unique bands. A detectable band intensity change has been observed in bottle gourd lines under different crossing periods. Thus, hybrid HBGH-35 at the F4 crossing period had a higher band intensity than both parental lines.

3.5. Multivariate Analysis of the Datasets

Multivariate principal component analysis (PCA), performed on all replicates of the four sets (crossing periods F1 to F4) and the control (F0) mixture of proteins, showed five statistically different groups, without overlapping, for sets 1 to 4 and the control (F0) (Figure 4).

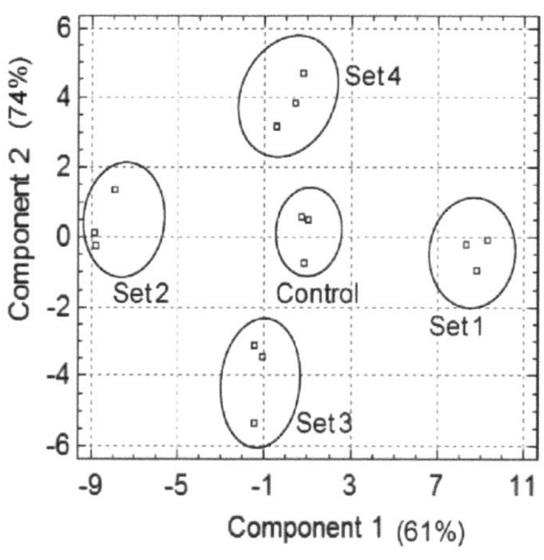

Figure 4. PCA score plot to assess the variance among the protein compositions of the four sets and the control protein mixtures (n = 3). Component−1 explains 62.6% of the variability of the data, whereas component 2 is responsible for 13.9% (the total accumulated percentage is 75%).

3.6. Protein Identification and Quantification

In the preliminary experiments, major differences in protein profiles were observed at F4 and F3 crossing periods. Label-free quantification (LFQ) via LC-MS/MS was performed in F4, F3, and F1 (control) crossing periods to detect the variably expressed proteins. Analysis of the LFQ results using MASCOT software revealed more than 100 differentially expressed proteins (DEPs), out of which 20 proteins were upregulated (Table 3) and 12 proteins were downregulated (Table 4), having fold changes of $\geq$1.5 and $\leq$0.6, respectively, which were considered for further analysis. The analysis further revealed some important proteins that play a role in seed reserve mobilization related to the source–sink relationship in bottle gourd. Tables 3 and 4 present a functional analysis of upregulated and downregulated proteins with differential abundance and fold change between genotypes HBGH35 and Pusa Naveen during different crossing periods (F3 and F4) compared to their control (F1). The DEPs were sorted based on their different molecular mechanisms, such as transporters, antioxidative enzymes, photosynthesis, signal transduction, biosynthetic pathways, nucleotide synthesis, carbohydrate metabolism, etc., with a higher protein score, protein coverage, and protein score match (PSM). Different bioinformatic software and tools designed for mass spectrometry-based protein identification and quantification were used to characterize the proteins based on their functions at cellular, biological, and molecular levels using Gene Ontology enrichment, PANTHER 8.0 software, and the KEGG pathway.

Among the upregulated proteins (Table 3), fold changes in ion transporters, such as H(+)-exporting diphosphatase, HATPase_c domain-containing protein, vacuolar proton pump subunit B, Cation_ATPase_N domain-containing protein, V-type proton ATPase subunit G, and V-ATPase 69 kDa subunit H(+)/Pi co-transporter, at F3 and F4 were more than 2.5 in HBGH35. The proteins for antioxidative enzymes, osmolytes, and molecular chaperones were highly upregulated in SSG 59-3 as compared to control fruits. Other upregulated proteins include metabolic interconversion enzymes, carbon metabolism, photosynthesis, DNA/RNA synthesis, chlorophyll synthesis, and oxidative phosphorylation. Among the downregulated proteins (Table 4), the proteins were from starch and sucrose metabolism, biosynthesis of secondary metabolites, carbon metabolism, gluconeogenesis, and amino acid biosynthesis, with a fold change of $\leq$0.5. The number of downregulated proteins in Pusa Naveen (♀) was higher compared to HBGH35.

Figure 5 compares the changes in differential protein expression under the different crossing periods. The expression of 77 and 13 proteins was significantly reduced and elevated, respectively, following the F1 crossing period (control). The abundance of 141 and 13 proteins was significantly lessened and increased, respectively, following the F4 crossing period, and the abundance of 19 and 13 proteins was significantly lessened and elevated, respectively, following the F3 crossing period.

Using Gene Ontology (GO) online PANTHER (Protein Analysis Through Evolutionary Relationships) 8.0 software, a total of 1861 pathway hits in HBGH-35 and 1006 pathway hits in Pusa Naveen were observed (Figure 6). PANTHER is an approach to outlining the genes and gene product properties that are shared across species. The PANTHER classification system (http://pantherdb.org/ 15 October 2022) has been designed to classify proteins (and their corresponding genes) to facilitate high-throughput analysis.

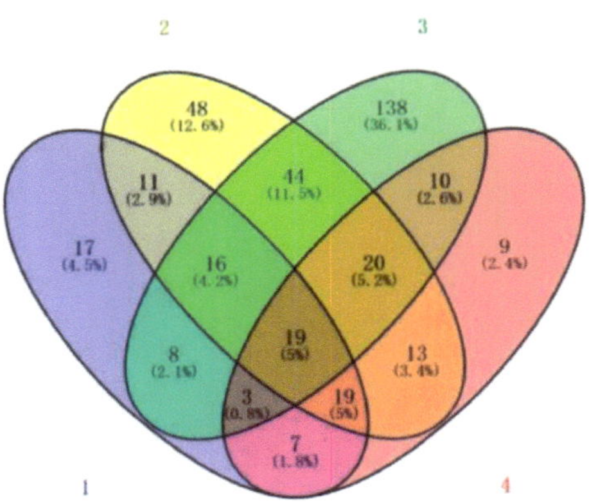

Figure 5. Venn diagram of the number of differentially expressed proteins identified in the bottle gourd proteome under different crossing periods.

Table 3. LFQ (MASCOT) list of upregulated proteins in HBGH-35 and Pusa Naveen (♀) at various crossing times.

S. No.	Accession	Protein Description	MW (kDa)	pI	Protein score	SC (%)	Peptides	PSM	Fold Change						KEGG Pathway
									HBGH35			Pusa Naveen (♀)			
									F1	F3	F4	F1	F3	F4	
Ion Transporters															
1.	A0A194YHK2	H(+)-exporting diphosphatase	79.7	5.33	467.22	13	3	168	1.45	4.83	6.15	1.21	2.44	2.65	Plant–pathogen interaction; protein processing in the endoplasmic reticulum
2.	C5YBL4	HATPase_c domain-containing protein	81.7	5.11	197.27	9	3	115	1.62	2.66	4.11	0.27	0.64	1.24	Phagosome; metabolic pathways; oxidative phosphorylation
3.	C5WP97	Cation_ATPase_N domain-containing protein	115.4	5.58	246.11	10	10	92	1.21	3.48	5.2	0.19	0.23	0.13	Phagosome; metabolic pathways; oxidative phosphorylation
4.	A0A194YNQ1	Vacuolar proton pump subunit B	54.2	5.36	998.47	17	17	392	1.2	3.42	3.05	0.28	0.84	1.02	Phagosome; metabolic pathways
5.	A0A1B6Q818	Glutathione transferase	24	6.52	582.35	7	7	193	1.34	4.26	3.55	1.02	0.51	0.36	Phagosome; oxidative phosphorylation; metabolic pathways
6.	C5YF66	V-type proton ATPase subunit G	12.3	6.13	18.57	2	2	7	2.67	3.46	4.9	1.64	2.64	2.32	Phagosome; metabolic pathways; oxidative phosphorylation
7.	C5YX05	V-type proton ATPase subunit C	42.7	5.78	446.02	13	6	170	2.13	5.94	6.54	1.24	3.91	4.09	Phagosome; metabolic pathways
Antioxidative enzymes, osmolytes, chaperons															
8.	A0A1B6PMT8	PEROXIDASE_4 domain-containing protein	38.3	8.19	372.99	11	11	155	1.23	3.23	4.35	1.2	2.1	2.8	Glutathione metabolism
9.	C5X6H6	L-ascorbate peroxidase	27.1	5.36	863.16	9	8	318	1.2	7.37	4.32	0.32	0.54	0.72	Arachidonic acid metabolism; glutathione metabolism
10	Q6JAG4	Glutathione peroxidase	18.4	7.08	326.46	7	6	122	1.17	3.32	5.11	1.2	1.08	2.34	Phagosome; metabolic pathways
11	C5WWX0	Glutathione reductase	59.3	7.56	458.25	14	14	209	2.04	4.23	6.23	1.17	2.43	3.01	Glutathione metabolism
12	A0A1B6QQQ9	Catalase	56.6	7.28	656.35	15	12	268	1.19	4.3	6.5	2.31	3.92	4.09	Scavenging
13	A0A1B6Q707	Delta-1-pyrroline-5-carboxylate synthase	78.3	6.42	193.73	11	10	83	1.04	6.64	4.8	1.24	2.27	3.01	Arginine and proline metabolism; biosynthesis of secondary metabolites; metabolic pathways
14	C5WSJ9	Proline dehydrogenase	53	7.69	2.48	1	1	1	4.64	6.54	6.64	1.07	2.64	3.64	Osmoregulation

Table 3. *Cont.*

S. No.	Accession	Protein Description	MW (kDa)	pI	Protein score	SC (%)	Peptides	PSM	Fold Change						KEGG Pathway
									HBGH35			Pusa Naveen (♀)			
									F1	F3	F4	F1	F3	F4	
15	C5Y3J4	LEA_2 domain-containing protein	24	9.11	42.13	2	2	20	1.35	3.77	4.51	1.03	2.62	4.43	Osmoregulation; molecular chaperons
16	A1E9W3	30S ribosomal protein S3, chloroplastic	25.9	9.74	102.5	5	5	51	1.2	3.2	4.6	1.3	2.2	3.3	Ribosome
17	A0A1B6PDG7	Diaminopimelate epimerase	37.8	6.54	106.48	6	6	46	2.11	4.29	5.22	2.08	3.01	2.32	Cysteine and methionine metabolism; metabolic pathways; biosynthesis of amino acids; sulfur metabolism; biosynthesis of secondary metabolites; carbon metabolism
18	A0A1B6P8M2	Pyruvate, phosphate dikinase	102.4	5.73	3300.97	44	44	1234	2.12	4.6	5.13	1.26	2.43	3.0	Protein export
19	C5YUG2	Starch synthase, chloroplastic/amyloplastic	103	6.49	101.91	6	6	32	2.39	2.75	3.05	1.14	3.04	4.2	Biosynthesis; carbohydrate metabolism
20	C5YD77	Glucose-6-phosphate 1-dehydrogenase	66.6	8.35	107.16	5	3	50	1.14	2.6	3.16	1.26	2.26	3.46	Glucose metabolism

Table 4. LFQ (MASCOT) list of downregulated proteins in HBGH35 and Pusa Naveen (♀) under different crossing periods.

S. No.	Accession	Protein Description	MW (kDa)	pI	Protein Score	SC (%)	Peptides	PSM	Fold Change						KEGG Pathway
									HBGH35			Pusa Naveen (♀)			
									F1	F3	F4	F1	F3	F4	
1.	A0A1Z5RLT7	Ribosomal_L16 domain-containing protein	20.8	10.18	192.37	6	3	82	−0.74	−0.72	−0.71	−0.81	−2.92	−0.01	Ribosome
2.	A1E9W3	30S ribosomal protein S3, chloroplastic	25.9	9.74	102.5	5	5	51	−0.1	−0.34	−0.11	-	-	-	Ribosome
3.	C5WYH8	Glutamyl-tRNA synthetase	81.2	7.44	317.85	11	4	133	−0.31	−0.2	−0.47	−0.51	−0.91	−0.61	Metabolic pathways; biosynthesis of secondary metabolites; aminoacyl-tRNA biosynthesis; porphyrin and chlorophyll metabolism

Table 4. *Cont.*

S. No.	Accession	Protein Description	MW (kDa)	pI	Protein Score	SC (%)	Peptides	PSM	Fold Change						KEGG Pathway
									HBGH35			Pusa Naveen (♀)			
									F1	F3	F4	F1	F3	F4	
4.	C5XFP1	Cysteine synthase	42.1	8.28	481.2	11	11	170	-	-	-	−0.33	0	−0.44	Cysteine and methionine metabolism; metabolic pathways; biosynthesis of amino acids; sulfur metabolism; biosynthesis of secondary metabolites; carbon metabolism
5.	C5XDW6	UDP-arabinopyranose mutase	40.8	7.05	212.84	7	2	103	-	-	-	−0.43	−0.14	−0.27	Amino sugar and nucleotide sugar metabolism
6.	C5YAY1	Protein kinase domain-containing protein	67.4	7.44	69.27	3	3	27	-	-	-	−0.35	−0.35	−0.16	Signal transduction
7.	C5Y2Z7	E1 ubiquitin-activating enzyme	116.7	5.36	629.61	22	2	258	-	-	-	−0.99	−0.17	−0.32	Ubiquitin-mediated proteolysis
8.	C5YDP0	40S ribosomal protein S8	24.9	10.39	390.36	6	2	155	-	-	-	−0.02	−0.91	−1	Ribosome
9.	C5XZJ2	Vacuolar protein sorting-associated protein 29	20.9	6.6	4.8	1	1	2	-	-	-	−6.64	−6.64	−0.3	Endocytosis
10.	C5WZ11	Glutathione transferase	25.7	7.56	95.7	3	1	36	-	-	-	−1.67	−6.64	−0.11	Glutathione metabolism
11.	C5XBD4	Phosphopyruvate hydratase	50.5	6.29	168.11	6	6	66	−0.05	−0.01	−0.16	−0.05	−1.57	−0.57	RNA degradation; carbon metabolism; glycolysis/Gluconeogenesis; biosynthesis of amino acids; metabolic pathways; biosynthesis of secondary metabolites
12.	C5Y9W4	14_3_3 domain-containing protein	29.6	4.81	959.55	14	9	347	−0.22	−0.72	−0.3	0-.32	−0.03	0.19	

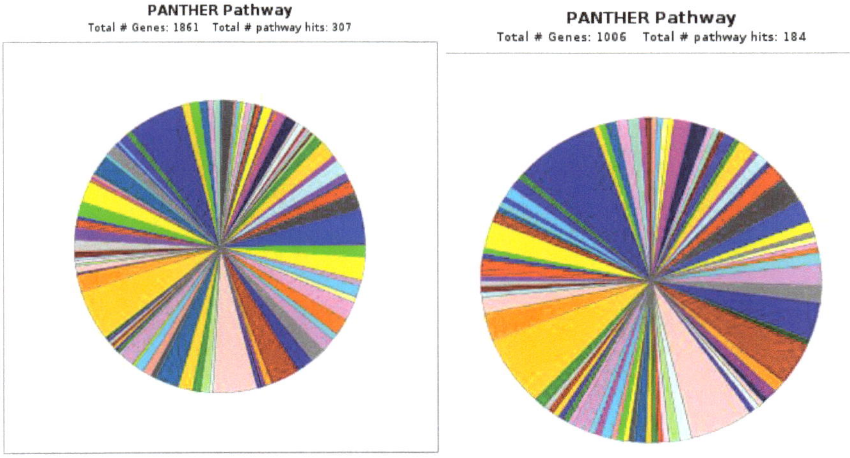

Figure 6. Gene Ontology classification of pathways using PANTHER in HBGH35 and Pusa Naveen (♀).

The functional characterization of identified proteins was based on Gene Ontology (GO) using the PANTHER 8.0 platform, which generated information regarding cellular localization and metabolic and biological processes. The classification of proteins was based on cellular components, molecular function, and biological processes (Figures 7–11), and to cluster them, we exploited the K-means clustering approach of the web tool STRING (https://string-db.org/ 17 October 2023). The obtained results are shown in Figure 12a (more abundant proteins) and Figure 12b (less abundant proteins), and Figure 12c shows gene co-occurrence across genomes. Gene ontology protein classification in HBGH35 and Pusa Naveen (♀) classify DEPs based on their molecular function (Figure 7). Around 1677 molecular function hits were observed across 2587 genes. About 62% of the proteins were involved in a catalytic role (GO: 0003824), which includes antioxidative enzymes, molecular chaperons, osmoprotectants, sugars, antioxidants, etc., and 25% of the proteins belonged to the binding protein class (GO: 0005488), which includes proteins that regulate abiotic stress conditions, signal transduction pathways, and post-transcriptional processes such as splicing regulation, mRNA transport, and mRNA translation modulation. Other proteins involved in molecular function include those involved in molecular adaptor activity (GO: 0060090), molecular transducer activity (GO: 0060089), structural molecule activity (GO: 0005198), and transporter activity (GO: 0005215). After identifying the molecular functions of the DEPs, further analysis was performed to determine the protein classes for the proteins involved in molecular function.

Around 1417 hits to biological processes were observed from 962 genes (Figure 8a) based on biological processes (Figure 8b) in HBGH-35, and the proteins were classified into 12 categories: those involved in cellular processes (44%), metabolic processes (35%), biological regulation (7%), localization (7%), response to stimulus (6%), biological phase (1.8%), reproductive processes (1.2%), signaling (1%), developmental processes (0.6%), reproduction (0.5%), multicellular organismal processes (0.4%), and immune system processes (0.4%) (Figure 8a). Majority of the proteins in Pusa Naveen (♀) were involved in the metabolic processes (23%), cellular processes (22%), biological regulation (17%), signaling (8%), response to stimulus (6%), localization (7%), developmental processes (7%), immune system response (2%), and multicellular organismal processes (0.1%) (Figure 8c).

The cellular component classification in HBGH35 and Pusa Naveen (♀) (Figure 9) revealed that in HBGH35, majority of proteins (42.5%) were found in the intracellular region, followed by cellular and anatomical activity (44.7%) and protein-containing complex (13.7%), while in Pusa Naveen (♀), four categories were observed: intracellular (43%), biosynthetic process (23%), cellular and anatomical activity (21%), and protein-containing complex (11%). Salinity under osmotic stress, salinity-induced cellular protein conversion to involvement in intracellular responses, which involved the metabolic and several signal transduction pathways, was followed by a cascade of intracellular signals via binding proteins and protein complex networks.

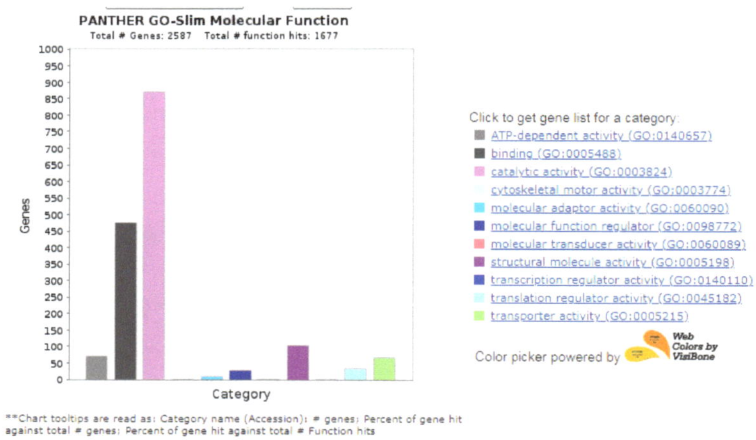

Figure 7. Gene Ontology classification of proteins in HBGH35 and Pusa Naveen (♀) based on their molecular function.

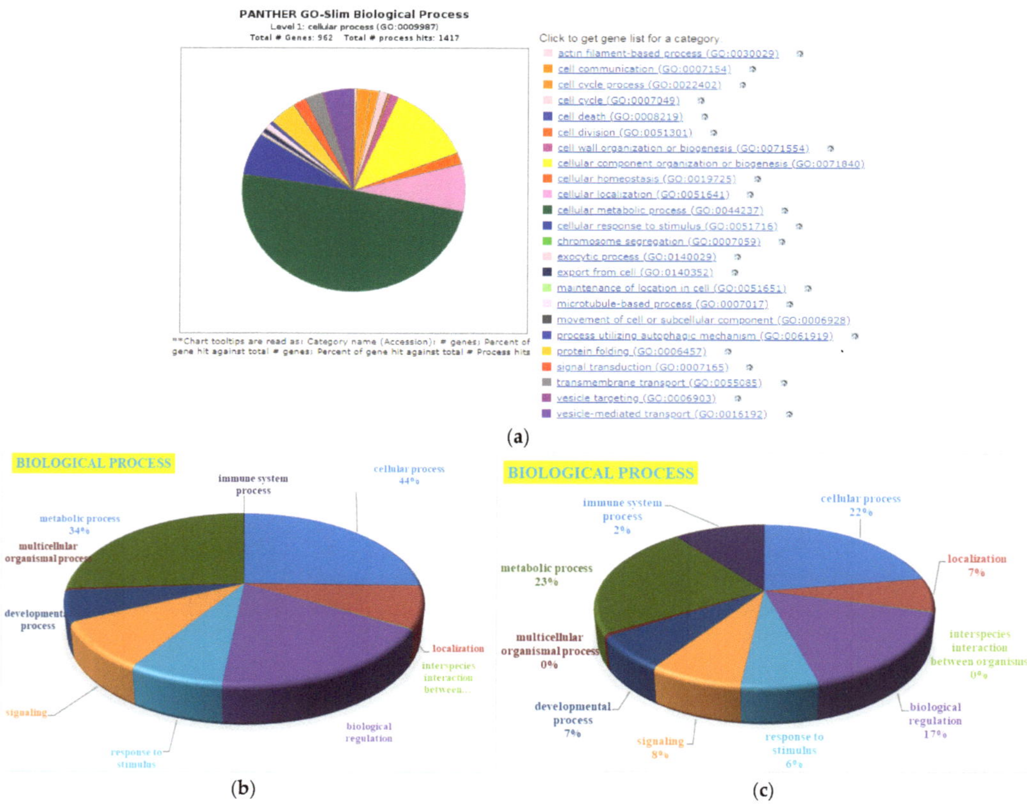

Figure 8. Gene Ontology classification of proteins with: (**a**) process hits, (**b**) HBGH35, and (**c**) Pusa Naveen (♀) based on their biological processes.

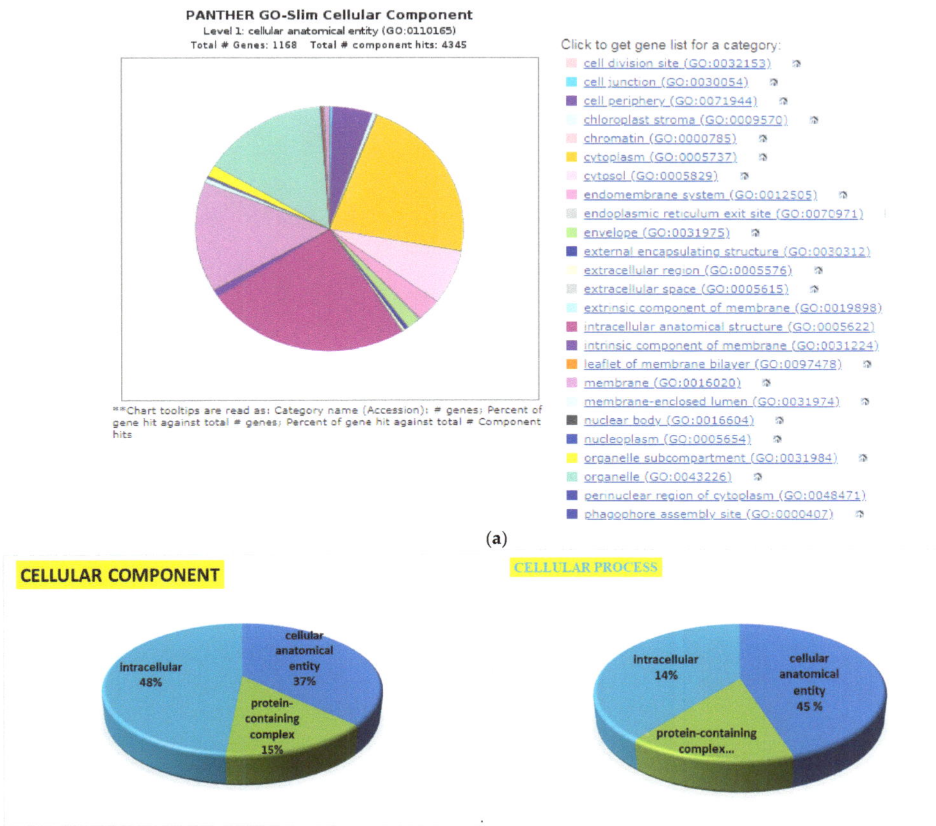

Figure 9. Gene Ontology classification of proteins: (**a**) process hits, (**b**) HBGH35, and (**c**) Pusa Naveen (♀) based on their cellular level.

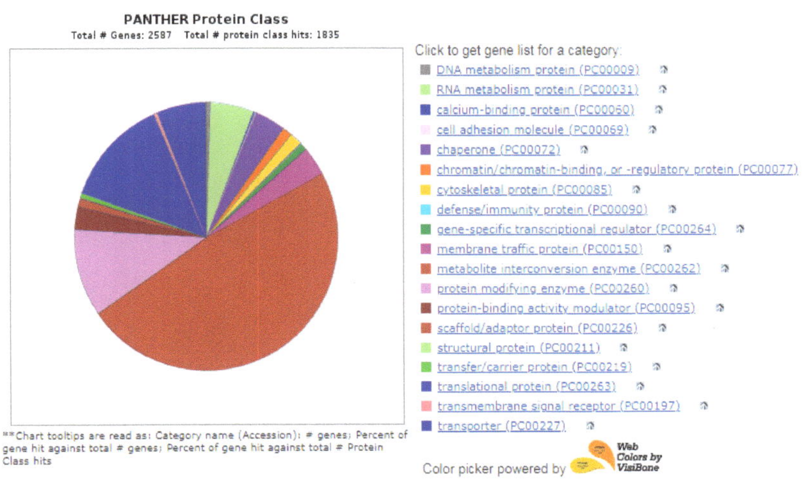

Figure 10. Gene Ontology classification of proteins in HBGH35 and Pusa Naveen (♀) based on their molecular function and metabolic interconversion enzymes.

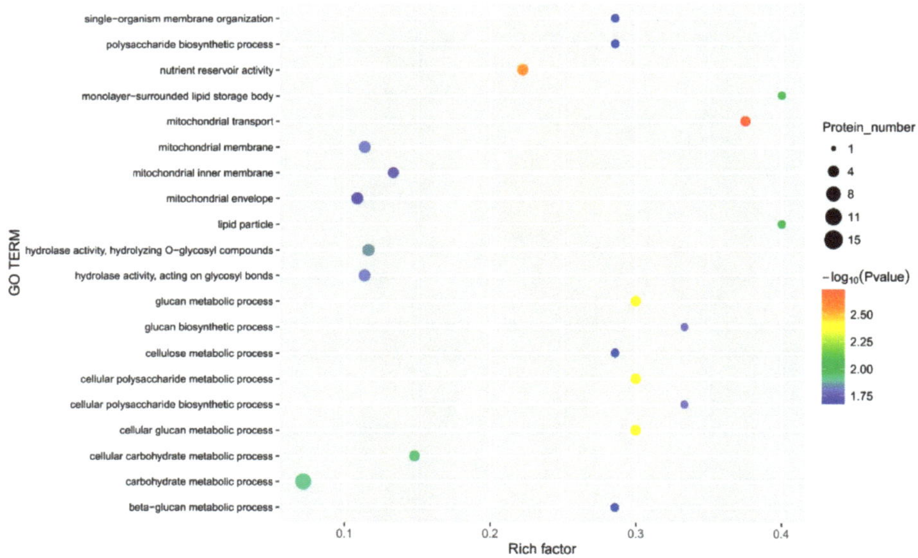

Figure 11. The functions of the main differential proteins in different crossing periods.

Among the metabolic interconversion enzymes (PC00262), Gene Ontology classification further classified 1835 proteins (Figure 10). Protein-modifying enzymes (PC00260) are the first category of metabolic interconversion enzymes, followed by transporters (PC00227) and translational proteins (PC00263). The oxidoreductase consists of alcohol dehydrogenase, catalase, choline oxidase, and enzymes involved in glycolysis, the TCA cycle, oxidative phosphorylation, and amino acid metabolism. The transferases include glutathione reductase, glutathione, glutathione S-transferases, and coenzyme A (CoA). Other categories of enzymes included were calcium-binding proteins (PC00060), chaperones (PC00072), membrane traffic proteins (PC00150), chromatin binding/regulatory proteins (PC00077), and RNA metabolism proteins (PC00031). The overall result of protein identification and the results of differential proteins are shown in Figure 11.

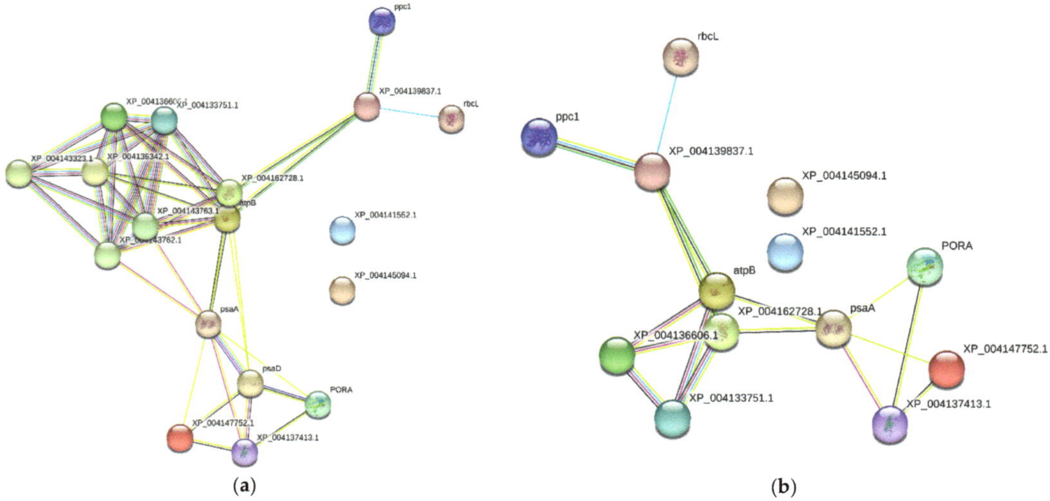

Figure 12. *Cont.*

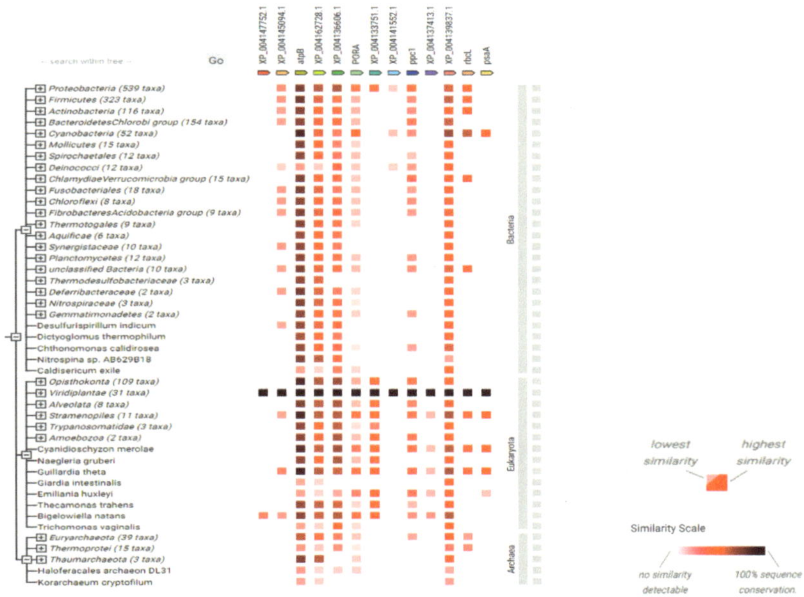

(c)

Figure 12. STRING analysis of (**a**) more abundant, (**b**) less abundant, and (**c**) gene co-occurrence patterns across genomes shows similarities.

4. Discussion

Due to climate change in the last decade, food production has been adversely affected worldwide and has become a significant threat to food security. Finding an appropriate crossing period that preserves high seed quality and yield may enhance high-quality seed production under different environmental conditions. It is indispensable because suitable crossing selection is essential for effective seed production with robust-quality seeds. Seed vigor determines the degree of aliveness and is depicted by the complicated interplay between genetic and environmental factors [30]. Bottle gourd grows well in areas with a rainfall range of 400–1500 mm per annum; however, moderate rather than excessive soil water is desired for a good harvest [31]. Therefore, bottle gourd is intolerant of waterlogging. Our results are well-supported by Grubben and Denton [32], who reported that bottle gourd grows well at warm temperatures (25–35 °C). The optimum germination temperature is between 20 and 25 °C. Flowering is highly sensitive to the photoperiod. Short days coupled with low night temperatures and high relative humidity promote male flowers' development, while the reverse promotes female flowers [31]. Not much information is available on the production of bottle gourd. Agronomic practices and physiological attributes that encourage more female flower production than male flower production might increase yields. Thus, our results might help to determine the optimal growth conditions to maximize fruit and seed sets.

At a temperature of 34 °C and a relative humidity of 87%, the F4 crossing period had maximum seed potential. The interaction plot between the bottle gourd parental lines and the hybrid revealed that both hybrid HBGH-35 and F4 crossing periods had maximum seed vigor I and seed vigor II. Our results are well-supported by several other researchers [33,34].

With the protein being hydrophobic, acetone reduced the protein's solubility and amplified its non-polar environment; thus, the proteins coagulated and did not disperse uniformly due to disturbance of the charge-to-mass ratio [7]. Protein precipitated with 2D-clean-up entirely and uniformly dissolved the pellets, and the protein concentration was also enhanced when quantified. The variations among extractable proteins were also reported by several authors [35–37], but no reports are available for comparative proteome analysis in bottle gourd. Henceforth, our results demonstrated that protein extraction using lysis buffer followed by its clean-up was the most efficient and reliable protocol for extracting protein from bottle gourd (leaves and seeds). Thus, standardization of extraction methods in bottle gourd holds potential in future understanding of the seed development process via identification of novel proteins involved in seed set.

The banding pattern was similar among all bottle gourd leaf proteins for a total of $\leq$29 protein bands at different crossing periods. Some bands were unique and specific markers for genotypes, which permitted their gel electropherograms' identification and characterization. Other researchers have also reported the genotypic variation among the extracted proteins based on their polypeptide composition [38].

Protein analysis using SDS-PAGE revealed a positive correlation between seed storage protein accumulation and their differential expression during different crossing periods in the current study. Furthermore, Capouchová et al. [38] used SDS-PAGE to investigate the proteome behavior in oat and sorghum genotypes. The authors reported that the higher proportion of proteins was characterized by prolamins and glutelins, approximately 75.02% and 50.63% of total seed storage proteins, respectively. This study is also supported by several other authors [36,39]. Srivastava et al. [40] and Punia et al. [7] conducted a molecular diversity pattern in bottle gourd using a RAPD marker and SDS-PAGE. Results showed that a total of 60 bands were observed, of which 60.29% were polymorphic and had a similarity coefficient of 0.68 on a similarity matrix. Our study may act as a baseline for detecting and identifying proteins under different crossing periods for quality seed production and might be exploited in molecular future breeding programs in developing bottle gourd cultivars with high yield and good kernel quality, with full vigor potential. Future insights about direct regulatory networks in bottle gourd would require comprehensive functional analysis of such proteins.

Proteomic studies have emerged as a new platform to unravel important relationships between protein abundance and plant stress acclimatization, allowing fast discovery and precise protein profiling under different abiotic stresses [2,41,42]. The leaves play a significant role in transporting water and minerals from the roots to the aerial parts. The electrophoretic profiling of proteins via SDS-PAGE provides a preliminary foundation for protein patterning under abiotic stress conditions. Label-free quantitative (LFQ) proteomics has allowed rapid identification, expression dynamics, and post-translational modifications (PTMs) of proteins [43]. Several reports on comparative proteomics have been explored in sorghum to study the abiotic stress responses and identify the prominent protein groups and protein classes as being salinity- or drought-responsive via several bioinformatics tools, such as Gene Ontology (GO), PANTHER, KEGG pathway, etc. [35,44,45]. Bandehagh et al. [46] studied salt-responsive proteins in canola leaves using a proteomic technique. The differentially expressed proteins involved various processes, including oxidative stress, energy production, electron transport signal transduction, translation, phosphate metabolic processes, and photosynthesis [47–49]. However, majority of these reports so far have relied on conventional protein extraction approaches [48] and gel-based protein abundance studies, including two-dimensional difference gel electrophoresis (2D-DIGE) or two-dimensional gel electrophoresis (2DE), followed by a matrix-assisted laser desorption/ionization-time of flight mass spectrometry (MALDI-TOF-MS) approach for individual protein identification. Nevertheless, consistent and reproducible results from 2DE or 2D-DIGE were challenging, and poor resolution of integral membrane proteins is another concern [23,50,51]. As a result, gel-free approaches in plant proteomics are now being used [52], which include liquid chromatography (LC) and ion-exchange chromatography, followed by de novo sequencing of the peptide fragments by MS/MS, and ultimately offer high-throughput analysis of the proteome profile, providing a snapshot of the major protein constituents.

5. Conclusions

The quantitative proteomics analysis of bottle gourd parental lines under different crossing periods revealed the mechanism of source–sink strength for superior seed quality. The standardization and comparative analysis of different protein extraction methods provided reproducible and high-quality SDS-PAGE results in both leaves and seeds at the F4 crossing period. The seed vigor followed a similar pattern as the differential protein expression by SDS-PAGE. Analysis of the LFQ results using MASCOT software revealed more than 100 differentially expressed proteins (DEPs), of which 20 proteins were upregulated and 12 proteins were downregulated, with fold changes of $\geq$1.5 and $\leq$0.6, respectively. The analysis further revealed some important proteins that play a role in seed reserve mobilization related to the source–sink relationship in bottle gourd. The DEPs were sorted based on their different molecular mechanisms, such as transporters, mineral deposition, antioxidative enzymes, photosynthesis, signal transduction, biosynthetic pathways, nucleotide synthesis, carbohydrate metabolism, etc., with a higher protein score, protein coverage, and protein score match (PSM). Future research aims to further understand the dynamic molecular interactions of protein candidates, especially those with specific bottle gourd seed attributes, and the modalities that enable this crop to be a valuable medicinal and horticultural crop with high nutritional value.

Author Contributions: Conceptualization, methodology, investigation, data curation, writing—original draft preparation, A.M. and H.P.; formal analysis, supervision, validation, V.S.M., D.S.D. and A.B.; writing—review and editing, J.T., M.N.A. and A.S. All authors have read and agreed to the published version of the manuscript.

Funding: This research received no external funding.

Institutional Review Board Statement: Not applicable.

Informed Consent Statement: Not applicable.

Data Availability Statement: Not applicable.

Acknowledgments: The authors are thankful to the Department of Vegetable Science, CCS HAU, and Hisar for providing the seeds of bottle gourd parental lines. The authors are also grateful to the Department of Seed Science and Technology for providing all the necessary facilities to research the field trials at the departmental farm and CCS Haryana Agricultural University, Hisar, for providing the required facilities. The authors would also like to extend their sincere appreciation to the Researchers Supporting Project (RSP2023R180) at King Saud University, Riyadh, Saudi Arabia.

Conflicts of Interest: The authors declare no conflict of interest.

References

1. Priya, K.; Deshpande, V.K.; Masuthi, D.K.; Haleshkumar, H.B. Influence of growth regulators and crossing period on flowering, seed yield and quality of chilli hybrid HCH-9646. *Int. J. Agric. Sci.* **2012**, *8*, 1–4.
2. Punia, H.; Tokas, J.; Malik, A.; Singh, S.; Phogat, D.S.; Bhuker, A.; Mor, V.S.; Rani, A.; Sheokand, R.N. Discerning morpho-physiological and quality traits contributing to salinity tolerance acquisition in sorghum [*Sorghum bicolor* (L.) Moench]. *South Afr. J. Bot.* **2021**, *140*, 409–418. [CrossRef]
3. Punia, H.; Tokas, J.; Malik, A.; Rani, A.; Gupta, P.; Kumari, A.; Mor, V.S.; Bhuker, A.; Kumar, S. Solar Radiation and Nitrogen Use Efficiency for Sustainable Agriculture. In *Resources Use Efficiency in Agriculture*; Springer: Berlin/Heidelberg, Germany, 2020; pp. 177–212.
4. Malik, A.; Mor, V.S.; Tokas, J.; Punia, H.; Malik, S.; Malik, K.; Sangwan, S.; Tomar, S.; Singh, P.; Singh, N. Biostimulant-Treated Seedlings under Sustainable Agriculture: A Global Perspective Facing Climate Change. *Agronomy* **2020**, *11*, 14. [CrossRef]
5. Yamasaki, Y.; Gao, F.; Jordan, M.C.; Ayele, B.T. Seed maturation associated transcriptional programs and regulatory networks underlying genotypic difference in seed dormancy and size/weight in wheat (*Triticum aestivum* L.). *BMC Plant Biol.* **2017**, *17*, 1–18. [CrossRef] [PubMed]
6. Doddagoudar, S.R.; Shekhargouda, M.; Khadi, B.M.; Eshanna, M.R.; Biradarpatil, N.K.; Vamadevaiah, H.M. Effect of crossing period on seed germination and seedling vigour of cotton hybrid DHH-543. *Crop. Res.* **2006**, *32*, 259.
7. Punia, H.; Tokas, J.; Bhadu, S.; Mohanty, A.K.; Rawat, P.; Malik, A. Satpal Proteome dynamics and transcriptome profiling in sorghum [Sorghum bicolor (L.) Moench] under salt stress. *3 Biotech* **2020**, *10*, 412. [CrossRef]

8. Bewley, J.D.; Black, M. *Physiology and Biochemistry of Seeds in Relation to Germination: 1 Development, Germination, and Growth*; Springer: Berlin/Heidelberg, Germany, 2014; ISBN 364266668X.
9. Kesari, V.; Rangan, L. Coordinated changes in storage proteins during development and germination of elite seeds of Pongamia pinnata, aversatile biodiesel legume. *AoB Plants* **2011**, *11*, 1–16. [CrossRef]
10. Punia, H.; Madan, S.; Malik, A.; Sethi, S.K. Stability analysis for quality attributes in durum wheat (*Triticum durum* L.) genotypes. *Bangladesh J. Bot.* **2019**, *48*, 967–972. [CrossRef]
11. Müntz, K.; Belozersky, M.A.; Dunaevsky, Y.E.; Schlereth, A.; Tiedemann, J. Stored proteinases and the initiation of storage protein mobilization in seeds during germination and seedling growth. *J. Exp. Bot.* **2001**, *52*, 1741–1752. [CrossRef]
12. Finnie, C.; Melchior, S.; Roepstorff, P.; Svensson, B. Proteome analysis of grain filling and seed maturation in barley. *Plant Physiol.* **2002**, *129*, 1308–1319. [CrossRef]
13. Santos, F.C.; Caixeta, F.; Clemente, A.C.S.; Pinho, E.V.; Rosa, S.D.V.F. Gene expression of antioxidant enzymes and coffee seed quality during pre- and post-physiological maturity. *Genet. Mol. Res.* **2014**, *13*, 10983–10993. [CrossRef] [PubMed]
14. Borém, F.M.; Coradi, P.C.; Saath, R.; Oliveira, J.A. Qualidade do café natural e despolpado após secagem em terreiro e com altas temperaturas. *Ciência E Agrotecnologia* **2008**, *32*, 1609–1615. [CrossRef]
15. Ahmad, P.; Abdel Latef, A.A.H.; Rasool, S.; Akram, N.A.; Ashraf, M.; Gucel, S. Role of proteomics in crop stress tolerance. *Front. Plant Sci.* **2016**, *7*, 1336. [CrossRef] [PubMed]
16. Tokas, J.; Punia, H.; Malik, A.; Sangwan, S.; Devi, S.; Malik, S. Growth performance, nutritional status, forage yield and photosynthetic use efficiency of sorghum [*Sorghum bicolor* (L.) Moench] under salt stress. *Range Manag. Agrofor.* **2021**, *42*, 59–70.
17. Mora, L.; Bramley, P.M.; Fraser, P.D. Development and optimisation of a label-free quantitative proteomic procedure and its application in the assessment of genetically modified tomato fruit. *Proteomics* **2013**, *13*, 2016–2030. [CrossRef] [PubMed]
18. Griffin, N.M.; Yu, J.; Long, F.; Oh, P.; Shore, S.; Li, Y.; Koziol, J.A.; Schnitzer, J.E. Label-free, normalized quantification of complex mass spectrometry data for proteomic analysis. *Nat. Biotechnol.* **2010**, *28*, 83–89. [CrossRef]
19. Sun, Q.; Zhang, N.; Wang, J.; Cao, Y.; Li, X.; Zhang, H.; Zhang, L.; Tan, D.X.; Guo, Y.D. A label-free differential proteomics analysis reveals the effect of melatonin on promoting fruit ripening and anthocyanin accumulation upon postharvest in tomato. *J. Pineal Res.* **2016**, *61*, 138–153. [CrossRef]
20. Huan, C.; An, X.; Yu, M.; Jiang, L.; Ma, R.; Tu, M.; Yu, Z. Effect of combined heat and 1-MCP treatment on the quality and antioxidant level of peach fruit during storage. *Postharvest Biol. Technol.* **2018**, *145*, 193–202. [CrossRef]
21. Brown, P.N.; Murch, S.J.; Shipley, P. Phytochemical diversity of cranberry (Vaccinium macrocarpon Aiton) cultivars by anthocyanin determination and metabolomic profiling with chemometric analysis. *J. Agric. Food Chem.* **2012**, *60*, 261–271. [CrossRef]
22. Chen, C.; Zhang, X.; Zhang, H.; Ban, Z.; Li, L.; Dong, C.; Ji, H.; Xue, W. Label-free quantitative proteomics to investigate the response of strawberry fruit after controlled ozone treatment. *RSC Adv.* **2019**, *9*, 676–689. [CrossRef]
23. Punia, H.; Tokas, J.; Malik, A.; Bajguz, A.; El-Sheikh, M.A.; Ahmad, P. Ascorbate–Glutathione Oxidant Scavengers, Metabolome Analysis and Adaptation Mechanisms of Ion Exclusion in Sorghum under Salt Stress. *Int. J. Mol. Sci.* **2021**, *22*, 13249. [CrossRef]
24. Punia, H.; Tokas, J.; Malik, A.; Sangwan, S. Characterization of phenolic compounds and antioxidant activity in sorghum [*Sorghum bicolor* (L.) Moench] grains. *Cereal Res. Commun.* **2021**, *49*, 343–353. [CrossRef]
25. Punia, H.; Tokas, J.; Mor, V.S.; Bhuker, A.; Malik, A.; Singh, N.; Alsahli, A.A.; Hefft, D.I. Deciphering Reserve Mobilization, Antioxidant Potential, and Expression Analysis of Starch Synthesis in Sorghum Seedlings under Salt Stress. *Plants* **2021**, *10*, 2463. [CrossRef] [PubMed]
26. Abdul-Baki, A.A.; Anderson, J.D. Vigor determination in soybean seed by multiple criteria 1. *Crop. Sci.* **1973**, *13*, 630–633. [CrossRef]
27. Suwanagul, A.; Deewathanawong, R. SDS-PAGE and Two-dimensional Electrophoresis Determination of Protein from Peel and Pulp Tissues of Ripening Durian Fruit. *Acta Hortic.* **2007**, *787*, 215–224. [CrossRef]
28. Ndimba, B.K.; Thomas, L.A.; Ngara, R. Sorghum 2-dimensional proteome profiles and analysis of Hsp70 expression under salinity stress. *Kasetsart. J. (Nat. Sci.)* **2010**, *44*, 768–775.
29. Toledo, F.; Arancibia-Avila, P.; Park, Y.-S.; Jung, S.-T.; Kang, S.-G.; Gu Heo, B.; Drzewiecki, J.; Zachwieja, Z.; Zagrodzki, P.; Pasko, P. Screening of the antioxidant and nutritional properties, phenolic contents and proteins of five durian cultivars. *Int. J. Food Sci. Nutr.* **2008**, *59*, 415–427. [CrossRef]
30. Li, F.X.; Zhou, Y.F.; Wang, Y.T.; Sun, L.; Bai, W.; Yan, T.; Xu, W.J.; Huang, R.D. Screening and identification of sorghum cultivars for alkali tolerance during germination. *Sci. Agric. Sin.* **2013**, *46*, 1762–1771.
31. Sagar, A. *Screening of Sorghum Germplasms for Salinity Tolerance Based on Morphophysiological and Biochemical Traits*; Bangladesh Agricultural University: Mymensingh, Bangladesh, 2017.
32. Qu, C.; Liu, C.; Gong, X.; Li, C.; Hong, M.; Wang, L.; Hong, F. Impairment of maize seedling photosynthesis caused by a combination of potassium deficiency and salt stress. *Environ. Exp. Bot.* **2012**, *75*, 134–141. [CrossRef]
33. Forcada, C.F.; Kodad, O.; Juan, T.; Estopañan, G. Genetic variability and pollen effect on the transmission of the chemical components of the almond kernel. *Span. J. Agric. Res.* **2011**, *9*, 781–789. [CrossRef]
34. Khalil, R.M.A. Molecular and biochemical markers associated with salt tolerance in some sorghum genotypes. *World Appl. Sci. J.* **2013**, *22*, 459–469. [CrossRef]
35. Ngara, R.; Ndimba, R.; Borch-Jensen, J.; Jensen, O.N.; Ndimba, B. Identification and profiling of salinity stress-responsive proteins in Sorghum bicolor seedlings. *J. Proteom.* **2012**, *75*, 4139–4150. [CrossRef] [PubMed]

36. Sathe, S.K.; Wolf, W.J.; Roux, K.H.; Teuber, S.S.; Venkatachalam, M.; Sze-Tao, K.W.C. Biochemical characterization of amandin, the major storage protein in almond (*Prunus dulcis* L.). *J. Agric. Food Chem.* **2002**, *50*, 4333–4341. [CrossRef] [PubMed]
37. Čolić, S.D.; Bakić, I.V.; Dabić Zagorac, D.; Natić, M.M.; Smailagić, A.T.; Pergal, M.V.; Pešić, M.B.; Milinčić, D.D.; Rabrenović, B.B.; Fotirić Akšić, M.M. Chemical Fingerprint and Kernel Quality Assessment in Different Grafting Combinations of Almond Under Stress Condition. *Sci. Hortic. (Amst.)* **2021**, *275*, 109705. [CrossRef]
38. Capouchová, I.; Petr, J.; Krejčířová, L. Protein composition of sorghum and oat grain and their suitability for gluten-free diet. *Žemdirbystė* **2006**, *93*, 271–284.
39. Radonić, A.; Thulke, S.; Mackay, I.M.; Landt, O.; Siegert, W.; Nitsche, A. Guideline to reference gene selection for quantitative real-time PCR. *Biochem. Biophys. Res. Commun.* **2004**, *313*, 856–862. [CrossRef] [PubMed]
40. Srivastava, D.; Khan, N.A.; Shamim, M.; Yadav, P.; Pandey, P.; Singh, K.N. Assessment of the genetic diversity in bottle gourd (*Lagenaria siceraria* [Molina] Standl.) genotypes using SDS-PAGE and RAPD Markers. *Natl. Acad. Sci. Lett.* **2014**, *37*, 155–161. [CrossRef]
41. Yin, H.; Chen, C.J.; Yang, J.; Weston, D.J.; Chen, J.G.; Muchero, W.; Ye, N.; Tschaplinski, T.J.; Wullschleger, S.D.; Cheng, Z.M.; et al. Functional genomics of drought tolerance in bioenergy crops. *Crit. Rev. Plant Sci.* **2014**, *33*, 205–224. [CrossRef]
42. Sharan, A.A.; Nikam, A.N.; Jaleel, A.; Tamhane, V.A.; Rao, S.P. Method for label-free quantitative proteomics for Sorghum bicolor L. Moench. *Trop. Plant Biol.* **2018**, *11*, 78–91. [CrossRef]
43. Barkla, B.; Vera-Estrella, R.; Pantoja, O. Progress and challenges for abiotic stress proteomics of crop plants. *Proteomics* **2013**, *13*, 1801–1815. [CrossRef]
44. Punia, H.; Tokas, J.; Malik, A.; Sangwan, S.; Rani, A.; Yashveer, S.; Alansi, S.; Hashim, M.J.; El-Sheikh, M.A. Genome-wide transcriptome profiling, characterization, and functional identification of NAC transcription factors in sorghum under salt stress. *Antioxidants* **2021**, *10*, 1605. [CrossRef] [PubMed]
45. Roy, S.K.; Kamal, A.H.M.; Kim, H.R.; Kwon, S.J.; Jang, H.Y.; Ko, J.H.; Kim, J.I.; Ko, T.S.; Xin, Z.; Woo, S.H. Proteome profiling of seed from inbred and mutant line of Sorghum (*Sorghum bicolor*). *Aust. J. Crop. Sci.* **2014**, *8*, 606–611.
46. Bandehagh, A.; Hosseini Salekdeh, G.; Toorchi, M.; Mohammadi, A.; Komatsu, K. Comparative proteomic analysis of canola leavesunder salinity stress. *Proteomics* **2011**, *11*, 1965–1975. [CrossRef] [PubMed]
47. Ahmad, P.; Tripathi, D.K.; Deshmukh, R.; Pratap Singh, V.; Corpas, F.J. *Revisiting the Role of ROS and RNS in Plants under Changing Environment*; Elsevier: Amsterdam, The Netherlands, 2019.
48. Mansoor, S.; Ali Wani, O.; Lone, J.K.; Manhas, S.; Kour, N.; Alam, P.; Ahmad, A.; Ahmad, P. Reactive Oxygen Species in Plants: From Source to Sink. *Antioxidants* **2022**, *11*, 225. [CrossRef]
49. Kohli, S.K.; Khanna, K.; Bhardwaj, R.; Abd_Allah, E.F.; Ahmad, P.; Corpas, F.J. Assessment of Subcellular ROS and NO Metabolism in Higher Plants: Multifunctional Signaling Molecules. *Antioxidants* **2019**, *8*, 641. [CrossRef]
50. Méchin, V.; Damerval, C.; Zivy, M. Total protein extraction with TCA-acetone. *Plant Proteomics Protocols.* **2007**, *355*, 1.
51. Westermeier, R.; Marouga, R. Protein detection methods in proteomics research. *Biosci. Rep.* **2005**, *25*, 19–32. [CrossRef]
52. Abdallah, C.; Dumas-Gaudot, E.; Renaut, J.; Sergeant, K. Gel-based and gel-free quantitative proteomics approaches at a glance. *Int. J. Plant Genomic.* **2012**, *2012*, 1–17. [CrossRef] [PubMed]

current issues in molecular biology

Review

Advances in Research on the Regulation of Floral Development by *CYC*-like Genes

Yuhong Chai [1,2,3,†], Hua Liu [2,†], Wendan Chen [2], Chenghu Guo [2], Haixia Chen [2], Xi Cheng [2], Dongliang Chen [2], Chang Luo [2], Xiumei Zhou [1,3,*] and Conglin Huang [2,*]

1 School of Horticulture and Landscape Architecture, Henan Institute of Science and Technology, Xinxiang 453003, China
2 Institute of Grassland, Flowers and Ecology, Beijing Academy of Agriculture and Forestry Sciences, Beijing 100097, China
3 Engineering Technology Research Center of Characteristic Horticultural Plant Development and Utilization, Henan Institute of Science and Technology, Xinxiang 453003, China
* Correspondence: zxm@hist.edu.cn (X.Z.); huangconglin@baafs.net.cn (C.H.)
† These authors contributed equally to this work.

Abstract: *CYCLOIDEA* (*CYC*)-like genes belong to the TCP transcription factor family and play important roles associated with flower development. The *CYC*-like genes in the CYC1, CYC2, and CYC3 clades resulted from gene duplication events. The CYC2 clade includes the largest number of members that are crucial regulators of floral symmetry. To date, studies on *CYC*-like genes have mainly focused on plants with actinomorphic and zygomorphic flowers, including *Fabaceae*, *Asteraceae*, *Scrophulariaceae*, and *Gesneriaceae* species and the effects of *CYC*-like gene duplication events and diverse spatiotemporal expression patterns on flower development. The *CYC*-like genes generally affect petal morphological characteristics and stamen development, as well as stem and leaf growth, flower differentiation and development, and branching in most angiosperms. As the relevant research scope has expanded, studies have increasingly focused on the molecular mechanisms regulating *CYC*-like genes with different functions related to flower development and the phylogenetic relationships among these genes. We summarize the status of research on the *CYC*-like genes in angiosperms, such as the limited research conducted on CYC1 and CYC3 clade members, the necessity to functionally characterize the *CYC*-like genes in more plant groups, the need for investigation of the regulatory elements upstream of *CYC*-like genes, and exploration of the phylogenetic relationships and expression of *CYC*-like genes with new techniques and methods. This review provides theoretical guidance and ideas for future research on *CYC*-like genes.

Keywords: *CYCLOIDEA* (*CYC*)-like gene; TCP gene family; CYC2 clade; floral symmetry; molecular regulatory mechanism; phylogeny

Citation: Chai, Y.; Liu, H.; Chen, W.; Guo, C.; Chen, H.; Cheng, X.; Chen, D.; Luo, C.; Zhou, X.; Huang, C. Advances in Research on the Regulation of Floral Development by *CYC*-like Genes. *Curr. Issues Mol. Biol.* **2023**, *45*, 2035–2059. https://doi.org/10.3390/cimb45030131

Academic Editor: Quan Zou

Received: 28 January 2023
Revised: 24 February 2023
Accepted: 27 February 2023
Published: 2 March 2023

1. Introduction

Cubas et al. first proposed the concept of the TCP transcription factor family, which is named according to the first letters of TEOSINTE BRANCHED 1 (TB1) in maize (*Zea mays*), CYCLOIDEA (CYC) in snapdragon (*Antirrhinum majus*), and PROLIFERATING CELL FACTOR 1 and 2 (PCF1 and PCF2) in rice (*Oryza sativa*) [1–5]. Genes encoding proteins with the TCP domain are involved in the regulation of angiosperm growth and development [6–9]. The TCP family members contain a highly conserved TCP domain, which forms a basic helix–loop–helix (bHLH) structure associated with DNA binding and protein dimerization [10,11]. TB1 is a major regulator of stem and lateral bud growth and the male flower formation of maize, rice, wheat, and other crops [12–14], whereas CYC controls the floral dorsal organ characteristics in snapdragon [1,15], and both PCF1 and PCF2 bind to the promoter of PROLIFERATING CELL NUCLEAR ANTIGEN (PCNA), which is crucial for DNA replication and repair, chromatin structure maintenance, chromosome isolation, and the cell cycle in rice [3]. According to their different domains, the members of the TCP family have been divided into the following two categories: TCP-P and TCP-C [16–18]. Moreover, TCP-C has been subdivided into the ECE (CYC/TB1) and CINCINNATA (CIN) clades [19,20].

The *CYC* genes belong to the ECE clade, which is unique to angiosperms [21,22]. In addition to the TCP and R domain sequences, *CYC* genes encode the glutamate–cysteine–glutamic acid (ECE) motif specific to core eudicots [23–25]. Phylogenetic analysis has indicated that *CYC* genes in angiosperms experienced two major gene duplication events, which led to the formation of the CYC1, CYC2, and CYC3 clades [26–28]. In different evolutionary lineages, gene duplication events occurred in each branch at different time points during evolution [29–35], as shown in Figure 1. A more thorough analysis of the CYC2 subgroup confirmed that they are key regulatory genes for the bilateral symmetry of flowers [36–38].

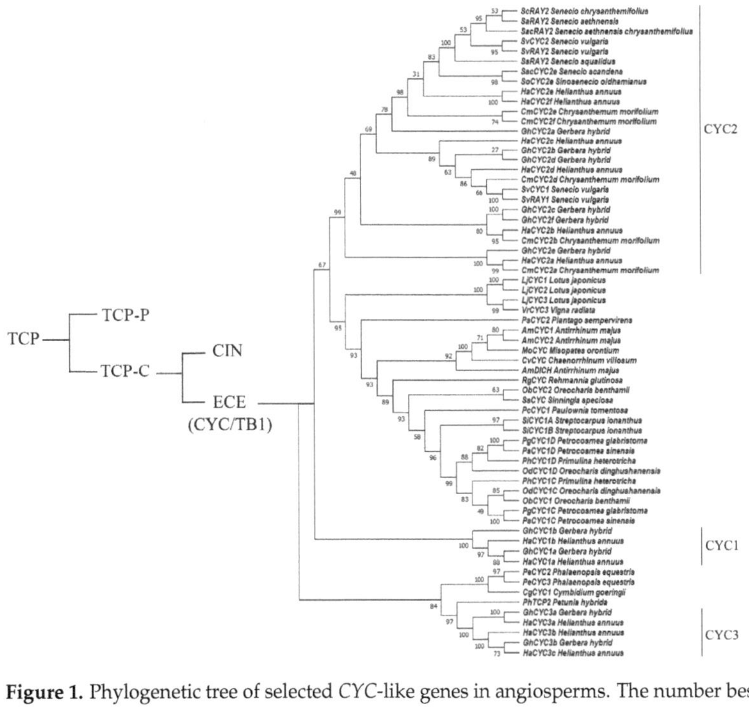

Figure 1. Phylogenetic tree of selected *CYC*-like genes in angiosperms. The number beside each node is the bootstrap support value.

Angiosperm flowers can be divided into the following three types according to the presence or absence of a plane of symmetry: (1) radially symmetrical flowers (actinomorphic) with multiple planes of symmetry, (2) bilaterally symmetrical flowers (zygomorphic) with only one plane of symmetry, and (3) asymmetrical flowers with no plane of symmetry [39–42]. Studies of fossils have revealed that angiosperm flowers were originally actinomorphic and that zygomorphic flowers arose during evolution [43–46]. Wild-type snapdragon, which has typical zygomorphic flowers, is a model plant for studying floral symmetry [47]. Early investigations on the molecular mechanism regulating snapdragon flower types demonstrated that zygomorphy was determined by *CYC*, *DICHOTOMA* (*DICH*), *RADIALIS* (*RAD*), and *DIVARICATA* (*DIV*) genes encoding regulators of the dorsoventral characteristics of flowers [1,4]. The functionally redundant *CYC* and *DICH* genes belong to the CYC2 clade and originated from gene duplication events associated with angiosperm evolution [48–50]. These genes are essential for the development of zygomorphic flowers in angiosperms, especially the *CYC* genes [45,51,52]. The results of studies on snapdragon compelled researchers to investigate the mechanism by which *CYC* genes regulate floral symmetry. In addition to snapdragon, the regulatory function of *CYC* genes has been explored in other insect-pollinated plants with zygomorphic flowers, including *Fabaceae*, *Gesneriaceae*, *Caprifoliaceae*, *Scrophulariaceae*, and *Malpighiaceae* species [53–57]. The results of these studies suggest that *CYC2* genes are the key genes for regulating the development of zygomorphic flowers.

This review summarizes the status of research and unresolved problems in *Fabaceae*, *Asteraceae*, *Scrophulariaceae*, *Gesneriaceae*, *Orchidaceae*, *Solanaceae*, and other families, while also proposing future directions for studies on *CYC*-like genes. As the three largest families of angiosperms, *Fabaceae*, *Asteraceae*, and *Orchidaceae* have been the main focus of research on *CYC*-like genes. In the *Fabaceae* species, *CYC*-like genes control floral symmetry, whereas the *CYC*-like genes in the *Asteraceae* species modulate changes to the capitulum morphology and regulate elongation of the showy corolla limb of the ray florets. In the *Scrophulariaceae* and *Gesneriaceae* species, *CYC*-like genes mainly affect the morphological characteristics of petals and stamen development. The *CYC*-like genes in the *Orchidaceae* species influence floral development or branching. The *CYC*-like genes in the *Solanaceae* species mainly affect axillary bud growth and development and stem growth, while also controlling branching, flower differentiation and development, and leaf size. These genes contribute to morphological changes to the dorsoventral floral symmetry in the *Caprifoliaceae* species. A systematic analysis of the functions and evolution of *CYC*-like genes may provide researchers and breeders with a theoretical basis for future research.

2. Progress in Research on *CYC*-like Genes in *Fabaceae*

The *Fabaceae* species are distributed worldwide. Because of their diversity in floral symmetry, legumes are suitable for exploring the evolution and underlying mechanism of floral symmetry [58]. Researchers have screened the *Fabaceae* for homologs of snapdragon *CYC* genes and then analyzed their functions to clarify the role of *CYC*-like genes in angiosperm floral development. The differences among the diverse species in terms of the *CYC*-like genes responsible for floral symmetry revealed a new regulatory system.

The duplication of *CYC* homologues gave rise to three copies of ECE clade genes in the TCP family in *Lotus Japonicus* [59]. In *L. japonicus*, both *LjCYC1* and *LjCYC2* mediate the development of asymmetrical inflorescences and flowers, and changes in the number of petals and in wing and keel morphology were observed in transgenic plants separately overexpressing *LjCYC1* and *LjCYC2* [59]. The asymmetrical expression pattern of *LjCYC2* is similar to that of the snapdragon *CYC* gene in the developing flower primordium [59]. However, *LjCYC2* is also expressed in the inflorescence primordium of *L. japonicus*, whereas the *CYC* gene is expressed only during floral primordium development in snapdragon [4].

Citerne et al. reported that the homologous genes of *CYC* in legumes can be divided into two major classes, ECE groups I and II, which are the result of an early duplication event [60]. ECE I can be further divided into two subclasses, IA and IB, which originated from duplication near or prior to the divergence of legumes. The *LEGCYC* genes in *Lupinus* are homologous to the regulatory gene *CYC* that controls the floral symmetry and paraxial floral organ characteristics of snapdragon and its related species [61]. Ree et al. suggested based on a molecular evolutionary analysis that positive selection has played a role in the evolution of the *LEGCYC1B* lineage, which is closely associated with floral morphological changes in *Lupinus*. Papilionoideae have strongly bilaterally symmetrical flowers, whereas *Cadia purpurea* flowers show radial symmetry associated with the expression of two *CYC* homologous genes (*LEGCYCs*) in the dorsal region of the flower (Figure 2) [58]. In addition, the expression pattern of one gene has expanded from the paraxial to the lateral and posterior regions of the corolla, which may result in reversion to evolutionarily regressive petal characters.

Wang et al. determined that the expression of three endogenous *LjCYC* genes is specifically inhibited by different RNAi transgenes [62]. A chimeric RNAi transgene containing *LjCYC1*- and *LjCYC2*-specific sequences down-regulated the expression of both endogenous genes. The effect of silencing the three *LjCYC* genes was mainly confined to the dorsal or lateral part of the petals, implying that the genes are associated with dorsal and lateral activities during the development of zygomorphic flowers [62]. Knockdown of the three *LjCYC* genes may result in wild-type petals that resemble ventral petals, complete organ internal (IN) asymmetry, and the lack of dorsoventral (DV) pathway-differentiated flowers. This suggests that DV asymmetry during the development of zygomorphic flowers is controlled by *LjCYC* genes, whereas floral organ IN asymmetry is independently determined by other genetic factors.

Figure 2. Flowers of *Cadia purpurea* and *Lupinus nanus* [60]. (**A**) Flower of *Cadia purpurea* with actinomorphic corolla. (**B**) Flowers of *Lupinus nanus* with zygomorphic corolla. Reprinted with permission from Ref. [60]. Copyright © 2023, Oxford University Press.

The mutation of *CYC2* in *Lathyrus odoratus* causes a change in dorsoventral petal type, resulting in a hooded (hdd) flower mutant with an epidermis and the pigmentation characteristic of a wing petal, and with a concave standard petal, the same as the lobed standard (*lst1*) mutant in *Pisum* [63]. Differences in *CYC* expression and activity may lead to differences in dorsal petal morphology in *Fabaceae*, and play a role in the negative regulation of petal edge growth in *Lathyrus*, mainly maintaining the flatness of the dorsal petal [64]. Interestingly, Ojeda et al. found that changes in the timing of *LjCYC2* expression during pollination of *Lotus* by bees and birds may be responsible for changes in flower petal micromorphology and size, whereas changes in the spatial distribution of gene expression had no effect on pollination [65].

Feng et al. determined that the upstream promoter regions of *GmCYC* genes vary in number and type of hormone response elements in *Glycine max* [66]. The expression of *GmCYC* genes is involved in different growth and developmental stages, induced by abscisic acid, brassinosteroids, aminocyclopropane–1–carboxylic acid, salicylic acid, and methyl jasmonate signals [66]. The *CYC*-like genes may have undergone multiple duplications and losses in different *Fabaceae* lineages and formed the distinct homologous clades CYC1 and CYC2, but the CYC3 clade was most likely lost [67]. The ancestors of *Papilionoideae* and *Caesalpinioideae* probably possessed two *CYC1* gene copies, but one of the copies was subsequently lost in *Papilionoideae* and was retained only in a few species of *Caesalpinioideae* [67]. The *CYC2* gene was replicated more frequently in *Papilionoideae* than in other legumes [67]. The diversity patterns of *CYC1* and *CYC2* genes are not associated with floral symmetry in non-papilionoid legumes, but the replication and functional differentiation of *CYC2* genes is necessary for floral symmetry in *Papilionoideae* [67].

The expression pattern of *VrCYC3*, which is homologous to *L. japonicus LjCYC3* and pea *PsCYC3*, differs from that of *VrCYC1* and *VrCYC2* in the dorsal, lateral, and ventral petals in mung bean (*Vigna radiata*) [68]. In addition, VrCYC3, which is localized to the nucleus, can induce transcription [68]. Moreover, it can interact with VrCYC1 and VrCYC2 in yeast cells, but this interaction is weakened by the deletion of two amino acid residues in its R domain [68]. This suggests that *LjCYC3/PsCYC3/VrCYC3* play a conserved role in determining the lateral petals shape, and the formation of symmetrical and asymmetrical flowers in *Fabaceae*.

3. Progress in Research on *CYC*-like Genes in *Asteraceae*

Asteraceae is the most highly evolved family of dicotyledonous plants with a complex inflorescence structure, termed a capitulum, that often consists of radially symmetrical disc florets and bilaterally symmetrical ray florets [69,70]. Researchers have cloned snapdragon *CYC* homologs in *Helianthus*, *Gerbera*, *Senecio*, *Chrysanthemum*, and other genera, which revealed the considerable abundance of these homologs in *Asteraceae*, many of which have diverse functions [71–75]. At the single-floret level, the *CYC* gene in the *Asteraceae* species uniquely regulates the elongation of the corolla limb of ray florets, which are critical for attracting pollinators [76].

3.1. CYC-like Genes of Helianthus

The radiate sunflower (*Helianthus annuus*) capitulum consists of bilaterally symmetrical sterile ray florets and radially symmetrical bisexual disc florets. Ten ECE clade members have been identified in the sunflower, and the spatiotemporal expression of these homologous genes varies [77]. The specific temporal expression of the different genes in diverse plant parts, including ray florets, disc florets, leaves, and roots, may maintain the complex sunflower inflorescence structure via coordinated expression.

The sunflower *tubular ray flower* (*turf*) mutant has hermaphroditic ray florets with an almost actinomorphic corolla. Fambrini et al. determined that this mutation was caused by the insertion of a TCP motif, a non-autonomous transposable element (TE) from the *CYC*-like gene *HaCYC2c*, named *Transposable element of turf1* (*Tetu1*) [78]. The excision of *Tetu1* can restore the wild-type phenotype or produce stable mutants, indicating that *HaCYC2c* is a key regulator of ray floret symmetry. A loss-of-function mutation to *HaCYC2c* can promote the transition of sterile florets to hermaphroditic florets, which reflects the importance of *CYC*-like genes for the inhibition of stamen development.

HaCYC2c was mutated in two independent *tubular-rayed* (*tub*) mutants, which apparently involved TE insertions, resulting in little or no expression and the formation of radially symmetrical ray florets, which are usually bilaterally symmetrical [79]. If *HaCYC2c* was inserted into the offspring, ray florets were more likely to replace disc florets at the center of the capitulum, whereas if *HaCYC2c* expression was inhibited, bilaterally symmetrical ray florets did not develop, and the capitulum comprised only disc florets (Figure 3).

Figure 3. Capitulum and florets of a wild-type sunflower plant and two mutants [79]. (**A**,**B**) Wild-type sunflower with disc and ray florets. (**C**,**D**) *dbl* mutant with disc and ray florets. (**E**,**F**) *tub* mutant with disc and ray florets. The florets from the disc center to the peripheral florets are arranged from left to right. Reprinted with permission from Ref. [79]. Copyright © 2023 Chapman et al.

Thus the *turf* and *tub* mutants are characterized by a transition from bilaterally symmetrical to radially symmetrical ray florets because of the insertion of TEs in *HaCYC2c* [80]. In the *dbl* or *Chrysanthemoides* (*Chry*) mutants, the insertion of *HaCYC2c* upstream of the coding region results in the ectopic expression of this gene and a transition from radially symmetrical disc florets to bilaterally symmetrical disc florets. The loss-of-function mutation to the *CYC*-like gene in sunflower *turf* mutants reportedly results in hermaphroditic tubular-like florets, which replace the normal sterile ray florets and the formation of a capitulum type that is not normally found in sunflower [81].

HaCYC2c was overexpressed after an insertion into the *HaNDUA2* promoter region to generate the sunflower long petal mutant (*lpm*) in which the abnormal elongation of the disc floret corolla and stamen abortion at an early stage of floral organ development was observed [82]. Furthermore, the floret symmetry changed from radial symmetry to bilateral symmetry, thus transforming the disc florets into ray florets. The overexpression of *HaCYC2c* and its control of *HaNDUA2* through transcriptional recognition may be an important regulatory node for floret type and functional differentiation in *Helianthus*, which was associated with maintaining the balance between the pollinator recruitment ability and the fertility of disc florets [82].

HdCYC2c and *HxmCYC2cB* belong to the CYC2 subclade in *Helianthus* [83]. *HdCYC2c* was differentially expressed in the different floret types of *Helianthus decapetalus*, with the expression level higher in the ray floret corolla than in the disc floret corolla [83]. In *Helianthus* × *multiflorus*, the insertion of TEs in *HxmCYC2cB* promoted the ectopic expression of *HxmCYC2cB* throughout the inflorescence, leading to the observed loss of actinomorphic florets and the production of ray florets [83]. Removal of a TE (CTEHM1) and epigenetic regulation of *HmCYC2c* expression resulted in two capitulum types of *Helianthus* × *multiflorus*, Meteor 1 and Meteor 2 [37]. The expression of *HmCYC2c* in the disc floret of Meteor 2 was significantly higher than that of Meteor 1. The CTEHM1 in *HmCYC2c* was truncated in Meteor 1, which showed the typical tubular corolla of *Helianthus*, whereas the remaining presence of CTEHM1 in *HmCYC2c* of Meteor 2 caused the largest corolla of disc florets to display the characteristics of a ray floret, resulting in an entirely radiate capitula not normally found in *Helianthus* [37].

Fambrini et al. isolated three CYC2 subclade genes (*HrCYC2c*, *HrCYC2d*, and *HrCYC2e*) associated with the identity of the *Helianthus* ray–floret (Hr), among which *HrCYC2c* played an important role in the initiation of the ray floret primordium [38]. The capitula of *HrCYC2c*-mutant homozygous dominant plants (*HrCYC2c/HrCYC2c*) and heterozygous dominant plants (*HrCYC2c/HrCYC2c-m*) initiated ray florets, whereas the recessive homozygous plants (*HrCYC2c-m/HrCYC2c-m*) did not develop ray florets [38].

3.2. CYC-like Genes of Gerbera Hybrida

The *CYC*-like homolog *GhCYC2a* is involved in the differentiation of *Gerbera hybrida* floret types, and its expression exhibits a gradient along the radial axis of the capitulum [71]. Specifically, *GhCYC2a* is expressed in the peripheral, bilaterally symmetrical ray florets, but not in the centermost disc florets, which are almost radially symmetrical and have more deeply incised corolla lobes. The overexpression of *GhCYC2a* results in disc florets acquiring a morphology similar to that of ray florets, whereas the inhibition of *GhCYC2a* expression leads to the development of limbs that are shorter than those of wild-type ray florets (Figure 4). This provided the first molecular evidence that a *CYC*-like TCP TF is involved in the definition of the capitulum of the *Asteraceae* species.

GhCYC2a collaborates with other *CYC*-like genes to participate in floret differentiation and ultimately determine the complex capitulum structure of *G. hybrida* [84]. *GhCYC2a* is specifically expressed in ray florets at an early developmental stage and is only activated in tubular flowers at an advanced developmental stage. *GhCYC2b* in *G. hybrida* and *HaCYC2d* and *HaCYC2c* in sunflowers belong to the CYC2 clade considered to be a strong candidate as regulators of ray–floret identity [74]. When *GhCYC2b* expression was inhibited in *G. hybrida*, the third type of transitional florets were shorter and the ray–floret corolla was five- or eight-lobed and radially symmetrical, which was in accordance with the findings of Broholm [84].

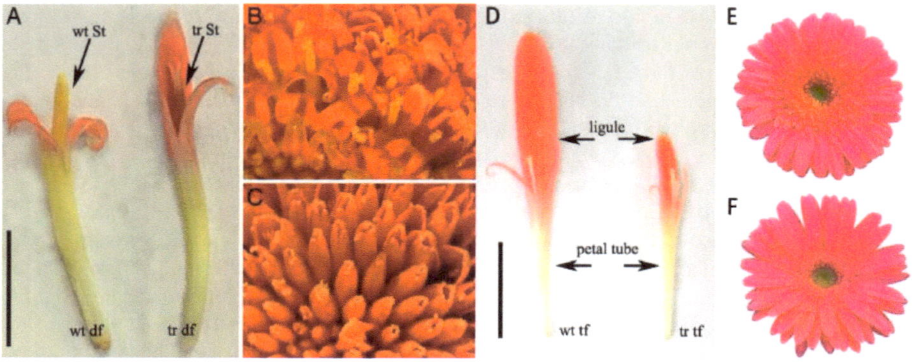

Figure 4. *G. hybrida* with *GhCYC2a* overexpression (**A**–**C**) and inhibition (**D**–**F**) [71]. (**A**) Disc florets (df) of wild-type (wt) *Gerbera hybrida* and transgenic (tr) *G. hybrida* with obvious phenotypic differences. St, stamen. (**B**) Pollen presentation on the style of the disc florets of wild-type *G. hybrida*. (**C**) Disc florets of transgenic *G. hybrida* lacking functional stamens. (**D**) Transitional floret (tf) of wild-type and transgenic *G. hybrida*. (**E**) Capitulum of wild-type *G. hybrida*. (**F**) Capitulum of genetically modified *G. hybrida*. Reprinted with permission from Ref. [71]. Copyright © 2023 by The National Academy of Sciences of the USA.

There is a substantial overlap in expression patterns among the CYC2 subclade genes (i.e., *GhCYC2a*, *GhCYC2b*, *GhCYC2c*, and *GhCYC2d*) in *G. hybrida* [74]. At the single-floret level, their expression domains in the corolla shifted spatially from the currently known dorsal pattern in bilaterally symmetrical flower species, which may have evolved after the origin of *Asteraceae* [74]. *GhCYC2a*, *GhCYC2b*, and *GhCYC2c* mediate the positioning in the proximal and distal axes of the capitulum, leading to ray floret differentiation, and also regulate ray–floret corolla growth by affecting cell proliferation until the corolla assumes its final size and shape [74]. In contrast, the expression of *GhCYC2d* may increase the floret initiation rate during the expansion of the capitulum, while the ectopic expression of *GhCYC2d* increases the floret density in the capitulum [74]. The upstream regulators of *GhCYC2b* (i.e., GhCIN1 and GhCIN2) are CINCINNATA-like homologous TCP proteins with unknown expression domains and functions, but are known to delay the development of marginal ray–floret primordia during early ontogeny [85]. In developing ray florets, the class E MADS-box TF GRCD5 activated *GhCYC2b* expression, whereas the class C MADS-box TF GAGA1 (upstream of *GhCYC2b*) contributed to stamen development.

3.3. *CYC-like Genes of Senecio*

Natural polymorphism of the capitulum in the *Senecio* species is due to the transfer of a set of regulatory genes containing the *RAY* locus from the diploid *Senecio squalidus* to the tetraploid *Senecio vulgaris* [86]. The *RAY* locus, which comprises a cluster of *CYC*-like genes expressed in the periphery of the inflorescence meristem that promote floral asymmetry and lead to increased outcrossing rates, has played a key role in the evolution of radiate capitulum types. The *CYC2*-like gene *RAY3* is initially uniformly expressed in ray florets during capitulum development, but at advanced stages is expressed only in the ventral corolla lobes of ray florets, resulting in the elongation of the ventral corolla limb in *S. vulgaris* [72]. The diversification of *CYC*-like genes has led to novel interactions, with *SvDIV1B* inhibiting *RAY3*, but potentially activating *RAY2*. The expression of *SvRAY1* may induce lateral cell division during the development of the *S. vulgaris* ray floret and, as a result, the morphology and arrangement of the ray floret cells change to some extent, thereby affecting the ray floret width [87]. The ray florets of *SvRAY1*-overexpressing plants were shorter and significantly broader than the wild-type ray florets.

3.4. CYC-like Genes of Chrysanthemum

Researchers have cloned the homologs of snapdragon *CYC* genes in several *Chrysanthemum* × *morifolium* cultivars and analyzed their expression and function [88–92]. Huang et al. identified six *CYC2* subclade *CmCYC2* genes (i.e., *CmCYC2a*, *CmCYC2b*, *CmCYC2c*, *CmCYC2d*, *CmCYC2e*, and *CmCYC2f*) in the *C.* × *morifolium* cultivar 'Maoxiangyu', wherein they mainly regulate the development of ray florets [88]. Compared with other *CYC* homologs, *CmCYC2*s in chrysanthemum may be similarly expressed or there may be distinct differences in expression patterns. The overexpression of *CmCYC2d* in wild-type *Arabidopsis thaliana* and the *tcp1* mutant showed that the vegetative growth of the transgenic lines was inhibited, the flowering period was delayed, and the petal size and arrangement were changed, making the originally radially symmetrical petals appear bilaterally symmetrical [88]. Furthermore the CmCYC2 proteins may form homodimers during flower organogenesis and participate in the regulation of ray and disc floret morphogenesis.

Chen et al. observed that the spatial expression patterns of six *Asteraceae CYC2*-like members are conserved throughout the family, and all of them influence capitulum development [93]. Both *CYC2c* and *CYC2g* are important for ray floret formation in *Chrysanthemum lavandulifolium*, whereas *CYC2d* inhibited the development of the dorsal corolla lobes and ray–floret stamens. The class A MADS-box genes interacted with *CYC2*-like genes potentially involved in processes associated with the formation of reproductive organs and the ray–floret corolla, especially corolla differentiation of the disc and ray florets in chrysanthemum [94]. The class B MADS-box gene *CDM19* may positively regulate the expression of the *CYC2*-like genes *CmCYC2c* and *CmCYC2d*, thereby modifying the floret symmetry in chrysanthemum [95].

Yang et al. cloned the *C.* × *morifolium* gene *CmTCP7*, which may be involved in the formation of floral buds as well as promote the growth of the corolla of the ray floret and participate in the formation of bilaterally symmetrical ray florets [96]. The *CYC*-like gene *CmCYC2* and *WUS*-like gene *CmWUS*, which were highly expressed in floral buds at the time of floral organ differentiation and in reproductive organs at advanced stages of development, coordinately regulate the development of *C.* × *morifolium* reproductive organs [89]. Furthermore, *CmCYC2* was highly expressed in the corolla of ray florets, which may promote ray floret growth and contribute to the formation of bilaterally symmetrical ray florets.

The expression of *CYC2b*, *CYC2d*, *CYC2e*, and *CYC2f* was differentially expressed in different types of *Chrysanthemum vestitum* ray florets, which confirmed the influence of *CYC*-like genes on floral morphology [90]. Yuan et al. observed that the ectopic expression of *CmCYC2* in the *Arabidopsis tcp1* mutant altered flower symmetry and flowering time, and the CmCYC2 TF may interact with or bind to the *CmCYC2* promoter to regulate floral symmetry development in *Chrysanthemum* [91]. Liu et al. cloned the *CYC2*-like gene *Cyc2CL* from *C.* × *morifolium* 'Pink Carpet' and revealed for the first time the variable shear pattern of a *CYC2*-like gene in chrysanthemum [92]. The transcription of *Cyc2CL* resulted in two mature mRNA sequences (*CyC2CL-1* and *CyC2CL-2*). Both transcripts were present at high levels in ray florets, but at very low levels in disc florets and inhibited the development of petals and stamens in *A. thaliana*.

The morphogenesis of the marginal florets in *Ajania* is interrupted shortly after the formation of the floral primordia, possibly because of the lack of expression of the *CYC2*-like gene *ClCYC2g* [97]. The decreased expression of *ClCYC2g* in *C. lavandulifolium* results in the gradual transformation of ray florets into disc florets (Figure 5). This transition may be associated with changes in pollination strategies under selective pressure. Zhang et al. identified four *ClCYC2*-like genes (i.e., *ClCYC2c*, *ClCYC2d*, *ClCYC2e*, and *ClCYC2f*), for which expression levels were significantly higher in ray florets than in disc florets of *C. lavandulifolium* [98].

Figure 5. Knockdown of *ClCYC2g* expression adversely affects the formation of symmetrical marginal florets in the radiate capitulum in chrysanthemum [97]. (**A**,**B**) Morphology of the capitulum of wild-type (WT) and transgenic *C. lavandulifolium.* (**C**) Ray and disc florets of wild-type *C. lavandulifolium.* (**D**) Ray-shaped florets with gradually increasing mutations. Orange and blue arrows indicate abnormal petals and stamens, respectively. (**E**,**F**) Expanded marginal florets of the transgenic lines. White and blue arrows indicate the bilabiate corolla with a deeply dentate limb apex and stamens, respectively. Reprinted with permission from Ref. [97]. Copyright © 2023 Society for Experimental Biology and John Wiley & Sons Ltd. (Hoboken, American).

3.5. *CYC-like Genes of Other Asteraceae Groups*

Bello et al. recovered eight major gene lineages in the highly derived genus *Anacyclus* (tribe *Anthemideae*) through phylogenetic reconstruction, comprising two *CYC1* genes, four *CYC2* genes, and two *CYC3* genes [21]. In *Anacyclus*, three *AcCYC2* genes are highly expressed in ray florets, and the expression patterns of four *AcCYC2* genes overlap in multiple organs, including the limb of ray florets, anthers, and ovule throughout development. Gene duplication events, as well as the subsequent subfunctionalization and neofunctionalization of *SEPALLATA*-like genes and *CYC*-like genes in *Asteraceae,* have been shown to be conducive to the identification of the floral meristem and the formation of key traits for floral differentiation in this large family [99]. Sun et al. identified five *CYC2*-like genes in several *Gaillardia* cultivars with different ray floret types [73]. Analyses of RNA re-sequencing results, quantitative real-time PCR (qRT-PCR) data, and the effects of gene silencing suggested that *CYC2c* is the main genetic factor affecting the formation of ray florets in *Gaillardia*.

4. Progress in Research on *CYC*-like Genes in *Lamiales*

Gene duplication, gene family retention, and tissue-specific expression of *CYC*-like genes are believed to have affected the evolution of corolla symmetry in *Lamiales* [100,101]. The *CYC*-like genes were differentially expressed in the higher core clades with high expression levels in adaxial petals, which had been widely replicated in *Lamiales* (including *Lamiaceae*, *Scrophulariaceae*, *Gesneriaceae*, *Oleaceae*, *Phrymaceae*, and many other families) [102,103]. The asymmetrical expression of *CYC*-like genes was not common but associated with the origin of bilaterally symmetrical corollas [104]. Changes to the *cis*-regulatory domain and the coding sequence of *CYC*-like genes may be critical for the symmetrical evolution of both sides of the corolla, with multiple selection mechanisms contributing to gene retention [100].

The expression pattern of *CYC2*-like genes has gradually evolved, and was widely expressed in the meristem of early-diverging *Lamiales* with a bilaterally symmetrical corolla, but limited in the meristem of core *Lamiales* and thus may be related to the origin of corolla bilateral symmetry [105,106]. The repeated loss of bilateral corolla symmetry is relatively frequent in *Lamiaceae*, which may be caused by different mechanisms and changes in floral symmetry-related genes, such as the loss of the CYC2 clade gene *Ml-CYC2A* in the genome and the contraction, expansion, or altered expression of *Cc-CYC2A* [107,108]. Sengupta and Hileman detected the significant enrichment of predicted autoregulatory sites in the 5′-terminal upstream noncoding region of *CYC*, the upstream regulator of floral zygomorphy in *Lamiales*. Their results suggest that the correlation between the autoregulation of *CYC* and the origin of zygomorphic flowers may be associated with zygomorphic flowers independently derived from eudicot lineages [40].

4.1. CYC-like Genes of Scrophulariaceae

Scrophulariaceae inflorescences are typically racemose, spicate, or cymose and often form a panicle [109]. The *CYC* gene associated with the regulation of floral symmetry was initially isolated from snapdragon, and its homologs in related species were subsequently cloned, including *Linaria vulgaris*, *Mohavea confertiflora*, *Veronica montana*, *Gratiola officinalis*, and *Torenia fournieri* [1,110–114]. These genes have diverse functions in *Scrophulariaceae*, but they primarily affect the morphological characteristics of petals and the development of stamens.

The wild-type snapdragon corolla comprises two dorsal lobes, one ventral lobe, and two lateral lobes. Snapdragon mutants have a semi-abnormal regular flower (semipeloric; *CYC* mutation) or an abnormal regular flower (peloric; *CYC/DICH* double mutation), which represent bilaterally symmetrical floral transitions into a radially symmetrical flower (Figure 6) [1,115]. In the classic *DICH* homozygous mutant, the ventral corolla lobes are more symmetrical than the wild-type ventral lobes and usually separate from each other because of a deep incision between the dorsal lobes [1].

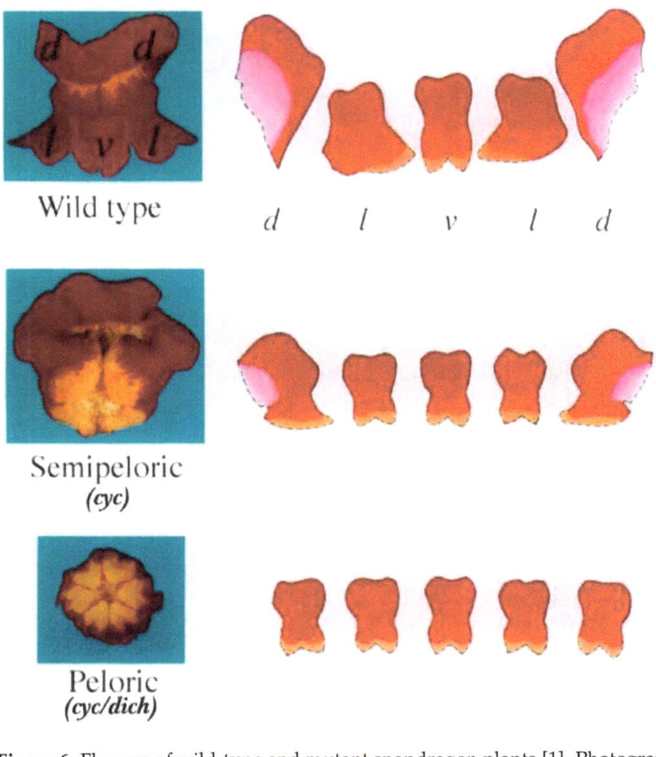

Figure 6. Flowers of wild-type and mutant snapdragon plants [1]. Photographs of the dorsal corolla lobe (d), lateral corolla lobe (l), and ventral corolla lobe (v) of the wild-type snapdragon flower are presented. The characteristics of the different corolla lobes are shown to the right of each flower. Reprinted with permission from Ref. [1]. Copyright © 2023, Nature Publishing Group.

Corley et al. determined that *AmCYC* was expressed in the dorsal corolla lobe, in which *AmRAD* was activated [116]. This ultimately led to the inhibited expression of *AmDIV* in the ventral and lateral lobes, and the formation of asymmetrical snapdragon flowers. Li et al. revealed that whole-genome duplication (WGD) and tandem replication had contributed to the expansion of the *CYC* gene family [15]. Both (*RAD*) and (*DIV*) controlled floral symmetry downstream of *CYC/DICH* and interacted with *DIV-RAD-INTERACTING FACTOR* (*DRIF*) [15]. The *DRIF* genes, which had homologous copies similar to *CYC/DICH*, were also located in the WGD-derived syntenic block [15]. These results further support the view that the key genetic factor regulating the asymmetry of snapdragon flowers was the result of a WGD event.

The peloric flowers (i.e., transitional from bilateral symmetry to radial symmetry) of an *L. vulgaris* mutant were the result of a spontaneous epigenetic mutation to the *CYC* allele [110,117]. The mutant harbored a defective *LCYC*, which is a *CYC* homolog. *LCYC* underwent a heritable modification (i.e., extensive methylation and transcriptional silencing) that was co-isolated with the mutant phenotype [117]. The mutant phenotype may be reversed during somatic development, which is associated with the demethylation of *LCYC* and the restoration of gene expression.

Hileman et al. identified the *M. confertiflora CYC* and *DICH* homologs, *McCYC1*, *McCYC2*, *McDICH1*, and *McDICH2*, of which expression levels increase from the stamen to the outermost floral whorl, which may be due to the change in the expression domain of the regulatory genes in the *CYC/DICH* pathway [111]. Changes to the *McCYC* and *McDICH* expression patterns result in new floral morphological traits, in that the two lateral stamens are aborted and show evidence of the adaxial corolla lobes' internal symmetry. The expression of *CYC*-like genes led to delayed growth or degradation of the adaxial floral organs, but it may also be associated with the loss of the adaxial floral organs [1,111].

A conservative floral symmetry gene network exists in *V. montana* and *G. officinalis*, in which *CYC*-like genes evolved after the gene duplication event, although the detailed genetic mechanisms of dorsal and ventral stamen abortion differ [114]. Specifically, *VmCYC1*, *GoCYC1*, and *GoCYC2* are only expressed in the dorsal region of the floral meristem and in developing flowers, in which expression patterns are independent of stamen abortion patterns, whereas the expression of *VmCYC2* and *GoCYC3* is mainly detected in vegetative and floral tissues.

A dorsally expressed *CYC*-like gene and the downstream target genes *RAD* and *DIV* are absent in *Plantago major* but are present in *Aragoa abietina* [53]. This *CYC*-like gene is expressed in all parts of the flower in *A. abietina*, including the dorsal, ventral, and lateral regions, similar to the expression of its homolog in the related species *Veronica serpyllifolia*. The duplication of *CYC*-like genes led to the evolution of radially symmetrical *A. abietina/P. major* flowers, and further disintegration of the symmetrical flower-related gene pathway led to the wind-pollination syndrome of *P. major* [66]. This model emphasizes the potential importance of gene loss in the evolution of important ecological traits.

Su et al. detected recent replication events of a *CYC*-like gene in *T. fournieri*, and functional analysis of two genes that show dorsal-specific expression, *TfCYC1* and *TfCYC2*, suggested the existence of a regulatory module integrating the dorsoventral pattern and asymmetric corolla pigmentation [112]. The ectopic expression of *TfCYC2* disrupts the asymmetrical corolla coloring pattern, resulting in a strongly dorsal flower, and the CYC–RAD module coordinates petal shape and corolla pigmentation. When *TfCYC2* expression was downregulated, the dorsal petal identity was lost. Diversified *CYC* genes evolved regulatory loops, and *TfCYC2* was directly bound to the regulatory region of the R2R3-MYB gene *TfMYB1*, resulting in asymmetric expression and ultimately the establishment of asymmetric pigmentation patterns [112]. Integration of the Ty1/Copia-like LTR retrotransposon TORE2 into the exon of *TfCYC2*, to generate the allele TfCYC2$^{\mathrm{TORE2}}$, inhibited the expression of *TfCYC2*, which is the main regulatory gene involved in anthocyanin pattern enrichment in *T. fournieri* [118]. The degree of pigmentation of the dorsal corolla lobe of *T. fournieri* is negatively correlated with *TfCYC2* expression.

4.2. CYC-like Genes of Gesneriaceae

The inflorescences of the *Gesneriaceae* species are usually double-flowered cymes or monochasia. The flowers are usually bilaterally symmetrical, but some species produce radially symmetrical flowers (e.g., *Tengia scopulorum*). In recent years, *CYC*-like genes have been isolated from several members of this family, including *Saintpaulia ionantha*, *Sinningia speciosa*, *Chirita heterotricha*, *Primulina heterotricha*, and *Petrocosmea glabristoma* [54,57,119,120]. These genes have the typical functions of *CYC*-like genes, which affect floral symmetry and stamen abortion.

The successive examination of the *CYC*-like genes in the *C. heterotricha*, *P. heterotricha*, and *Petrocosmea* species showed that their expression in developing flowers is regulated by various mechanisms [54,119,120]. First, the promoter sequences of the *CYC* homologs *ChCYC1C* and *ChCYC1D* in *C. heterotricha* were isolated. Subsequent analysis indicated the genes may have evolved automatic regulatory loops to maintain expression during the establishment of bilaterally symmetrical flowers [119]. The *RAD*-like gene *ChRAD* may be directly targeted by *ChCYC1* as part of a regulatory network. Next, the expression and function of two *CYC2* genes (*CYC1C* and *CYC1D*) in *P. heterotricha* were analyzed, which revealed positive self-regulatory and cross-regulatory effects [120]. This mechanism may lead to the independent formation of bilaterally symmetrical flowers, which is associated with plant–insect co-evolution and the adaptive radiation of angiosperms. Finally, changes in the dorsal corolla lobe size of *P. glabristoma* and *Petrocosmea sinensis* were determined to be mainly mediated by the expression and differentiation of *CYC1C* and *CYC1D*, and the changes in the petal shape were associated with the expression-level changes to the *CIN*-like TCP gene *CIN1* (Figure 7) [54]. Highly redundant homologous genes with the same expression patterns and interspecific differences in expression may be controlled by markedly different regulatory pathways, because natural selection may have resulted in diverse regulatory modifications rather than sequence changes to key developmental genes to generate morphological diversity [54].

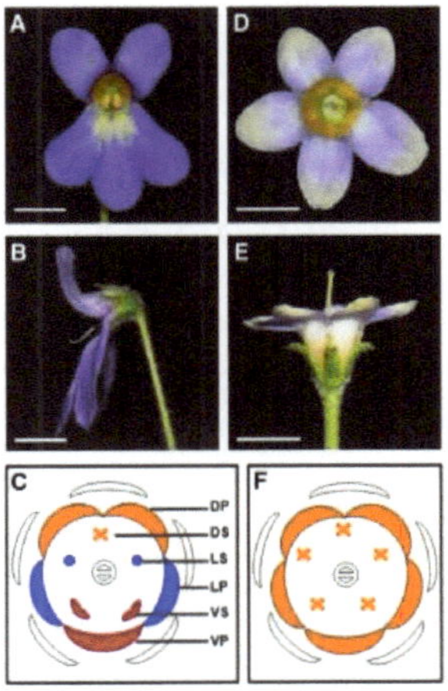

Figure 7. Front and side view and floral diagram of normal and peloric flowers of *Petrocosmea sinensis* [54]. (**A–C**) Front and side view and floral diagram of *P. sinensis* normal flowers, which have a typical bilaterally symmetrical corolla. (**D–F**) Front and side view and floral diagram of *P. sinensis* peloric flowers, which have a radially symmetrical corolla. DP, dorsal corolla lobes; DS, dorsal stamens; LS, lateral stamens; LP, lateral corolla lobes; VS, ventral stamens; VP, ventral corolla lobe. Reprinted with permission from Ref. [54]. Copyright © 2023, Oxford University Press.

Hsu et al. reported that in *S. speciosa*, the dorsal corolla lobes are bent outward, the midvein of the lateral corolla lobes is asymmetrical, and the expansion of the ventral area of the corolla is closely related to the *SsCYC* genotype [57]. Expression shifts of the

CYC-like genes *SiCYC* and *SiCYC1B*, which show dorsal-specific expression in the wild-type *S. ionantha*, led to two completely different reversals of radial symmetry, namely dorsalized actinomorphic (DA) and ventralized actinomorphic (VA) peloria, which may be controlled by upstream *trans*-acting factors or epigenetic regulation [121]. *SiCYC* and *SiCYC1B* were metastasized with an ectopically extended expression on all corolla lobes in DA, whereas their dorsal-specific expression was greatly reduced in VA [121]. The main highly expressed copies of *SiCYC* were constrained by purification selection, whereas selection of the low-expression helper gene *SiCYC1B* was relaxed after duplication [121]. Heterologous expression of *SiCYC* in *A. thaliana* was characterized by delayed corolla growth owing to limited cell proliferation [121].

CYC-like gene duplication events have occurred at least five times in the evolutionary history of *Gesneriaceae* [51]. Three copies of *CYC*-like genes in the actinomorphic *Conandron ramondioides* were not expressed in the corolla, whereas the zygomorphic species *Hemiboea bicornuta* and *Lysionotus pauciflorus* retained a *CYC1* copy (i.e., *GCYC1C* and *GCYC1D*, respectively) expressed in dorsal corolla lobes [51]. Selective relaxation after the duplication of *CYC1* created evolutionary diversification, in which multiple copies retained the effect of random differentiation affecting the dorsal-specific expression of genes associated with floral symmetry changes [51]. The promoter region of *CpCYC* is a key determinant of its specific expression in the dorsal corolla lobe of *Chirita pumila*, where the LEAFY element may directly activate and regulate *CpCYC* to form a bilaterally symmetrical flower [122].

Yang et al. determined that the ortholog of *LjCYC1* in *S. ionantha* is highly expressed in the root, leaf, peduncle, calyx, petal, stamen, and pistil of transgenic *S. ionantha* plants [123]. Two flower-type variations were observed in T_1 transgenic plants. The first was the change in floral symmetry. Specifically, radially symmetrical wild-type flowers were replaced by bilaterally symmetrical flowers or flowers with obvious differences between the dorsal and ventral corolla lobes. The second variation involved floral organ morphology (e.g., a lobe incision towards the base of the corolla, and stamen, pistil, and calyx petalization). Liu et al. functionally characterized the *CYC*-like flower symmetry-related gene *CpCYC* in *C. pumila* [124]. By transforming plants with a RNAi:CpCYC vector, vertically radially symmetrical flowers were obtained, implying that *CpCYC* determines the establishment of zygomorphy and the horizontal plane of flowers. The insertion of a *CpCYC* promoter:GUS vector into *C. pumila* confirmed that the *CpCYC* promoter was active in dorsal corolla lobes, dorsal/lateral staminodes, and pedicels.

4.3. CYC-like Genes of Phrymaceae

The *Phrymaceae* species have bilaterally symmetrical bisexual flowers borne in spikes at the top of the stem and in the upper leaf axils. To date, there has been relatively little research on the *CYC*-like genes in this family. The flowers of the *Phrymaceae* species *Diplacus pictus* have distinct dorsal, ventral, and lateral corolla lobes. The expression and function of *CYC* genes may vary between *D. pictus* and snapdragon [125]. The *CYC*-like gene *DpCYC* is expressed in a narrow part of the upper lip of the dorsal corolla lobe. The novel upturned abaxial corolla lobe of *D. pictus* may be associated with the localized expression of *DpCYC* on the upper surface of this structure.

5. Progress in Research on *CYC*-like Genes in *Orchidaceae*

Orchidaceae is a large family, second in species number only to *Asteraceae* and the largest family of monocotyledons [126,127]. The bisexual flowers of the members of this family are generally bilaterally symmetrical [128]. The median tepal is typically modified into the labellum, which is often in the abaxial position because the inflorescence droops or the pedicel twists [129]. Many *CYC*-like genes in *Phalaenopsis* and *Cattleya trianae* have been identified and observed to influence floral or inflorescence development [130–132].

Lin et al. observed that the ECE clade gene *PeCIN8* is highly expressed in the late ovule developmental stage, with overlapping expression on the 16th day after the pollination of *Phalaenopsis equestris* flowers [130]. Hence *PeCIN8* may be crucial for orchid ovule development because of its regulatory effects on cell division. In addition, Liu et al. recently identified 10 ECE clade genes (*CgCINs* and *CgCYCs*) from the *Cymbidium goeringii* genome and observed that flower-specific gene expression may be associated with the regulation of flower development [131].

The *CYC*-like genes are single, short, low-expressed copies in *Orchidoideae* [132]. Madrigal et al. performed a phylogenetic analysis of the TCP genes in *C. trianae* and observed that the TCP genes were resolved into three major clades with multiple gene duplication events detected [132]. Among these genes, the *CYC*-like genes are single short copies with low expression, and their homogeneous expression in the labial and lateral tepals suggested that they contribute little to bilateral perianth formation.

6. Progress in Research on *CYC*-like Genes in *Solanaceae*

The *Solanaceae* is dominated by species with radially symmetrical flowers, but the early-diverging clades often show bilateral flower symmetry [133]. At present, few studies have explored the influence of *CYC*-like genes on flower development in the *Solanaceae*. *Petunia hybrida* is an annual herb that produces solitary flowers in leaf axils [134]. Its funnel-shaped corolla is white or purple and may be variously striped. There is considerable diversity in the flower types of *P. hybrida*. Research on *Petunia CYC*-like genes of the ECE clade showed that they may be mainly associated with the growth and development of axillary buds, while also modulating stem growth and the development of branches, flowers, and leaves (e.g., size) [135–137].

Zou et al. isolated the *P. hybrida CYC1* clade genes *PhTCP3* and *PhTCP4*, the *CYC2* clade gene *PhTCP2*, and the *CYC3* clade gene *PhTCP1*, which were predominantly associated with axillary bud growth and development [135]. These four genes were highly homologous to the ECE clade genes from tomato (*Solanum lycopersicum*), gerbera, grape (*Vitis vinifera*), poplar (*Populus*), *A. thaliana*, and other plant species and exhibited tissue-specific expression patterns, and the expression levels in the axillary buds were significantly higher than those in the other tissues analyzed [135]. The overexpression and silencing of *PhTCP1* led to decreased and increased branching, respectively, suggestive of the regulatory effects of *PhTCP1* on branch formation and development.

Zhang et al. conducted the first genome-wide spatiotemporal expression profile and promoter analysis of the petunia ECE clade *PaTCP* genes, and reported that these genes may play an important role in the various developmental processes of petunia through multiple hormonal pathways, especially in petal development and the formation of petal size [136]. Sengupta and Hileman observed that the *CYC* homologous genes positively regulate the *RAD* homologous genes in tomato, which was similar to previous findings in snapdragon [137].

7. Progress in Research on *CYC*-like Genes in Other Angiosperms

To date, there have been only a few studies on *CYC*-like genes in other angiosperm families. These investigations have indicated the genes are primarily expressed in floral organs and affect the establishment of floral symmetry and stamen fertility [138–142]. Thus future research needs to be extended to these less-studied plant families.

7.1. CYC-like Genes of Brassicaceae

The flowers of *Brassicaceae* are borne in racemes of many small flowers, which are actinomorphic in most species but zygomorphic in a few species [143,144]. Cubas et al. were the first researchers to identify a *CYC2* lineage gene, *AtTCP1*, in *A. thaliana* that regulates the symmetrical development of floral organs [52]. Arabidopsis flowers are actinomorphic, and *AtTCP1* is briefly expressed specifically in the dorsal region of the floral primordium, which suggests that *AtTCP1* does not play a key role in floral organ differentiation because there is no flower-specific direct transcriptional autoregulation or other expression pattern that persists until advanced stages of flower differentiation [52].

The corolla of the genus *Iberis* is zygomorphic, with two small adaxial petals and two large abaxial petals [145]. Busch et al. identified a *CYC* homolog from *Iberis amara*, *IaTCP1*, for which the temporary alteration of expression is important for the control of zygomorphic corolla formation [146]. The timing of *IaTCP1* expression differed from that of *AtTCP1* and other *CYC* homologs. Specifically, *IaTCP1* expression was lacking early in asymmetric petal development, but the gene was strongly differentially expressed in the corolla during advanced asymmetric petal development. In addition, *IaTCP1* activated the expression of many more genes in adaxial petals than it inhibited [147]. These findings suggest that asymmetric corolla formation in *I. amara* may be caused by the strong expression of the *CYC2* clade gene in the small ventral petals and weak expression in the large dorsal petals.

Busch et al. revealed through phylogenetic reconstruction that the zygomorphic genera *Iberis*, *Calepina*, and *Teesdalia* belong to a major *Brassicaceae* lineage [148]. Zygomorphy is most evident in *Iberis*, but less so in *Calepina* and *Teesdalia*, with an expression-dependent positive correlation between the strength of the difference in *CYC2* expression and the degree of zygomorphy [148]. This study suggested that zygomorphy evolved through the heterochronic expression of *CYC2*, from paraxial expression in the ancestral floral meristem to accumulation of paraxial *CYC2* transcripts late in petal development in the *Brassicaceae*.

7.2. *CYC-like Genes of Dipsacales*

Howarth and Donoghue identified three major ECE clade *CYC*-like genes (*DipsCYC1*, *DipsCYC2*, and *DipsCYC3*) in the ancestors of *Dipsacales*, as well as additional duplication events involving genes in this clade [26]. The *DipsCYC1* gene was not involved in subsequent replication events and may not be expressed in flower tissues. In contrast, *DipsCYC2* and *DipsCYC3* had similar duplication patterns in several clades. In the *Caprifoliaceae* species *Lonicera morrowii*, *DipsCYC2B* was expressed in the four dorsal petals, but not in the ventral petal, whereas *DipsCYC3B* was expressed in the flowers and petal primordia, with the peak expression level in the ventral petal [26].

Berger et al. compared the expression patterns of six *CYC*-like genes in the dorsal, lateral, and ventral petals of the inner and outer florets of the capitulum of *Knautia macedonica* and observed that *CYC*-like genes duplicated many times were differentially expressed among the petal types and the inner and outer florets [55]. The formation of bilateral symmetry may be regulated by a dorsoventral expression gradient. In addition, dorsoventral morphological specificity was associated with significant changes in ventral petal gene expression.

7.3. *CYC-like Genes of Zingiberales*

Bartlett and Specht showed that gene replication unique to the *Zingiberales*, including one replication event involving the ECE clade gene *TBL*, preceded the diversification of commelinid monocotyledons [142]. The changes in *TBL* expression were associated with the evolutionary changes in floral symmetry and stamen abortion. In addition, *ZinTBL1a* was expressed in the posterior (adaxial) stamen lip of *Heliconia stricta* (*Heliconiaceae*) and the anterior (abaxial) stamen lip of *Costus spicatus* (*Costaceae*) [141]. The *ZinTBL2* expression level was highest in the anterior sepals of *H. stricta* and the posterior fertile stamens of *C. spicatus*. These findings indicate that the ECE clade genes were repeatedly recruited in the evolutionary process, which accelerated the evolution of bilaterally symmetrical flowers.

Canna indica (*Cannaceae*) has noticeably asymmetrical flowers, in which the stamens develop into petal-like staminodes or are aborted (semi-fertile stamens) at an early developmental stage associated with three ECE clade homologs (*CiTBL1a*, *CiTBL1b-1*, and *CiTBL1b-2*) (Figure 8) [142]. The overexpression of *CiTBL* genes in *A. thaliana* resulted in dwarfism, the production of small petals and relatively few stamens, as well as mature flowers with altered symmetry, which provided evidence of the involvement of ECE clade homologs in the development of asymmetrical *C. indica* flowers.

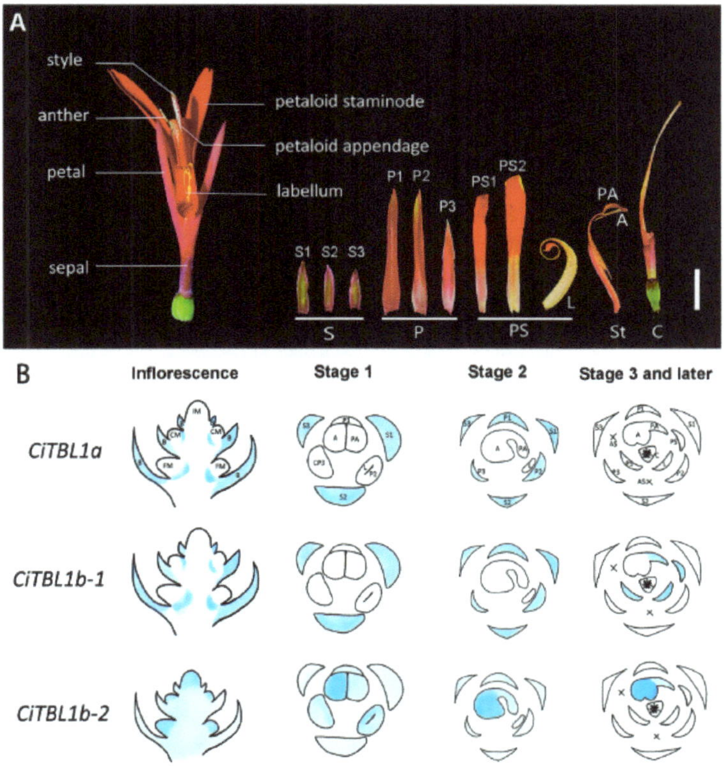

Figure 8. Morphology of *Canna indica* flowers and expression of *CiTBL* genes [142] (Yu et al., 2020). (**A**) Flower morphology. A: anthers; C: carpel; L: labellum; P: petals; PA: petaloid appendage; PS: petaloid staminodes; S: sepals; St: staminodes. (**B**) Expression of *CiTBL1a*, *CiTBL1b-1*, and *CiTBL1b-2* in young inflorescences and flowers at different developmental stages. Gene expression sites are indicated in blue, with the intensity of the coloration reflecting the expression level. AS: abortive staminodes; B: primary bracts; CM: meristem of monochasium; CP: common primordium of the petal and stamen; FM: floral meristem; IM: inflorescence meristem. Reprinted with permission from Ref. [142]. Copyright © 2023 Frontiers Media S.A.

7.4. CYC-like Genes of Ranunculales

Ranunculaceae underwent an evolutionary transition from actinomorphy to zygomorphy, with the accumulation of as many as four copies of the *CYC*-like gene *RanaCyL* in zygomorphic species [149,150]. The *RanaCyL* homologous genes are expressed early in flower bud development, and the expression duration varies with species and gene class. In actinomorphic species, at most one *RanaCyL* paralog is expressed late in flower development, whereas in zygomorphic species, all paralogs are expressed, constituting a species-specific recognition code for the perianth [149].

Fumariaceae and *Papaveraceae* show morphological diversity in flower symmetry and inflorescence structure, which may be related to the duplication and functional diversity of *CYC*-like genes [151]. Damerval et al. reported that the homologous genes of *CYC* in *Papaveraceae*, *PAPACYL1*, and *PAPACYL2*, which are ECE clade members, were expressed during the flower development of all three species studied and were specifically expressed in the outer petals of the two species with asymmetrical flowers [152].

The *CYC*-like *CYL* genes of the *Papaveraceae* species *Eschscholzia california* and *Cysticapnos vesicaria* have highly diverse expression patterns and functions [153]. The silencing of *EscaCYL1* enhances the control of bud branching, whereas *PapaCYL* genes promote germination and growth of stamens. In addition, *CyveCYL* genes are involved in the regulation of floral symmetry and perianth development of *Cysticapnos* by regulating B-class floral–organ identity genes to determine sepal and petal characteristics.

7.5. CYC-like Genes of other Families

An ECE clade gene, *CcCYC*, is not expressed in the radially symmetrical perianth of *Tradescantia pallida*, but is expressed asymmetrically in the bilaterally symmetrical perianth of *Commelina communis* and *Commelina dianthifolia* [138]. These observations were related to genes that were recruited in parallel through the independent evolution of flower bilateral symmetry in the early stage of floral development. The *Actinodium cunninghamii* (*Myrtaceae*) capitulum consists of a pseudanthium, with a ray flower that is not a single flower but a branch with a short bud that flowers occasionally; its proximal branch is also similar to the ray flower [139]. The changes in the expression of *CYC*-like genes in the pseudanthium modulated ray flower structures and branching patterns. This gene expression pattern is similar to that observed in the distantly related *Asteraceae* species, indicating that flowering plants seem to have recruited *CYC*-like genes at least twice in their evolutionary history for the development of heterotypic inflorescences.

Horn et al. showed that the ECE clade gene *CYCL* is present in basal angiosperms and *Magnoliaceae* species [140]. In *Aristolochia*, *CYCL* was involved in the differentiation of the perianth and the mushroom pseudo-structure, but did not participate in the process mediating the formation of zygomorphic flowers. Only when the TCP domain of the *Aristolochia CYCL* gene was replaced by the CYC2 domain could the functionally similar gene be obtained. The differentiation and evolution of the ECE lineage led to significant changes in the coding region and the *cis*-regulatory elements, which ultimately established *CYC2* as a key gene regulating floral zygomorphy in dicotyledons. Pabon-Mora et al. reported that ECE clade genes may also be involved in cell division in leaves, pistils, and ovules [154]. Specifically, *CYC*-like genes maintain differential expansion of the perianth by promoting cell division in the distal and ventral extremities during middle and late flower development in *Aristolochia fimbriata*.

Zhang and co-workers confirmed that the expression of the *CYC2*-like genes *CYC2A* and *CYC2B* was associated with the floral symmetry of *Malpighiaceae*, and that relaxation of their conserved expression and expansion to a wider floral area (including the dorsal stamen) were related to the development of dorsoventral heteranthery in *Hiptage benghalensis* and contributed to the elaborated androecium, which is essential for adaptation to the new pollination strategy [56,155]. Berger et al. compared the corolla shape of *Fedia graciliflora* expressing the wild-type or knocked-out *CYC2*-like gene *FgCYC2A* using canonical variable analysis, and observed that gene knockout resulted in significant changes in flower shape, which affected the position of the dorsal lobe relative to the lateral lobe and led to more radially symmetrical flowers [156].

Radially symmetric *Rhododendron taxifolium* and bilaterally symmetric *Rhododendron beyerinckianum* have four and five *CYC*-like genes from shared tandem duplications, respectively [28]. The *CYC*-like genes are expressed in the longer dorsal petals and stamens, and are highly expressed in the pistil of *R. beyerinckianum*, whereas in *R. taxifolium* the orthologs are either ubiquitously expressed, have been lost from the genome, or are weakly expressed [28]. As the main regulatory factor for the growth of differentiated organs in *Rhododendron*, *CYC*-like genes did not regulate the expression of *RAD*-like genes, which revealed a certain deviation from the typical floral symmetry-related gene regulatory network of asterids [28].

Three *CYC*-like genes (*CamCYC1*, *CamCYC2*, and *CamCYC3*) in *Campanulaceae* have undergone dynamic changes in replication and loss, including the first instance of the loss of *CamCYC2* in a bilaterally symmetrical group [27]. The *CamCYC1* gene was included in duplication events in the radially symmetrical *Campanuloideae* species, whereas *CamCYC2* was duplicated but *CamCYC3* was lost at an early stage of divergence, in the bilaterally symmetrical and inverted *Lobelioideae* species [27]. In addition, the bilaterally symmetrical and non-inverted *Cyphioideae* species lost *CamCYC2*, but replicated *CamCYC3* [27]. The late expression of *CamCYC2* along the dorsoventral axis of the inverted flower was confirmed, and was not regulated by external factors, such as gravity [27].

8. Outlook

Researchers have conducted systematic and detailed studies on the *CYC*-like genes of many angiosperm families, such as *Fabaceae*, *Asteraceae*, *Scrophulariaceae*, *Gesneriaceae*, and *Orchidaceae*. However there are still many issues regarding the function and evolution of *CYC*-like genes that require exploration in greater detail.

8.1. Conduct Systematic Functional and Evolutionary Research, Especially Regarding CYC1 and CYC3 Clade Members

The *CYC*-like genes have extensive and important roles affecting plant development [26,29,30]. The current relevant research has mainly focused on the *CYC2* clade, which is primarily associated with the regulation of floral symmetry, with less research conducted on the *CYC1* and *CYC3* clades [53–57,112]. Therefore, the functions and evolution of *CYC1* and *CYC3* genes should be investigated, to expand our understanding of the contributions of *CYC*-like genes to the growth and development of angiosperms.

8.2. Functionally Characterize the CYC-like Genes in More Plant Groups

Through developmental biology, genetics, and evolutionary genetics, scientists revealed that the ancestors of *CYC* in core eudicots were expressed in the dorsal flower organs, thus affecting floral symmetry [19,26,112,157]. The *CYC* genes are expressed in the ventral floral organs of several monocot groups (*Zingiberaceae*, *Alstroemeriaceae*, and *Commelinaceae*) [138,141,158]. In *Alstroemeriaceae*, Hoshino et al. observed that the *CYC*-like genes *AaTCP1*, *AmTCP1*, *ApTCP1*, and *ApTCP2*, which belong to the ECE clade, are involved in the development of floral asymmetry and the identity of ventral floral organs in *Alstroemeria aurea*, *Alstroemeria magenta*, and *Alstroemeria pelegrina* with bilaterally symmetrical flowers [158]. In addition, *AaTCP1* transcripts were specifically accumulated in flower buds and located at the paraxial perianth base of *A. aurea*. These results reflect the complexity of the *CYC* expression pattern in angiosperms. Additional research on these genes and their regulatory effects on floral symmetry will require the inclusion of more plant groups.

8.3. Investigate the Regulatory Elements Upstream of CYC-like Genes

Increasing numbers of studies have isolated and analyzed the phylogenetic relationships, expression patterns, and functions of *CYC*-like homologs in different angiosperm groups, but there has been minimal research on the upstream regulatory elements. Yang and co-workers determined that the bilateral symmetry of the flowers in *Gesneriaceae* may have involved the evolution of an automatic regulatory loop for the *CYC*-like gene [119,120]. In the *double-flowered* (*dbl*) sunflower mutant, *HaCYC2c* inserted into the promoter region is usually expressed specifically in wild-type ray florets, but not throughout the capitulum, possibly resulting in the inability to observe radially symmetrical flowers [79]. CmWUS can bind to the *cis*-acting element TAAT in the *CmCYC3a* promoter in yeast, and activate the expression of resistance genes, while also regulating floral symmetry and flower organ development together with ECE TFs in chrysanthemum [89]. The chrysanthemum TF CmCYC2c can bind to the *cis*-acting element of *CmCYC2f* to activate its expression, but it can also form heterodimers with CmCYC2c-2, CmCYC2d, and CmCYC2e, which may participate in the regulation of floral organ symmetry [88]. The spatiotemporal expression patterns and functions of *CYC*-like genes in different flower organs in different taxonomic groups are diverse, which may be related to changes in the upstream regulatory elements, ultimately resulting in a variety of angiosperm flower types. Therefore, the regulatory elements upstream of *CYC*-like genes must be studied, which will help to clarify the evolution and functional differentiation of these genes in angiosperms.

8.4. Study the Phylogenetic Relationships and Expression of CYC-like Genes with New Techniques and Methods

In addition to traditional methods for verifying gene functions, methylation analyses and other technical methods should be used to study the apparent modifications to *CYC*-like genes. Zhang et al. used qRT-PCR and bisulfite sequencing PCR techniques to determine the expression patterns and DNA methylation patterns of *CYC2*-like genes in two types of chrysanthemum florets, thereby providing new epigenetic-related insights into the formation of the capitulum in *Asteraceae* [98]. Sun et al. confirmed that *CYC2c* is the main factor influencing the *Gaillardia* ray–floret phenotype by applying RNA resequencing technology as well as qRT-PCR and gene-silencing methods [73]. In future studies, additional new technologies and methods including comparative genomics can be used to elucidate the role of *CYC*-like genes in angiosperm floral development, which will lead to new ideas for future research on the evolution and development of angiosperms.

9. Method

We randomly selected *CYC*-like genes from the published literature to construct a phylogenetic tree (Figure 1) using MEGA 11 software. The nucleotide sequences of the homologous *CYC* genes were downloaded from NCBI GenBank (https://www.ncbi.nlm.nih.gov/) (accessed on 9 October 2022) and aligned with ClustalW. The phylogenetic tree was constructed with MEGA 11 software using the maximum likelihood method [159]. The accession numbers of the sequence data used to construct the phylogenetic tree are present in the Supplemental Table S1. To assess support for the topology of the tree, a bootstrap analysis with 1000 replications was performed. In the phylogenetic tree, the CYC1, CYC2, and CYC3 clades are labeled with reference to the literature, whereas other *CYC*-like genes do not have an explicit classification at present [21,37,74–76].

Supplementary Materials: The following supporting information can be downloaded at: https://www.mdpi.com/article/10.3390/cimb45030131/s1, Supplementary Table S1. The accession numbers of the genes used to construct the phylogenetic tree.

Author Contributions: Conceptualization, H.L.; methodology, X.Z.; software, Y.C., W.C., C.G. and H.C.; writing—original draft preparation, Y.C. and H.L.; writing—review and editing, Y.C., H.L. and C.H.; funding acquisition, X.C., C.L. and D.C. All authors have read and agreed to the published version of the manuscript.

Funding: This work was supported by the National Natural Science Foundation of China (31901354), Beijing Innovation Consortium of Agriculture Research System (BAIC09-2022), and the Innovation Foundation of the Beijing Academy of Agriculture and Forestry Sciences (KJCX20200112).

Institutional Review Board Statement: Not applicable.

Data Availability Statement: No confidential/unpublished data have been used in this article.

Conflicts of Interest: The authors declare that the research was conducted in the absence of any commercial or financial relationships that could be construed as potential conflicts of interest.

References

1. Luo, D.; Carpenter, R.; Vincent, C.; Copsey, L.; Coen, E. Origin of floral asymmetry in *Antirrhinum*. *Nature* **1996**, *383*, 794–799. [CrossRef]
2. Doebley, J.; Stec, A.; Hubbard, L. The evolution of apical dominance in maize. *Nature* **1997**, *386*, 485–488. [CrossRef]
3. Kosugi, S.; Ohashi, Y. PCF1 and PCF2 specifically bind to cis elements in the rice proliferating cell nuclear antigen gene. *Plant Cell* **1997**, *9*, 1607–1619. [CrossRef] [PubMed]
4. Luo, D.; Carpenter, R.; Copsey, L.; Vincent, C.; Clark, J.; Coen, E. Control of organ asymmetry in flowers of *Antirrhinum*. *Cell* **1999**, *99*, 367–376. [CrossRef] [PubMed]
5. Cubas, P.; Lauter, N.; Doebley, J.; Coen, E. The TCP domain: A motif found in proteins regulating plant growth and development. *Plant J.* **1999**, *18*, 215–222. [CrossRef] [PubMed]
6. Manassero, N.G.; Viola, I.L.; Welchen, E.; Gonzalez, D.H. TCP transcription factors: Architectures of plant form. *Biomol. Concepts* **2013**, *4*, 111–127. [CrossRef]

7. Nicolas, M.; Torres-Perez, R.; Wahl, V.; Cruz-Oro, E.; Rodriguez-Buey, M.L.; Zamarreno, A.M.; Martin-Jouve, B.; Garcia-Mina, J.M.; Oliveros, J.C.; Prat, S.; et al. Spatial control of potato tuberization by the TCP transcription factor BRANCHED1b. *Nat. Plants* **2022**, *8*, 281–294. [CrossRef]
8. Damerval, C.; Claudot, C.; Le Guilloux, M.; Conde, E.S.N.; Brunaud, V.; Soubigou-Taconnat, L.; Caius, J.; Delannoy, E.; Nadot, S.; Jabbour, F.; et al. Evolutionary analyses and expression patterns of TCP genes in Ranunculales. *Front. Plant Sci.* **2022**, *13*, 1055196. [CrossRef]
9. Sharma, P.; Kumar, P.; Fri, N.; Ijast, I. Genome wide analysis and identification of TCP gene family in Wheat (*Triticum aestivum* L.). *Int. J. Appl. Sci. Technol.* **2022**, *8*, 19–35.
10. Parapunova, V.; Busscher, M.; Busscher-Lange, J.; Lammers, M.; Karlova, R.; Bovy, A.G.; Angenent, G.C.; de Maagd, R.A. Identification, cloning and characterization of the tomato TCP transcription factor family. *BMC Plant Biol.* **2014**, *14*, 157. [CrossRef]
11. Baulies, J.L.; Bresso, E.G.; Goldy, C.; Palatnik, J.F.; Schommer, C. Potent inhibition of TCP transcription factors by miR319 ensures proper root growth in Arabidopsis. *Plant Mol. Biol.* **2022**, *108*, 93–103. [CrossRef] [PubMed]
12. Igartua, E.; Contreras-Moreira, B.; Casas, A.M. TB1: From domestication gene to tool for many trades. *J. Exp. Bot.* **2020**, *71*, 4621–4624. [CrossRef] [PubMed]
13. De Souza, M.T.; Van Es, S.W.; Hernandez-Pinzon, I.; Kirschner, G.K.; Van Der Wal, F.; Da, S.S.; Busscher-Lange, J.; Angenent, G.C.; Moscou, M.; Immink, R.; et al. The TCP transcription factor HvTB2 heterodimerizes with VRS5 and controls spike architecture in barley. *Plant Reprod.* **2022**, *35*, 205–220. [CrossRef] [PubMed]
14. Dixon, L.E.; Pasquariello, M.; Boden, S.A. TEOSINTE BRANCHED1 regulates height and stem internode length in bread wheat. *J. Exp. Bot.* **2020**, *71*, 4742–4750. [CrossRef] [PubMed]
15. Li, M.; Zhang, D.; Gao, Q.; Luo, Y.; Zhang, H.; Ma, B.; Chen, C.; Whibley, A.; Zhang, Y.; Cao, Y.; et al. Genome structure and evolution of *Antirrhinum majus* L. *Nat. Plants* **2019**, *5*, 174–183. [CrossRef] [PubMed]
16. Steiner, E.; Livne, S.; Kobinson-Katz, T.; Tal, L.; Pri-Tal, O.; Mosquna, A.; Tarkowska, D.; Mueller, B.; Tarkowski, P.; Weiss, D. The Putative O-Linked N-Acetylglucosamine Transferase SPINDLY Inhibits Class I TCP Proteolysis to Promote Sensitivity to Cytokinin. *Plant Physiol.* **2016**, *171*, 1485–1494. [CrossRef]
17. Busch, A.; Deckena, M.; Almeida-Trapp, M.; Kopischke, S.; Kock, C.; Schussler, E.; Tsiantis, M.; Mithofer, A.; Zachgo, S. MpTCP1 controls cell proliferation and redox processes in Marchantia polymorpha. *New Phytol.* **2019**, *224*, 1627–1641. [CrossRef]
18. Spears, B.J.; McInturf, S.A.; Collins, C.; Chlebowski, M.; Cseke, L.J.; Su, J.; Mendoza-Cozatl, D.G.; Gassmann, W. Class I TCP transcription factor AtTCP8 modulates key brassinosteroid-responsive genes. *Plant Physiol.* **2022**, *190*, 1457–1473. [CrossRef]
19. Cubas, P. Floral zygomorphy, the recurring evolution of a successful trait. *Bioessays* **2004**, *26*, 1175–1184. [CrossRef]
20. Martín-Trillo, M.; Cubas, P. TCP genes: A family snapshot ten years later. *Trends Plant Sci.* **2010**, *15*, 31–39. [CrossRef]
21. Bello, M.A.; Cubas, P.; Alvarez, I.; Sanjuanbenito, G.; Fuertes-Aguilar, J. Evolution and expression patterns of CYC/TB1 genes in Anacyclus: Phylogenetic insights for floral symmetry genes in Asteraceae. *Front. Plant Sci.* **2017**, *8*, 589. [CrossRef] [PubMed]
22. Levin, K.A.; Boden, S.A. A new branch of understanding for barley inflorescence development. *J. Exp. Bot.* **2020**, *71*, 6869–6871. [CrossRef] [PubMed]
23. Cubas, P. Role of TCP genes in the evolution of morphological characters in angiosperms. In *Role of TCP Genes in the Evolution of Key Morphological Characters in Angiosperms*; Cronk, Q.C.B., Hawkins, J., Bateman, R.M., Eds.; Taylor & Francis: London, UK, 2002; pp. 247–266.
24. Howarth, D.G.; Donoghue, M.J. Duplications in *CYC*-like genes from Dipsacales correlate with floral form. *Int. J. Plant Sci.* **2005**, *166*, 357–370. [CrossRef]
25. Citerne, H.L.; Le Guilloux, M.; Sannier, J.; Nadot, S.; Damerval, C. Combining phylogenetic and syntenic analyses for understanding the evolution of TCP ECE genes in eudicots. *PLoS ONE* **2013**, *8*, e74803. [CrossRef] [PubMed]
26. Howarth, D.G.; Donoghue, M.J. Phylogenetic analysis of the "ECE" (CYC/TB1) clade reveals duplications predating the core eudicots. *Proc. Natl. Acad. Sci. USA* **2006**, *103*, 9101–9106. [CrossRef] [PubMed]
27. Tong, J.; Knox, E.B.; Morden, C.W.; Cellinese, N.; Mossolem, F.; Zubair, A.S.; Howarth, D.G. Duplication and expression patterns of *CYCLOIDEA*-like genes in Campanulaceae. *Evodevo* **2022**, *13*, 5. [CrossRef]
28. Ramage, E.; Soza, V.L.; Yi, J.; Deal, H.; Chudgar, V.; Hall, B.D.; Di Stilio, V.S. Gene Duplication and Differential Expression of Flower Symmetry Genes in Rhododendron (Ericaceae). *Plants* **2021**, *10*, 1994. [CrossRef]
29. Carlson, S.E.; Howarth, D.G.; Donoghue, M.J. Diversification of *CYCLOIDEA*-like genes in Dipsacaceae (Dipsacales): Implications for the evolution of capitulum inflorescences. *BMC Evol. Biol.* **2011**, *11*, 325. [CrossRef]
30. Busch, A.; Zachgo, S. Flower symmetry evolution: Towards understanding the abominable mystery of angiosperm radiation. *Bioessays* **2009**, *31*, 1181–1190. [CrossRef]
31. Panchy, N.; Lehti-Shiu, M.; Shiu, S.H. Evolution of Gene Duplication in Plants. *Plant Physiol.* **2016**, *171*, 2294–2316. [CrossRef]
32. Karaaslan, E.S.; Wang, N.; Faiss, N.; Liang, Y.; Montgomery, S.A.; Laubinger, S.; Berendzen, K.W.; Berger, F.; Breuninger, H.; Liu, C. Marchantia TCP transcription factor activity correlates with three-dimensional chromatin structure. *Nat. Plants* **2020**, *6*, 1250–1261. [CrossRef] [PubMed]
33. Ferrero, L.V.; Gastaldi, V.; Ariel, F.D.; Viola, I.L.; Gonzalez, D.H. Class I TCP proteins TCP14 and TCP15 are required for elongation and gene expression responses to auxin. *Plant Mol. Biol.* **2021**, *105*, 147–159. [CrossRef] [PubMed]
34. Tabarelli, M.; Malnoy, M.; Janik, K. Chasing Consistency: An Update of the TCP Gene Family of Malus x Domestica. *Genes* **2022**, *13*, 1696. [CrossRef] [PubMed]

35. Sinou, C.; Cardinal-McTeague, W.; Bruneau, A. Testing generic limits in Cercidoideae (Leguminosae): Insights from plastid and duplicated nuclear gene sequences: Phylogeny and generic limits in Cercidoideae. *Taxon* **2020**, *69*, 67–86. [CrossRef]
36. Fambrini, M.; Pugliesi, C. *CYCLOIDEA*-2 Clade Genes: Key Players in the Control of Floral Symmetry, Inflorescence Architecture, and Reproductive Organ Development. *Plant Mol. Biol. Rep.* **2017**, *35*, 20–36. [CrossRef]
37. Fambrini, M.; Pugliesi, C. Presence/absence of a CACTA transposon in the CYC2c gene of two genotypes of Helianthus × multiflorus cv. "Meteor" characterized by a radiate inflorescence with different shape of disk flower corollas. *Biologia* **2019**, *74*, 1675–1686. [CrossRef]
38. Fambrini, M.; Bernardi, R.; Pugliesi, C. Ray flower initiation in the Helianthus radula inflorescence is influenced by a functional allele of the HrCYC2c gene. *GENESIS* **2020**, *58*, e23401. [CrossRef]
39. Bukhari, G.; Zhang, J.; Stevens, P.F.; Zhang, W. Evolution of the process underlying floral zygomorphy development in pentapetalous angiosperms. *Am. J. Bot.* **2017**, *104*, 1846–1856. [CrossRef]
40. Sengupta, A.; Hileman, L.C. Novel Traits, Flower Symmetry, and Transcriptional Autoregulation: New Hypotheses From Bioinformatic and Experimental Data. *Front. Plant Sci.* **2018**, *9*, 1561. [CrossRef]
41. Kalisz, S.; Ree, R.H.; Sargent, R.D. Linking floral symmetry genes to breeding system evolution. *Trends Plant Sci.* **2006**, *11*, 568–573. [CrossRef]
42. Endress, P.K. Evolution of floral symmetry. *Curr. Opin. Plant Biol.* **2010**, *4*, 86–91. [CrossRef]
43. Coen, E.; Nugent, J.; Luo, D.; Bradley, D.; Cubas, P.; Chadwick, M.; Copsey, L.; Carpenter, R. Evolution of floral symmetry. *Philos. Trans. R. Soc. B* **1995**, *350*, 35–38. [CrossRef]
44. Zhang, J.; Stevens, P.F.; Zhang, W. Evolution and development of inflorescences and floral symmetry in Solanaceae. *Am. J. Bot.* **2022**, *109*, 746–767. [CrossRef] [PubMed]
45. Jiang, Y.; Moubayidin, L. Floral symmetry: The geometry of plant reproduction. *Emerg. Top. Life Sci.* **2022**, *6*, 259–269. [CrossRef] [PubMed]
46. Rosin, F.M.; Kramer, E.M. Old dogs, new tricks: Regulatory evolution in conserved genetic modules leads to novel morphologies in plants. *Dev. Biol.* **2009**, *332*, 25–35. [CrossRef] [PubMed]
47. Otero, A.; Fernandez-Mazuecos, M.; Vargas, P. Evolution in the Model Genus Antirrhinum Based on Phylogenomics of Topotypic Material. *Front. Plant Sci.* **2021**, *12*, 631178. [CrossRef]
48. Nicolas, M.; Cubas, P. The Role of TCP Transcription Factors in Shaping Flower Structure, Leaf Morphology, and Plant Architecture. In *Plant Transcription Factors*; Gonzalez, D.H., Ed.; Academic Press: Boston, MA, USA, 2016; pp. 249–267.
49. Preston, J.C.; Hileman, L.C.; Cubas, P. Reduce, reuse, and recycle: Developmental evolution of trait diversification. *Am. J. Bot.* **2011**, *98*, 397–403. [CrossRef]
50. Preston, J.C.; Hileman, L.C. Developmental genetics of floral symmetry evolution. *Trends Plant Sci.* **2009**, *14*, 147–154. [CrossRef]
51. Hsin, K.T.; Lu, J.Y.; Moller, M.; Wang, C.N. Gene duplication and relaxation from selective constraints of GCYC genes correlated with various floral symmetry patterns in Asiatic Gesneriaceae tribe Trichosporeae. *PLoS ONE* **2019**, *14*, e210054. [CrossRef] [PubMed]
52. Cubas, P.; Coen, E.; Zapater, J.M. Ancient asymmetries in the evolution of flowers. *Curr. Biol.* **2001**, *11*, 1050–1052. [CrossRef]
53. Preston, J.C.; Martinez, C.C.; Hileman, L.C. Gradual disintegration of the floral symmetry gene network is implicated in the evolution of a wind-pollination syndrome. *Proc. Natl. Acad. Sci. USA* **2011**, *108*, 2343–2348. [CrossRef]
54. Yang, X.; Zhao, X.G.; Li, C.Q.; Liu, J.; Qiu, Z.J.; Dong, Y.; Wang, Y.Z. Distinct Regulatory Changes Underlying Differential Expression of TEOSINTE BRANCHED1-CYCLOIDEA-PROLIFERATING CELL FACTOR Genes Associated with Petal Variations in Zygomorphic Flowers of *Petrocosmea* spp. of the Family Gesneriaceae. *Plant Physiol.* **2015**, *169*, 2138–2151. [CrossRef] [PubMed]
55. Berger, B.A.; Thompson, V.; Lim, A.; Ricigliano, V.; Howarth, D.G. Elaboration of bilateral symmetry across *Knautia macedonica* capitula related to changes in ventral petal expression of *CYCLOIDEA*-like genes. *Evodevo* **2016**, *7*, 8. [CrossRef] [PubMed]
56. Zhang, W.; Kramer, E.M.; Davis, C.C. Differential Expression of *CYC2* Genes and the Elaboration of Floral Morphologies in *Hiptage*, an Old World Genus of Malpighiaceae. *Int. J. Plant Sci.* **2016**, *177*, 551–558. [CrossRef]
57. Hsu, H.C.; Wang, C.N.; Liang, C.H.; Wang, C.C.; Kuo, Y.F. Association between Petal Form Variation and *CYC2*-like Genotype in a Hybrid Line of *Sinningia speciosa*. *Front. Plant Sci.* **2017**, *8*, 558. [CrossRef] [PubMed]
58. Citerne, H.L.; Pennington, R.T.; Cronk, Q.C. An apparent reversal in floral symmetry in the legume Cadia is a homeotic transformation. *Proc. Natl. Acad. Sci. USA* **2006**, *103*, 12017–12020. [CrossRef]
59. Feng, X.; Zhao, Z.; Tian, Z.; Xu, S.; Luo, Y.; Cai, Z.; Wang, Y.; Yang, J.; Wang, Z.; Weng, L.; et al. Control of petal shape and floral zygomorphy in *Lotus japonicus*. *Proc. Natl. Acad. Sci. USA* **2006**, *103*, 4970–4975. [CrossRef]
60. Citerne, H.L.; Luo, D.; Pennington, R.T.; Coen, E.; Cronk, Q.C. A phylogenomic investigation of CYCLOIDEA-like TCP genes in the Leguminosae. *Plant Physiol.* **2003**, *131*, 1042–1053. [CrossRef]
61. Ree, R.H.; Citerne, H.L.; Lavin, M.; Cronk, Q.C. Heterogeneous selection on LEGCYC paralogs in relation to flower morphology and the phylogeny of Lupinus (Leguminosae). *Mol. Biol. Evol.* **2004**, *21*, 321–331. [CrossRef]
62. Wang, J.; Wang, Y.; Luo, D. *LjCYC* genes constitute floral dorsoventral asymmetry in *Lotus japonicus*. *J. Integr. Plant Biol.* **2010**, *52*, 959–970. [CrossRef]
63. Woollacott, C.; Cronk, Q. The hooded mutant of Lathyrus odoratus (Fabaceae) is associated with a cycloidea gene mutation. *Botany* **2017**, *96*, 47–55. [CrossRef]

64. Woollacott, C.; Wang, L.; Beyer, S.; Walus, K.; Cronk, Q. *CYCLOIDEA* gene activity, local growth and curvature in the dorsal petal of Lathyrus odoratus (Fabaceae). *Bot. Lett.* **2018**, *166*, 64–69. [CrossRef]
65. Ojeda, D.I.; Jaen-Molina, R.; Santos-Guerra, A.; Caujape-Castells, J.; Cronk, Q. Temporal, but not spatial, changes in expression patterns of petal identity genes are associated with loss of papillate conical cells and the shift to bird pollination in Macaronesian Lotus (Leguminosae). *Plant Biol.* **2017**, *19*, 420–427. [CrossRef]
66. Feng, Z.; Liu, N.; Zhang, G.; Xu, S.; Gong, Y. Identification of Branch Related *CYL* Gene in Soybean and Analysis of Hormone Expression Pattern in Vegetable Soybean Species. *Mol. Plant Breed.* **2019**, *17*, 4865–4872. [CrossRef]
67. Zhao, Z.; Hu, J.; Chen, S.; Luo, Z.; Luo, D.; Wen, J.; Tu, T.; Zhang, D. Evolution of CYCLOIDEA-like genes in Fabales: Insights into duplication patterns and the control of floral symmetry. *Mol. Phylogenet. Evol.* **2019**, *132*, 81–89. [CrossRef]
68. Li, X.; Sun, M.; Jia, Y.; Qiu, D.; Peng, Q.; Zhuang, L. Genetic control of the lateral petal shape and identity of asymmetric flowers in mungbean (*Vigna radiata* L.). *Front. Plant Sci.* **2022**, *13*, 996239. [CrossRef]
69. Zoulias, N.; Duttke, S.; Garces, H.; Spencer, V.; Kim, M. The Role of Auxin in the Pattern Formation of the Asteraceae Flower Head (Capitulum). *Plant Physiol.* **2019**, *179*, 391–401. [CrossRef]
70. Hernandez, L.F.; Palmer, J.H. Regeneration of the sunflower capitulum after cylindrical wounding of the receptacle. *Am. J. Bot.* **1988**, *75*, 1253–1261. [CrossRef]
71. Broholm, S.K.; Tahtiharju, S.; Laitinen, R.A.; Albert, V.A.; Teeri, T.H.; Elomaa, P. A TCP domain transcription factor controls flower type specification along the radial axis of the *Gerbera* (*Asteraceae*) inflorescence. *Proc. Natl. Acad. Sci. USA* **2008**, *105*, 9117–9122. [CrossRef]
72. Garces, H.M.; Spencer, V.M.; Kim, M. Control of Floret Symmetry by *RAY3, SvDIV1B*, and *SvRAD* in the Capitulum of *Senecio vulgaris*. *Plant Physiol.* **2016**, *171*, 2055–2068. [CrossRef]
73. Sun, P.; Bao, Y.; Zhu, Y.; Huang, N.; Wang, X.; Wu, Z. Possible role of the *CYC2c* gene in the cornflower-like ray floret phenotype of Gaillardia cultivars. *J. Plant Res.* **2022**, *135*, 465–472. [CrossRef]
74. Tahtiharju, S.; Rijpkema, A.S.; Vetterli, A.; Albert, V.A.; Teeri, T.H.; Elomaa, P. Evolution and diversification of the *CYC/TB1* gene family in Asteraceae—A comparative study in Gerbera (Mutisieae) and sunflower (Heliantheae). *Mol. Biol. Evol.* **2012**, *29*, 1155–1166. [CrossRef]
75. Juntheikki-Palovaara, I.; Tahtiharju, S.; Lan, T.; Broholm, S.K.; Rijpkema, A.S.; Ruonala, R.; Kale, L.; Albert, V.A.; Teeri, T.H.; Elomaa, P. Functional diversification of duplicated *CYC2* clade genes in regulation of inflorescence development in *Gerbera hybrida* (Asteraceae). *Plant J.* **2014**, *79*, 783–796. [CrossRef]
76. Spencer, V.; Kim, M. Re"CYC"ling molecular regulators in the evolution and development of flower symmetry. *Semin. Cell Dev. Biol.* **2018**, *79*, 16–26. [CrossRef] [PubMed]
77. Chapman, M.A.; Leebens-Mack, J.H.; Burke, J.M. Positive selection and expression divergence following gene duplication in the sunflower *CYCLOIDEA* gene family. *Mol. Biol. Evol.* **2008**, *25*, 1260–1273. [CrossRef] [PubMed]
78. Fambrini, M.; Salvini, M.; Pugliesi, C. A transposon-mediate inactivation of a *CYCLOIDEA*-like gene originates polysymmetric and androgynous ray flowers in *Helianthus annuus*. *Genetica* **2011**, *139*, 1521–1529. [CrossRef] [PubMed]
79. Chapman, M.A.; Tang, S.; Draeger, D.; Nambeesan, S.; Shaffer, H.; Barb, J.G.; Knapp, S.J.; Burke, J.M. Genetic analysis of floral symmetry in Van Gogh's sunflowers reveals independent recruitment of *CYCLOIDEA* genes in the Asteraceae. *PLoS Genet.* **2012**, *8*, e1002628. [CrossRef] [PubMed]
80. Fambrini, M.; Salvini, M.; Basile, A.; Pugliesi, C. Transposon-dependent induction of Vincent van Gogh's sunflowers: Exceptions revealed. *Genesis* **2014**, *52*, 315–327. [CrossRef]
81. Mizzotti, C.; Fambrini, M.; Caporali, E.; Masiero, S.; Pugliesi, C. A *CYCLOIDEA* -like gene mutation in sunflower determines an unusual floret type able to produce filled achenes at the periphery of the pseudanthium. *Botany* **2015**, *93*, 171–181. [CrossRef]
82. He, Z.; Zeng, W.; Chen, W.; Wu, Y.; Wen, G.; Chen, X.; Wang, Q.; Zhou, J.; Li, Y.; Yang, Z.; et al. HaCYC2c regulating the heteromorphous development and functional differentiation of florets by recognizing HaNDUA2 in sunflower. *Plant Cell Rep.* **2022**, *41*, 1025–1041. [CrossRef]
83. Fambrini, M.; Bellanca, M.; Costa, M.M.; Usai, G.; Cavallini, A.; Pugliesi, C. Ligulate inflorescence of Helianthus x multiflorus, cv. Soleil d'Or, correlates with a mis-regulation of a CYCLOIDEA gene characterised by insertion of a transposable element. *Plant Biol.* **2018**, *20*, 956–967. [CrossRef] [PubMed]
84. Broholm, S. The Role of MADS and TCP Transcription Factors in Gerbera Hybrida Flower Development. Ph.D. Thesis, University of Helsinki, Helsinki, Finland, 2009.
85. Zhao, Y.; Broholm, S.K.; Wang, F.; Rijpkema, A.S.; Lan, T.; Albert, V.A.; Teeri, T.H.; Elomaa, P. TCP and MADS-Box Transcription Factor Networks Regulate Heteromorphic Flower Type Identity in *Gerbera hybrida*. *Plant Physiol.* **2020**, *184*, 1455–1468. [CrossRef] [PubMed]
86. Kim, M.; Cui, M.L.; Cubas, P.; Gillies, A.; Lee, K.; Chapman, M.A.; Abbott, R.J.; Coen, E. Regulatory genes control a key morphological and ecological trait transferred between species. *Science* **2008**, *322*, 1116–1119. [CrossRef] [PubMed]
87. Chen, K.; Po, C.; Hao, Y.; Feng, L.; Zhou, J.; Luan, S.; Liu, L.; Li, F.; Yuan, S.; Cui, M. Cloning and functional analysis of *CYCLOIDEA(CYC)*-like *SvRAY1* gene from *Senecio vulgaris*. *J. Zhejiang AF Univ.* **2021**, *38*, 1153–1160. [CrossRef]
88. Huang, D.; Sun, M.; Yuan, C.; Cheng, T.; Wang, J.; Zhang, Q. Isolation and functional analysis of *CYC2d* orthologous genes from several plants of the tribe Anthemideae. *J. Beijing For. Univ.* **2017**, *39*, 63–71. [CrossRef]

89. Yang, Y.; Sun, M.; Yuan, C.; Han, Y.; Zheng, T.; Cheng, T.; Wang, J.; Zhang, Q. Interactions between *WUSCHEL* and *CYC2*-like Transcription Factors in Regulating the Development of Reproductive Organs in *Chrysanthemum morifolium*. *Int. J. Mol. Sci.* **2019**, *20*, 1276. [CrossRef]
90. Pu, Y.; Huang, H.; Wen, X.; Lu, C.; Zhang, B.; Gu, X.; Qi, S.; Fan, G.; Wang, W.; Dai, S. Comprehensive transcriptomic analysis provides new insights into the mechanism of ray floret morphogenesis in chrysanthemum. *BMC Genom.* **2020**, *21*, 728. [CrossRef]
91. Yuan, C.; Huang, D.; Yang, Y.; Sun, M.; Cheng, T.; Wang, J.; Pan, H.; Zhang, Q. CmCYC2-like transcription factors may interact with each other or bind to the promoter to regulate floral symmetry development in Chrysanthemum morifolium. *Plant Mol. Biol.* **2020**, *103*, 159–171. [CrossRef]
92. Liu, H.; Sun, M.; Pan, H.; Cheng, T.; Wang, J.; Zhang, Q. Two *Cyc2CL* transcripts (*Cyc2CL-1* and *Cyc2CL-2*) may play key roles in the petal and stamen development of ray florets in chrysanthemum. *BMC Plant Biol.* **2021**, *21*, 105. [CrossRef]
93. Chen, J.; Shen, C.Z.; Guo, Y.P.; Rao, G.Y. Patterning the Asteraceae Capitulum: Duplications and Differential Expression of the Flower Symmetry *CYC2*-Like Genes. *Front. Plant Sci.* **2018**, *9*, 551. [CrossRef]
94. Wen, X.; Qi, S.; Huang, H.; Wu, X.; Zhang, B.; Fan, G.; Yang, L.; Hong, Y.; Dai, S. The expression and interactions of ABCE-class and *CYC2*-like genes in the capitulum development of *Chrysanthemum lavandulifolium* and *C.* × *morifolium*. *Plant Growth Regul.* **2019**, *88*, 205–214. [CrossRef]
95. Kironji, G.S. Patterning of the *Chrysanthemum* Inflorescence Roles of B-Class and *CYC2* Subclade Genes. Ph.D. Thesis, Nanjing Agricultural University, Nanjing, China, 2019.
96. Yang, Y.; Yang, Y.J.; Yuan, C.Q.; Zhang, Q.X. Cloning and Expression Analysis of CmTCP7 in Chrysanthemum morifolium. *Acta Bot. Boreali-Occident. Sin.* **2019**, *39*, 595–602. [CrossRef]
97. Shen, C.Z.; Chen, J.; Zhang, C.J.; Rao, G.Y.; Guo, Y.P. Dysfunction of *CYC2g* is responsible for the evolutionary shift from radiate to disciform flowerheads in the *Chrysanthemum* group (Asteraceae: Anthemideae). *Plant J.* **2021**, *106*, 1024–1038. [CrossRef] [PubMed]
98. Zhang, F.; Lu, C.; Qi, S.; Dai, S. Difference Analysis of *ClCYC2*-Like Genes Expression and DNA Methylation between the Two Types of Florets in *Chrysanthemum lavandulifolium*. *J. Plant Growth Regul.* **2022**, *41*, 1316–1330. [CrossRef]
99. Elomaa, P.; Zhao, Y.; Zhang, T. Flower heads in Asteraceae-recruitment of conserved developmental regulators to control the flower-like inflorescence architecture. *Hortic. Res.* **2018**, *5*, 36. [CrossRef]
100. Zhong, J.; Kellogg, E.A. Duplication and expression of *CYC2*-like genes in the origin and maintenance of corolla zygomorphy in Lamiales. *New Phytol.* **2014**, *205*, 852–868. [CrossRef]
101. Ha, Y.H.; Kim, C.; Choi, K.; Kim, J.H. Molecular Phylogeny and Dating of Forsythieae (Oleaceae) Provide Insight into the Miocene History of Eurasian Temperate Shrubs. *Front. Plant Sci.* **2018**, *9*, 99. [CrossRef]
102. Smyth, D.R. Evolution and genetic control of the floral ground plan. *New Phytol.* **2018**, *220*, 70–86. [CrossRef] [PubMed]
103. Ajani, Y.; Jamzad, Z.; Claßen-Bockhoff, R. Floral biology in the endemic Iranian Salvia majdae—Implications for taxonomy, character evolution and conservation. *Flora* **2021**, *287*, 151986. [CrossRef]
104. Wessinger, C.; Hileman, L. Parallelism in Flower Evolution and Development. *Annu. Rev. Ecol. Evol. Syst.* **2020**, *51*, 387–408. [CrossRef]
105. Zhong, J.; Kellogg, E.A. Stepwise evolution of corolla symmetry in CYCLOIDEA2-like and RADIALIS-like gene expression patterns in Lamiales. *Am. J. Bot.* **2015**, *102*, 1260–1267. [CrossRef] [PubMed]
106. Poulin, V.; Amesefe, D.; Gonzalez, E.; Alexandre, H.; Joly, S. Testing candidate genes linked to corolla shape variation of a pollinator shift in Rhytidophyllum (Gesneriaceae). *PLoS ONE* **2022**, *17*, e267540. [CrossRef] [PubMed]
107. Zhong, J.; Preston, J.C.; Hileman, L.C.; Kellogg, E.A. Repeated and diverse losses of corolla bilateral symmetry in the Lamiaceae. *Ann. Bot.* **2017**, *119*, 1211–1223. [CrossRef] [PubMed]
108. Hileman, L.C.; Cubas, P. An expanded evolutionary role for flower symmetry genes. *J. Biol.* **2009**, *8*, 90. [CrossRef] [PubMed]
109. Neel, M.C. Conservation implications of the reproductive ecology of Agalinis acuta (Scrophulariaceae). *Am. J. Bot.* **2002**, *89*, 972–980. [CrossRef] [PubMed]
110. Cubas, P.; Vincent, C.; Coen, E. An epigenetic mutation responsible for natural variation in floral symmetry. *Nature* **1999**, *401*, 157–161. [CrossRef] [PubMed]
111. Hileman, L.C.; Kramer, E.M.; Baum, D.A. Differential regulation of symmetry genes and the evolution of floral morphologies. *Proc. Natl. Acad. Sci. USA* **2003**, *100*, 12814–12819. [CrossRef]
112. Su, S.; Xiao, W.; Guo, W.; Yao, X.; Xiao, J.; Ye, Z.; Wang, N.; Jiao, K.; Lei, M.; Peng, Q.; et al. The CYCLOIDEA-RADIALIS module regulates petal shape and pigmentation, leading to bilateral corolla symmetry in *Torenia fournieri* (Linderniaceae). *New Phytol.* **2017**, *215*, 1582–1593. [CrossRef]
113. Clark, J.I.; Coen, E.S. The cycloidea gene can respond to a common dorsoventral prepattern in Antirrhinum. *Plant J.* **2002**, *30*, 639–648. [CrossRef]
114. Preston, J.C.; Kost, M.A.; Hileman, L.C. Conservation and diversification of the symmetry developmental program among close relatives of snapdragon with divergent floral morphologies. *New Phytol.* **2009**, *182*, 751–762. [CrossRef]
115. Gubitz, T.; Caldwell, A.; Hudson, A. Rapid molecular evolution of CYCLOIDEA-like genes in Antirrhinum and its relatives. *Mol. Biol. Evol.* **2003**, *20*, 1537–1544. [CrossRef] [PubMed]
116. Corley, S.B.; Carpenter, R.; Copsey, L.; Coen, E. Floral asymmetry involves an interplay between TCP and MYB transcription factors in *Antirrhinum*. *Proc. Natl. Acad. Sci. USA* **2005**, *102*, 5068–5073. [CrossRef] [PubMed]

117. Theissen, G. Evolutionary developmental genetics of floral symmetry: The revealing power of Linnaeus' monstrous flower. *Bioessays* **2000**, *22*, 209–213. [CrossRef]
118. Kondo, M.; Tanikawa, N.; Nishijima, T. Mutation of CYCLOIDEA expands variation of dorsal-ventral flower asymmetry expressed as a pigmentation pattern in Torenia fournieri cultivars. *Hortic. J.* **2020**, *89*, 481–487. [CrossRef]
119. Yang, X.; Cui, H.; Yuan, Z.; Wang, Y. Significance of consensus *CYC*-binding sites found in the promoters of both *ChCYC* and *ChRAD* genes in *Chirita heterotricha* (*Gesneriaceae*). *J. Syst. Evol.* **2010**, *48*, 249–256. [CrossRef]
120. Yang, X.; Pang, H.B.; Liu, B.L.; Qiu, Z.J.; Gao, Q.; Wei, L.; Dong, Y.; Wang, Y.Z. Evolution of double positive autoregulatory feedback loops in *CYCLOIDEA2* clade genes is associated with the origin of floral zygomorphy. *Plant Cell* **2012**, *24*, 1834–1847. [CrossRef]
121. Hsu, H.J.; He, C.W.; Kuo, W.H.; Hsin, K.T.; Lu, J.Y.; Pan, Z.J.; Wang, C.N. Genetic analysis of floral symmetry transition in African violet suggests the involvement of trans-acting factor for CYCLOIDEA expression shifts. *Front. Plant Sci.* **2018**, *9*, 1008. [CrossRef]
122. Liu, J.; Jie, W.; Yang, X.; Wang, Y. Regulatory pathways of CYC-like genes in patterning floral zygomorphy exemplified in Chirita pumila. *J. Syst. Evol.* **2020**, *59*, 567–580. [CrossRef]
123. Yang, M.; Xu, J.; Pang, J. Flower Shape Changes of African Violets Caused by *LjCYC1* Gene in *Lotus japonicus*. *Acta Horticulturae Sinica* **2020**, *47*, 708–716. [CrossRef]
124. Liu, J.; Wang, J.J.; Wu, J.; Wang, Y.; Liu, Q.; Liu, F.P.; Yang, X.; Wang, Y.Z. An Optimized Transformation System and Functional Test of CYC-Like TCP Gene CpCYC in Chirita pumila (Gesneriaceae). *Int. J. Mol. Sci.* **2021**, *22*, 4544. [CrossRef]
125. Ferraro, B.J. Examining the roles of *CYCLOIDEA*, *RADIALIS* and *DIVARICATA* in driving the evolution of flower shape Californian *Diplacus pictus* (Curran ex Greene) Nesom (*Phrymaceae*). Master's Thesis, California State University, Long Beach, CA, USA, 2014.
126. Lucibelli, F.; Valoroso, M.C.; Aceto, S. Radial or bilateral? The molecular basis of floral symmetry. *Genes* **2020**, *11*, 395. [CrossRef] [PubMed]
127. Lucibelli, F.; Valoroso, M.; Theißen, G.; Nolden, S.; Mondragon Palomino, M.; Aceto, S. Extending the toolkit for beauty: Differential co-expression of DROOPING LEAF-like and class B MADS-Box genes during Phalaenopsis flower development. *Int. J. Mol. Sci.* **2021**, *22*, 7025. [CrossRef] [PubMed]
128. Madrigal, Y.; Pabon-Mora, N.; Alzate, J. Unraveling the genetic basis of floral symmetry variation in monocots with an emphasis on Asparagales. In Proceedings of the World Orchid Conference 2017, Guayaquil, Ecuador, 8–12 November 2017.
129. Klepikova, A.V.; Kasianov, A.S.; Ezhova, M.A.; Penin, A.A.; Logacheva, M.D. Transcriptome atlas of Phalaenopsis equestris. *PeerJ* **2021**, *9*, e12600. [CrossRef]
130. Lin, Y.F.; Chen, Y.Y.; Hsiao, Y.Y.; Shen, C.Y.; Hsu, J.L.; Yeh, C.M.; Mitsuda, N.; Ohme-Takagi, M.; Liu, Z.J.; Tsai, W.C. Genome-wide identification and characterization of TCP genes involved in ovule development of *Phalaenopsis equestris*. *J. Exp. Bot.* **2016**, *67*, 5051–5066. [CrossRef] [PubMed]
131. Liu, D.K.; Zhang, C.; Zhao, X.; Ke, S.; Li, Y.; Zhang, D.; Zheng, Q.; Li, M.H.; Lan, S.; Liu, Z.J. Genome-wide analysis of the TCP gene family and their expression pattern in Cymbidium goeringii. *Front. Plant Sci.* **2022**, *13*, 1068969. [CrossRef]
132. Madrigal, Y.; Alzate, J.F.; Pabon-Mora, N. Evolution and expression patterns of TCP genes in Asparagales. *Front. Plant Sci.* **2017**, *8*, 9. [CrossRef]
133. Zhang, J.; Stevens, P.; Zhang, W. Evolution of floral zygomorphy in androecium and corolla in Solanaceae. *J. Syst. Evol.* **2017**, *55*, 581–590. [CrossRef]
134. Shimada, A.; Kimura, Y. Influence of glyphosate on flower morphogenesis and pigmentation in Petunia hybrida. *Z. Naturforsch. C* **2006**, *61*, 578–582. [CrossRef]
135. Zou, S.; Wang, H.; Yu, Y.; Ma, J.; Guo, Y.; Li, M. Studies on molecular cloning, expression pattern of ECE Clade TCP genes in Petunia. *Acta Hortic. Sin.* **2013**, *40*, 307–316. [CrossRef]
136. Zhang, S.; Zhou, Q.; Chen, F.; Wu, L.; Liu, B.; Li, F.; Zhang, J.; Bao, M.; Liu, G. Genome-Wide Identification, Characterization and Expression Analysis of TCP Transcription Factors in Petunia. *Int. J. Mol. Sci.* **2020**, *21*, 6594. [CrossRef]
137. Sengupta, A.; Hileman, L.C. A CYC-RAD-DIV-DRIF interaction likely pre-dates the origin of floral monosymmetry in Lamiales. *Evodevo* **2022**, *13*, 3. [CrossRef] [PubMed]
138. Preston, J.C.; Hileman, L.C. Parallel evolution of TCP and B-class genes in Commelinaceae flower bilateral symmetry. *Evodevo* **2012**, *3*, 6. [CrossRef] [PubMed]
139. Classen-Bockhoff, R.; Ruonala, R.; Bull-Herenu, K.; Marchant, N.; Albert, V.A. The unique pseudanthium of *Actinodium* (Myrtaceae)—Morphological reinvestigation and possible regulation by *CYCLOIDEA*-like genes. *Evodevo* **2013**, *4*, 8. [CrossRef] [PubMed]
140. Horn, S.; Pabon-Mora, N.; Theuss, V.S.; Busch, A.; Zachgo, S. Analysis of the CYC/TB1 class of TCP transcription factors in basal angiosperms and magnoliids. *Plant J.* **2015**, *81*, 559–571. [CrossRef] [PubMed]
141. Bartlett, M.E.; Specht, C.D. Changes in expression pattern of the *teosinte branched1*-like genes in the Zingiberales provide a mechanism for evolutionary shifts in symmetry across the order. *Am. J. Bot.* **2011**, *98*, 227–243. [CrossRef] [PubMed]
142. Yu, Q.; Tian, X.; Lin, C.; Specht, C.D.; Liao, J. Expression and function studies of *CYC/TB1*-like genes in the asymmetric flower *Canna* (Cannaceae, Zingiberales). *Front. Plant Sci.* **2020**, *11*, 580576. [CrossRef] [PubMed]
143. Iljinska, A.P. The range of the morphological features of Brassicaceae s. l.: Inflorescence, flower. *Ukr. Bot. J.* **2015**, *72*, 122–134. [CrossRef]
144. Nikolov, L.A. Brassicaceae flowers: Diversity amid uniformity. *J. Exp. Bot.* **2019**, *70*, 2623–2635. [CrossRef]

145. Reichling, J.; Saller, R. Iberis amara L. (Bittere Schleifenblume)—Profil einer Heilpflanze. *Complement. Med. Res.* **2002**, *9*, 21–33. [CrossRef]
146. Busch, A.; Zachgo, S. Control of corolla monosymmetry in the Brassicaceae Iberis amara. *Proc. Natl. Acad. Sci. USA* **2007**, *104*, 16714–16719. [CrossRef]
147. Busch, A.; Horn, S.; Zachgo, S. Differential transcriptome analysis reveals insight into monosymmetric corolla development of the crucifer Iberis amara. *BMC Plant Biol.* **2014**, *14*, 285. [CrossRef] [PubMed]
148. Busch, A.; Horn, S.; Muhlhausen, A.; Mummenhoff, K.; Zachgo, S. Corolla monosymmetry: Evolution of a morphological novelty in the Brassicaceae family. *Mol. Biol. Evol.* **2012**, *29*, 1241–1254. [CrossRef] [PubMed]
149. Jabbour, F.; Cossard, G.; Le Guilloux, M.; Sannier, J.; Nadot, S.; Damerval, C. Specific duplication and dorsoventrally asymmetric expression patterns of Cycloidea-like genes in zygomorphic species of Ranunculaceae. *PLoS ONE* **2014**, *9*, e95727. [CrossRef] [PubMed]
150. Zalko, J.; Frachon, S.; Morel, A.; Deroin, T.; Espinosa Moreno, F.; Xiang, K.; Wang, W.; Zhang, W.; Lang, S.; Dixon, L.; et al. Floral organogenesis and morphogenesis of Staphisagria (Ranunculaceae): Implications for the evolution of synorganized floral structures in Delphinieae. *Int. J. Plant Sci.* **2020**, *182*, 59–70. [CrossRef]
151. Kolsch, A.; Gleissberg, S. Diversification of CYCLOIDEA-like TCP genes in the basal eudicot families Fumariaceae and Papaveraceae s.str. *Plant Biol.* **2006**, *8*, 680–687. [CrossRef]
152. Damerval, C.; Le Guilloux, M.; Jager, M.; Charon, C. Diversity and evolution of CYCLOIDEA-like TCP genes in relation to flower development in Papaveraceae. *Plant Physiol.* **2007**, *143*, 759–772. [CrossRef]
153. Zhao, Y.; Pfannebecker, K.; Dommes, A.B.; Hidalgo, O.; Becker, A.; Elomaa, P. Evolutionary diversification of CYC/TB1-like TCP homologs and their recruitment for the control of branching and floral morphology in Papaveraceae (basal eudicots). *New Phytol.* **2018**, *220*, 317–331. [CrossRef]
154. Pabon-Mora, N.; Madrigal, Y.; Alzate, J.F.; Ambrose, B.A.; Ferrandiz, C.; Wanke, S.; Neinhuis, C.; Gonzalez, F. Evolution of Class II TCP genes in perianth bearing Piperales and their contribution to the bilateral calyx in Aristolochia. *New Phytol.* **2020**, *228*, 752–769. [CrossRef]
155. Zhang, W.; Steinmann, V.W.; Nikolov, L.; Kramer, E.M.; Davis, C.C. Divergent genetic mechanisms underlie reversals to radial floral symmetry from diverse zygomorphic flowered ancestors. *Front. Plant Sci.* **2013**, *4*, 302. [CrossRef]
156. Berger, B.A.; Ricigliano, V.A.; Savriama, Y.; Lim, A.; Thompson, V.; Howarth, D.G. Geometric morphometrics reveals shifts in flower shape symmetry and size following gene knockdown of CYCLOIDEA and ANTHOCYANIDIN SYNTHASE. *BMC Plant Biol.* **2017**, *17*, 205. [CrossRef]
157. Hileman, L.C. Trends in flower symmetry evolution revealed through phylogenetic and developmental genetic advances. *Philos. Trans. R. Soc. B* **2014**, *369*, 20130348. [CrossRef] [PubMed]
158. Hoshino, Y.; Igarashi, T.; Ohshima, M.; Shinoda, K.; Murata, N.; Kanno, A.; Nakano, M. Characterization of CYCLOIDEA-like genes in controlling floral zygomorphy in the monocotyledon Alstroemeria. *Sci. Hortic.* **2014**, *169*, 6–13. [CrossRef]
159. Letunic, I.; Bork, P. Interactive Tree Of Life (iTOL): An online tool for phylogenetic tree display and annotation. *Bioinformatics* **2007**, *23*, 127–128. [CrossRef] [PubMed]

current issues in molecular biology

MDPI

Article

Genome-Wide Identification of DUF26 Domain-Containing Genes in Dongxiang Wild Rice and Analysis of Their Expression Responses under Submergence

Cheng Huang †, Dianwen Wang †, Hongping Chen, Wei Deng, Dazhou Chen, Ping Chen * and Jilin Wang *

Rice National Engineering Research Center (Nanchang), Rice Research Institute, Jiangxi Academy of Agricultural Sciences, Nanchang 330200, China; chenghuang@webmail.hzau.edu.cn (C.H.); dianwen1989@126.com (D.W.); 13970920363@139.com (H.C.); dw19710330@163.com (W.D.); cdz288@163.com (D.C.)

* Correspondence: cp87090379@163.com (P.C.); wangjilin1982@163.com (J.W.); Tel.: +86-185-7906-9996 (P.C.); +86-133-8753-2293 (J.W.)

† These authors contributed equally to this work.

Abstract: The DUF26 domain-containing protein is an extracellular structural protein, which plays an important role in signal transduction. Dongxiang wild rice (*Oryza rufipogon* Griff.) is the northernmost common wild rice in China. Using domain analysis, 85 DUF26 domain-containing genes were identified in Dongxiang wild rice (DXWR) and further divided into four categories. The DUF26 domain-containing genes were unevenly distributed on chromosomes, and there were 18 pairs of tandem repeats. Gene sequence analysis showed that there were significant differences in the gene structure and motif distribution of the DUF26 domain in different categories. Motifs 3, 8, 9, 13, 14, 16, and 18 were highly conserved in all categories. It was also found that there were eight plasmodesmata localization proteins (PDLPs) with a unique motif 19. Collinearity analysis showed that DXWR had a large number of orthologous genes with wheat, maize, sorghum and zizania, of which 17 DUF26 domain-containing genes were conserved in five gramineous crops. Under the stress of anaerobic germination and seedling submergence treatment, 33 DUF26 domain-containing genes were differentially expressed in varying degrees. Further correlation analysis with the expression of known submergence tolerance genes showed that these DUF26 domain-containing genes may jointly regulate the submergence tolerance process with these known submergence tolerance genes in DXWR.

Keywords: Dongxiang wild rice (*Oryza rufipogon* Griff.); DUF26 domain; expression mode; submergence tolerance

Citation: Huang, C.; Wang, D.; Chen, H.; Deng, W.; Chen, D.; Chen, P.; Wang, J. Genome-Wide Identification of DUF26 Domain-Containing Genes in Dongxiang Wild Rice and Analysis of Their Expression Responses under Submergence. *Curr. Issues Mol. Biol.* **2022**, *44*, 3351–3363. https://doi.org/10.3390/cimb44080231

Academic Editor: Shimeles Tilahun

Received: 15 June 2022
Accepted: 25 July 2022
Published: 27 July 2022

1. Introduction

Signal transduction mechanisms exist widely in plants and other eukaryotes, and are involved in the regulation of cell functions, the coordination of cell–cell, and the exchange of information between cells and the environment. Proteins with extracellular domains and large gene families encoding secretory proteins in plants play an important role in sensing environmental changes and development through signal transduction mechanisms [1–3]. Receptor like kinase (RLK) with an extracellular domain is involved in signal sensing, while intracellular kinase domain transduces signal to substrate protein. RLK plays an important role in stress response, hormone signal transduction, cell wall monitoring, and plant development.

The DUF26 domain (PF01657) belongs to the extracellular domain, and its core contains a conserved cysteine motif (C-8X-C-2X-C), which exists in three plant proteins [4,5]. The first is a cysteine rich receptor like secreted protein (CRRSP). The most typical CRRSP is Gnk2 from Ginkgo biloba leaves, which has an antifungal activity as mannose binding lectin in vitro [5]. Two maize CRRSPs have also been shown to bind mannose and participate in the defense against fungal pathogens [6]. The second is a cysteine rich receptor kinase (CRK) that has a typical DUF26 structure in the extracellular region, forming a large RLK subgroup in plants and playing a role in the response of Arabidopsis to stress [7–13]. The third DUF26 domain-containing protein is plasmodesmata localized protein (PDLP). PDLP contains two DUF26 domains and a transmembrane helix in its extracellular domain, but lacks a kinase domain. They are related to plasmodesmata and participate in intercellular signal transduction, pathogen response, systemic signal transduction, and callose deposition control [14,15]. However, the specific biochemical function of the DUF26 domain-containing gene in plants is still unclear.

Rice is one of the most important food crops in the world, and nearly half of the world's population lives on rice [16]. Submergence stress is one of the most serious natural disasters facing mankind, and has become an important factor limiting the stable yield and yield increase of rice [17]. DXWR is a common wild rice with the highest latitude and the northernmost distribution, which is rich in disease and insect resistance genes and submergence, cold, drought, and barren resistance genes [18]. Previous studies have shown that the DUF26 domain-containing gene plays a role in signal transduction and stress response. However, the research on the molecular mechanism of the DUF26 domain-containing gene response to submergence stress in DXWR is still very limited. The whole genome recognition and characterization of DUF26 domain-containing genes in DXWR is helpful to better explore the molecular mechanism of biotic or abiotic stress resistance. The high-quality sequencing of DXWR has been completed, which provides an excellent opportunity for us to identify and dissect the DUF26 domain-containing genes of DXWR. In this study, the DUF26 domain-containing gene in DXWR were comprehensively studied including phylogeny, chromosome location, gene structure, protein motif, evolution, expression pattern analysis, and correlation analysis of submergence stress. In general, the systematic analysis of DUF26 domain-containing proteins and their expression regulation mode under submergence tolerance has laid the foundation for further exploring the function of DUF26 domain-containing genes in DXWR.

2. Materials and Methods

2.1. Identification of DUF26 Domain-Containing Genes in DXWR

The sequence information of DXWR comes from the gene annotation website (http://www.ricegermplasmgenome.com/, accessed on 20 May 2022), and the hidden Markov model of DUF26 domain-containing genes (PF01657) was used to search and screen candidate genes in hmmer3 [19,20]. Then, it was passed through the NCBI conservative domain database (CDD; https://www.ncbi.nlm.nih.gov/cdd, accessed on 26 May 2022) and SMART database (http://smart.embl-heidelberg.de/, accessed on 12 June 2022) to confirm the integrity of the protein domain [21,22]. On this basis, pkinase_Tyr (PF07714) searched for proteins containing kinase domains, and identified candidate protein domains through the PFAM, SMART, and NCBI-CDD protein databases. The isoelectric point and molecular weight of the DUF26 domain-containing genes were obtained from the ExPASY website [23]. The online analysis software Wolf Psort (http://wolfpsort.hgc.jp/, accessed on 16 June 2022) was used to predict the subcellular localization of the DUF26 domain-containing protein in DXWR.

2.2. Analysis of the Main Characteristics of DUF26 Domain-Containing Genes in DXWR

Using MEGA11, the phylogenetic tree of DXWR was constructed by neighbor joining (NJ) and bootstrap repeats set to 1000 [24]. The distribution of all DUF26 domain- containing genes on the chromosome of DXWR was analyzed and visualized by TBtools [25]. The characteristic motif of the DUF26 domain-containing gene was determined by MEME (http://meme-suite.org/tools/meme, accessed on 12 June 2022), the base sequence number was 20, and the sequence of site distribution was 0 or 1 occurrence for each sequence [26].

2.3. *Repeat Events and Collinearity Analysis of DUF26 Domain-Containing Genes*

The identification of serial repeat events was analyzed by the multilinear analysis tool MCscanX [27]. The genome data of Wheat (Triticum_aestivum.IWGSC.52), Sorghum (Sorghum_bicolor.Sorghum_bicolor_NCBIv3.52), and Maize (Zea_mays.Zm-B73- REFERENCE-NAM-5.0.52) were all from the Ensembl Plant genome website (https://plants.ensembl.org/index.html, accessed on 20 June 2022) and Zizania (Zlat_genome_v1) was downloaded from the website (https://ngdc.cncb.ac.cn/, accessed on 20 June 2022). The colinear relationship between DXWR and the gramineous crops was analyzed by JCVI [28].

2.4. *Expression Data of DUF26 Domain-Containing Genes under Submergence*

The expression data of all genes were from the transcriptome data of submergence stress tolerance (unpublished), in which AG is anaerobic germination, SS is seedling submergence, and CK is normal growth. Anaerobic germination and normal germination treatment: the dxwr seeds were shelled, then germinated in 20 cm deep and 5 mm normal water, respectively, and then the embryo tissues germinated for 2D and 4D were taken successively. Submergence treatment at the seedling stage: the seedlings were normally grown to the leaf stage (6LS), and all of them were submerged in the deep water of 50 cm. The aboveground tissues of 0, 3, and 5 days of submergence were taken, respectively. The aboveground tissues included the stem base area of 5 mm in the area of the internode, node, stem tip, and leaf base. All of the above materials were grown in incubators under the conditions of 28 °C normal temperature, 14 h of light, and 10 h of darkness. All samples were quickly frozen in liquid nitrogen and stored at −80 °C until the RNA was extracted (in order to avoid the difference of gene expression between samples caused by circadian rhythm, submergence treatment, and sample collection were conducted at 4 pm).

2.5. *Correlation Analysis between DUF26 Domain-Containing Genes and Submergence Tolerance Genes*

In order to predict the function of DUF26 domain-containing genes, submergence tolerant genes with known functions include *RAmy3D* (*LOC_Os08g36910*), *CIPK14/15* (*LOC_Os12g02200/LOC_Os11g02240*), *G3PDH* (*LOC_Os02g38920*), *MPK3* (*LOC_ Os03g17700*), *OsERF66/67* (*LOC_Os03g22170/LOC_Os07g47790*), *PPDK* (*LOC_ Os03g31750*), *SLR1* (*LOC_Os03g49990*), *SLRL1* (*LOC_ Os01g45860*), and *SnRK1A* (*LOC_ Os05g45420*). The correlation analysis between the DUF26 domain-containing genes and submergence tolerant genes for functional identification was calculated using the cor function in R language and default Pearson correlation coefficient.

3. Results

3.1. *Identification of DUF26 Domain-Containing Genes in DXWR*

Hmmer3 was used to search all of the DUF26 domain-containing genes in the local DXWR database. After de-redundancy, candidate DUF26 domain-containing genes were obtained. Finally, 85 DUF26 domain-containing genes were identified, of which 10 did not contain the pkinase-cysteine rich receptor like secretory proteins (CRRSPs) of the Tyr domain and 24 plasmodesmata localization proteins (PDLPs), 51 of which contain pkinase-cysteine rich receptor protein kinase (CRKs) of the Tyr domain. The cysteine rich receptor protein kinases (CRKs) can be further divided into two categories: the cysteine rich receptor protein kinases (sdCRKs) containing a single DUF26 domain and the cysteine rich receptor protein kinases (ddCRKs) containing double DUF26 domains are 6 and 45, respectively (Table S1). In addition, the physical and chemical properties of 85 DUF26 domain-containing genes including the molecular weight, isoelectric point, gene location on chromosome, and subcellular localization were also analyzed (Table S1).

3.2. *Chromosomal Localization of DUF26 Domain-Containing Genes in DXWR*

The chromosomal mapping analysis showed that 85 DUF26 domain-containing genes were present on all chromosomes except for chromosome 9. The results showed that the DUF26 domain-containing genes were unevenly distributed on the chromosomes, with only one gene on Chr6 and 35 genes on Chr7. It is worth noting that those belonging to the same group in the phylogenetic tree were not mapped on the same chromosome, but scattered in different positions on the same chromosome or exist on different chromosomes (Figure 1). In addition, the genome collinearity analysis of DXWR showed that 18 pairs of DUF26 domain-containing genes were tandem repeats, indicating that tandem repeats played an important role in the expansion of DUF26 domain-containing genes (Figure 1).

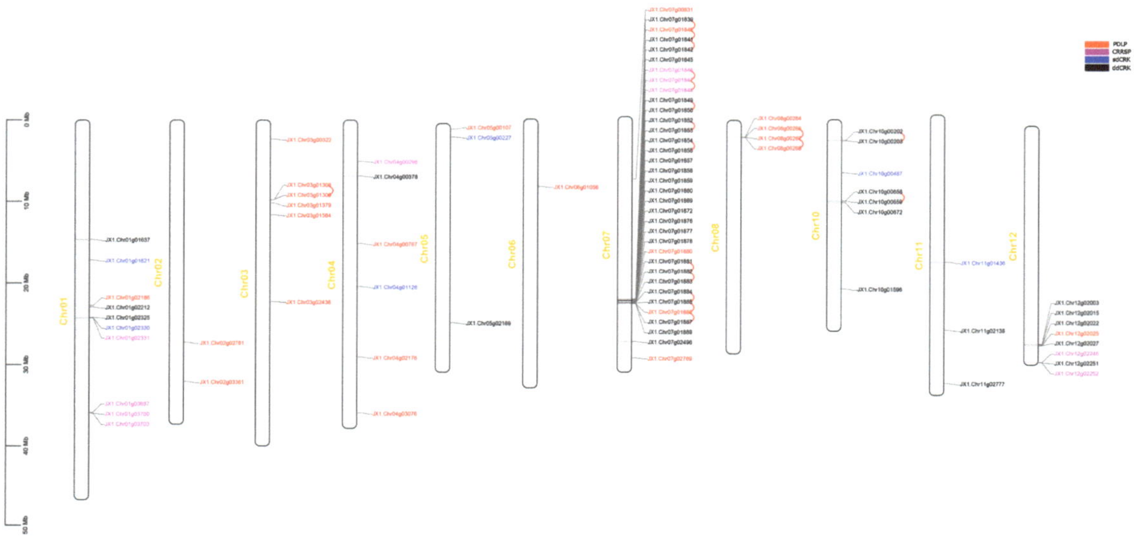

Figure 1. The distribution of DUF26 domain-containing genes on chromosomes. Different colors represent different types of DUF26 containing genes, in which red represents PDLPs, magenta represents CRRSPs, blue represents sdCRKs, black represents ddCRKs, and the red connecting line represents tandem repeat gene pairs.

3.3. *Motifs and Structural Composition of DUF26 Domain-Containing Genes in DXWR*

The 85 DUF26 domain-containing genes were divided into four groups, and the phylogenetic tree was constructed by using amino acid sequences. At the same time, we analyzed the conserved motifs of the DUF26 domain-containing genes to explore various biological functions of the protein domain. Twenty conserved motifs were identified, among which Motifs 3, 8, 9, 13, 14, 16, and 18 were conserved in four groups of DUF26 domain-containing genes (Figure 2). Motifs 11 and 20 only appeared in three other DUF26 domain-containing genes except for CRRSPs. Meanwhile, 47 of the 51 cysteine rich receptor protein kinases (CRKs) contained conserved Motif 1, 2, 4, 5, 6, 7, 10, 12, 15, and 17. In addition, Motif 19 was found only in eight plasmodesmata localization proteins (PDLPs). Moreover, the DUF26 domain-containing gene was also analyzed, and it was found that the location of this domain was highly consistent with that of the motif. Specifically, the DUF26 domain-containing genes in members of the same class had similar conserved motifs, and the motifs of the members of different classes were quite different. Whether these motif composition differences affect their biological functions needs to be further studied through biological experiments (Figure 2). At the same time, the exon/intron patterns of 85 DUF26 domain-containing genes were also studied (Figure 2). The results showed that the number of exons of the DUF26 domain-containing gene ranged from 1 to 11, and 74 of them contained at least two exons. The gene structure of the DUF26 domain-containing in the CRKs group was more complex than that of PDLPs and CRRSPs.

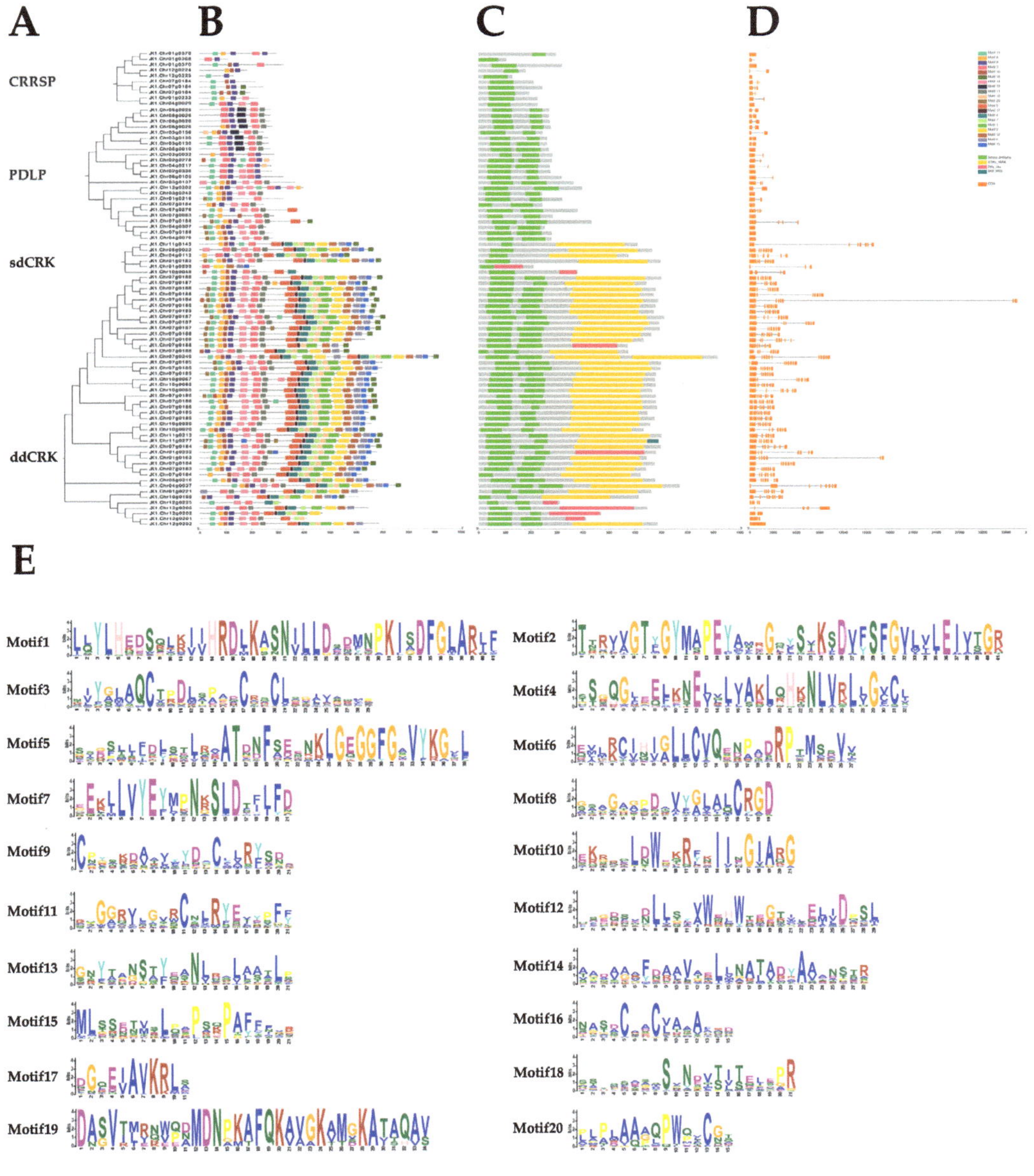

Figure 2. The phylogenetic relationship, structure, domain, and conserved motif analysis of DUF26 domain-containing genes in DXWR. (**A**) The phylogenetic tree was constructed based on the full-length sequences of DXWR DUF26 domain-containing proteins using MEGA12. (**B**) The distribution of motifs of the MAPK proteins. The conserved motifs of DUF26 domain-containing proteins were determined by MEME and visualized by TBtools. (**C**) Domain analysis of 85 DUF26 domain-containing proteins. Green represents DUF26 domain (stress antifung). Yellow represents the sTKc_IRAK kinase domain. Magenta represents the PKc_like domain. Blue green represents the duf3404 domain. (**D**) The exon–intron structures of the DUF26 domain-containing genes. Orange boxes indicate exons. (**E**) The sequence logo of the DUF26 domain-containing proteins motifs. The height of each amino acid represents the relative frequency of the amino acid at that position.

3.4. Collinearity Analysis of DUF26 Domain-Containing Genes in DXWR among Gramineous Species

The collinearity between species provides a reference for studying the evolution of gene families and gene functions. Therefore, the collinearity of the DUF26 domain-containing genes between DXWR and four other gramineous plants was analyzed (Figure 3). The collinearity analysis showed that DXWR and wheat had the most homologous genes, with 64 pairs, followed by zizania (41 pairs), sorghum (27 pairs), and maize (23 pairs). Orthologous DUF26 domain-containing genes in five gramineous crops (Table S2). The 17 domain-containing genes containing *JX1.Chr01g02212*, *JX1 Chr02g02781*, *JX1 Chr02g03361*, *JX1 Chr03g01308*, *JX1 Chr05g00107*, *JX1 Chr05g00227*, *JX1 Chr06g01056*, *JX1 Chr07g01839*, *JX1 Chr07g01841*, *JX1 Chr07g01842*, *JX1 Chr07g01845*, *JX1 Chr07g01847*, *JX1 Chr07g01849*, *JX1 Chr07g01853*, *JX1 Chr07g01869*, *JX1 Chr10g01596*, and *JX1.Chr12g02022* were highly conserved among the five gramineous species. Among them, 11 genes had three or more homologous genes in wheat. These highly conserved homologous genes are of great significance in exploring the relationship between species and predicting gene function.

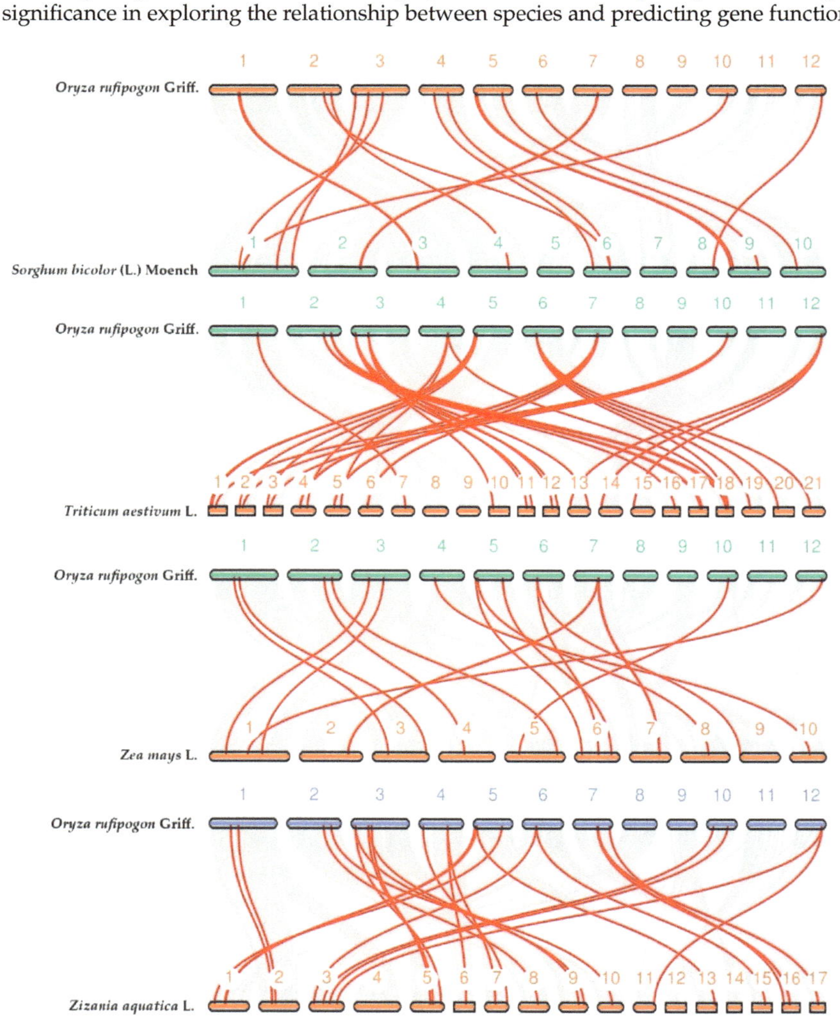

Figure 3. The collinearity analysis of the DXWR and gramineaes plants. Arabic numerals indicate chromosome numbers. The red line highlights the collinearity gene pairs containing the DUF26 domain-containing genes among different species.

3.5. Expression Analysis of DUF26 Domain-Containing Genes in DXWR under Submergence

In order to explore the response of the DUF26 domain-containing gene in DXWR to submergence, the transcriptome data of the DUF26 domain-containing genes before and after submergence germination and seedling submergence treatment were studied. The results showed that 33 out of the 85 DUF26 domain-containing genes in DXWR responded to submergence (Figure 4). A total of 50% (12/24) of PDLPs were induced to express by submergence including *JX1.Chr06g01056* and *JX1.Chr07g01880*, which were down-regulated after submerged germination and seedling submergence treatment, while the other 10 PDLPs were significantly upregulated after 2 days of submerged germination, but the upregulation multiple decreased after 4 days of submerged germination. The expression of *JX1.Chr03g 01309*, *JX1.Chr03g01308*, and *JX1.Chr04g03076* was upregulated after 3 and 5 days of seedling submergence. At the same time, 42% (19/42) of ddCRKs responded to submergence. *JX1.Chr07g01882*, *JX1.Chr10g00202*, *JX1.Chr07g01881*, and *JX1.Chr07g02496* were downregulated in both the submerged germination and seedling submergence treatments, while *JX1.Chr07g01884* was upregulated in both the submergence treatments. In addition, there were 10 ddCRKs with different responses under the two submergence treatments, for example, *JX1.Chr07g01857* and *JX1.Chr11g02777* were upregulated after submergence treatment at the seedling stage, but downregulated at the germination stage.

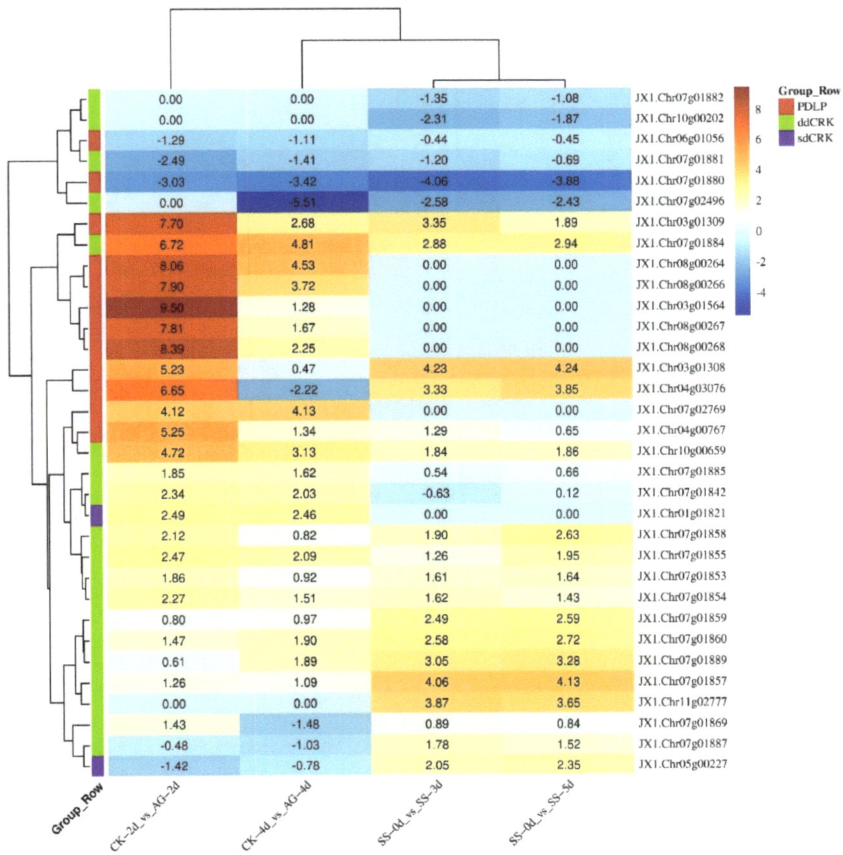

Figure 4. The expression profile of the DUF26 domain-containing genes after different periods of submergence, in which AG is anaerobic germination, SS is seedling submergence, and CK is normal growth. The heatmap was generated by taking log2 fold of the FPKM ratio.

3.6. Correlation Analysis between DUF26 Domain-Containing Genes and Submergence Tolerant Response Genes in DXWR

In order to further study whether the DUF26 domain-containing genes are involved in the response of DXWR to submergence, the correlation between 33 DUF26 domain-containing genes differentially expressed in submergence germination and seedling submergence treatment and other known genes involved in the response of rice to submergence were analyzed. The expression data of all genes were used under submergence stress (Table S3). The results showed that 11 DUF26 domain-containing genes including *JX1.Chr07g01842*, *JX1.Chr01g01821*, *JX1.Chr08g00268*, *JX1.Chr03g01564*, and *JX1.Chr03g01564* were significantly positively correlated with the expression levels of SnRK1A and G3PDH. In addition, 13 DUF26 domain-containing genes including *JX1.Chr07g01885*, *JX1. Chr07g01857*, and *JX1.Chr07g01860* were significantly positively correlated with the expression levels of *CIPK14/15*, *PPDK*, *OsERF66/67*, *RAmy3D*, and *SLRL1* (Figure 5). In conclusion, the correlation analysis showed that the 33 DUF26 do main-containing genes were involved in the response of DXWR to submergence and were likely to be related to known submergence tolerance genes.

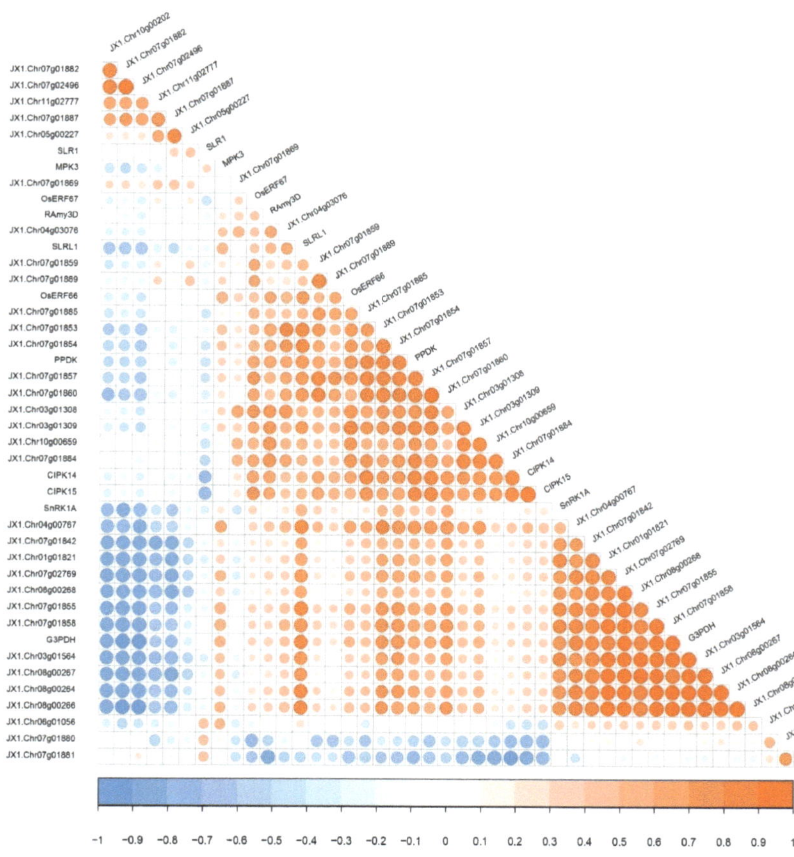

Figure 5. The correlation analysis of the DUF26 domain-containing genes and submergence tolerance genes in DXWR.

4. Discussion

The DUF26 domain-containing genes are an ancient family in plants, which mainly regulate plant growth, development, and respond to various biological and abiotic stresses through signal transduction. DUF26 domain-containing genes have been identified in a variety of plants, where a total of 108, 79, and 48 DUF26 domain-containing genes have been identified in Arabidopsis, rice, and maize, respectively [29]. We identified 85 DUF26 domain-containing genes in the genome of DXWR (Table S1). According to the differences in domains, they were divided into four categories, in which the number of CRRSPs, PDLPs, sdCRKs, and ddCRKs were 10, 24, 6, and 45, respectively. Tandem repeat genes play a role in processes that require rapid adaptation such as adaptation to the environment, pathogen response, and secondary metabolism [30]. It was found that the DUF26 domain-containing genes in DXWR were mainly clustered on chromosomes, and there were 18 pairs of tandem repeat genes, which played an important role in the expansion of its members. For example, *JX1.Chr08g00266*, *JX1.Chr08g00267*, and *JX1.Chr08g00268* are three genes derived from tandem repeats. Interestingly, they are significantly upregulated during anaerobic germination, indicating that these three genes may participate in the anaerobic germination process together.

Orthologous genes are highly conserved among species, and they may be preserved due to their similar important functions in the process of evolution [31]. As a hexaploid plant, wheat has experienced two polyploidization events. Theoretically, one DXWR gene corresponds to three wheat homologous genes [32]. For example, *JX1.Chr07g01849* and 11 other genes have three homologous genes in wheat, indicating that these genes are very conservative in the evolution of wheat. In addition, *JX1.Chr07g01882* and *JX1.Chr12g02022* have six homologous genes in wheat, indicating that these two genes have been expanded in the process of wheat evolution. Rice, sorghum, and maize have experienced at least two whole genome duplication events before the differentiation of Gramineae [32]. Theoretically, there are duplicate copies of DUF26 domain-containing genes among different species. For example, *JX1.Chr07g01849* and *JX1.Chr06g01056* have two lineal homologous genes in sorghum and maize, respectively. Zizania has good characters of deep water tolerance, and there is a significant collinearity between the zizania and rice genome [33]. There are 35 DUF26 domain-containing genes in DXWR that are collinear with zizania, of which about 23% (8/35) have two orthologs genes in zizania. In the collinearity analysis with wheat, maize, sorghum, and zizania, 17 DUF26 domain-containing genes were found to be highly conserved in five gramineous crops, indicating that they may play similar functions. For example, *JX1.Chr03g01308* is highly conserved in five gramineous crops, and it is significantly upregulated during the anaerobic germination and seedling submergence of DXWR, suggesting that this gene may play a similar role in the submergence tolerance of these species. In addition, about 47% (40/85) of the DUF26 domain-containing genes in DXWR have no corresponding ortholog genes in other gramineous crops, and these genes may play a unique role in the evolution of DXWR and be preserved. For example, *JX1.Chr04g00767*, a unique DUF26 domain-containing gene in DXWR, was significantly upregulated during anaerobic germination and seedling flooding, suggesting that it may be the reason why DXWR has strong submergence tolerance.

The analysis of gene structure shows that there were similar intron/exon arrangement patterns and numbers among the genes located in the same group of intron DUF26 domain, and their conservative domains also had similar numbers and position orders, indicating that the members of the subfamily were relatively conservative in the process of evolution, but there are great structural differences among members of different groups, where different cleavage patterns may play an important role in the functional differentiation of DUF26 domain-containing genes. Protein sequence analysis showed that all DUF26 domain-containing genes contained highly conserved Motif 3, 8, 9, 13, 14, 16, and 18, and had similar motif structures among the same group of DUF26 domain-containing genes. Similarly, the unique motif provides an idea for studying the function of DUF26 domain-containing genes. For example, Motif 19 only appeared in eight PDLPs. It is noteworthy that these eight genes were significantly upregulated during anaerobic germination. Therefore, it is speculated that Motif 19 may play an important role in the anaerobic germination of DXWR, which still needs to be verified by subsequent experiments.

There are two strategies in the process of rice submergence tolerance: the first one is to avoid submergence, which shows coleoptile elongation in the anaerobic germination stage, rapid stem elongation in the seedling stage, and the second one is to stop growth in the seedling stage [34]. Prior to this study, there was no report on the role of the DUF26 domain-containing gene in submergence tolerance. This study confirmed that 33 of the 85 DUF26 domain-containing genes in DXWR responded to anaerobic germination or seedling submergence to varying degrees. Subsequent correlation analysis showed that these response genes may have the same expression pattern as the known submergence tolerant genes. Protein kinase CIPK14/15 plays an important role in the tolerance of rice to hypoxia stress. SnRK1A is an important intermediary in the sugar signal cascade. Under submergence conditions, CIPK15 links the hypoxia signal with the SnRK1A dependent sugar signal cascade, regulates the production of sugar and energy, and enables rice to grow under submergence conditions [35,36]. Similarly, G3PDH plays a key role in glycolysis [37]. MPK3 is induced by submergence and plays a key role in its adaptation to submergence conditions [38]. *JX1.Chr07g01842*, *JX1.Chr01g01821*, *JX1.Chr08g00268*, *JX1.Chr03g01564*, *JX1.Chr08g00267*, *JX1.Chr08g00264*, *JX1.Chr08g00266*, *JX1.Chr07g02769*, *JX1.Chr07g01855*, and *JX1.Chr07g01858* are the 10 genes highly correlated with the *SnRK1A* and *G3PDH* expression regulation of the sugar signal channel, and also positively correlated with *CIPK14/15* and *MPK3*. Interestingly, these genes were highly upregulated during anaerobic germination, indicating that they may participate in the anaerobic germination process of DXWR by affecting glucose metabolism. Pyruvate phosphate double kinase gene (*PPDK*) is a response gene to hypoxia stress. Hypoxia stress induces the expression of PPDK in the rice roots, leaf sheaths, and shoots of etiolated seedlings [39]. Under anoxic conditions, the α-amylase gene (*RAmy3D*) is upregulated, which is very important for germination and growth under hypoxia [40–42]. *OsERF66/67* is stable under hypoxia. The overexpression of *OsERF66/67* will lead to the increased expression of anaerobic survival genes, indicating that they are responsible for submergence mediated transcriptional regulation in submergence tolerance strategies [43]. Two GRAS proteins SLR1/SLRL1 are inhibitors of GA signal, which mediate submergence tolerance by limiting gibberellin response in rice [44–46]. *JX1.Chr07g01885*, *JX1.Chr07g01857*, *JX1.Chr07g01860*, *JX1.Chr07g01859*, *JX1.Chr07g01889*, *JX1.Chr03g01308*, *JX1.Chr03g01309*, *JX1.Chr07g0188*, *JX1.Chr10g0065*, *JX1.Chr04g00767*, *JX1.Chr07g01853*, and *JX1.Chr07g01854* had a highly similar expression and regulation pattern with *PPDK*, *OsERF66/67*, *RAmy3D*, and *SLRL1*, except for *JX1.Chr07g01885*, which was upregulated only in anaerobic germination, and the rest were upregulated in the process of seedling submergence. These results suggest that these genes may play an important role in the submergence tolerance of DXWR.

5. Conclusions

In this study, 85 DUF26 domain-containing genes were identified from the DXWR genome. These genes were unevenly distributed on 12 chromosomes and had the characteristics of cluster arrangement. According to the composition of the conserved domains, the DUF26 domain-containing genes in DXWR can mainly be divided into four categories: CRRSPs, PDLPs, sdCRKs, and ddCRKs. The protein sequence analysis showed that DUF26 domain-containing genes in the same group had a similar structure and motif composition. The collinearity analysis with gramineous crops showed that DUF26 domain-containing genes were preserved and lost during evolution, and some genes with special functions were preserved among species. Transcriptome data analysis showed that 33 DUF26 domain-containing genes responded to anaerobic germination and seedling submergence to varying degrees. Subsequent correlation analysis showed that the involvement of these genes in the submergence tolerance process may be related to the known submergence tolerance genes. The study of the DXWR DUF26 domain-containing gene family in the organization, structure, evolution, and expression level under submergence conditions is conducive to the functional analysis of DUF26 domain-containing genes, and lays an important foundation for a better understanding of the molecular mechanism of submergence tolerance in DXWR.

Supplementary Materials: The following supporting information can be downloaded at: https://www.mdpi.com/article/10.3390/cimb44080231/s1.

Author Contributions: C.H., D.W., P.C. and J.W. designed the research. Formal analysis, C.H., D.W. and H.C.; Funding acquisition, P.C. and H.C.; Investigation, J.W., D.C. and H.C.; Data curation, D.W. and W.D.; Project administration, P.C. and J.W.; Resources, D.W., H.C. and W.D. wrote the manuscript; Review and editing, P.C., J.W., D.W., C.H., H.C. and D.C. All authors have read and agreed to the published version of the manuscript.

Funding: This research was jointly funded by the National Natural Science Foundation of China (31960076); the Project of Discovery of Favorable Genes of Wild Rice and Breeding of Green and Efficient Varieties of Jiangxi Province (20213AAF01001); the Jiangxi Technological Innovation Guidance Program (20212BDH81023); the Jiangxi Modern Agricultural Scientific Research Collaborative Innovation Special Project (JXXTCXBSJJ202204); and the Doctoral Startup Fund Project of Jiangxi Academy of Agricultural Sciences (20191CBS002).

Institutional Review Board Statement: Not applicable.

Informed Consent Statement: Not applicable.

Data Availability Statement: Not applicable.

Acknowledgments: We are grateful to Wukai Wang and Jianwei Zheng for assistance in the data analysis.

Conflicts of Interest: The authors declare no conflict of interest.

References

1. Nakamura, S.; Suzuki, T.; Kawamukai, M.; Nakagawa, T. Expression analysis of Arabidopsis thaliana small secreted protein genes. *Biosci. Biotechnol. Biochem.* **2012**, *76*, 436–446. [CrossRef]
2. Agrawal, G.K.; Jwa, N.S.; Lebrun, M.H.; Job, D.; Rakwal, R. Plant secretome: Unlocking secrets of the secreted proteins. *Proteomics* **2010**, *10*, 799–827. [CrossRef]
3. Tavormina, P.; De Coninck, B.; Nikonorova, N.; De Smet, I.; Cammue, B.P. The Plant Peptidome: An Expanding Repertoire of Structural Features and Biological Functions. *Plant Cell* **2015**, *27*, 2095–2118. [CrossRef] [PubMed]
4. Miyakawa, T.; Miyazono, K.; Sawano, Y.; Hatano, K.; Tanokura, M. Crystal structure of ginkbilobin-2 with homology to the extracellular domain of plant cysteine-rich receptor-like kinases. *Proteins* **2009**, *77*, 247–251. [CrossRef] [PubMed]
5. Miyakawa, T.; Hatano, K.; Miyauchi, Y.; Suwa, Y.; Sawano, Y.; Tanokura, M. A secreted protein with plant-specific cysteine-rich motif functions as a mannose-binding lectin that exhibits antifungal activity. *Plant Physiol.* **2014**, *166*, 766–778. [CrossRef] [PubMed]
6. Ma, L.S.; Wang, L.; Trippel, C.; Mendoza-Mendoza, A.; Ullmann, S.; Moretti, M.; Carsten, A.; Kahnt, J.; Reissmann, S.; Zechmann, B.; et al. The Ustilago maydis repetitive effector Rsp3 blocks the antifungal activity of mannose-binding maize proteins. *Nat. Commun.* **2018**, *9*, 1711. [CrossRef]
7. Lee, D.S.; Kim, Y.C.; Kwon, S.J.; Ryu, C.M.; Park, O.K. The Arabidopsis Cysteine-Rich Receptor-Like Kinase CRK36 Regulates Immunity through Interaction with the Cytoplasmic Kinase BIK1. *Front. Plant Sci.* **2017**, *8*, 1856. [CrossRef]
8. Yeh, Y.H.; Chang, Y.H.; Huang, P.Y.; Huang, J.B.; Zimmerli, L. Enhanced Arabidopsis pattern-triggered immunity by overexpression of cysteine-rich receptor-like kinases. *Front. Plant Sci.* **2015**, *6*, 322. [CrossRef]
9. Wrzaczek, M.; Brosche, M.; Salojarvi, J.; Kangasjarvi, S.; Idanheimo, N.; Mersmann, S.; Robatzek, S.; Karpinski, S.; Karpinska, B.; Kangasjarvi, J. Transcriptional regulation of the CRK/DUF26 group of receptor-like protein kinases by ozone and plant hormones in Arabidopsis. *BMC Plant Biol.* **2010**, *10*, 95. [CrossRef]
10. Idanheimo, N.; Gauthier, A.; Salojarvi, J.; Siligato, R.; Brosche, M.; Kollist, H.; Mahonen, A.P.; Kangasjarvi, J.; Wrzaczek, M. The Arabidopsis thaliana cysteine-rich receptor-like kinases CRK6 and CRK7 protect against apoplastic oxidative stress. *Biochem. Biophys. Res. Commun.* **2014**, *445*, 457–462. [CrossRef] [PubMed]
11. Chen, K.; Fan, B.; Du, L.; Chen, Z. Activation of hypersensitive cell death by pathogen-induced receptor-like protein kinases from Arabidopsis. *Plant Mol. Biol.* **2004**, *56*, 271–283. [CrossRef]
12. Chen, K.; Du, L.; Chen, Z. Sensitization of defense responses and activation of programmed cell death by a pathogen-induced receptor-like protein kinase in Arabidopsis. *Plant Mol. Biol.* **2003**, *53*, 61–74. [CrossRef]
13. Acharya, B.R.; Raina, S.; Maqbool, S.B.; Jagadeeswaran, G.; Mosher, S.L.; Appel, H.M.; Schultz, J.C.; Klessig, D.F.; Raina, R. Overexpression of CRK13, an Arabidopsis cysteine-rich receptor-like kinase, results in enhanced resistance to Pseudomonas syringae. *Plant J.* **2007**, *50*, 488–499. [CrossRef] [PubMed]
14. Caillaud, M.C.; Wirthmueller, L.; Sklenar, J.; Findlay, K.; Piquerez, S.; Jones, A.; Robatzek, S.; Jones, J.; Faulkner, C.; Birch, P.J.P.P. The Plasmodesmal Protein PDLP1 Localises to Haustoria-Associated Membranes during Downy Mildew Infection and Regulates Callose Deposition. *PLoS Pathog.* **2014**, *10*, e1004496. [CrossRef]

15. Lim, G.H.; Shine, M.B.; Delorenzo, L.; Yu, K.; Cui, W.; Navarre, D.; Hunt, A.; Lee, J.Y.; Kachroo, A.; Kachroo, P.J.C.H.; et al. Plasmodesmata Localizing Proteins Regulate Transport and Signaling during Systemic Acquired Immunity in Plants. *Cell Host Microbe* **2016**, *19*, 541–549. [CrossRef] [PubMed]
16. Sasaki, T.; Burr, B. International Rice Genome Sequencing Project: The effort to completely sequence the rice genome. *Curr. Opin. Plant Biol.* **2000**, *3*, 138–141. [CrossRef]
17. Bailey-Serres, J.; Lee, S.C.; Brinton, E. Waterproofing crops: Effective flooding survival strategies. *Plant Physiol.* **2012**, *160*, 1698–1709. [CrossRef]
18. Ma, X.; Han, B.; Tang, J.; Zhang, J.; Cui, D.; Geng, L.; Zhou, H.; Li, M.; Han, L. Construction of chromosome segment substitution lines of Dongxiang common wild rice (*Oryza rufipogon* Griff.) in the background of the japonica rice cultivar Nipponbare (*Oryza sativa* L.). *Plant Physiol. Biochem.* **2019**, *144*, 274–282. [CrossRef] [PubMed]
19. Potter, S.C.; Luciani, A.; Eddy, S.R.; Park, Y.; Lopez, R.; Finn, R.D. HMMER web server: 2018 update. *Nucleic Acids Res.* **2018**, *46*, W200–W204. [CrossRef]
20. Mistry, J.; Chuguransky, S.; Williams, L.; Qureshi, M.; Salazar, G.A.; Sonnhammer, E.L.L.; Tosatto, S.C.E.; Paladin, L.; Raj, S.; Richardson, L.J.; et al. Pfam: The protein families database in 2021. *Nucleic Acids Res.* **2021**, *49*, D412–D419. [CrossRef]
21. Marchler-Bauer, A.; Bryant, S.H. CD-Search: Protein domain annotations on the fly. *Nucleic Acids Res.* **2004**, *32*, W327–W331. [CrossRef] [PubMed]
22. Letunic, I.; Khedkar, S.; Bork, P. SMART: Recent updates, new developments and status in 2020. *Nucleic Acids Res.* **2021**, *49*, D458–D460. [CrossRef] [PubMed]
23. Artimo, P.; Jonnalagedda, M.; Arnold, K.; Baratin, D.; Csardi, G.; de Castro, E.; Duvaud, S.; Flegel, V.; Fortier, A.; Gasteiger, E.; et al. ExPASy: SIB bioinformatics resource portal. *Nucleic Acids Res.* **2012**, *40*, W597–W603. [CrossRef] [PubMed]
24. Tamura, K.; Stecher, G.; Kumar, S. MEGA11: Molecular Evolutionary Genetics Analysis Version 11. *Mol. Biol. Evol.* **2021**, *38*, 3022–3027. [CrossRef] [PubMed]
25. Chen, C.; Chen, H.; Zhang, Y.; Thomas, H.R.; Frank, M.H.; He, Y.; Xia, R. TBtools: An Integrative Toolkit Developed for Interactive Analyses of Big Biological Data. *Mol. Plant* **2020**, *13*, 1194–1202. [CrossRef]
26. Bailey, T.L.; Johnson, J.; Grant, C.E.; Noble, W.S. The MEME Suite. *Nucleic Acids Res.* **2015**, *43*, W39–W49. [CrossRef]
27. Wang, Y.; Tang, H.; Debarry, J.D.; Tan, X.; Li, J.; Wang, X.; Lee, T.H.; Jin, H.; Marler, B.; Guo, H.; et al. MCScanX: A toolkit for detection and evolutionary analysis of gene synteny and collinearity. *Nucleic Acids Res.* **2012**, *40*, e49. [CrossRef]
28. Tang, H.; Bowers, J.E.; Wang, X.; Ming, R.; Alam, M.; Paterson, A.H. Synteny and collinearity in plant genomes. *Science* **2008**, *320*, 486–488. [CrossRef]
29. Vaattovaara, A.; Brandt, B.; Rajaraman, S.; Safronov, O.; Veidenberg, A.; Luklova, M.; Kangasjarvi, J.; Loytynoja, A.; Hothorn, M.; Salojarvi, J.; et al. Mechanistic insights into the evolution of DUF26-containing proteins in land plants. *Commun. Biol.* **2019**, *2*, 56. [CrossRef]
30. Rogers, R.L.; Shao, L.; Thornton, K.R. Tandem duplications lead to novel expression patterns through exon shuffling in Drosophila yakuba. *PLoS Genet.* **2017**, *13*, e1006795. [CrossRef]
31. Wu, S.; Han, B.; Jiao, Y. Genetic Contribution of Paleopolyploidy to Adaptive Evolution in Angiosperms. *Mol. Plant* **2020**, *13*, 59–71. [CrossRef] [PubMed]
32. Ren, R.; Wang, H.; Guo, C.; Zhang, N.; Zeng, L.; Chen, Y.; Ma, H.; Qi, J. Widespread Whole Genome Duplications Contribute to Genome Complexity and Species Diversity in Angiosperms. *Mol. Plant* **2018**, *11*, 414–428. [CrossRef] [PubMed]
33. Yan, N.; Yang, T.; Yu, X.T.; Shang, L.G.; Guo, D.P.; Zhang, Y.; Meng, L.; Qi, Q.Q.; Li, Y.L.; Du, Y.M.; et al. Chromosome-level genome assembly of Zizania latifolia provides insights into its seed shattering and phytocassane biosynthesis. *Commun. Biol.* **2022**, *5*, 36. [CrossRef]
34. Dwivedi, S.K.; Kumar, S.; Bhakta, N.; Singh, S.K.; Rao, K.K.; Mishra, J.S.; Singh, A.K. Improvement of submergence tolerance in rice through efficient application of potassium under submergence-prone rainfed ecology of Indo-Gangetic Plain. *Funct. Plant Biol.* **2017**, *44*, 907–916. [CrossRef]
35. Kurusu, T.; Hamada, J.; Nokajima, H.; Kitagawa, Y.; Kiyoduka, M.; Takahashi, A.; Hanamata, S.; Ohno, R.; Hayashi, T.; Okada, K.; et al. Regulation of microbe-associated molecular pattern-induced hypersensitive cell death, phytoalexin production, and defense gene expression by calcineurin B-like protein-interacting protein kinases, OsCIPK14/15, in rice cultured cells. *Plant Physiol.* **2010**, *153*, 678–692. [CrossRef] [PubMed]
36. Lu, C.A.; Lin, C.C.; Lee, K.W.; Chen, J.L.; Huang, L.F.; Ho, S.L.; Liu, H.J.; Hsing, Y.I.; Yu, S.M. The SnRK1A protein kinase plays a key role in sugar signaling during germination and seedling growth of rice. *Plant Cell* **2007**, *19*, 2484–2499. [CrossRef]
37. Kollmer, I.; Novak, O.; Strnad, M.; Schmulling, T.; Werner, T. Overexpression of the cytosolic cytokinin oxidase/dehydrogenase (CKX7) from Arabidopsis causes specific changes in root growth and xylem differentiation. *Plant J.* **2014**, *78*, 359–371. [CrossRef]
38. Singh, P.; Sinha, A.K. A Positive Feedback Loop Governed by SUB1A1 Interaction with MITOGEN-ACTIVATED PROTEIN KINASE3 Imparts Submergence Tolerance in Rice. *Plant Cell* **2016**, *28*, 1127–1143. [CrossRef]
39. Moons, A.; Valcke, R.; Van Montagu, M. Low-oxygen stress and water deficit induce cytosolic pyruvate orthophosphate dikinase (PPDK) expression in roots of rice, a C3 plant. *Plant J.* **1998**, *15*, 89–98. [CrossRef]
40. Park, M.; Yim, H.K.; Park, H.G.; Lim, J.; Kim, S.H.; Hwang, Y.S. Interference with oxidative phosphorylation enhances anoxic expression of rice alpha-amylase genes through abolishing sugar regulation. *J. Exp. Bot.* **2010**, *61*, 3235–3244. [CrossRef]

41. Xiong, M.; Yu, J.; Wang, J.; Gao, Q.; Huang, L.; Chen, C.; Zhang, C.; Fan, X.; Zhao, D.; Liu, Q.Q.; et al. Brassinosteroids regulate rice seed germination through the BZR1-RAmy3D transcriptional module. *Plant Physiol.* **2022**, *189*, 402–418. [CrossRef] [PubMed]
42. Toyofuku, K.; Umemura, T.; Yamaguchi, J. Promoter elements required for sugar-repression of the RAmy3D gene for alpha-amylase in rice. *FEBS Lett.* **1998**, *428*, 275–280. [CrossRef]
43. Lin, C.C.; Chao, Y.T.; Chen, W.C.; Ho, H.Y.; Chou, M.Y.; Li, Y.R.; Wu, Y.L.; Yang, H.A.; Hsieh, H.; Lin, C.S.; et al. Regulatory cascade involving transcriptional and N-end rule pathways in rice under submergence. *Proc. Natl. Acad. Sci. USA* **2019**, *116*, 3300–3309. [CrossRef] [PubMed]
44. Ayano, M.; Kani, T.; Kojima, M.; Sakakibara, H.; Kitaoka, T.; Kuroha, T.; Angeles-Shim, R.B.; Kitano, H.; Nagai, K.; Ashikari, M. Gibberellin biosynthesis and signal transduction is essential for internode elongation in deepwater rice. *Plant Cell Environ.* **2014**, *37*, 2313–2324. [CrossRef] [PubMed]
45. Fukao, T.; Bailey-Serres, J. Submergence tolerance conferred by Sub1A is mediated by SLR1 and SLRL1 restriction of gibberellin responses in rice. *Proc. Natl. Acad. Sci. USA* **2008**, *105*, 16814–16819. [CrossRef] [PubMed]
46. Itoh, H.; Shimada, A.; Ueguchi-Tanaka, M.; Kamiya, N.; Hasegawa, Y.; Ashikari, M.; Matsuoka, M. Overexpression of a GRAS protein lacking the DELLA domain confers altered gibberellin responses in rice. *Plant J.* **2005**, *44*, 669–679. [CrossRef]

current issues in molecular biology

Article

Genome-Wide Identification and Expression Analysis of the Aquaporin Gene Family in *Lycium barbarum* during Fruit Ripening and Seedling Response to Heat Stress

Wei He [1,†], Mingyu Liu [1,†], Xiaoya Qin [2], Aihua Liang [3,4], Yan Chen [1], Yue Yin [2], Ken Qin [2,*] and Zixin Mu [1,3,4,*]

1 College of Life Sciences, Northwest A&F University, Yangling, Xianyang 712100, China
2 National Wolfberry Engineering Research Center, Ningxia Academy of Agriculture and Forestry Sciences, Yinchuan 750002, China
3 College of Life Sciences & Technology, Tarim University, Alar 843301, China
4 State Key Laboratory Breeding Base for the Protection and Utilization of Biological Resources in Tarim Basin Co-Funded by Xinjiang Corps and the Ministry of Science and Technology, Alaer 843300, China
* Correspondence: qinken7@163.com (K.Q.); muzx@nwsuaf.edu.cn (Z.M.)
† These authors contributed equally to this work.

Abstract: Plant–water relations mediated by aquaporins (AQPs) play vital roles in both key plant growth processes and responses to environmental challenges. As a well-known medicinal and edible plant, the harsh natural growth habitat endows *Lycium* plants with ideal materials for stress biology research. However, the details of their molecular switch for water transport remain unclear. In the present work, we first identified and characterized AQP family genes from *Lycium* (*L.*) *barbarum* at the genome scale and conducted systemic bioinformatics and expression analyses. The results showed that there were 38 *Lycium barbarum* AQPs (LbAQPs) in *L. barbarum*, which were classified into four subfamilies, including 17 LbPIP, 9 LbTIP, 10 LbNIP, and 2 LbXIP. Their encoded genes were unevenly distributed on all 12 chromosomes, except chromosome 10. Three of these genes encoded truncated proteins and three genes underwent clear gene duplication events. Cis-acting element analysis indicated that the expression of *LbAQPs* may be mainly regulated by biotic/abiotic stress, phytohormones and light. The qRT-PCR assay indicated that this family of genes presented a clear tissue-specific expression pattern, in which most of the genes had maximal transcript levels in roots, stems, and leaves, while there were relatively lower levels in flowers and fruits. Most of the *LbAQP* genes were downregulated during *L. barbarum* fruit ripening and presented a negative correlation with the fruit relative water content (RWC). Most of their transcripts presented a quick and sharp upregulation response to heat stress following exposure of the 2-month-old seedlings to a 42 °C temperature for 0, 1, 3, 12, or 24 h. Our results proposed that *LbAQPs* were involved in *L. barbarum* key development events and abiotic stress responses, which may lay a foundation for further studying the molecular mechanism of the water relationship of *Lycium* plants, especially in harsh environments.

Keywords: *Lycium barbarum*; aquaporins; genome-wide; water relations; fruit ripening; heat stress

Citation: He, W.; Liu, M.; Qin, X.; Liang, A.; Chen, Y.; Yin, Y.; Qin, K.; Mu, Z. Genome-Wide Identification and Expression Analysis of the Aquaporin Gene Family in *Lycium barbarum* during Fruit Ripening and Seedling Response to Heat Stress. *Curr. Issues Mol. Biol.* **2022**, *44*, 5933–5948. https://doi.org/10.3390/cimb44120404

Academic Editor: Vijai Bhadauria

Received: 19 October 2022
Accepted: 10 November 2022
Published: 28 November 2022

1. Introduction

Plant water transport and related molecular components are responsive to an extremely wide array of environmental and hormonal signals, which is an essential process throughout their life cycle [1]. It is well known that the plant water balance can usually be broken down by two key physiological processes, namely, expansion growth and stress response [2,3]. In both cases, ensuring water transport to the growth centers or avoiding water loss means that maintaining water homoeostasis is vital for whole-plant development and survival [1]. This is especially true under global climate change, in which plants are subjected to incorporated drought and heat stress, which further deteriorates the plant water status compared with the exertion of single environmental factors. Maintaining plant water balance, therefore, is becoming increasingly indispensable for both sustainable agriculture and ecosystems.

Aquaporins (AQPs), also called water channels or major intrinsic proteins (MIPs), are characterized by six transmembrane domains that together facilitate the transport of water and a variety of low-molecular-weight solutes. The greatest numbers of AQP isoforms are present in plants versus animals and microorganisms, implying their indispensable role in plant sessile living and environmental response [4,5]. The increase in AQP number during the evolution of plants from aquatic to terrestrial and from lower to higher further highlighted the importance of AQPs in the adaptation of higher plants to land life [6,7]. Plant AQPs are present in the plasma and intracellular membranes of most plant cells and play central roles in various physiological processes by ensuring cell-to-cell water transport and, to a lesser extent, single-cell osmotic regulation [8]. These processes include stomatal and leaf movements, seed dormancy and germination, plant growth and development, CO_2 fixation, nutrient allocation and toxicity, ROS detoxification and signaling, whole-plant water transport and transpiration, plant reproduction, and abiotic and biotic stress responses [6,9–12]. Through genome sequencing technology, AQP gene families have been comprehensively identified in approximately 50 various plant species, covering algae, mosses, lycophytes, monocots, and dicots [6,11,12].

AQPs are generally classified into five subfamilies in higher plants on the basis of their localizations and amino acid sequences. These include plasma membrane intrinsic proteins (PIPs), tonoplast intrinsic proteins (TIPs), nodulin 26-like intrinsic proteins (NIPs), small basic intrinsic proteins (SIPs), and uncategorized (X) intrinsic proteins (XIPs). The two types of PIPs are PIP1 s and PIP2s, which have differences in amino acid length, amino acid substitution, and water permeability. TIP AQPs exist in more divergent forms than PIPs, in which they are further divided into five isoforms, TIP1, TIP2, TIP3, TIP4, and TIP5, on the basis of their sequence homologies. NIP proteins are also present in divergent forms, as they are divided into five subgroups (NIP1, NIP2, NIP3, NIP4, and NIP). In *Arabidopsis*, NIP isoforms include AtNIP1;1, AtNIP1;2, AtNIP2;1, AtNIP3;1, AtNIP4;1, and AtNIP4;2. XIP is generally found in plasma membranes and is known to be permeable to the largest uncharged solutes, such as urea, boric acid, glycerol, and H_2O_2 [6]. In addition, the entire XIP subfamily is absent in both Arabidopsis and rice, presenting its high variation among plant species.

Aquaporins certainly exert a crucial function in alleviating abiotic stress by transporting water and other small molecules to maintain cellular homeostasis [9,10]. Although AQP exists as a large protein family in plants, an increasing number of genetic assays have demonstrated that a single AQP gene can orchestrate the whole plant function, implying its potential application in crop improvement when facing extreme climate challenges [9,13–15]. Taken together, in-depth research on AQP function and regulatory mechanisms will provide a breakthrough point to reveal the molecular mechanism underlying plant water balance. Identification and characterization of AQP family genes in more plant species, including medicinal plants, ex. *Lycium* plants, are, therefore, indispensable both for basic research and human health care practices.

Goji *Lycium* of the Solanaceae family contains ~80 species, representing important food plants and enjoying a reputation as one of the world's most economically and medicinally valuable fruit crops [16–19]. It has been reported that *Lycium* species present a fragmented distribution pattern at high altitudes in the subtropics to temperate regions but are absent in tropical regions. Correspondingly, the temperature requirement for these *Lycium* species is consistent with its geographical distribution—they are sensitive to heat stress and, therefore, significantly reduce yield and fruit quality under high temperature. Referring to a previous study, in addition to the visible wilting in morphology, *L. barbarum* will also show obvious changes in transcription and metabolism when exposed to heat stress [20].

Compared with other Solanaceae, *Lycium* plants adapt to strong light and arid and saline environments [16–19]. In China, *Lycium* species are also applied as pioneer trees in vegetation restoration and saline–alkali land improvement, in which they have important ecological value [17]. Their natural growth habitat endows *Lycium* plants with ideal materials for stress biology research; however, the continuing lack of a genome sequence of this genus has severely impeded advances in this field. Recently, a genome sequence for one *Lycium* plant, *Lycium* (*L.*) *barbarum*, was released, which undoubtedly will greatly impel studies in *Lycium* biology [18].

To explore the water relationships of *Lycium* plants both under key growth processes and for the abiotic stress response, in the present work, we first identified and characterized AQP family genes from *L. barbarum* at a genome scale and then conducted systematic bioinformatics analysis, including assessments of gene structures, conserved domains, phylogenetic analysis, and cis-elements in promoters, as well as a transcript profiling assay. Our results highlight the importance of these gene families in fresh fruit ripening and seedling heat stress responses and, therefore, lay an invaluable foundation for the in-depth elucidation of the acclimation mechanisms of *Lycium* plants to extreme environments, especially in terms of water transport.

2. Materials and Methods

2.1. Plant Materials

The different tissues, including roots, stems, leaves, and flowers, as well as the five ripening stages of *Lycium* (L.) *barbarum* (Ningqi 7, N7), were collected from three of the 5-year-old trees at the Wolfberry (Lycium) Germplasm Repository of Ningxia, Academy of Agriculture and Forestry Sciences, Ningxia Hui Autonomous Region, China (38°080′ N, 106°090′ E and altitude 1100 m). To assay fruit ripening dynamics, fruits were sampled at five different stages (S1, S2, S3, S4, and S5) under their natural state, as described by Cao et al. (2021) [18]. Within 30 days before sampling, the mean high temperature and low temperature of Yinchuan were 27.50 °C and 12.37 °C, respectively, while the field was regularly irrigated to keep soil moisture suitable. In addition to rational field management, healthy fruits in good condition were selected after sampling and used for experiments. Our field studies were conducted in accordance with local legislation and appropriate permissions.

2.2. Heat Stress Assay

Uniform clonal seedlings that were grown in a greenhouse for approximately 5 weeks were transported to a growth chamber. After acclimation to the artificial environment (25 °C) for one week, the seedlings were divided into two parts. Whereas one part was still left in the same growth chamber as the control, the other part was transported to a 42 °C growth chamber for heat stress exposure. The seedlings were subjected to heat stress for 0, 1, 3, 6, 12, and 24 h. The leaves were sampled at each time point and divided into three biological replicates, then immediately frozen by liquid nitrogen for RNA extraction.

2.3. Relative Water Content Determination of *L. barbarum* Fresh Fruits

After weighing their fresh weight, the fruits (5~6) at different ripening stages (three biological replicates per stage) were transferred from their aluminum boxes to the corresponding centrifuge tubes for water soaking. The amount of water added to each tube was consistent and the fruits were completely immersed. After 24 h, the fruits were taken out, the surface moisture was removed, and the fruits were transferred into their corresponding aluminum boxes to weigh the saturated weight. Then, the aluminum boxes containing the chopped fruits were transferred to an oven for drying. The samples were first dried at 60~80 °C for 2~3 h, ensuring that the tissue became brittle and dry, and then dried at 100~105 °C for 1~2 h. After the difference between the two weights was less than 0.002 g, the weight was denoted as the dry weight. The RWC was calculated by the formula RWC (%) = (fresh weight − dry weight) × 100/(saturated weight − dry weight).

2.4. Identification and Chromosomal Location of LbAQP Genes

To identify the putative AQP proteins in *L. barbarum* (LbAQPs), the 35 AtAQPs in the A. thaliana genome were downloaded from the TAIR database (https://www.arabidopsis.org/) and were then used as queries to BLAST search the *L. barbarum* genome (https://www.ncbi.nlm.nih.gov/genome/81199?genome_assembly_id=1656998, acessed on 20 August 2022) with an E-value of e^{-10}. Then, preliminary amino acid sequences that may have the function of LbAQPs were obtained according to the homology of AtAQPs. On this basis, conservative AQP domains obtained from the PFAM database (http://pfam-legacy.xfam.org/) were blast searched against these candidate sequences. Finally, amino acid sequences without conserved AQP domains and redundant sequences were manually removed.

The molecular weights (MWs), isoelectric points (pIs) and grand average hydropathicity (GRAVY) values of the LbAQP proteins were analyzed with ProtParam (http://web.expasy.org/protparam/) [21].

We retrieved the genome annotation files (for internal use only) from the *L. barbarum* genome database of NCBI and summarized their physical positions into a graph using TBtools software v1.098774, in which the chromosome numbers and positions of each sequence in the genome were indicated.

2.5. Classification of LbAQP Protein Members and Construction of a Phylogenetic Tree

The phylogenetic tree was constructed by MEGAX software from a ClustalX alignment of related amino acid sequences (bootstrap replicates = 1000) using the maximum likelihood method. The ML tree was formatted for visualization by the Chiplot website (https://www.chiplot.online/).

2.6. Structure and Conserved Motif Analysis of LbAQPs

Sequence alignments between selected sequences and the genome were carried out according to the GFF format genome annotation files obtained from the genome database in NCBI, and the intron-exon structure information of these genes was generated. TBtools software was used to draw the structure map. Conserved motif analysis was performed in the classic mode of Multiple Em for Motif Elicitation (MEME, https://meme-suite.org/meme/), where the number of motifs was set to 15, the E-value was set to e^{-10}, and the other settings were consistent with the default parameters.

2.7. Promoter Cis-Element Analysis of LbAQPs

The promoter regions 1 kb upstream of the corresponding genes were analyzed, and the cis-elements were predicted by the PlantCARE database (https://bioinformatics.psb.ugent.be/webtools/plantcare/html/). Subsequently, according to the results of the PlantCARE calculations, we recorded the numbers of cis-elements in these sequences and summarized this information into a figure for subsequent analysis.

2.8. Quantitative Real-Time PCR (qRT-PCR)

Total RNA was isolated from the indicated samples using TRIzol reagent (Invitrogen), and its quality was determined by a NanoDrop (Thermo Scientific, Wilmington, DE, USA). First-strand cDNAs were synthesized from 0.5 μg of RNA using the Super-Script III First-Strand Synthesis SuperMix Kit (Invitrogen, Grand Island, NY, USA). qRT-PCR was conducted using 0.01 μg of the cDNA on a LightCycler 480 Instrument System (Roche, Diagnostics GmbH, Mannheim, Germany) with KAPA SYBR FAST qPCR Master Mix and with an initial denaturing step at 95 °C for 5 min, followed by 55 cycles of 95 °C for 10 s, 60 °C for 20 s, and 72 °C for 5 s. The fold changes in the relative expression levels were analyzed via the $2^{-\Delta\Delta CT}$ method using the LbACTIN1 gene as an internal control. The gene-specific primers used for qRT-PCR are listed in Table S1. All experimental results were performed with three biological and technical replicates.

2.9. Statistical Analysis

Statistical analyses were performed using SPSS version 19.0. Parameter differences among various ripening stages of fruits were determined using one-way ANOVA with appropriate post hoc analysis.

3. Results

3.1. Characterization of the LbAQP Gene Family

To identify *LbAQP* family genes in the *L. barbarum* genome, the 35 Arabidopsis AQP protein sequences and the MIP PF00230 conserved domain were employed as queries to search against the *L. barbarum* genome database using the BlastP program. Forty-seven genes were identified by homologous alignment, and nine genes were eliminated by conserved domain and amino acid site analysis. Finally, 38 full-length genes encoding potential *LbAQPs* were identified and named according to their sequence similarity and phylogenies with both individual AtAQP proteins and SlAQP proteins (Table 1). While three of these genes (*LbPIP1;5*, *LbPIP2;7*, and *LbNIP4;3*) encoded severely truncated proteins, three genes experienced clear gene duplication events (*LbTIP2;2*, *LbNIP1;2*, and *LbNIP6;1*). In-depth analysis of the identified 38 *LbAQPs* was performed to determine their CDS, TMHMM, isoelectric point (pI), MW, and grand average of hydropathicity (GRAVY). Most of the LbAQPs consisted of CDSs ranging in length from 740 to 900 base pairs (bp).

Table 1. Detailed information on 38 aquaporin (AQP) genes of *Lycium* (*L.*) *barbarum* and their encoded proteins.

Gene Name	Gene ID	CDS Length (bp)	Size (aa)	Size (aa)	PI	TMHMM	GRAVY
LbPIP1;1	Lba07g01359	861	287	30.777	8.31	6	0.409
LbPIP1;2	Lba06g02040	855	285	30.784	8.64	6	0.318
LbPIP1;3	Lba04g00268	858	286	30.694	7.69	6	0.375
LbPIP1;4	Lba09g02427	861	287	30.866	7.68	6	0.422
LbPIP1;5	Lba09g02372	492	164	17.854	9.91	3	0.746
LbPIP1;6	Lba01g01166	858	286	30.627	9.10	6	0.416
LbPIP2;1	Lba03g00307	849	283	30.223	8.21	6	0.501
LbPIP2;2	Lba03g00306	849	283	30.168	6.94	6	0.496
LbPIP2;4	Lba08g00456	861	287	30.732	6.37	6	0.537
LbPIP2;5	Lba08g01659	855	285	30.425	8.24	6	0.408
LbPIP2;6	Lba06g00111	861	287	30.624	8.57	6	0.590
LbPIP2;7	Lba01g02704	528	176	18.901	5.16	4	0.545
LbPIP2;8	Lba06g03476	852	284	30.435	9.28	6	0.486
LbPIP2;9	Lba08g01171	849	283	30.150	9.24	6	0.512
LbPIP2;10	Lba10g01287	852	284	30.421	8.83	6	0.545
LbPIP2;11	Lba12g01911	810	270	28.867	8.82	6	0.522
LbPIP2;12	Lba05g00065	1029	343	37.897	6.40	5	0.285
LbTIP1;1	Lba01g02671	753	251	25.712	5.16	6	0.740
LbTIP2;1	Lba07g01176	744	248	25.043	6.15	7	0.988
LbTIP2;2	Lba03g02891	1476	492	49.630	5.51	15	1.009
LbTIP2;3	Lba01g01487	750	250	25.220	5.35	6	0.919
LbTIP2;4	Lba01g02017	744	248	24.988	5.66	7	0.952
LbTIP3;1	Lba01g02391	774	258	27.175	6.70	6	0.621
LbTIP3;2	Lba03g01980	780	260	27.752	8.07	6	0.523
LbTIP4;1	Lba04g01403	741	247	25.883	6.01	7	0.866
LbTIP5;1	Lba03g01262	768	256	26.549	8.58	6	0.709
LbNIP1;2	Lba12g01378	1614	538	57.114	8.43	11	0.431
LbNIP2;1	Lba11g02541	753	251	26.644	9.20	5	0.381
LbNIP3;1	Lba01g02539	1041	347	37.614	8.45	6	0.386
LbNIP3;2	Lba07g01497	810	270	29.075	9.30	5	0.564
LbNIP4;1	Lba12g02217	834	278	29.332	5.92	7	0.698
LbNIP4;2	Lba05g02002	2859	953	105.105	7.34	5	-0.546
LbNIP4;3	Lba12g00430	441	147	15.815	9.61	3	0.678
LbNIP4;4	Lba12g00428	837	279	29.702	7.55	6	0.702
LbNIP5;1	Lba09g02501	1167	389	40.914	6.11	5	0.249
LbNIP6;1	Lba03g02537	1866	622	64.687	8.48	12	0.421
LbXIP1;2	Lba08g01139	975	325	34.597	7.66	7	0.698
LbXIP1;6	Lba06g03416	972	324	34.684	7.04	6	0.737

In addition, nine sequences were longer than 900 bp, among which seven sequences were longer than 1000 bp and one sequence was longer than 2000 bp. The predicted proteins ranged in length from 147 to 953 amino acids, with 15.815 to 105.105 kDa MW. The pI values varied between 5.16 and 9.91. The GRAVY values were also detected through bioinformatics analysis and were positive, except for LbNIP4;2, and ranged from −0.546 to 1.009. The conserved transmembrane domains (TMDs) values ranged from 3 to 15. The TMDs were predicted using TMHMM version 2.0 (http://www.cbs.dtu.dk/services/TMHMM/).

3.2. Physical Distribution of LbAQPs on L. barbarum Chromosomes

We further analyzed the *LbAQP* gene location on the *L. barbarum* chromosomes (Chr). It was shown that the *LbAQP*-encoded genes were unevenly distributed on 11 out of 12 *L. barbarum* chromosomes, with the exception of chr 02. In detail, there were seven, six, two, two, four, three, four, three, one, one, and five genes located on chromosomes 01, 03, 04, 05, 06, 07, 08, 09, 10, 11, and 12, respectively (Figure 1).

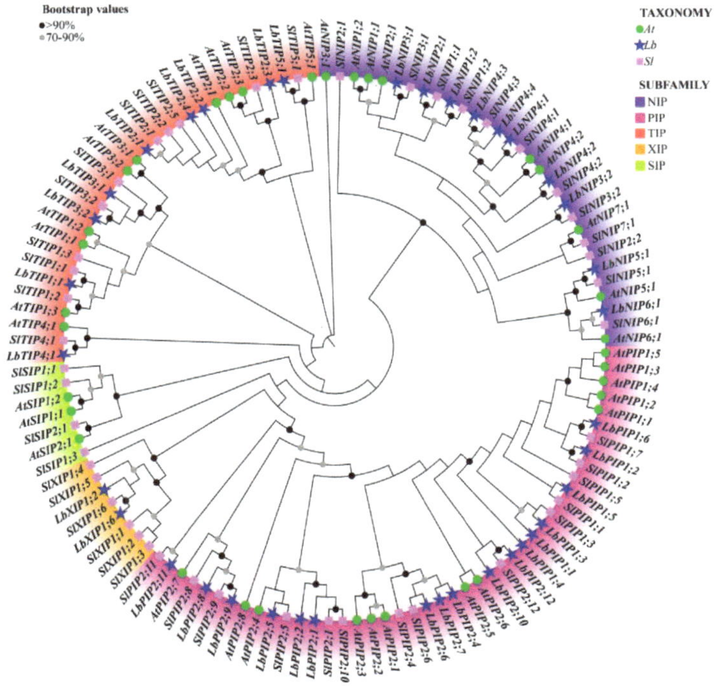

Figure 1. Phylogenetic analysis of the 38 *Lycium barbarum* AQPs (LbAQPs) with *Arabidopsis* and tomato homologs. Deduced amino acid sequences were aligned using ClustalX, and a phylogenetic tree was constructed using the bootstrap maximum likelihood tree (1000 replicates) method and MEGAX software. The full-length AQP protein sequences, including 38 members from *Lycium barbarum* (Lb), 35 members from *Arabidopsis thaliana* (At), and 47 members from *Solanum lycopersicum* (Sl), were classified into PIP, TIP, NIP, XIP, and SIP subfamilies, respectively. The branches of different classes have altered colors, and each represents a different subfamily.

3.3. Phylogenetic Comparison of LbAQP Proteins of L. barbarum, Tomato, and Arabidopsis

To study the evolutionary characteristics of genes and the evolutionary relationships among AQP proteins, we performed cross-genus phylogenetic analysis with 38 LbAQPs, along with 35 and 45 AQP proteins from *Arabidopsis* and tomato, respectively (Figure 2). According to the known *Arabidopsis* and tomato AQP families, LbAQPs can be divided into four distinct subfamilies: LbPIPs, LbTIPs, LbNIPs, and LbXIPs. There are 17 *LbPIP*, 9 *LbTIP*, 10 *LbNIP*, and 2 *LbXIP* in the *L. barbarum* genome, respectively. In contrast to

Arabidopsis and other *Solanaceae* species [21–23], there are no LbSIPs found in the *L. barbarum* genome. The *17 LbPIPs* can be further subdivided into two subgroups, 6 *LbPIP1* and 11 *LbPIP2*. The comparative phylogenetic trees of AQP subfamilies among *Arabidopsis*, tomato, and *L. barbarum* demonstrated that AQP was highly species-specific. We also found that the proteins in the same or adjacent classification groups belonged to the same subfamily, which further confirmed the phylogenetic tree analysis.

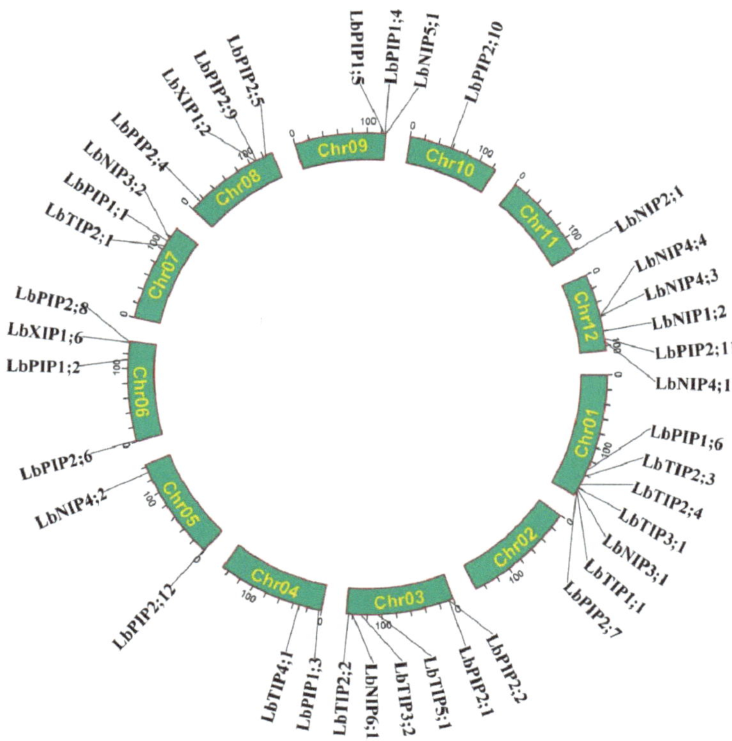

Figure 2. Chromosomal distribution of *LbAQP* genes.

3.4. Gene Structure, Conserved Motifs, and Phylogenetic-Tree-Based Classification of LbAQPs

Multisequence alignment was performed to explore the evolutionary relationships of all 38 *LbAQP* genes (Figure 3A). A maximum likelihood tree was constructed by comparing related amino acid sequences. We also used MEME software to analyze the conserved motifs of LbAQPs with motif-specific sequences. Full-length protein sequence analysis identified a total of 10 motifs (Figure 3B). The number of conserved motifs in each LbAQP varied between three and eight. Although the conserved motifs of the 38 LbAQP genes were different in composition, they all contained motif 2, motif 3, and motif 4 and they were arranged in the same order, with motif 2 first, then motifs 3 and 4. Motifs 2, 3, and 4 existed in most LbAQPs, representing the characteristic structures of AQP. Motif analyses showed that most of the motifs were specific to subfamilies. Most members of the LbPIPs contained motifs 2, 3, 4, 5, 6, 8, 9, and 10, with the exception of LbPIP1;5, LbPIP2;7, and PIP2;12. All members of the LbTIPs, except LbTIP5;1, contained motifs 1, 2, 3, 4, and 7. While motifs 1, 2, 3, 4, and 5 appeared in two LbXIPs, LbNIPs contained motifs 1, 2, 3, and 4, except for LbNIP4;3. There was a clear duplication of motif 1 in subfamilies LbXIPs and LbNIPs, with two copies present in their genes in these two subfamilies.

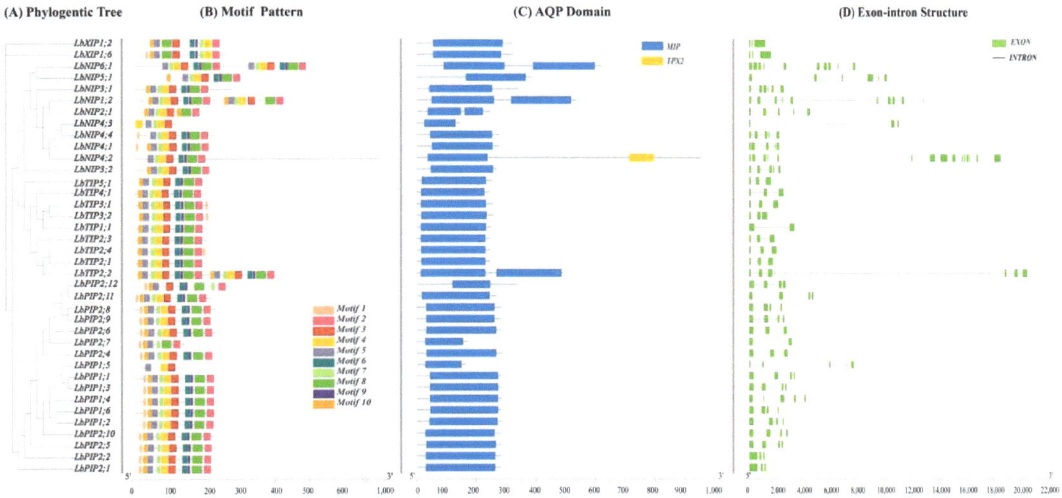

Figure 3. Phylogenic relationships, gene structure conserved domains, and conserved motif analyses. (**A**) A phylogenetic tree of *LbAQP* proteins was constructed with MEGAX software. (**B**) Conserved motif distribution of *LbAQP* proteins. The conserved motifs identified with MEME are displayed with boxes in different colors. A total of 10 motifs were identified. The scale at the bottom shows the length of the protein. (**C**) Predicted conserved structural domains of *LbAQP* proteins. Gray lines represent the length of each protein sequence, and conserved domains are indicated by colored boxes. (**D**) Exon-intron structures of the *LbAQP* genes. Gene structures were analyzed with Tbtools software. Exons and introns are indicated with green boxes and gray lines, respectively. The lengths of the exons and introns (bp) are indicated on the *x*-axis. The combined figure was illustrated with Tbtools software.

While 34 members out of 38 LbAQPs have one conserved MIP domain (Figure 3C), *LbTIP2;2*, *LbNIP1;2*, and *LbNIP6;1* have two similar MIP domains due to gene duplication. In addition to containing one MIP domain, *LbNIP4;2* also contained a targeting protein for Xklp2 (TPX2) domain.

To better understand the corresponding gene structure, the exon-intron structures of 38 *LbAQPs* were analyzed by the Pfam database based on amino acid sequences (Figure 3D). Our study revealed that the number and length of introns were significantly different among different subfamilies in LbAQPs, in which the number ranged from one to five introns. Eleven *LbPIP* genes had three introns, while four genes (*LbPIP2;1*, *LbPIP2;2*, *LbPIP2;4*, and *LbPIP2;6*) had two introns, one gene (LbPIP1;4) had four introns, and one gene (*LbPIP2;7*) had one intron. Seven LbTIP genes had two introns, *LbTIP1;1* had one, and *LbTIP2;2* had three. Six LbNIP genes had four introns, *LbNIP3;1* and *LbNIP5;1* had five introns, and *LbNIP4;3* had two introns. Both *LbXIP* genes had two introns. The lowest number of introns was observed in *LbTIPs* (~two) and *LbXIPs* (two), followed by *LbPIPs* (~three) and NIPs (~4). The varied number of introns among the *LbAQPs* contributed to the variations in gene length (Figure 3).

Multiple sequence alignment of the DNA-binding domains of 38 LbAQP proteins revealed the conserved amino acid sequences of AQP (Figure 4). Most of the LbAQPs contained dual NPA motifs, except *LbPIP2;7*, *LbPIP2;12*, *LbPIP1;5*, and *LbNIP4;3*, which were found to harbor a single NPA motif. Both the first and second NPA motifs were found to be conserved in the LbPIP and LbTIP subfamilies. While LbNIP2;1 and LbNIP5;1 showed alanine to serine substitutions, *LbNIP4;3* showed alanine to valine substitutions in the first NPA motif. It was found that valine substituted alanine in the second NPA motif for both NIP5;1 and NIP6;1. Two LbXIPs were not conserved in the first NPA motif, in which valine substituted alanine for *LbXIP1;2*, while serine substituted asparagine and valine substituted alanine for *LbXIP1;6*. Therefore, both the first and the second NPA motifs were not conserved for the LbNIP subfamily.

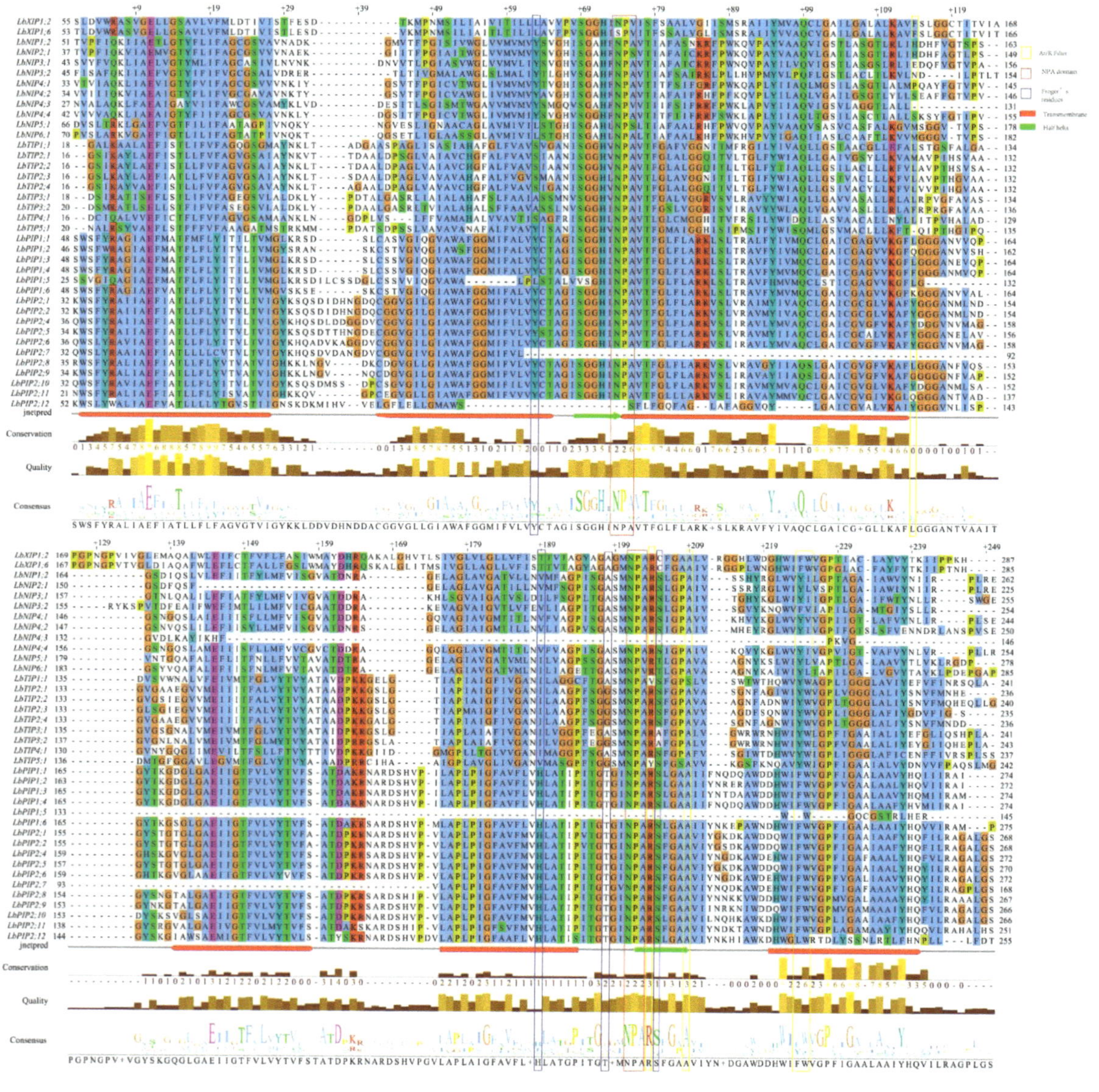

Figure 4. Protein sequence alignment of LbAQPs. Conserved transmembrane domains (TM1-6) and amino acids at NPA domains, ar/R selectivity filters, and Froger's residues identified in LbAQPs.

3.5. Analyzing Cis-Elements in the LbAQP Promoters

Promoter cis-acting elements are important binding regions of transcription initiation factors and play an important role in regulating gene expression. To further explore the possible biological functions of the 38 LbAQPs, the presence of cis-acting elements in the 1 kb upstream promoter regions of the corresponding genes was predicted using the PlantCARE database (Figure 5). This result indicated that, except for the core promoter elements, such as TATA-box and CAAT-box (data not shown), some unique cis-acting elements related to hormones, stress resistance, and tissue and organ development were identified, which are speculated to play regulatory roles in the activation and induction of *LbAQP* expression. The identified phytohormone responsiveness elements included ABA response-related element (ABRE), TGACG-motif (cis-acting regulatory element involved in MeJA responsiveness), Eth-responsive element (ERE), P-box and GA-responsive element (GARE) motif, TCA-element (SA-acting element), and TGA-element (auxin-responsive element). The biotic/abiotic stress response elements included MYC, MYB, stress response element (STRE), LTR (low-temperature control element), wound-responsive element (WRE3), W-box, TC-rich repeats (cis-acting element involved in defense and stress responsiveness), DRE core (drought and osmotic stress induction element), anaerobic induction regulatory element (ARE), and MYB-binding site (MBS, drought-responsive element), which may be related to the tolerance and response mechanisms of plants to biotic or abiotic stress. The others included G-box (light-responsive element), Box-4 (part of a conserved DNA module involved in light response), TCT-motif (part of a light-responsive element), AAGAA-motif (cis-elements for oxidative defense pathway), and circadian (cis-acting regulatory element involved in circadian control). Among these, the more abundant cis-elements in the LbAQP promoters were MYC, MYB, ABRE, G-box, ARE, STRE, and Box-4, reflecting the vital functions of these genes in abiotic stress response and light response.

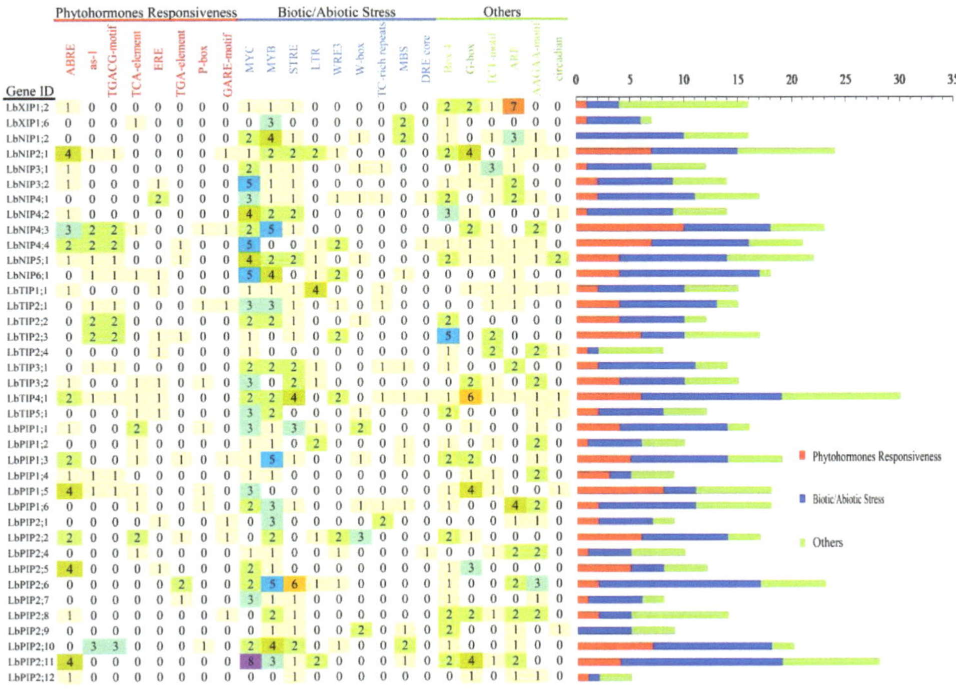

Figure 5. The cis-acting elements on the putative promoters of the *LbAQP* genes. The identified cis-acting elements were mainly divided into three major categories: phytohormone responsiveness, biotic/abiotic stress, and others.

3.6. Tissue-Specific Expression

To determine the tissue-specific expression of this family of genes, a quantitative real-time PCR (qRT-PCR) assay was conducted in various tissues of *L. barbarum*, including roots, stems, leaves, flowers, and fruits (Figure 6). Five genes were specifically expressed in roots (*LbNIP3;1*, *LbTIP2;3*, *LbTIP2;4*, *LbTIP2;2*, and *LbPIP1;4*). Although they were also expressed in other tissues, the genes with the highest transcripts in roots are *LbNIP4;2*, *LbPIP1;3*, *LbPIP1;1*, *LbPIP2;6*, *LbPIP2;4*, *LbNIP1;2*, and *LbNIP2;1*. The genes *LbTIP3;1* and

LbTIP3;2 were mainly expressed in fruits. Eleven genes had maximal transcript levels in stems versus other tissues, which included *LbNIP5;1*, *LbNIP6;1*, *LbTIP4;1*, *LbTIP2;1*, *LbXIP1;6*, *LbPIP1;6*, *LbPIP1;2*, *LbPIP2;8*, *LbPIP2;11*, *LbPIP2;7*, and *LbPIP2;10*. The genes that were highly coexpressed in stems and leaves were *LbTIP2;1*, *LbXIP1;2*, *LbPIP2;7*, *LbPIP2;10*, *LbPIP2;1*, *LbNIP6;1*, and *LbPIP2;2*, while the genes that were highly coexpressed in roots and stems were *LbNIP5;1*, *LbNIP2;1*, *LbNIP1;2*, and *LbPIP2;6*. The genes that were coexpressed in roots, stems, and leaves were *LbNIP1;2* and *LbPIP2;7*. Meanwhile, the gene that was specifically expressed in leaves was *LbPIP2;5* and the genes with the highest transcript levels in leaves were *LbXIP1;2* and *LbPIP2;12*. The most highly expressed genes in flowers were *LbNIP4;3*, *LbNIP4;1*, and *LbTIP5;1*, although many genes had their lowest transcript levels in flowers vs. other tissues, such as *LbPIP1;3*, *LbPIP1;1*, *LbPIP2;9*, and *LbPIP2;6*. Taken together, the overlapping and preferential expression patterns of these *LbAQPs* might confer *Lycium* to conduct distinct ABA responses in specific tissues.

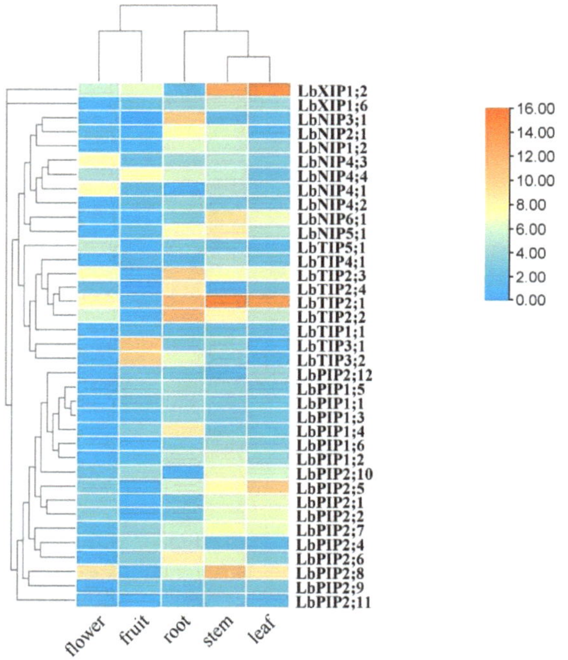

Figure 6. Tissue-specific expression of *LbAQP* genes. Tissue-specific expression of *LbAQPs* was determined by quantitative real-time PCR (qRT-PCR) in leaves, young roots, stems, ripening fruits, and flowers with gene-specific primers. qRT-PCR was performed in triplicate, and the fold changes were analyzed via the $2^{-\Delta\Delta CT}$ method using the *LbACTIN1* gene as an internal control. Values are the means of three independent experiments.

3.7. Expression Profiles of LbAQPs during Fruit Ripening

Fruit development and ripening are complex processes that undergo dramatic physiological changes, including a changing water status [24–26]. To determine whether the expression of *LbAQPs* is sensitive to developmental cues, the transcriptional levels of 38 *LbAQPs* were analyzed using qRT-PCR with gene-specific primers (Figure 7). Within the *LbAQP* gene families, 24 genes were downregulated and three genes (*LbNIP4;1*, *LbNIP4;3*, and *LbNIP4;4*) were upregulated during *L. barbarum* fruit ripening. While the upregulated genes all belonged to the LbNIP subfamily, the downregulated genes were distributed in all four subfamilies. Ten *LbAQP* genes presented irregular expression patterns, in which the transcripts of four genes first increased and then decreased, whereas those of six genes first decreased and then increased during fruit ripening. Among the downregulated genes, most were significantly expressed in the S1 stage, presenting a negative correlation with the fruit relative water content (RWC).

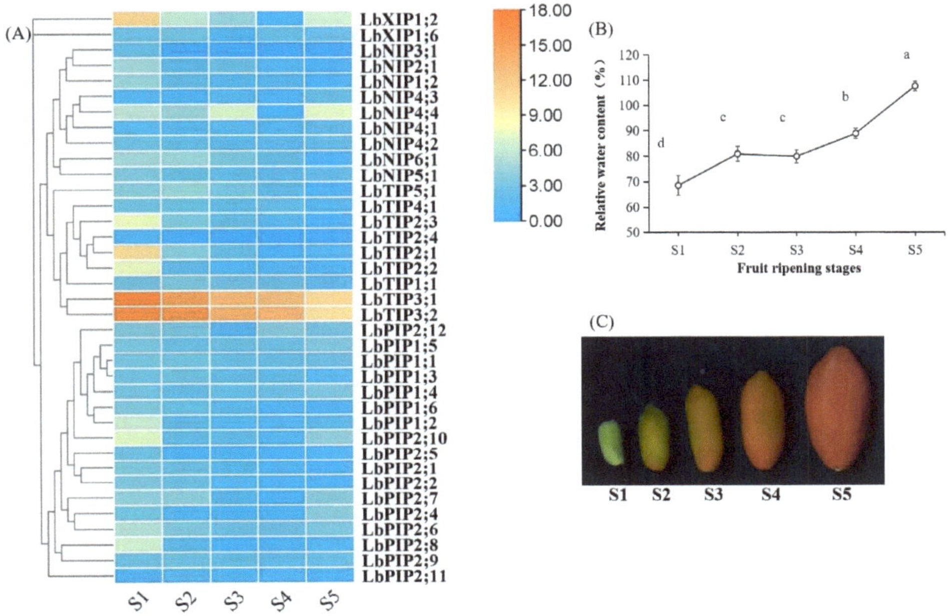

Figure 7. Development-dependent expression profiling during fruit ripening. (**A**) The transcript abundance of *LbAQPs* during fruit ripening was analyzed at five different developmental stages under their natural state. The fold changes in the relative expression levels were analyzed via the $2^{-\Delta\Delta CT}$ method using the *LbACTIN1* gene as an internal control. Values are the means of three independent experiments. (**B**) Relative water content (%) of the fruits for the above five ripening stages. Values represent the mean ± SEM, n = 3 (biological replicates). Means with different letters are significantly different ($p < 0.01$; one-sided ANOVA). (**C**) Phenotypes of the five representative ripening stages (S1, S2, S3, S4, and S5) of *L. barbarum* fruits.

3.8. Expression Profiles of LbAQPs in Response to Heat Stress

Whereas most of the *LbAQP* genes were significantly upregulated by heat stress, *LbNIP5;1*, *LbXIP1;2*, and *LbPIP1;6* were downregulated (Figure 8). For the upregulated genes, most of their transcript abundance presented first increasing and then decreasing expression patterns, while several were maintained at constant levels (*LbNIP4;2* and *LbTIP1;1*), several continuously increased (*LbNIP3;1*, *LbTIP5;1*, and *LbXIP1;6*), and several had irregular changes during 24 h of heat stress. Compared with the control values, approximately half of the gene transcripts were upregulated by over fivefold. Among them, *LbNIP3;1*, *LbTIP3;1*, *LbTIP2;4*, and *LbNIP4;4* were upregulated by over 15-, 30-, 40-, and 70-fold, respectively. Six genes (*LbNIP4;1*, *LbNIP4;4*, *LbTIP3;1*, *LbTIP3;2*, *LbPIP2;9*, and *LbPIP2;11*) achieved maximal transcription at 1 h, and 13 genes peaked at 3 h, which can be considered earlier response genes. The others were part of the later response genes.

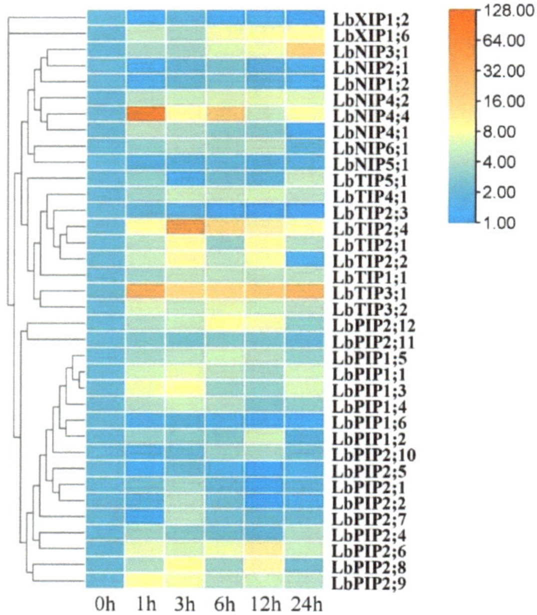

Figure 8. Expression profiling responses to heat stress. The expression levels of *LbAQPs* were measured in the leaves of 2-month-old *L. barbarum* subjected to 42 °C for 0, 1, 3, 12, or 24 h. The fold changes in the relative expression levels were analyzed via the $2^{-\Delta\Delta CT}$ method using the *LbACTIN1* gene as an internal control. Values are the means of three independent experiments.

4. Discussion

Thirty-eight homologs of *Arabidopsis* AQPs were identified in the *L. barbarum* genome, indicating this gene family conservation in higher plants [6]. Certainly, it should be noted that clear genus-specific features exist. Within *Solanaceae* species, 45, 47, and 76 AQPs have been identified in tomato (*Solanum lycopersicum*), potato (*Solanum tuberosum*), and tobacco (*Nicotiana tabacum*) genomes, respectively [21–23]. Moreover, all five subfamilies of AQPs that have been identified thus far were present in the above three Solanaceae species, whereas SIP was not found in the *L. barbarum* genome. Interestingly, within the 38 *LbAQP* genes, both truncation and duplication events occurred. It is increasingly recognized that not all plant AQP subfamilies are good water channels. Within the NIPs (mainly), as well as some PIPs and XIPs, transported substrates include metalloids, protonated organic acids, or metal complexes [27]. The polymorphism of LbAQPs may reflect the multifunctionality of this family of proteins and may also reflect the best acclimation of *Lycium* plants to extreme environments, reversing other Solanaceae species. Of course, it cannot be ruled out that now we only have a relatively coarse reference genome of *L. barbarum*. In the future, with the further improvement of the genome map, the number of AQP genes for *Lycium* plants at the whole-genome level may also vary, as has been shown in maize. Moreover, it should be highlighted that different species within *Lycium* plants may have varied AQP gene numbers, similar to what has been observed in cotton, olive trees, and *Linum* species. Pangenome exploration will bring interesting results in the future.

Developing fruits are strong sink organs, and the accumulation of sugars in them causes a negative water potential, which endows AQPs with pivotal roles at both the tissue and cellular levels [3,28,29]. An attention study was conducted in tomato fruits, in which regulation of AQP expression can clearly modify fruit quality (e.g., size, flavor, nutrition, and firmness) by the method of deficit irrigation-derived water scarcity [3,28,29]. It has been shown in grapes that the discharge of surplus phloem water may be required for normal grape ripening [25]. Our present work showed that most of the AQP genes identified in

L. barbarum were downregulated during fruit ripening and presented a negative correlation with fruit RWC. This means that most of them were significantly involved in fruit growth, while several took part in fruit ripening. Considering the coexpression manner of most of the genes and the great change levels within different ripening stages, it can be speculated that at least some AQPs identified here as expressed in fruits are necessary for water transport during fruit development.

With global climate change, extreme temperatures are becoming increasingly frequent, posing serious threats to plant growth and food production [24]. Since they have evolutionarily acclimated to cold and cool local environments for a long time [19], the main threat to current and future global production and quality of *Lycium* plants is heat stress. In the present work, we found that most of the LbAQP transcripts presented quick and sharp responses to heat stress following seedling exposure to a 42 °C temperature, indicating the potential targets of this protein family for engineering the heat tolerance of *Lycium* species. Considering the rapidity and recoverability of the transcriptional response pattern, it is speculated that LbAQPs may also be involved in the heat stress signaling pathway, as has been determined in other physiological processes [1,14]. Among them, *LbNIP4;4* and *LbTIP3;1* may play primary functions, followed by *LbTIP2;4*, *LbTIP2;1*, *LbTIP2;2*, *LbPIP1;3*, *LbPIP2;8*, and *LbPIP1;1*.

Except for transport water, it is increasingly recognized that AQPs have emerged as central membrane targets of environmental and hormonal signaling pathways acting on plant–water relations [1,27,30,31]. Considering that functional compounds, as pharmacodynamic components, are generally secondary metabolites formed by medicinal plants against stress, more studies, including molecular genetics and systems biology approaches, are now needed to comprehend how LbAQPs and secondary metabolites interact during fruit ripening and *Lycium* plants respond to abiotic stress.

5. Conclusions

Thirty-eight LbAQP genes were first identified and characterized from the *L. barbarum* genome, which fell into four subfamilies, including 17 LbPIP, 9 LbTIP, 10 LbNIP, and 2 LbXIP. There were no SIP subfamily genes found in *Lycium* plants, unlike in other Solanaceae. The transcript profiling showed that their expression presented clear tissue-, developmental-, and stressor-specific patterns. The rapidity and recoverability of the transcriptional response highlight the potential roles of this protein family in regulating *L. barbarum* fruit ripening and the heat stress response. These findings also suggest that LbPYLs might be good candidates for future biotechnological use to enhance *Lycium* resistance to drought and hot environments. Our results lay a foundation for further studying the molecular mechanism of the water relationship of *Lycium* plants, especially for the two above key physiological processes.

Supplementary Materials: The following supporting information can be downloaded at: https://www.mdpi.com/article/10.3390/cimb44120404/s1.

Author Contributions: Z.M. and K.Q. designed the research. W.H. and Y.C. performed the experiments. W.H. and Y.Y. conducted the data analyses. K.Q. and X.Q. conducted the field management work. Z.M., M.L. and A.L. wrote the manuscript. All the authors have read and approved the manuscript. All authors have read and agreed to the published version of the manuscript.

Funding: This work was jointly supported by the Key Research and Development Project foundation of Ningxia province of China (No. 2020BFH03005), the Foreign Science and Technology Cooperation Project of Ningxia Academy of Agriculture and Forestry Sciences (No. DW-X-2020009).

Institutional Review Board Statement: Not applicable.

Informed Consent Statement: Not applicable.

Data Availability Statement: The original data presented in the study are included in the article/Supplementary Material, further inquiries can be directed to the corresponding author.

Conflicts of Interest: The authors declare that the research was conducted in the absence of any commercial or financial relationship that could be construed as a potential conflict of interest.

References

1. Maurel, C.; Tournaire-Roux, C.; Verdoucq, L.; Santoni, V. Hormonal and environmental signaling pathways target membrane water transport. *Plant Physiol.* **2021**, *187*, 2056–2070. [CrossRef] [PubMed]
2. Wang, Y.; Zhao, Z.; Liu, F.; Sun, L.; Hao, F. Versatile roles of aquaporins in plant growth and development. *Int. J. Mol. Sci.* **2020**, *21*, 9485. [CrossRef] [PubMed]
3. Li, H.; Zhang, X.; Hou, X.; Du, T. Developmental and water deficit-induced changes in hydraulic properties and xylem anatomy of tomato fruit and pedicels. *J. Exp. Bot.* **2021**, *72*, 2741–2756. [CrossRef] [PubMed]
4. Fox, A.R.; Maistriaux, L.C.; Chaumont, F. Toward understanding of the high number of plant aquaporin isoforms and multiple regulation mechanisms. *Plant Sci.* **2017**, *264*, 179–187. [CrossRef] [PubMed]
5. Laloux, T.; Junqueira, B.; Maistriaux, L.C.; Ahmed, J.; Jurkiewicz, A.; Chaumont, F. Plant and mammal aquaporins: Same but different. *Int. J. Mol. Sci.* **2018**, *19*, 521. [CrossRef] [PubMed]
6. Maurel, C.; Boursiac, Y.; Luu, D.T.; Santoni, V.; Shahzad, Z.; Verdoucq, L. Aquaporins in plants. *Physiol. Rev.* **2015**, *95*, 1321–1358. [CrossRef]
7. Bezerra-Neto, J.P.; de Araújo, F.C.; Ferreira-Neto, J.R.C.; da Silva, M.D.; Pandolfi, V.; Aburjaile, F.F.; Sakamoto, T.; de Oliveira Silva, R.L.; Kido, E.A.; Barbosa Amorim, L.L.; et al. Plant aquaporins: Diversity, evolution and biotechnological applications. *Curr. Protein Pept. Sci.* **2019**, *20*, 368–395. [CrossRef]
8. Ahmed, S.; Kouser, S.; Asgher, M.; Gandhi, S.G. Plant aquaporins: A frontward to make crop plants drought resistant. *Physiol. Plant.* **2021**, *172*, 1089–1105. [CrossRef]
9. Srivastava, A.K.; Penna, S.; Nguyen, D.V.; Tran, L.S. Multifaceted roles of aquaporins as molecular conduits in plant responses to abiotic stresses. *Crit. Rev. Biotechnol.* **2016**, *36*, 389–398. [CrossRef]
10. Kapilan, R.; Vaziri, M.; Zwiazek, J.J. Regulation of aquaporins in plants under stress. *Biol. Res.* **2018**, *51*, 4. [CrossRef]
11. Deshmukh, R.K.; Sonah, H.; Belanger, R.R. Plant Aquaporins: Genome-wide identification, transcriptomics, proteomics, and advanced analytical tools. *Front. Plant Sci.* **2016**, *7*, 1896. [CrossRef] [PubMed]
12. Guo, Z.; Ma, D.; Li, J.; Wei, M.; Zhang, L.; Zhou, L.; Zhou, X.; He, S.; Wang, L.; Shen, Y.; et al. Genome-wide identification and characterization of aquaporins in mangrove plant Kandelia obovata and its role in response to the intertidal environment. *Plant Cell Environ.* **2022**, *45*, 1698–1718. [CrossRef] [PubMed]
13. Britto, D.T.; Kronzucker, H.J. From aquaporin to ecosystem: Plants in the water cycle. *J. Plant Physiol.* **2018**, *227*, 1–2. [CrossRef] [PubMed]
14. Feng, Z.J.; Liu, N.; Zhang, G.W.; Niu, F.G.; Xu, S.C.; Gong, Y.M. Investigation of the AQP family in soybean and the promoter activity of TIP2;6 in heat stress and hormone responses. *Int. J. Mol. Sci.* **2019**, *20*, 262. [CrossRef] [PubMed]
15. Reddy, P.S.; Dhaware, M.G.; Sivasakthi, K.; Divya, K.; Nagaraju, M.; Sri Cindhuri, K.; Kavi Kishor, P.B.; Bhatnagar-Mathur, P.; Vadez, V.; Sharma, K.K. Pearl Millet aquaporin gene PgPIP2;6 improves abiotic stress tolerance in transgenic tobacco. *Front. Plant Sci.* **2022**, *13*, 820996. [CrossRef]
16. Potterat, O. Goji (Lycium barbarum and L. chinense): Phytochemistry, pharmacology and safety in the perspective of traditional uses and recent popularity. *Planta Med.* **2010**, *76*, 7–19. [CrossRef]
17. Yao, R.; Heinrich, M.; Weckerle, C.S. The genus Lycium as food and medicine: A botanical, ethnobotanical and historical review. *J. Ethnopharmacol.* **2018**, *212*, 50–66. [CrossRef]
18. Cao, Y.L.; Li, Y.L.; Fan, Y.F.; Li, Z.; Yoshida, K.; Wang, J.Y.; Ma, X.K.; Wang, N.; Mitsuda, N.; Kotake, T.; et al. Wolfberry genomes and the evolution of Lycium (Solanaceae). *Commun. Biol.* **2021**, *4*, 671. [CrossRef]
19. Gong, H.; Rehman, F.; Ma, Y.; Zeng, S.; Yang, T.; Huang, J.; Li, Z.; Wu, D.; Wang, Y. Germplasm resources and strategy for genetic breeding of Lycium species: A Review. *Front. Plant Sci.* **2022**, *13*, 802936. [CrossRef]
20. Qin, X.; Qin, B.; He, W.; Chen, Y.; Yin, Y.; Cao, Y.; An, W.; Mu, Z.; Qin, K. Metabolomic and Transcriptomic Analyses of Lycium barbarum L. under Heat Stress. *Sustainability* **2022**, *14*, 12617. [CrossRef]
21. Reuscher, S.; Akiyama, M.; Mori, C.; Aoki, K.; Shibata, D.; Shiratake, K. Genome-wide identification and expression analysis of aquaporins in tomato. *PLoS ONE* **2013**, *8*, e79052. [CrossRef] [PubMed]
22. Venkatesh, J.; Yu, J.W.; Park, S.W. Genome-wide analysis and expression profiling of the Solanum tuberosum aquaporins. *Plant Physiol. Biochem.* **2013**, *73*, 392–404. [CrossRef]
23. De Rosa, A.; Watson-Lazowski, A.; Evans, J.R.; Groszmann, M. Genome-wide identification and characterisation of aquaporins in Nicotiana tabacum and their relationships with other Solanaceae species. *BMC Plant Biol.* **2020**, *20*, 266. [CrossRef] [PubMed]
24. Ripoll, J.; Urban, L.; Staudt, M.; Lopez-Lauri, F.; Bidel, L.P.; Bertin, N. Water shortage and quality of fleshy fruits–making the most of the unavoidable. *J. Exp. Bot.* **2014**, *65*, 4097–4117. [CrossRef] [PubMed]
25. Zhang, Y.; Keller, M. Discharge of surplus phloem water may be required for normal grape ripening. *J. Exp. Bot.* **2017**, *68*, 585–595. [CrossRef]
26. Song, S.; Zhang, D.; Ma, F.; Xing, W.; Huang, D.; Wu, B.; Chen, J.; Chen, D.; Xu, B.; Xu, Y. Genome-wide identification and expression analyses of the aquaporin gene family in Passion fruit (Passiflora edulis), revealing PeTIP3-2 to be involved in drought Stress. *Int. J. Mol. Sci.* **2022**, *23*, 5720. [CrossRef]

27. Tyerman, S.D.; McGaughey, S.A.; Qiu, J.; Yool, A.J.; Byrt, C.S. Adaptable and multifunctional ion-conducting aquaporins. Annu. *Rev. Plant. Biol.* **2021**, *72*, 703–736. [CrossRef]
28. Hou, X.; Zhang, W.; Du, T.; Kang, S.; Davies, W.J. Responses of water accumulation and solute metabolism in tomato fruit to water scarcity and implications for main fruit quality variables. *J. Exp. Bot.* **2020**, *71*, 1249–1264. [CrossRef]
29. Hou, X.; Li, H.; Zhang, W.; Yao, Z.; Wang, Y.; Du, T. Water transport in fleshy fruits: Research advances, methodologies, and future directions. *Physiol. Plant.* **2021**, *172*, 2203–2216. [CrossRef]
30. Chen, Q.; Liu, R.; Wu, Y.; Wei, S.; Wang, Q.; Zheng, Y.; Xia, R.; Shang, X.; Yu, F.; Yang, X.; et al. ERAD-related E2 and E3 enzymes modulate the drought response by regulating the stability of PIP2 aquaporins. *Plant Cell* **2021**, *33*, 2883–2898. [CrossRef]
31. Jing, W.K.; Li, Y.H.; Zhang, S.; Zhou, X.F.; Gao, J.P.; Ma, N. Aquaporin, beyond a transporter. *Horticult. Plant J.* **2022**. [CrossRef]

current issues in molecular biology

MDPI

Article

Identification of Candidate Genes Associated with Pulp Color by Transcriptomic Analysis of 'Huaxiu' Plum (*Prunus salicina* Lindl.) during Fruit-Ripening

Gang Wang, Wenxin Weng, Zhanhui Jia, Jiyu Zhang, Tao Wang and Jiping Xuan *

Jiangsu Key Laboratory for the Research and Utilization of Plant Resources, Institute of Botany, Jiangsu Province and Chinese Academy of Sciences, Nanjing 210014, China
* Correspondence: xuanjiping@cnbg.net

Abstract: The plum (*Prunus salicina* Lindl.) is one of the traditional and economically important stone fruit trees in China. Anthocyanins are important pigments in plums. However, little is known about the molecular mechanisms underlying anthocyanin accumulation in plum fruits, which has hindered research on the molecular mechanism of its utilization. Our research shows that the chlorophyll content was gradually decreased and the contents of anthocyanin and flavonoid increased during the coloring process of the pulp in 'Huaxiu' plums (*P. salicina*). Then, the RNA-Seq technique was used to analyze the transcriptome of pulp color changes with three different stages (yellow, orange, and red) in the 'Huaxiu' plum (*P. salicina*). A total of 57,119 unigenes with a mean length of 953 bp were generated, and 61.6% of them were annotated to public databases. The Gene Ontology (GO) database assigned 21,438 unigenes with biological process, cellular components, and molecular function. In addition, 32,146 unigenes were clustered into 25 categories for functional classification by the COG database, and 7595 unigenes were mapped to 128 KEGG pathways by the KEGG pathway database. Of these, 1095 (YS-versus-OS), 4947 (YS-versus-RS), and 3414 (OS-versus-RS) genes were significantly expressed differentially between two coloration stages. The GO and KEGG pathway enrichment analysis revealed that 20 and 1 differentially expressed genes (DEG) are involved in flavonoid biosynthesis and anthocyanin biosynthesis, respectively. Finally, we mainly identified three structural genes as candidate genes. The transcriptome information in this study provide a basis for further studies of pulp colors in plum and contribute to our understanding of the molecular mechanisms underlying anthocyanin biosynthesis in pulp.

Keywords: *Prunus salicina*; pulp color; transcriptome; RNA-Seq; DGE; anthocyanin and flavonoid biosynthesis

Citation: Wang, G.; Weng, W.; Jia, Z.; Zhang, J.; Wang, T.; Xuan, J. Identification of Candidate Genes Associated with Pulp Color by Transcriptomic Analysis of 'Huaxiu' Plum (*Prunus salicina* Lindl.) during Fruit-Ripening. *Curr. Issues Mol. Biol.* **2022**, *44*, 6368–6384. https://doi.org/10.3390/cimb44120434

Academic Editors: Shimeles Tilahun and Vijai Bhadauria

Received: 4 November 2022
Accepted: 12 December 2022
Published: 15 December 2022

Publisher's Note: MDPI stays neutral with regard to jurisdictional claims in published maps and institutional affiliations.

1. Introduction

The plum (*Prunus salicina* L.) is an economically important stone fruit tree in China. The fruit can be eaten fresh or processed into many value-added products, such as juice, fermented wine, and preserves [1]. When the fruit matures, the pulp color can be green, yellow–green, yellow, red, or purple based on the difference cultivar of the plum. The fruit color is mainly determined by the level of anthocyanin accumulation [2]. Anthocyanins have been extensively studied, not only because they result in the production of red, purple, and black pigments, but also in the contexts of their diverse roles in UV protection and pathogen defense, as well as their nutritional value in the human diet [3].

Anthocyanins are formed by the combination of anthocyanidin and one or more carbohydrates through glycosides bonds which then stably exist in plant cells, which make the fruit appear bright colors [4]. Anthocyanins are synthesized through the flavonoid pathway which has been well-studied in many plants. Many structural genes of the flavonoid pathway have been isolated and identified as the critical gene for anthocyanin biosynthesis in plants, such as chalcone synthase (CHS), chalcone isomerase (CHI), flavonoid

3′-hydroxylase (F3′H), anthocyanidin synthase (ANS), and UDP-glucose: flavonoid 3-O-glucosyltransferase (UFGT) [4–7]. In addition, transcription factors play important roles in modulating anthocyanin biosynthetic pathway activity and color changes, including MYB proteins, basic helix–loop–helix (bHLH) proteins, and WD40 proteins [8]. These regulators form an MYB–bHLH–WD40 (MBW) complex that binds to promoters and regulates the expression of structural genes of the anthocyanin biosynthetic pathway [9,10]. The role of MBW in anthocyanin biosynthesis has been elucidated in fruit trees such as apple [11], pears [12,13], grape [14], strawberry [15], Chinese bayberry [16], kiwifruit [17], litchi [18], blueberry [19], sweet cherry [20] peach [21,22], plum [1], and fig [23]. In addition, other regulatory factors also affect anthocyanin biosynthesis via interaction with the MBW complex or by modulating the transcription of structural genes directly [1]. A SQUAMOSA promoter-binding protein-like (SPL) transcription factor (*PpSPL1*) has been shown to inhibit the expression of anthocyanin biosynthetic genes and negatively regulate anthocyanin accumulation through the destabilization of MBW [24]. MADS-box and NAC transcription factors also are reported to be involved in the regulation of anthocyanin accumulation [25,26].

In recent years, RNA sequencing (RNA-Seq) transcriptome analysis has been extensively used for the identification of functional genes in fruit trees, including plum. The genome of Chinese plum 'Sanyueli' and the chloroplast genome sequence of 'Wushan plum' (*P. salicina*) have been published [27,28]. The transcriptome analysis of 'Siyueli'-pollinated and 'Yinhongli'-pollinated fruits revealed 2762 and 1018 differentially expressed genes (DEGs) involved in the response to different pollen sources [29]. Recently reported are the fruit skin transcriptome assembly of 'Angeleno' and 'Lamoon' Japanese plum cultivars with different skin color [30], and the transcriptomic changes during fruit-ripening in the red-fleshed plum cultivar 'Furongli' (*P. salicina*) [1]. During the ripening and development of plum fruits, chlorophyll degradation, the content of organic acids decreases, while anthocyanins accumulate rapidly in the colored varieties, and this phenomenon is especially obvious in the late stage of plum fruit-ripening [31]. In this study, as a first step towards understanding gene expression during fruit-ripening in plum, the three mature stages of 'Huaxiu' plum fruits with pulp colors of yellow, orange, and red were collected for transcriptome sequencing. This study presents the results of a comprehensive analysis of transcriptome data from fruits at three representative developmental stages. The results will help us to understand the gene expression difference and explore the molecular mechanisms of the biosynthetic pathways in secondary metabolites in plum.

2. Materials and Methods

2.1. Plant Materials

Fruits of 'Huaxiu' plum (*P. salicina*) were collected from 6-year-old field-grown trees in an orchard in Donghai County, Jiangsu Province, China. In July 2007, a natural-bud mutation for early maturing fruit of 'Qiuji' was found by a farmer and named as 'Huaxiu'. According to our observations, 'Huaxiu' characteristics are mid-early ripening (fruits mature in late-July in this area, about 108 days after flowering), where the skin is dark purple and the pulp orange–red. These mature fruit samples were divided into three different stages based on pulp color and termed the 'Yellow stage' (YS), 'Orange stage' (OS), 'Red stage' (RS), respectively.

During the fruit ripening periods, three biological replicates were collected per sample, each with 20 fruits randomly collected from two trees in order to decrease background variation. Then, three pulps were peeled and sliced into appropriate pieces after measuring the weight and diameter, and then immediately frozen in liquid nitrogen for the determination of chlorophyll, carotenoid, anthocyanins, and flavonoids. Two biological replicated samples were used at each stage for transcriptome sequencing due to the background and limited yield of fruits.

2.2. Total Anthocyanin, Flavonoid, and Chlorophyll Measurements

Total anthocyanin levels were determined according to the HPLC system [32]. Approximately 1.5 g of the sample was ground to a fine powder in liquid nitrogen and extracted with a 1% HCl-methanol solution at 4 °C for 24 h to obtain the extract by filtration. Then, 20 μL of the samples were injected into a C18 column (Agilent Zorbax Eclipse SB-C18, 4.6 mm × 100 mm, 1.8 μm, Santa Clara, CA, USA). The binary solvent system was 5% formic acid in water as mobile phase A and methanol as mobile phase B. The gradient elution was 20% B at 0–30 min, 40% B at 30–40 min, 100% B at 40 min. The flow rate was kept at 10 mL/min, and the column temperature was maintained at 40 °C. The chlorophyll and carotenoids levels were measured according to the absorbance at 663 nm, 645 nm, and 470 nm (A_{663}, A_{645}, and A_{470}) [33]. The flavonoid levels were measured as previously described [34]. Each sample comprised three biological replicates and the analysis represented the results of three independent experiments.

2.3. RNA Extraction, cDNA Library Construction, and Sequencing

Total RNA from approximately 100 mg of frozen fruits of YS, OS, and RS was extracted using the TRIzol 1 Reagent (Invitrogen, Waltham, MA, USA) according to manufacturer's protocol. Genomic DNA was eliminated by using RNase-free DNase I (TaKaRa, Dalian, China) and then RNA integrity and purity was confirmed on 1% agarose gels and NanoDrop™ 2000 spectrophotometer (NanoDrop Technologies, Wilmington, DE, USA). All RNA extracts showed a 260/280 nm ratio from 1.9 to 2.2. Approximately 25 μg total RNA of no less than 600 ng/μL concentration was used for the cDNA library construction. The steps of the cDNA library construction and the subsequent sequencing using an Illumina HiSeq™ 4000 (San Diego, CA, USA) were performed by staff at Gene Denovo Biotechnology Co. (Guangzhou, China).

2.4. De Novo Assembly and Functional Annotation

First, the raw data were filtered to remove reads with unknown sequencing 'N' and low-quality reads with more than 50% bases with a quality value ≤5. Then, mixing the clean reads from the three samples, the de novo assembly was performed using the assembly software Trinity (Trinityrnaseq_r2013_08_14 version) to obtain transcript sequences, and the longest transcript in each gene was taken as a single gene cluster (unigene) [35]. To annotate unigenes sequences of 'Huaxiu' plums, Blastx search (E-value < 10^{-5}) was used to search against the NCBI nonredundant protein database [36] (NR, http://www.ncbi.nlm.nih.gov/refseq/about/nonredundantproteins, last accessed on 23 August 2022), the SwissProt protein database (http://www.expasy.ch/sprot, last accessed on 24 August 2022), the Kyoto Encyclopedia of Genes and Genomes database [37] (KEGG, http://www.genome.jp/kegg, last accessed on 24 August 2022), and the Clusters of Orthologous Groups of proteins database [38] (COG, http://www.ncbi.nlm.nih.gov/COG, last accessed on 24 August 2022). These annotation databases were used for comparisons with homologous genes. GO function classification was performed using Blast2GO [39] (http://www.blast2go.com/b2ghome, last accessed on 14 September 2022) with E ≤ 10^{-5}, the distribution of the GO functional classifications of the unigenes was plotted using WEGO software [40], which allowed categorization into three different GO terms, including molecular function, cellular component, and biological process. The KEGG pathway annotation was performed using the Blast software against the KEGG database [41].

2.5. Expression Analysis

After assembly, unigene expression levels were quantified using fragments per kilo base of transcript per million mapped reads (FPKM). FPKM values were calculated using RSEM (RNA-Seq by Expectation Maximization) [42]. Differentially expressed genes (DEGs) analysis of two samples was performed using the DEG Seq R package [43]. The differentially expressed unigenes between two samples were screened using a false discovery rate (FDR), which is used to determine *p*-value thresholds in multiple testing [44]. The significance of the DEGs were determined based on a threshold of FDR <0.05 and absolute log2 fold changes $\geq$1 in a comparison as significant DEGs. GO enrichment analysis provided all GO terms that significantly enriched in DEGs compared to the genome background and filtered the DEGs that corresponded to biological functions. KEGG pathway enrichment analysis identified significantly enriched metabolic pathways or signal transduction pathways in DEGs compared with the whole genome background. DEGs were subsequently mapped to the database for the GO and KEGG pathway enrichment analysis with the p-value < 0.05 using the software GOatools [45] (https://github.com/tanghaibao/GOatools, last accessed on 16 September 2022).

Twelve DEGs involved in flavonoid biosynthesis were selected for validation by real-time quantitative RT-PCR (qRT-PCR). Total RNA was extracted and qRT-PCR was performed with a method modified as previously described. The primer sequences used for qRT-PCR are listed in Table S1. The relative gene expression level was calculated according to the $2^{-\Delta\Delta Ct}$ method. To visualize the relative expression levels data, YS was normalized as "1". Three biological and three technical replicates were performed in these experiments.

2.6. Statistical Analysis

Statistical analysis of variance (ANOVA) with Duncan's new multiple range test was carried out to compare cultivar mean values using SPSS Version 16.0 (Chicago, IL, USA). The significance level was set to $p < 0.01$. Graph Pad Prism version 6.0 (Graph Pad Software, San Diego, CA, USA) and Photoshop CS5 (Microsoft, Redmond, WA, USA and Adobe, San Jose, CA, USA) were used for graph plotting.

3. Results

3.1. Total Anthocyanin, Chlorophyll, Carotenoids, and Flavonoid Levels

As indicated in Figure 1A, the pulp color of 'Huaxiu' plums changed from yellow to red during ripening and these coincided with changes in the abundance of total anthocyanins, chlorophyll, carotenoids, and flavonoid in the pulp (Figure 1B). We determined that the total anthocyanins content of 'Huaxiu' plums increased from 6.45×10^{-3} mg/g FW at the pulp yellow stage (YS) to 81.52×10^{-3} mg/g FW at the pulp red stage (RS) as the ripening pulp color proceeded. The chlorophyll content was decreased from 8.94×10^{-3} mg/g at YS to 5.4×10^{-3} mg/g at RS, with a constant increase of the carotenoids content from 2.39×10^{-3} mg/g at YS to 5.61×10^{-3} mg/g at RS. The levels of flavonoids increased continuously during the change of the pulp color, and the contents were 1.59 mg/g at YS, 1.76 mg/g at OS, and 3.79 mg/g at RS, respectively.

3.2. Transcriptome Sequencing and De Novo Assembly

Three cDNA libraries were constructed from the total RNA of 'Huaxiu' plums pulp at yellow, orange, and red stage, we obtained 56,019,255, 58,211,832, and 55,985,990 raw reads, respectively, which consist of 8,380,734,292 bp, 8,713,283,611 bp, and 8,379,336,001 bp (Table S2). In addition, the Q20 of all pulps exceeded 97% and the GC content was around 46% (Table S2).

After the removal of adaptor sequences, ambiguous reads, and low-quality reads for the assembly, a total of 57,119 unigenes consisting of 54,447,493 bp have been acquired. The percentage GC of all the unigenes was 41.07%, and the average length of unigene was 953 bp, the largest and smallest unigenes were 16,373 and 201 bp in length, respectively, and the N50 value was 1997 bp (Table 1). The results show that the sequences of unigenes are mostly concentrated between 200 and 399 bp and 400 and 799 bp, accounting for 46.9% and 20.5% of the total unigenes, respectively (Figure 2A and Table S3).

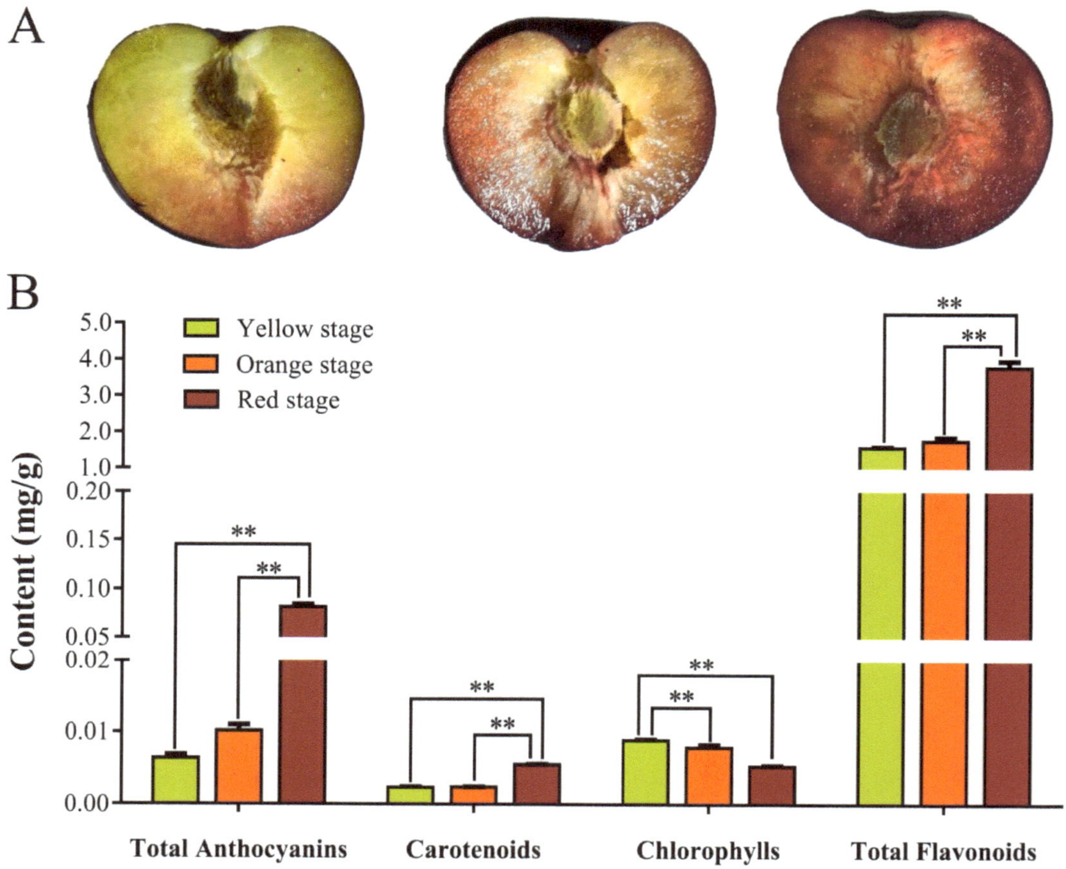

Figure 1. The phenotype of 'Huaxiu' plums fruits and pigment contents in the pulp. (**A**): Images of 'Huaxiu' plums fruits pulp at different coloration stages. (**B**): Contents of total anthocyanin, chlorophylls, carotenoids, and total flavonoids in the pulp of three coloration stages. The vertical bars represent the standard error of triplicate experiments. ** indicates significant differences in comparison to stems at $p < 0.01$.

Table 1. Summary of sequencing and de novo assembly results of unigenes.

Type	Unigene
Genes Num	57,119
Total assembled bases	54,447,493
Percent GC (%)	41.07
Max length (bp)	16,373
Min length (bp)	201
Average length (bp)	953
N50	1997

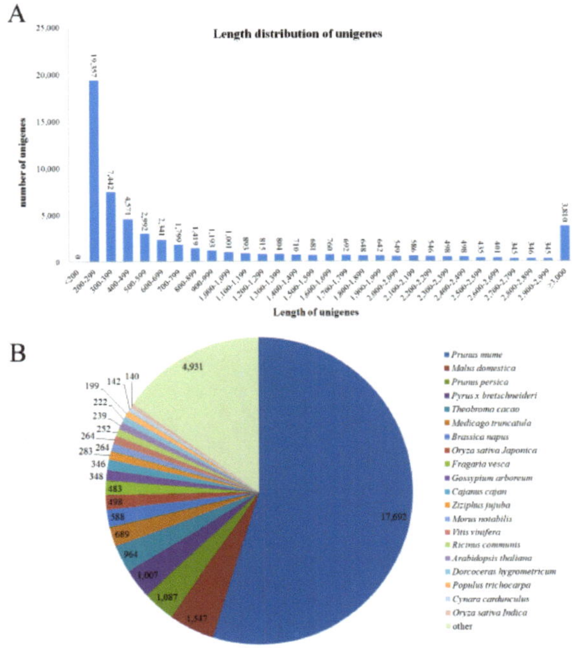

Figure 2. The distribution of the unigenes in 'Huaxiu' plums fruit transcriptome. (**A**): Sequence length distribution of the unigenes in 'Huaxiu' plums fruit transcriptome. The x-axis indicates unigene length interval from 200 bp to >3000 bp. The y-axis indicates the number of unigenes of each sequence length. (**B**): Results of species distribution of 'Huaxiu' plums unigenes top BLAST hits for all homologous sequences in NR (NCBI nonredundant) database. Different species are indicated by different colors.

3.3. Functional Annotation of Unigenes

To annotate the transcriptome of 'Huaxiu' plums, 57,119 unigenes were searched against four public databases (NR, SwissProt, KOG, and KEGG) with a cutoff E-value of 10^{-5} (Figure S1 and Table S4). The functional annotation results showed only 61.6% of the unigenes (35,181) were identified (Table S4). Different unigenes have been matched in different databases, and 32,185 unique sequences were annotated with reference to the NR database, while 2996 unigenes were annotated using the other databases (Table S4).

The species distribution of 'Huaxiu' plum unigenes in the NR database is the best match, and showed the top matches with *Prunus mume* (55.0%), followed by *Malus domestica* (4.8%), *Prunus persica* (3.4%), *Pyrus bretschneideri* (3.1%), *Theobroma cacao* (3.0%), *Medicago truncatula* (2.1%), *Brassica napus* (1.8%), *Oryza sativa Japonica* (1.5%), *Fragaria vesca* (1.5%), *Gossypium arboreum* (1.1%), and *Cajanus cajan* (1.1%) (Figure 2B and Table S5).

3.4. Functional Classification of Unigenes

Functional classification of unigenes by GO assignments resulted in the successful annotation of 21,438 unigenes. These unigenes were classified into three main GO categories, including biological process, molecular function, and cellular components (Figure 3, Table S6). Under the biological process category, metabolic process (12,176), cellular process (11,287), and single-organism process (8730) were the most dominant subcategories; the molecular function category mainly consisted of binding (11,852) and catalytic activity (10,764); the cellular components category mainly consisted of cell (8896), cell part (8895), organelle (7031), and membrane (4084).

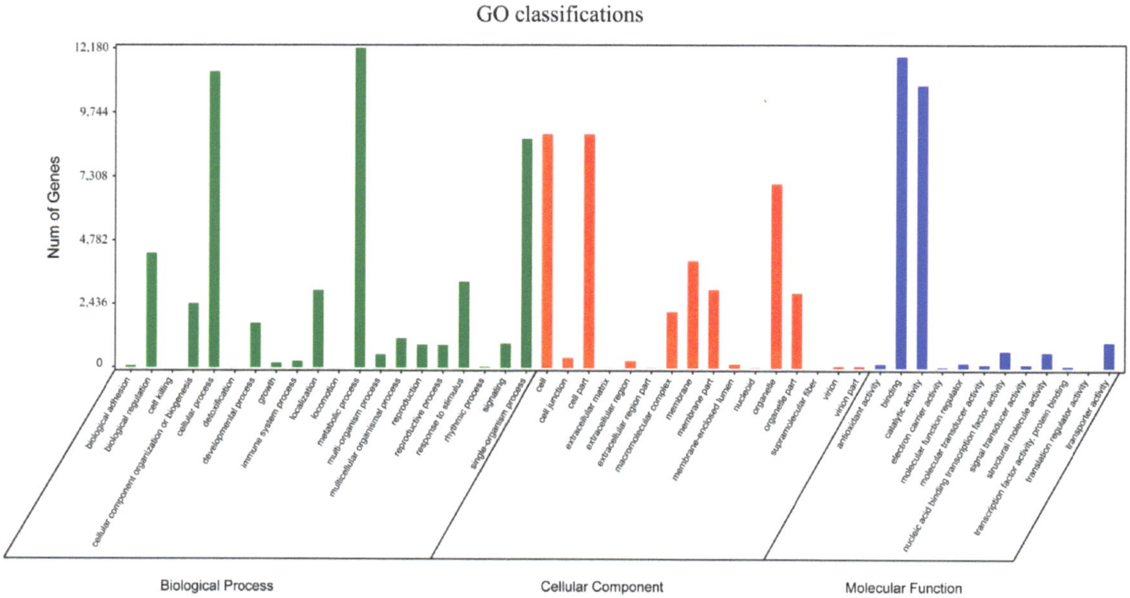

Figure 3. Gene ontology (GO) term level 2 categories of unigenes of the 'Huaxiu' plums (*P. salicina*).

The validity of the plum transcriptome were further evaluated by KOG annotation, whereby 32,146 unigenes sequences were clustered into 25 KOG categories (Figure 4, Table S7). Among the 25 categories, which were alphabetized, the cluster of R (general function prediction only, 6176, 19.2%), group O (post-translational modification, protein turnover, and chaperones, 3697, 11.5%), and group T (signal transduction mechanisms, 3380, 10.5%) represented the three most abundant groups in the data. The group T (cell motility, 20, 0.1%) represented the smallest group.

To better understand biological pathways of 'Huaxiu' plum fruits' pulp color change, all unigenes were searched in the KEGG database. A total of 7595 unigenes sequences were mapped onto 128 KEGG pathways in our results (Table S8). The pathways with the five highest unigenes representations were those related with the ribosome (ko03010; 644; 8.48%), carbon metabolism (ko01200; 629; 8.28%), biosynthesis of amino acids (ko01230; 517; 6.81%), protein processing in endoplasmic reticulum (ko04141; 387; 5.1%), and plant–pathogen interaction (ko04626; 322; 4.24%). Among these, carotenoids biosynthesis (ko00906; 45; 0.59%), flavonoid biosynthesis (ko00941; 44; 0.58%), and anthocyanin biosynthesis (ko00942; 1; 0.01%) appeared to be the smallest groups.

3.5. Differential Gene Expression between Three Stages of Pulp Coloration

To study unigene expression during different stages, the reads were mapped to the assembled transcriptome. A total of 56,570 unigenes (99%) were detected at all three stages. Among them, 50,987, 49,626, and 49,619 unigenes were detected in three pulp discoloration stages, respectively. In the results, 1095 transcripts were expressed differentially in yellow pulp compared to orange pulp, and among these genes, 632 were upregulated and 463 were downregulated (Figure 5 and Table S9). A total of 4947 shared DEGs were identified during the comparison between yellow and red, and among these DEGs, 1434 were unregulated and 3513 were down-regulated (Figure 5 and Table S11). Moreover, 3414 differentially expressed unigenes (864 unregulated and 2550 down regulated) were found between orange and red (Figure 5 and Table S10).

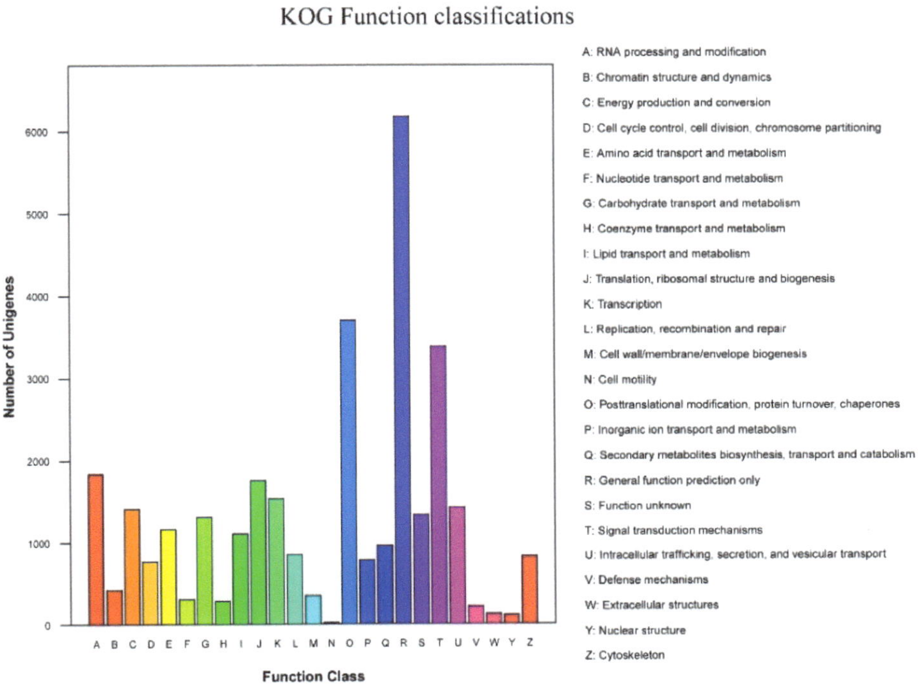

Figure 4. KOG functional classification of 'Huaxiu' plums (*P. salicina*) unigenes.

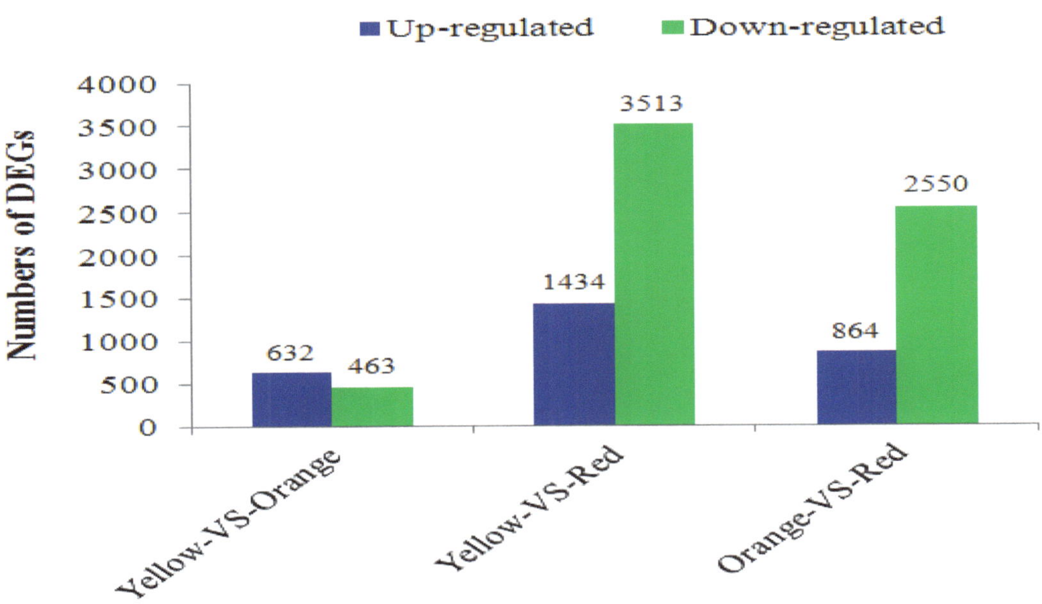

Figure 5. The numbers of DEGs in comparisons of the YS-vs.-OS, OS-vs.-RS, and YS-vs.-RS fruit samples.

The significant DEGs were represented in the three GO categories using the GO database (Figure 6). The results showed that five pathway (single-organism process, metabolic process, cellular process, catalytic activity, and binding) genes were most significantly enriched in the GO category. To annotate the functions of the unigene, we conducted a pathway-enrichment analysis of the DEGs in YS-vs.-OS, OS-vs.-RS, and YS-vs.-RS based on the KEGG database. According to the p-value statistics of the KEGG pathway enrichment, the top 20 pathways (p-values < 0.05) with the number of DEGs in the three periods are listed (Figure S2, Table S12). In YS-vs.-OS, the results showed that 218 DEGs were significantly enriched and were associated with 84 pathways, and among them, 10 pathways indicated the most significant enrichment, which mainly includes phenylpropanoid biosynthesis (18, 8.26%), starch and sucrose metabolism (21, 9.63%), plant hormone signal transduction (14, 6.42%), and plant–pathogen interaction (15, 6.88%) (Figure S2A, Table S12-1). In OS-vs.-RS, 580 DEGs were significantly enriched in 113 pathways, of which 19 pathways indicated the most significant enrichment. The following pathways are the most important in OS-vs.-RS: starch and sucrose metabolism (51, 8.79%), plant hormone signal transduction (39, 6.72%), plant–pathogen interaction (45, 7.76%), and phenylpropanoid biosynthesis (29, 5.00%) (Figure S2B, Table S12-2). In YS-vs.-RS, 750 DEGs were significantly enriched in 117 pathways, of which 22 pathways indicated the most significant enrichment. The following pathways are the most important in YS-vs.-RS: starch and sucrose metabolism (72, 9.6%), plant hormone signal transduction (55, 7.33%), plant–pathogen interaction (62, 8.27%), and pentose and glucuronate interconversions (29, 3.87%) (Figure S2C, Table S12-3).

3.6. Candidate Genes Related to Anthocyanin and Flavonoid Pathways

To identify unigenes involved in anthocyanin and flavonoid biosynthesis, KEGG functional enrichment was analyzed to characterize the functions of differentially expressed unigenes (Table 2). A total of 20 genes involved in anthocyanin and flavonoid biosynthesis were differentially expressed during ripening of 'Huaxiu' plums and only 1 DEG (Unigene0032408) was associated with the anthocyanin biosynthetic pathway, which was significantly upregulated during OS-vs.-RS. A total of 6, 12, and 17 DEGs were annotated as being involved in flavonoid biosynthesis in YS-vs.-OS, OS-vs.-RS, and YS-vs.-RS, respectively, with 3 DEGs common to the three sample groups. These structural genes of the flavonoid and anthocyanin biosynthetic pathways were significantly upregulated, including beta-D-glucosyl crocetin beta-1, 6-glucosyltransferase-like (Unigene0032408), CHS (Unigene0001957, Unigene0043265, Unigene0043266), CHI (Unigene0026465), DFR (Unigene0012798), LDOX (Unigene0001105), F3H (Unigene0002347, Unigene0040558), F3′H (Unigene0027142), and leucoanthocyanidin reductase-like (Unigene0025372), while other genes were downregulated. Structural genes CHS (Unigene0001957), CHS (Unigene0043265), DFR (Unigene0012798), and leucoanthocyanidin reductase-like (Unigene0025372) were the most significant DEGs in YS-vs-RS group, which showed 3.44-, 5.25-, 3.47-, and 5.33-fold upregulation, respectively. In addition, two genes (Unigene0002731 and Unigene0039938) were downregulated by 4.09- and 2.89-fold in YS-vs-RS, respectively.

To validate the RNA-Seq results, we selected 12 DEGs (flavonoid biosynthetic pathway genes) (Supplementary Figure S6) and analyzed their expression levels in YS, OS, and RS using RT-qPCR (Figure S3). The expression level trends of these genes were consistent with the changes in the abundance as detected in the RNA-Seq data.

Table 2. Pathways associated with candidate differentially expressed genes involved in anthocyanin biosynthesis and flavonoid biosynthesis in 'Huaxiu' plums.

Pathway	DEGs	KO ID	Description	$\log_2$ (OS/YS)	*p*-Value	$\log_2$ (RS/OS)	*p*-Value	$\log_2$ (RS/YS)	*p*-Value
Anthocyanin biosynthesis	Unigene0032408	ko00942	Beta-D-glucosyl crocetin beta-1,6-glucosyltransferase-like	—	—	1.31	1.01×10^{-24}	—	—
Flavonoid biosynthesis	Unigene0001957	ko00941	Chalcone synthase (CHS)	1.79	3.19×10^{-43}	1.65	1.83×10^{-64}	3.44	5.17×10^{-144}
	Unigene0043265	ko00941	Chalcone synthase (CHS)	3.33	1.04×10^{-33}	1.92	2.01×10^{-34}	5.25	1.47×10^{-172}
	Unigene0043266	ko00941	Chalcone synthase (CHS)	—	—	1.90	0.0006	1.79	0.0004
	Unigene0026465	ko00941	Chalcone isomerase (CHI)	—	—	1.69	0.0006	1.46	2.26×10^{-43}
	Unigene0012798	ko00941	Dihydroflavonol 4-reductase (DFR)	1.49	4.61×10^{-7}	1.98	6.68×10^{-20}	3.47	1.09×10^{-55}
	Unigene0001105	ko00941	Leucoanthocyanidin dioxygenase (LDOX)	1.32	4.43×10^{-27}	—	—	1.89	1.90×10^{-61}
	Unigene0002347	ko00941	Flavonol synthase/flavanone 3-hydroxylase-like (FS/F3′H)	—	—	1.70	5.48×10^{-14}	2.37	1.86×10^{-32}
	Unigene0040558	ko00941	Flavanone 3-hydroxylase (F3H)	—	—	—	—	1.54	1.16×10^{-56}
	Unigene0027142	ko00941	Flavonoid 3′-hydroxylase(F3′H)	—	—	1.19	1.22×10^{-38}	1.00	5.62×10^{-30}
	Unigene0022847	ko00940	Flavonoid 3′,5′-methyltransferase-like (F3′5′H)	—	—	—	—	−1.64	0.0035
	Unigene0023053	ko00940	Flavonoid 3′,5′-methyltransferase-like(F3′5′H)	—	—	—	—	−1.58	2.62×10^{-7}
	Unigene0025278	ko00941	Leucoanthocyanidin reductase-like	—	—	−1.13	3.50×10^{-5}	−1.03	1.09×10^{-5}
	Unigene0025372	ko00941	Leucoanthocyanidin reductase-like isoform X1	—	—	3.82	6.66×10^{-7}	5.33	3.46×10^{-11}
	Unigene0039938	ko00941	Shikimate O-hydroxycinnamoyltransferase-like	—	—	—	—	−2.89	0.0059
	Unigene0039940	ko00941	Shikimate O-hydroxycinnamoyltransferase-like	−1.07	5.02×10^{-15}	—	—	−1.45	7.34×10^{-31}
	Unigene0039941	ko00941	Shikimate O-hydroxycinnamoyltransferase-like	—	—	−1.25	0.0005	−1.91	9.24×10^{-11}
	Unigene0002731	ko00941	Shikimate O-hydroxycinnamoyltransferase-like	—	—	−3.75	0.0054	−4.09	0.0012
	Unigene0037918	ko00941	Acylsugar acyltransferase 3-like	1.26	0.0001	—	—	—	—
	Unigene0019559	ko00941	3,5-dihydroxybiphenyl synthase-like	—	—	−1.09	1.99×10^{-7}	—	—

Note: —: The gene expression level that had no significant differences.

A

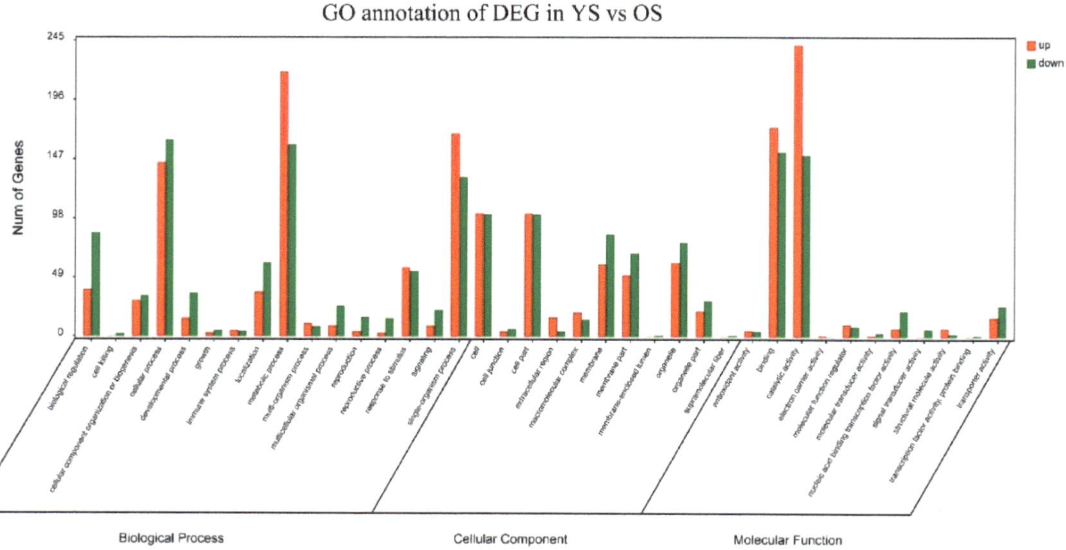

B

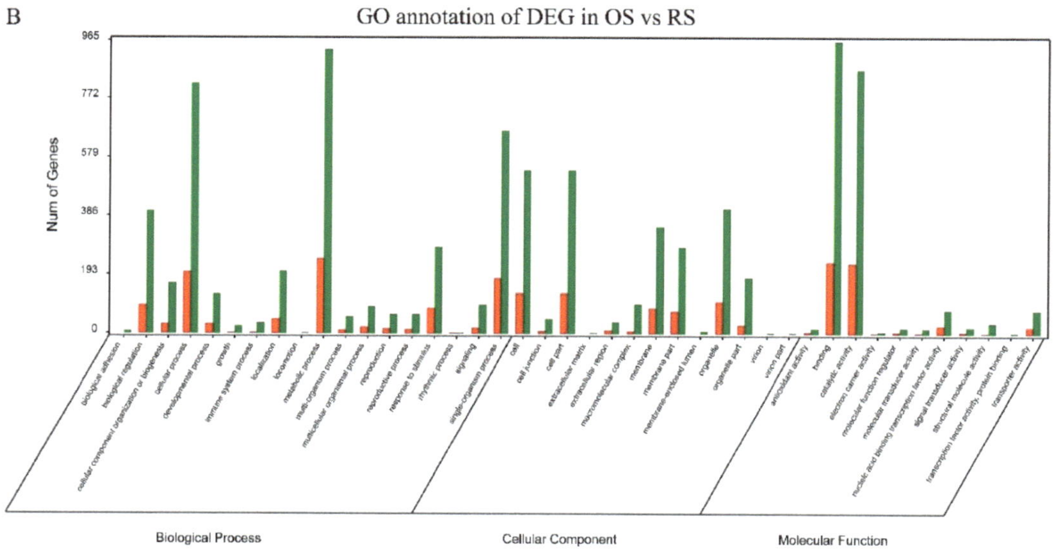

Figure 6. *Cont.*

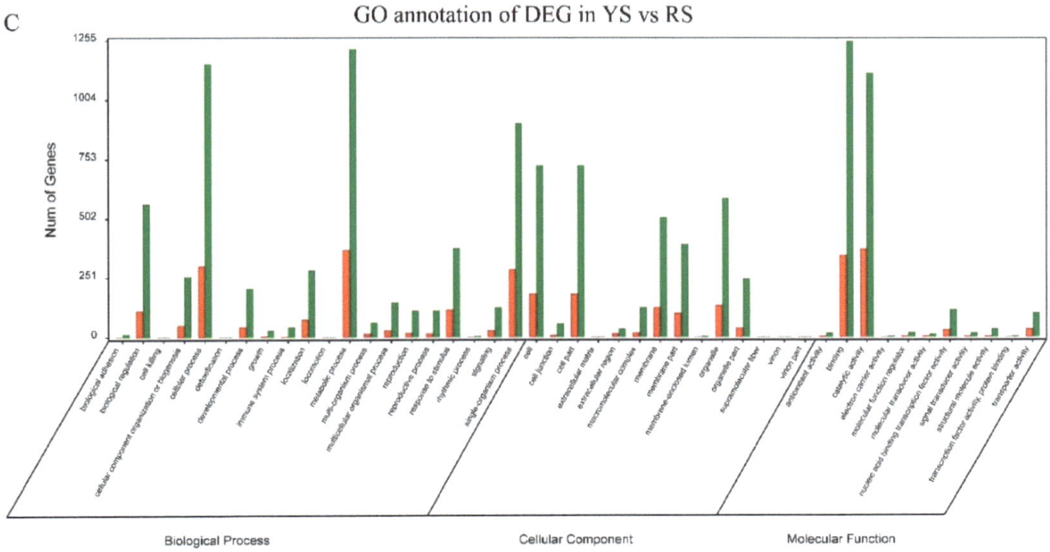

Figure 6. GO annotation of up- and downregulated expression genes of 'Huaxiu' plums (*P. salicina*) unigenes. (**A**): GO annotation in YS-vs.-OS; (**B**): GO annotation in OS-vs.-RS; (**C**): GO annotation in YS-vs.-RS.

4. Discussion

The color of the pulp cannot only be attractive in appearance, but also has higher nutritional value [1]. Therefore, it is very important to elucidate the genetic mechanism of pulp color regulation. It is generally believed the color of the pulp is determined by the comprehensive performance of chlorophyll, carotenoids, anthocyanins, and other pigment substances. Anthocyanin biosynthesis is fairly complex and is associated with flavonoids [46]. In our study, it was found that, during the coloring process of the pulp of 'Huaxiu' plums, the chlorophyll was gradually degraded and the synthesis of anthocyanin increased (Figure 1). When the comprehensive performance of anthocyanins and carotenoids made the pulp orange, then anthocyanins were synthesized in large amounts, the color of carotenoids was masked by anthocyanins, and the pulp was red. The main objective of this study was to identify the genes involved in anthocyanin biosynthesis in plums.

In this study, we have performed transcriptome sequencing of the stages of three fruit pulp colors of 'Huaxiu' plums with the use of advanced high-throughput Illumina RNA-Seq technology. The results not only enrich the gene information of *P. salicina*, but also can help explore the molecular genetics and biochemical characteristics of *Prunus salicina* Lindl. and its related species with the generated transcriptome data. In total, 57,119 unigenes were assembled, with a mean length of 953 bp (Table 1), which is comparable to 944 bp for sweet cherry (*P. avium* L.) [47], and with the longer to the previously reported other species such as 872 bp for cultivar 'Furongli' (*P. salicina* L.) [1] and 531 bp for Chinese bayberry (*Myrica rubra*) [48]. Approximately 61.6% of the unigenes were annotated to public databases (NR, Swiss-Prot, GO, COG, and KEGG), which means that more than one-third of the unigenes have no apparent homologs, with similar results seen in other no model plant species [49]. The unannotated unigenes could be plum-specific genes with novel functions, which may be related with some unique biosynthesis processes and pathways in the results. Furthermore, the annotated unigenes of *P. salicina* L. indicated the highest homology to those of *Prunus mume* (55.0%), followed by *Malus domestica* (4.8%) and *Prunus persica* (3.4%) (Figure 2B and Table S4), which may indicate the evolutionary relationship among these species. In spite of a large number of unigene sequences that indicated no matches, many of the unigenes were still assigned to a wide range of GO and KEGG classifications. The KEGG function annotation analysis showed that 7595 unigenes were involved in 128 biosynthesis process. The largest number of unigenes was associated with the ribosome, carbon metabolism, and biosynthesis of amino acids. However, the smallest number of unigenes was associated with anthocyanin biosynthesis, which have only one matching unigene. All of these data contribute to the study of the metabolic and biosynthesis mechanisms in *P. salicina*.

The RNA-Seq analysis revealed that the numbers of DEGs differed at the coloration stages, and we identified 1095, 3414, and 4947 DEGs between yellow and orange, orange and red, and yellow and red stages, respectively (Figure 5). More DEGs were detected at the yellow and red stage than at the yellow and orange stage, suggesting greater changes in the pulp color during the final ripening stage. Anthocyanin, the most important metabolite in flavonoid production, is an essential nutritional component in *P. salicina* fruits and their products. In our study, the flavonoid pathways were significantly enriched in the KEGG pathway. We identified many DEGs between different stages of the pulp color involved in anthocyanin and flavonoid biosynthesis, which mainly were structural genes, including CHS, CHI, DFR, F3H, F3′H, and LDOX, and were significantly upregulated during the pulp color of yellow vs. red color stages (Table 2). These observations agree well with those qRT-PCR results mentioned above. This is in accordance with findings for other fruits. To date, most of the structural genes in the anthocyanin biosynthetic pathway are upregulated during the fruit development of red/green skin color mutations of pear [50]. Similarly, coordinated expression changes of ANS, DFR1, F3H, F3′H, and UFGT have also been demonstrated in differently colored plum, sweet cherry, Chinese bayberries, and other plants [47,51].

CHS is considered to be a key enzyme in the anthocyanin biosynthesis of *Rosaceous* plants, which have diverse functions such as defense against pathogens and pigment biosynthesis. CHS proteins have been found responsible for the red coloration in crabapple cultivars [52], and the CHS protein of Japanese morning glory was also found to enhance both flavonoid production and flower pigmentation [53]. In the present study, we found that two CHSs (Unigene0001957 and Unigene0043265) were significantly upregulated at each stage, with the highest expression in the red stage. However, they do not correlate exactly with the increased concentration of anthocyanin content and total flavonoids during 'Huaxiu' plum coloration (Figure 1). This was probably due to the complicated composition of flavonoids. DFR and LDOX/ANS are late anthocyanin biosynthetic genes. F3H was one of three main enzymes in the primary phases of the flavonoid pathway. Transcript levels of F3H were greater in RS than YS, and the upregulation of F3H genes in RS indicates that it contributed to the accumulation of anthocyanin, which led to the red pigmentation in the plums. The expression of F3H also proved that it could make apples red [40]. In this study, these two structural genes (Unigene0012798 and Unigene0001105) showed the highest expression levels in the red stage, with the highest anthocyanin concentration (Figure 1, Table 2), which is consistent with findings for apple skin [54]. Thus, we believe that these genes may be play an important role in regulating anthocyanin biosynthesis in 'Huaxiu' plum fruit pulp.

The red hue of plant organs is caused due to anthocyanins, and the accumulation of these pigments is also regulated by transcription factors (TFs). In Rosaceae species, MYBs play a critical role as key transcription factors for all of the anthocyanin biosynthetic pathway genes or for the regulation of single key genes in fruit and flower color formation, particularly MYB10 genes, which are responsible for part of the natural variation in anthocyanin colors [10]. The bHLH proteins and NAC proteins have also been reported to be involved in anthocyanin synthesis [1]. The exact roles of these candidate transcription factor should be investigated in further studies.

5. Conclusions

The pulp of plums has been the focus of studies associated with fruit taste, quality, and nutrition. In the current study, we used RNA-Seq to analyze changes in the transcriptome with the pulp color of 'Huaxiu' plums. A total of 57,119 unigenes with a mean length of 953 bp were generated, and 61.6% of them were annotated to public databases. This study provides a large collection of DEG associated with plum fruit maturation processes. In addition, candidate genes involved in anthocyanin biosynthesis and flavonoid biosynthesis were identified in the transcriptome dataset. Further studies are needed to determine whether the identified candidate genes are related to anthocyanin biosynthesis in plums. This study has enriched the genetic dates of plums and provided an important platform for studying pulp-ripening processes in plums, especially anthocyanin and flavonoid biosynthesis, which will lay the foundation for further functional genomics and molecular metabolic mechanism studies on plums.

Supplementary Materials: The following supporting information can be downloaded at: https://www.mdpi.com/article/10.3390/cimb44120434/s1, Figure S1: The Venn diagram of unigene numbers annotated by BLASTx against four public databases; Figure S2: Pathway enrichment (top 20) of differentially expressed genes between the two colors stages of pulp; Figure S3: Expression analysis of DEG by qRT-PCR; Table S1: Primer sequences of unigenes used for quantitative RT-PCR verification; Table S2: Summary of read numbers based on the RNA-Seq data from the pulp of 'Huaxiu' pump during the yellow, orange and red stages of maturation; Table S3: The length distribution of the assembled unigenes; Table S4: Summary of functional annotations for the assembled unigenes; Table S5: The species distribution of unigene BLAST results against the nr database; Table S6: Summary of level2 GO term classification for the 'Huaxiu' plums (*P. salicina*) transcriptome; Table S7: Summary of KOG functional classification for the 'Huaxiu' plums (*P. salicina*) unigenes; Table S8: Summary of KEGG pathways enrichment involved in the 'Huaxiu' plums (*P. salicina*) unigenes; Table S9: Summary of significance DEGs in YS-vs.-OS; Table S10: Summary of significance DEGs in OS-vs.-RS; Table S11: Summary of significance DEGs in YS-vs.-RS; Table S12: Summary of pathway enrichment of DEGs between the two colors stages of pulp (top 20).

Author Contributions: Conceptualization, J.X. and G.W.; methodology, W.W. and J.Z.; software, T.W.; validation, W.W.; investigation, J.X., W.W., G.W. and Z.J.; resources, J.X., W.W., G.W. and Z.J.; data curation, W.W. and G.W.; writing—original draft preparation, G.W.; writing—review and editing, G.W and J.X.; project administration, J.X. All authors have read and agreed to the published version of the manuscript.

Funding: This research was funded by the National Science and Technology Program (2014BAD16B04), partly supported by the Independent Innovation of Agricultural Science and Technology of Jiangsu Province (CX(21)3046), and the Natural Science Foundation of Jiangsu Province of China (BK20210161).

Institutional Review Board Statement: Not applicable.

Informed Consent Statement: Not applicable.

Data Availability Statement: All relevant data are within the paper and its Supplementary Materials.

Acknowledgments: We are grateful to an orchard in Donghai County (Jiangsu Province, China) for providing experimental materials.

Conflicts of Interest: The authors declare no conflict of interest.

Abbreviations

ANS	anthocyanidin synthase
bHLH	basic helix–loop–helix
CHS	chalcone synthase
COG	cluster of orthologous groups of proteins
CH	chalcone isomerase

DFR	dihydroflavonol 4-reductase
DGE	differentially expressed genes
F3H	flavanone 3-hydroxylase
F3'H	flavonoid 3'-hydroxylase
FDR	false discovery rate
FPKM	fragments per kilobase of transcript per million mapped reads
GO	gene ontology
KEGG	Kyoto Encyclopedia of Genes and Genomes
LDOX	leucoanthocyanidin dioxygenase
NR	NCBI nonredundant protein database
OS	orange stage
qRT-PCR	quantitative real-time PCR
RNA-Seq	RNA sequencing
RS	red stage
RSEM	RNA-Seq by expectation maximization
UFGT	UDP-glucose: flavonoid 3-O-glucosyltransferase
YS	yellow stage

References

1. Fang, Z.Z.; Zhou, D.R.; Ye, X.F.; Jiang, C.C.; Pan, S.L. Identification of Candidate Anthocyanin-Related Genes by Transcriptomic Analysis of 'Furongli' Plum (*Prunus salicina* Lindl.) during Fruit Ripening Using RNA-Seq. *Front. Plant Sci.* **2016**, *7*, 1338. [CrossRef] [PubMed]
2. Jaakola, L. New insights into the regulation of anthocyanin biosynthesis in fruits. *Trends Plant Sci.* **2013**, *18*, 477–483. [CrossRef] [PubMed]
3. Hooshmand, S.; Arjmandi, B.H. Viewpoint: Dried plum, an emerging functional food that may effectively improve bone health. *Ageing Res. Rev.* **2009**, *8*, 122–127. [CrossRef] [PubMed]
4. Tanaka, Y.; Sasaki, N.; Ohmiya, A. Biosynthesis of plant pigments: Anthocyanins, betalains and carotenoids. *Plant J. Cell Mol. Biol.* **2008**, *54*, 733–749. [CrossRef]
5. Han, Y.; Vimolmangkang, S.; Soria-Guerra, R.E.; Korban, S.S. Introduction of apple ANR genes into tobacco inhibits expression of both CHI and DFR genes in flowers, leading to loss of anthocyanin. *J. Exp. Bot.* **2012**, *63*, 2437–2447. [CrossRef]
6. Rafique, M.Z.; Carvalho, E.; Stracke, R.; Palmieri, L.; Herrera, L.; Feller, A.; Malnoy, M.; Martens, S. Nonsense Mutation Inside Anthocyanidin Synthase Gene Controls Pigmentation in Yellow Raspberry (*Rubus idaeus* L.). *Front. Plant Sci.* **2016**, *7*, 1892. [CrossRef]
7. Griesser, M.; Hoffmann, T.; Bellido, M.L.; Rosati, C.; Fink, B.; Kurtzer, R.; Aharoni, A.; Munoz-Blanco, J.; Schwab, W. Redirection of flavonoid biosynthesis through the down-regulation of an anthocyanidin glucosyltransferase in ripening strawberry fruit. *Plant Physiol.* **2008**, *146*, 1528–1539. [CrossRef]
8. Feller, A.; Machemer, K.; Braun, E.L.; Grotewold, E. Evolutionary and comparative analysis of MYB and bHLH plant transcription factors. *Plant J. Cell Mol. Biol.* **2011**, *66*, 94–116. [CrossRef]
9. Xu, W.; Dubos, C.; Lepiniec, L. Transcriptional control of flavonoid biosynthesis by MYB-bHLH-WDR complexes. *Trends Plant Sci.* **2015**, *20*, 176–185. [CrossRef]
10. Zhao, L.; Gao, L.; Wang, H.; Chen, X.; Wang, Y.; Yang, H.; Wei, C.; Wan, X.; Xia, T. The R2R3-MYB, bHLH, WD40, and related transcription factors in flavonoid biosynthesis. *Funct. Integr. Genom.* **2013**, *13*, 75–98. [CrossRef]
11. Chagne, D.; Lin-Wang, K.; Espley, R.V.; Volz, R.K.; How, N.M.; Rouse, S.; Brendolise, C.; Carlisle, C.M.; Kumar, S.; De Silva, N.; et al. An ancient duplication of apple MYB transcription factors is responsible for novel red fruit-flesh phenotypes. *Plant Physiol.* **2013**, *161*, 225–239. [CrossRef]
12. Li, L.; Ban, Z.J.; Li, X.H.; Wu, M.Y.; Wang, A.L.; Jiang, Y.Q.; Jiang, Y.H. Differential expression of anthocyanin biosynthetic genes and transcription factor PcMYB10 in pears (*Pyrus communis* L.). *PLoS ONE* **2012**, *7*, e46070. [CrossRef]
13. Cui, D.; Zhao, S.; Xu, H.; Allan, A.C.; Zhang, X.; Fan, L.; Chen, L.; Su, J.; Shu, Q.; Li, K. The interaction of MYB, bHLH and WD40 transcription factors in red pear (*Pyrus pyrifolia*) peel. *Plant Mol. Biol.* **2021**, *106*, 407–417. [CrossRef]
14. Hichri, I.; Deluc, L.; Barrieu, F.; Bogs, J.; Mahjoub, A.; Regad, F.; Gallois, B.; Granier, T.; Trossat-Magnin, C.; Gomes, E.; et al. A single amino acid change within the R2 domain of the VvMYB5b transcription factor modulates affinity for protein partners and target promoters selectivity. *BMC Plant Biol.* **2011**, *11*, 117. [CrossRef]
15. Schaart, J.G.; Dubos, C.; Romero De La Fuente, I.; van Houwelingen, A.; de Vos, R.C.H.; Jonker, H.H.; Xu, W.; Routaboul, J.M.; Lepiniec, L.; Bovy, A.G. Identification and characterization of MYB-bHLH-WD40 regulatory complexes controlling proanthocyanidin biosynthesis in strawberry (*Fragaria x ananassa*) fruits. *New Phytol.* **2013**, *197*, 454–467. [CrossRef]
16. Cao, Y.; Jia, H.; Xing, M.; Jin, R.; Grierson, D.; Gao, Z.; Sun, C.; Chen, K.; Xu, C.; Li, X. Genome-Wide Analysis of MYB Gene Family in Chinese Bayberry (*Morella rubra*) and Identification of Members Regulating Flavonoid Biosynthesis. *Front. Plant Sci.* **2021**, *12*, 691384. [CrossRef]

17. Fraser, L.G.; Seal, A.G.; Montefiori, M.; McGhie, T.K.; Tsang, G.K.; Datson, P.M.; Hilario, E.; Marsh, H.E.; Dunn, J.K.; Hellens, R.P.; et al. An R2R3 MYB transcription factor determines red petal colour in an *Actinidia* (kiwifruit) hybrid population. *BMC Genom.* **2013**, *14*, 28. [CrossRef]
18. Lai, B.; Du, L.N.; Liu, R.; Hu, B.; Su, W.B.; Qin, Y.H.; Zhao, J.T.; Wang, H.C.; Hu, G.B. Two LcbHLH Transcription Factors Interacting with LcMYB1 in Regulating Late Structural Genes of Anthocyanin Biosynthesis in Nicotiana and Litchi chinensis During Anthocyanin Accumulation. *Front. Plant Sci.* **2016**, *7*, 166. [CrossRef]
19. Zhao, M.; Li, J.; Zhu, L.; Chang, P.; Li, L.; Zhang, L. Identification and Characterization of MYB-bHLH-WD40 Regulatory Complex Members Controlling Anthocyanidin Biosynthesis in Blueberry Fruits Development. *Genes* **2019**, *10*, 496. [CrossRef]
20. Jin, W.; Wang, H.; Li, M.; Wang, J.; Yang, Y.; Zhang, X.; Yan, G.; Zhang, H.; Liu, J.; Zhang, K. The R2R3 MYB transcription factor PavMYB10.1 involves in anthocyanin biosynthesis and determines fruit skin colour in sweet cherry (*Prunus avium* L.). *Plant Biotechnol. J.* **2016**, *14*, 2120–2133. [CrossRef]
21. Rahim, M.A.; Busatto, N.; Trainotti, L. Regulation of anthocyanin biosynthesis in peach fruits. *Planta* **2014**, *240*, 913–929. [CrossRef]
22. Tuan, P.A.; Bai, S.; Yaegaki, H.; Tamura, T.; Hihara, S.; Moriguchi, T.; Oda, K. The crucial role of PpMYB10.1 in anthocyanin accumulation in peach and relationships between its allelic type and skin color phenotype. *BMC Plant Biol.* **2015**, *15*, 280. [CrossRef] [PubMed]
23. Wang, Z.; Cui, Y.; Vainstein, A.; Chen, S.; Ma, H. Regulation of Fig (*Ficus carica* L.) Fruit Color: Metabolomic and Transcriptomic Analyses of the Flavonoid Biosynthetic Pathway. *Front. Plant Sci.* **2017**, *8*, 1990. [CrossRef] [PubMed]
24. Gou, J.Y.; Felippes, F.F.; Liu, C.J.; Weigel, D.; Wang, J.W. Negative regulation of anthocyanin biosynthesis in *Arabidopsis* by a miR156-targeted SPL transcription factor. *Plant Cell* **2011**, *23*, 1512–1522. [CrossRef]
25. Jaakola, L.; Poole, M.; Jones, M.O.; Kamarainen-Karppinen, T.; Koskimaki, J.J.; Hohtola, A.; Haggman, H.; Fraser, P.D.; Manning, K.; King, G.J.; et al. A SQUAMOSA MADS box gene involved in the regulation of anthocyanin accumulation in bilberry fruits. *Plant Physiol.* **2010**, *153*, 1619–1629. [CrossRef] [PubMed]
26. Zhou, H.; Lin-Wang, K.; Wang, H.; Gu, C.; Dare, A.P.; Espley, R.V.; He, H.; Allan, A.C.; Han, Y. Molecular genetics of blood-fleshed peach reveals activation of anthocyanin biosynthesis by NAC transcription factors. *Plant J. Cell Mol. Biol.* **2015**, *82*, 105–121. [CrossRef]
27. Fang, Z.Z.; Lin-Wang, K.; Dai, H.; Zhou, D.R.; Jiang, C.C.; Espley, R.V.; Deng, C.; Lin, Y.J.; Pan, S.L.; Ye, X.F. The genome of low-chill Chinese plum "Sanyueli" (*Prunus salicina* Lindl.) provides insights into the regulation of the chilling requirement of flower buds. *Mol. Ecol. Resour.* **2022**, *22*, 1919–1938. [CrossRef]
28. Fang, B.; Xu, Y.; Yu, J. The complete chloroplast genome sequence of *Prunus salicina* 'Wushan plum'. *Mitochondrial DNA Part B Resour.* **2021**, *6*, 1243–1244. [CrossRef]
29. Deng, L.; Wang, T.; Hu, J.; Yang, X.; Yao, Y.; Jin, Z.; Huang, Z.; Sun, G.; Xiong, B.; Liao, L.; et al. Effects of Pollen Sources on Fruit Set and Fruit Characteristics of 'Fengtangli' Plum (*Prunus salicina* Lindl.) Based on Microscopic and Transcriptomic Analysis. *Int. J. Mol. Sci.* **2022**, *23*, 12959. [CrossRef]
30. Gonzalez, M.; Maldonado, J.; Salazar, E.; Silva, H.; Carrasco, B. De novo transcriptome assembly of 'Angeleno' and 'Lamoon' Japanese plum cultivars (*Prunus salicina*). *Genom. Data* **2016**, *9*, 35–36. [CrossRef]
31. Li, Q.; Chang, X.X.; Wang, H.; Brennan, C.S.; Guo, X.B. Phytochemicals Accumulation in Sanhua Plum (*Prunus salicina* L.) during Fruit Development and Their Potential Use as Antioxidants. *J. Agric. Food Chem.* **2019**, *67*, 2459–2466. [CrossRef]
32. Muthukrishnan, S.D.; Kaliyaperumal, A.; Subramaniyan, A. Identification and determination of flavonoids, carotenoids and chlorophyll concentration in *Cynodon dactylon* (L.) by HPLC analysis. *Nat. Prod. Res.* **2015**, *29*, 785–790. [CrossRef]
33. Macias-Sanchez, M.D.; Mantell Serrano, C.; Rodriguez Rodriguez, M.; Martinez de la Ossa, E.; Lubian, L.M.; Montero, O. Extraction of carotenoids and chlorophyll from microalgae with supercritical carbon dioxide and ethanol as cosolvent. *J. Sep. Sci.* **2008**, *31*, 1352–1362. [CrossRef]
34. Larit, F.; Leon, F.; Benyahia, S.; Cutler, S.J. Total Phenolic and Flavonoid Content and Biological Activities of Extracts and Isolated Compounds of *Cytisus villosus* Pourr. *Biomolecules* **2019**, *9*, 732. [CrossRef]
35. Grabherr, M.G.; Haas, B.J.; Yassour, M.; Levin, J.Z.; Thompson, D.A.; Amit, I.; Adiconis, X.; Fan, L.; Raychowdhury, R.; Zeng, Q.; et al. Full-length transcriptome assembly from RNA-Seq data without a reference genome. *Nat. Biotechnol.* **2011**, *29*, 644–652. [CrossRef]
36. Pruitt, K.D.; Tatusova, T.; Maglott, D.R. NCBI reference sequences (RefSeq): A curated non-redundant sequence database of genomes, transcripts and proteins. *Nucleic Acids Res.* **2007**, *35*, D61–D65. [CrossRef]
37. Altermann, E.; Klaenhammer, T.R. PathwayVoyager: Pathway mapping using the Kyoto Encyclopedia of Genes and Genomes (KEGG) database. *BMC Genom.* **2005**, *6*, 60. [CrossRef]
38. Tatusov, R.L.; Fedorova, N.D.; Jackson, J.D.; Jacobs, A.R.; Kiryutin, B.; Koonin, E.V.; Krylov, D.M.; Mazumder, R.; Mekhedov, S.L.; Nikolskaya, A.N.; et al. The COG database: An updated version includes eukaryotes. *BMC Bioinform.* **2003**, *4*, 41. [CrossRef]
39. Conesa, A.; Gotz, S.; Garcia-Gomez, J.M.; Terol, J.; Talon, M.; Robles, M. Blast2GO: A universal tool for annotation, visualization and analysis in functional genomics research. *Bioinformatics* **2005**, *21*, 3674–3676. [CrossRef]
40. Ye, J.; Zhang, Y.; Cui, H.; Liu, J.; Wu, Y.; Cheng, Y.; Xu, H.; Huang, X.; Li, S.; Zhou, A.; et al. WEGO 2.0: A web tool for analyzing and plotting GO annotations, 2018 update. *Nucleic Acids Res.* **2018**, *46*, W71–W75. [CrossRef]
41. Kanehisa, M.; Furumichi, M.; Tanabe, M.; Sato, Y.; Morishima, K. KEGG: New perspectives on genomes, pathways, diseases and drugs. *Nucleic Acids Res.* **2017**, *45*, D353–D361. [CrossRef] [PubMed]

42. Li, B.; Dewey, C.N. RSEM: Accurate transcript quantification from RNA-Seq data with or without a reference genome. *BMC Bioinform.* **2011**, *12*, 323. [CrossRef] [PubMed]
43. Wang, L.; Feng, Z.; Wang, X.; Wang, X.; Zhang, X. DEGseq: An R package for identifying differentially expressed genes from RNA-seq data. *Bioinformatics* **2010**, *26*, 136–138. [CrossRef] [PubMed]
44. Chen, X. False discovery rate control for multiple testing based on discrete *p*-values. *Biom. J. Biom. Z.* **2020**, *62*, 1060–1079. [CrossRef] [PubMed]
45. Klopfenstein, D.V.; Zhang, L.; Pedersen, B.S.; Ramirez, F.; Warwick Vesztrocy, A.; Naldi, A.; Mungall, C.J.; Yunes, J.M.; Botvinnik, O.; Weigel, M.; et al. GOATOOLS: A Python library for Gene Ontology analyses. *Sci. Rep.* **2018**, *8*, 10872. [CrossRef]
46. Luo, X.; Cao, D.; Li, H.; Zhao, D.; Xue, H.; Niu, J.; Chen, L.; Zhang, F.; Cao, S. Complementary iTRAQ-based proteomic and RNA sequencing-based transcriptomic analyses reveal a complex network regulating pomegranate (*Punica granatum* L.) fruit peel colour. *Sci. Rep.* **2018**, *8*, 12362. [CrossRef]
47. Wei, H.; Chen, X.; Zong, X.; Shu, H.; Gao, D.; Liu, Q. Comparative transcriptome analysis of genes involved in anthocyanin biosynthesis in the red and yellow fruits of sweet cherry (*Prunus avium* L.). *PLoS ONE* **2015**, *10*, e0121164. [CrossRef]
48. Feng, C.; Chen, M.; Xu, C.J.; Bai, L.; Yin, X.R.; Li, X.; Allan, A.C.; Ferguson, I.B.; Chen, K.S. Transcriptomic analysis of Chinese bayberry (*Myrica rubra*) fruit development and ripening using RNA-Seq. *BMC Genom.* **2012**, *13*, 19. [CrossRef]
49. Hou, D.Y.; Shi, L.C.; Yang, M.M.; Li, J.; Zhou, S.; Zhang, H.X.; Xu, H.W. De novo transcriptomic analysis of leaf and fruit tissue of *Cornus officinalis* using Illumina platform. *PLoS ONE* **2018**, *13*, e0192610. [CrossRef]
50. Yang, Y.; Yao, G.; Yue, W.; Zhang, S.; Wu, J. Transcriptome profiling reveals differential gene expression in proanthocyanidin biosynthesis associated with red/green skin color mutant of pear (*Pyrus communis* L.). *Front. Plant Sci.* **2015**, *6*, 795. [CrossRef]
51. Drogoudi, P.; Pantelidis, G. Phenotypic Variation and Peel Contribution to Fruit Antioxidant Contents in European and Japanese Plums. *Plants* **2022**, *11*, 1338. [CrossRef]
52. Tai, D.; Tian, J.; Zhang, J.; Song, T.; Yao, Y. A *Malus* crabapple chalcone synthase gene, McCHS, regulates red petal color and flavonoid biosynthesis. *PLoS ONE* **2014**, *9*, e110570. [CrossRef]
53. Kang, J.H.; McRoberts, J.; Shi, F.; Moreno, J.E.; Jones, A.D.; Howe, G.A. The flavonoid biosynthetic enzyme chalcone isomerase modulates terpenoid production in glandular trichomes of tomato. *Plant Physiol.* **2014**, *164*, 1161–1174. [CrossRef]
54. Li, W.F.; Ning, G.X.; Mao, J.; Guo, Z.G.; Zhou, Q.; Chen, B.H. Whole-genome DNA methylation patterns and complex associations with gene expression associated with anthocyanin biosynthesis in apple fruit skin. *Planta* **2019**, *250*, 1833–1847. [CrossRef]

current issues in molecular biology

MDPI

Article

Characterization of Highbush Blueberry (*Vaccinium corymbosum* L.) Anthocyanin Biosynthesis Related MYBs and Functional Analysis of *VcMYB* Gene

Yongyan Zhang [1,2,†], Dingquan Huang [1,3,†], Bin Wang [1], Xuelian Yang [4], Huan Wu [1], Pengyan Qu [2], Li Yan [4], Tao Li [5], Chunzhen Cheng [2,*] and Dongliang Qiu [1,*]

1 College of Horticulture, Fujian Agriculture and Forestry University, Fuzhou 350002, China
2 College of Horticulture, Shanxi Agricultural University, Jinzhong 030801, China
3 College of Agriculture, Fujian Agriculture and Forestry University, Fuzhou 350002, China
4 College of Agriculture, Guizhou University, Guiyang 550025, China
5 Fruit Research Institute, Fujian Academy of Agricultural Sciences, Fuzhou 350013, China
* Correspondence: ld0532cheng@sxau.edu.cn (C.C.); qiudl1970@fafu.edu.cn (D.Q.)
† These authors contributed equally to this work.

Abstract: As one of the most important transcription factors regulating plant anthocyanin biosynthesis, MYB has attracted great attentions. In this study, we identified fifteen candidate anthocyanin biosynthesis related MYB (ABRM) proteins, including twelve R2R3-MYBs and three 1R-MYBs, from highbush blueberry. The subcellular localization prediction results showed that, with the exception of VcRVE8 (localized in chloroplast and nucleus), all of the blueberry ABRMs were nucleus-localized. The gene structure analysis revealed that the exon numbers of the blueberry *ABRM* genes varied greatly, ranging between one and eight. There are many light-responsive, phytohormone-responsive, abiotic stress-responsive and plant growth and development related *cis*-acting elements in the promoters of the blueberry *ABRM* genes. It is noteworthy that almost all of their promoters contain light-, ABA- and MeJA-responsive elements, which is consistent with the well-established results that anthocyanin accumulation and the expression of *MYBs* are influenced significantly by many factors, such as light, ABA and JA. The gene expression analysis revealed that *VcMYB*, *VcMYB6*, *VcMYB23*, *VcMYBL2* and *VcPH4* are expressed abundantly in blueberry fruits, and *VcMYB* is expressed the highest in the red, purple and blue fruits among all blueberry *ABRMs*. *VcMYB* shared high similarity with functionally proven *ABRMs* from many other plant species. The gene cloning results showed that *VcMYB* had three variable transcripts, but only the transient overexpression of *VcMYB-1* promoted anthocyanin accumulation in the green fruits. Our study can provide a basis for future research on the anthocyanin biosynthesis related *MYBs* in blueberry.

Keywords: MYB; blueberry; anthocyanin; gene expression; expression regulation

Citation: Zhang, Y.; Huang, D.; Wang, B.; Yang, X.; Wu, H.; Qu, P.; Yan, L.; Li, T.; Cheng, C.; Qiu, D. Characterization of Highbush Blueberry (*Vaccinium corymbosum* L.) Anthocyanin Biosynthesis Related MYBs and Functional Analysis of *VcMYB* Gene. *Curr. Issues Mol. Biol.* **2023**, *45*, 379–399. https://doi.org/10.3390/cimb45010027

Academic Editors: Shimeles Tilahun and Vijai Bhadauria

Received: 2 December 2022
Revised: 22 December 2022
Accepted: 31 December 2022
Published: 3 January 2023

1. Introduction

Anthocyanins, which are natural polyphenols that are widely found in many foods, possess many biological and health-beneficial abilities, such as anti-inflammatory, visual-improving and so on [1–4]. All of these abilities are achieved through their high antioxidant activity [5,6] and some other biological effects, such as their anti-proliferative effect [7], hepatoprotective effect [8] and anti-depressant behavior [9].

Blueberries are one of the richest sources of anthocyanins among common fruits [10], and the bio-availabilities of blueberry anthocyanins have been extensively investigated [11,12]. During the last two decades, a large number of studies have been conducted on anthocyanin metabolism [13–22], and some anthocyanin biosynthesis related structural genes (such as *cinnamic acid 4-hydroxylase* (*C4H*) [13], *flavanone-3-hydroxylase* (*F3H*) [14], *dihydroflavonol reductase* (*DFR*) [15], *anthocyanidin synthase* (*ANS*) [16]) and regulatory transcriptional factors (such as genes encoding MYBs and bHLHs [17–22]) were identified or functionally verified in blueberry.

The MYB transcription factor plays an important role in regulating the spatio-temporal expression of anthocyanin biosynthesis related structural genes and anthocyanin accumulation, and is one of the most important and widely investigated transcription factors involved in anthocyanin biosynthesis regulation [17,22]. The first plant MYB gene, *C1*, was discovered from maize [23], whose encoded protein was required for anthocyanin biosynthesis in the aleurone layer [24]. Since then, *MYBs* have been isolated from many plant species. Most MYB proteins contain conserved MYB domain/domains at their N-terminus. According to the number of their MYB domains, MYBs can be further classified into four types, i.e., 1R-MYB/MYB-related, R2R3-MYB, 3R-MYB and 4R-MYB. It is noteworthy that most of the reported anthocyanin biosynthesis related MYBs (ABRMs) were R2R3-MYBs, followed by 1R-MYB/MYB-related MYBs [25]. In blueberry, most of the reported anthocyanin metabolism regulatory transcription factors were MYBs [26]. For example, the up-regulated expression of *VcMYB21* and *VcR2R3MYB* was confirmed to be associated with the UV-B-induced anthocyanin accumulation in blueberry pericarp [27]; VcMYB21 played a negative regulatory role in anthocyanin accumulation [28]; the transient overexpression of blueberry *MYBA* in an *Antirrhinum majus MYB* mutant restored the anthocyanin accumulation, and its co-expression with a heterologous *bHLH* could induce anthocyanin accumulation in tobacco leaves [29]; the pink fruit mutation phenotype of 'Pink Lemonade' was caused by the mutation of *MYB1* [20]; the expression of *VcUFGT* in highbush blueberry was positively regulated by VcMYBA1 and negatively regulated by VcMYBC2 [30].

Although some studies have been conducted on blueberry *ABRMs* [31,32], their applications are restricted as almost all these reported *ABRMs* were obtained based on *de novo* assembled transcriptome data. The publication of the draft genome of blueberry [33,34] greatly facilitated the genome-wide identification of the functional and regulatory genes involved in anthocyanin biosynthesis [35]. Based on the genome data, Wang et al. [31] identified a total of 229 *MYB* members that could be further divided into 23 subfamilies, but they did not focus on the anthocyanin biosynthesis related members; Zhao et al. [17] identified 11 MYBs from blueberry fruits through homologous protein searching, using Arabidopsis, apple, grape and strawberry MYBs belonging to the MBW complex as queries. Recently, many *ABRMs* have been identified and functionally approved in various plant species, such as Arabidopsis [36], *Helianthus tuberosus* [37], monkeyflower [38], *Eutrema salsugineum* [39], *Freesia hybrida* [40], grape hyacinth [41], apple [42] and so on. In this study, for the exploration and characterization of blueberry ABRMs, we identified the blueberry ABRMs by homologous searches against the blueberry protein data provided by the highbush blueberry genome project using the reported ABRMs from some other plant species as queries, characterized their sequences, and investigated their corresponding genes' expression patterns in blueberry fruits at five different ripening stages, based on our previously obtained transcriptome data and quantitative real time PCR (qRT-PCR) analysis. Moreover, the function of *VcMYB* (VaccDscaff1486-snap-gene-0.3), whose encoded protein showed high similarity with ten functionally proved ABRMs from other plant species (including Arabidopsis AtMYB114, AtMYB90, AtMYB75 and AtMYB113, *H. tuberosus* HtMYB2, monkeyflower PELAN, *E. salsugineum* EsMYB90, *F. hybrida* FhPAP1, grape hyacinth MaAN2 and apple MdMYB10), was further studied by transient overexpression in young blueberry fruits. The results obtained in this study will provide a foundation for the functional analysis and applications of blueberry anthocyanin biosynthesis related *MYB* genes, and will lay the foundations for research on the high-anthocyanin aimed blueberry breeding in the future.

2. Materials and Methods

2.1. Plant Materials

The plant materials used in this study were the fruits of four-year-old southern highbush blueberry 'FL03' at five different stages (green fruit (GF), pink fruit (PiF), red fruit (RF), purple fruit (PF) and blue fruit (BF)) [22]. After harvesting, the fruits were quickly taken back to the laboratory, washed with distilled water three times, immediately frozen in liquid nitrogen after draining the fruit surface with sterilized filter paper, and stored in the refrigerator at −80 °C for further use.

2.2. Identification of ABRM Proteins in Blueberry

The gDNA, cDNA and protein sequence files of the blueberries were downloaded from https://www.vaccinium.org/analysis/49 (accessed on 3 March 2021). To identify ABRM proteins in the blueberries, homologous protein sequence alignment was performed against the blueberry protein data, using recently reported ABRM protein sequences as queries with e-value $\leq 1 \times 10^{-5}$ and similarity >50% as criteria. The screened sequences with the highest bit score were selected as candidate MYBs and were named according to their homologous proteins sharing the highest similarity with them. One exception is that one candidate MYB (VaccDscaff1486-snap-gene-0.3) was named as VcMYB to distinguish it from blueberry VcMYBA (MH105054) [32]. A phylogenetic tree was constructed by Maximum Likelihood method using MEGAX (Possion mode, complete deletion, and 1000 bootstrap values) to show the relationships among the blueberry ABRMs and the reported ABRMs from some other plants.

2.3. Bioinformatic Analysis of Blueberry ABRM Genes and Their Encoded Proteins

The physicochemical properties, conserved motifs (motif number set as 10) and domains, as well as the subcellular localization of the blueberry MYB proteins, were analyzed according to the method described by Zhang et al. [22]. Gene structure analysis was performed using GSDS (http://gsds.gao-lab.org/, accessed on 3 March 2021). The ClustalW program in MEGA 6.06 software was used for multiple protein sequence alignment, and the alignment results were incorporated into MEGA 6.06 for the construction of the phylogenetic tree using the neighbor-joining method under the criteria of the Poisson model, complete deletion and bootstrap value = 1000. The 2000 bp sequences to upstream the start codon of the *MYBs* were extracted from the blueberry genome database and considered as promoter sequences, and were then subjected to *cis*-acting element analysis using PlantCARE (http://bioinformatics.psb.ugent.be/webtools/plantcare/html/, accessed on 3 March 2021).

2.4. Gene Expression Analysis

In our previous study, we sequenced the transcriptome of 'FL03' blueberry fruits at five different ripening stages (GF, PiF, RF, PF and BF). To study the expression patterns of these identified blueberry *ABRMs*, their FPKM (Fragments Per Kilobase of exon model per Million mapped fragments) values were first extracted from the transcriptome data and transformed into $\text{Log}_2(\text{FPKM} + 1)$ for heatmap drawing using TBtools software [43].

To validate the expression of the blueberry *ABRMs*, four highly expressed *ABRMs*, including *VcMYB6*, *VcMYB23*, *VcMYBL2* and *VcMYB*, were selected and subjected to quantitative real time PCR analysis. Primers were designed according to their CDS sequences using Primer 3.0 (Table 1). A Trizol RNA Extraction Kit (TaKaRa, Dalian, China) was used to isolate the total RNA from the blueberry fruits at five different ripening stages. Then, high quality RNA was used for cDNA synthesis using TransScript All-in-One First-Strand cDNA Synthesis SuperMix for qPCR (One-Step gDNA Removal) kit (TransGen, Beijing China). The qRT-PCR reactions were performed on a Bio-Rad CFX96^{TM} real-time quantitative fluorescent PCR instrument using a TB Green Premix Ex Taq II (Tli RNaseH Plus; TaKaRa, Dalian, China) kit according to the manual. The relative expression levels of the selected blueberry *ABRM* genes in the fruits at different ripening stages were calculated using the $2^{-\Delta\Delta Ct}$ method with *GAPDH* (Genbank ID: AY123769) as the internal reference gene [22]. Three biological and three technical replications were made for the qRT-PCR analysis of the selected genes.

Table 1. Information for the primers used in this study.

Target Gene	Primer Name	Primer Sequence	Target Length (bp)	Annealing Temperature (°C)	Applications
VcMYB	VcMYB-F VcMYB-R	ATGGACATAGTTCCATTGGGAGTGA TAAAATATCCCAAAGGTCCACATTGTC	798	59	Gene cloning
	VcMYB-InF	ACGGGGGACTCTAGAG-GATCCATGGACATAGTTCC-ATTGGGAGTGA	786/704/568	60	Vector construction
	VcMYB-InR	GCTCACCATCGCTGCACTAGTTAAAA-TATCCCAAAGGTCCACATTGTC			
	VcMYB-qF VcMYB-qR	TCCATTGGGAGTGAGAAAGG CAATCCTGCCCTGTAAGGAA	115	60	qRT-PCR
VcMYB6	VcMYB6-qF VcMYB6-qR	CTCTCCTCAGGTGGAGCATC TTCCTCTTGAGCGTGGAGTT	164	60	qRT-PCR
VcMYB2	VcMYBL2-qF VcMYBL2-qR	TCAAAATCCACGTCCCTCTC CATTCTCCGCTAGCTTGGTC	92	60	qRT-PCR
VcMYB23	VcMYB23-qF VcMYB23-qR	TGTTGGGAAACAGATGGTCA TTTCAAGTGGGTGTGCCATA	89	60	qRT-PCR
GAPDH	GAPDH-qF GAPHD-qR	ACTACCATCCACTCTATCACCG AACACCTTACCAACAGCCTTG	116	59	qRT-PCR

2.5. Gene Cloning, Vector Construction and Transient Overexpression Analysis

A RevertAid First-strand cDNA synthesis Kit (Thermo Scientific, Shanghai, China) was used to synthesize the cDNA for gene cloning. The primers for *VcMYB* gene amplification were designed using Primer 3 (Table 1). The 25 μL PCR system contained 1 μL cDNA, 1 μL each of the forward and reverse primer, 12.5 μL 2 × Green mix and 9.5 μL ddH_2O. The PCR conditions were set as follows: predenaturation at 95 °C for 3 min; 35 cycles of denaturation at 95 °C for 30 s, annealing at 59 °C for 30 s, and extension at 72 °C for 2.5 min; and final extension at 72 °C for 8 min. The electrophoresis detection results showed that there were three variable transcripts for *VcMYB*. The PCR products of the three transcripts were separately gel extracted, ligated into an 18-T vector and transformed into *Escherichia coli* DH5a. Positive clones were selected and sent to Beijing Liuhe Huada Gene Technology Co., Ltd. (Beijing, China) for sequencing verification. According to the sequencing results, the primers for the *VcMYB* gene vector construction were further designed and used for the amplification of the three variable transcripts (Table 1). The cloned fragments were ligated into the pBI123 vector using the Ready-to-use Seamless Cloning Kit (Sangon Biotech, Shanghai, China), and transformed into *Agrobacterium tumefaciens* GV3101. *A. tumefaciens* GV3101 carrying empty pBI123 (as control), pBI123-*VcMYB-1*, pBI123-*VcMYB-2* and pBI123-*VcMYB-3* vectors were cultured in the dark at 180 rpm and 28 °C until OD_{600} = 0.6~0.8 they were centrifuged at 5000 rpm for 4 min, resuspended using an MES solution (2-(N-Morpholino) and ethanesulfonic acid hydrate) solution (containing 10 mM MES, 10 mM $MgCl_2 \cdot 6H_2O$ and 200 μM acetosyringone) to a final concentration of OD_{600} = 0.6. Then, they were injected into the green fruits (at about 15 days post flowering (dpf), to further confirm the function of *VcMYB-1*, green fruits at 20 dpf were also used) of the 'Legacy' blueberry, cultured in the dark for 1 day and removed to normal light for fruit color change observation. At three days post treatment, the colour parameters (L^*, a^* and b^* values) of the fruit pericarps were measured using a CR8 Portable Colorimeter (Shenzhen Threenh Technology Co., Ltd., Shenzhen, China), and the anthocyanin was extracted using the acidified ethanol [44] from the blueberry fruit pericarps. The anthocyanin content in the blueberry pericarps was measured and calculated according to the method of Zhuang et al. [45].

3. Results

3.1. The Identified Blueberry ABRMs

Through homologous protein searching, fifteen candidate blueberry ABRM proteins were identified (Table 2). The phylogenetic analysis results revealed that all of these blueberry ABRM proteins shared a close relationship with the query homologous proteins from other plants (Supplementary Figure S1). Among them, VcMYB shared high similarity, up to 93.40%, with the reported blueberry anthocyanin biosynthesis regulatory VcMYBA [32]. Moreover, it shared more than 50% sequence similarities with ten known ABRMs, including Arabidopsis AtMYB114 (74.36%), *H. tuberosus* HtMYB2 (62.25%), monkeyflower PELAN (59.87%), *E. salsugineum* EsMYB90 (59.35%), Arabidopsis AtMYB90 (PAP2) (58.97%) and AtMYB75(PAP1) (58.71%), *F. hybrida* FhPAP1 (56.86%), grape hyacinth MaAN2 (56.86%), Arabidopsis AtMYB113 (56.77%) and apple MdMYB10 (54.36%). VcMYB12 shared the highest similarity with Arabidopsis AtMYB12 (66.86%), followed by AtMYB12 and AtMYB111. The similarities of VcMYB123 with kiwifruit AcMYB123 and apple MYB9 were both higher than 70% (72.33% and 71.43%, respectively). VcMYBL2 shared the highest similarity with apple MdMYBL2 (65.29%), followed by eggplant SmelMYBL1 (51.61%). VcMYB6 shared the highest similarity with apple MdMYB6 (55.26%). VcPL shared the highest similarity with rice OsPL (75.19%). VcMYB24L shared the highest similarity with apple MdMYB24L (75.00%). VcRVE8 shared the highest similarity to *Pyrus bretschneideri* PbRVE8 (68.22%). VcCPC shared the highest similarity with Arabidopsis AtCPC (57.41%). VcMYBC1, VcMYB23, VcMYB340, VcPH4, VcMYBATV and VcMYB85 shared the highest similarity with kiwifruit AaMYBC1, apple MdMYB23, sweet potato IbMYB340, citrus CsPH4, tomato SlMYBATV and millet SiMYB85, respectively.

3.2. Physiobiochemical Properties and Sequence Characteristics of Blueberrry ABRMs

The identified blueberry ABRM proteins consisted of between 73 and 365 aa, with molecular weights ranging between 8407.08 and 40,095.85 Da, and isoelectric points (pI) ranging between 4.31 and 9.72. Their instability coefficients ranged between 36.10 and 79.67. All of these proteins were predicted to be hydrophilic proteins. The subcellular localization prediction results showed that all of these blueberry ARBMs were nucleus-localized, with the exception of VcRVE8 (localized to both chloroplast and nucleus) (Table 3). By analyzing the amino acid sequences, we found that VcMYB12, VcMYB85, VcMYB340, VcPL, VcMYB6, VcMYB23, VcMYBC1, VcMYB24L, VcMYB123, VcMYB, VcPH4 and VcMYBL2 have conserved R2 and R3 domains, indicating that they belong to the R2R3-MYB (2R-MYB). The remaining three ARBMs belong to 1R-MYB, of which VcMYBATV and VcCPC contain a R3 domain, and VcRVE8 contains a R1/2 domain (Figure 1).

3.3. Conserved Motifs in Blueberry ABRM Proteins and Gene Structures of Their Corresponding Genes

In total, we identified five conserved motifs from the blueberry ABRM proteins (Figure 2A). Among these motifs, Motif1 was found in all of the ABRMs. The R2R3 type blueberry ABRMs all contained Motif1~3, VcMYB23 contained two Motif3, and VcMYB contained an extra Motif5. VcPCP and VcMYBATV contained only one Motif1, VcRVE8 contained only one Motif1 and one Motif4.

Table 2. The identified blueberry ABRM proteins.

Protein Name	Gene ID	Homologous Protein Name	Homologous Gene ID	Similarity (%)	References
VcMYB	VaccDscaff1486-snap-gene-0.3	VcMYBA	MH105054	93.40	[32]
		AtMYB114	At1G66380	74.36	[36]
		HtMYB2	MN887536	62.25	[37]
		PELAN	KJ011144	59.87	[38]
		EsMYB90	XP_006391393	59.35	[39]
		AtMYB90(PAP2)	At1G66390	58.97	[36]
		AtMYB75(PAP1)	At1G56650	58.71	[36]
		FhPAP1	MT210093	56.86	[40]
		MaAN2	KY781168	56.86	[41]
		AtMYB113	At1G66370	56.77	[36]
		MdMYB10	EU51829.2	54.36	[42]
VcMYB12	VaccDscaff43-snap-gene-6.43	AtMYB12	At2G47460	66.86	[46]
		AtMYB11	At3G62610	52.17	[46]
		AtMYB111	At5G49330	50.19	[46]
VcMYB123	VaccDscaff34-augustus-gene-10.31	AcMYB123	MH643776	72.33	[47]
		MdMYB9	MDP0000210851	71.43	[48]
VcMYBL2	VaccDscaff28-augustus-gene-197.19	MdMYBL2	NP_001281006.1	65.29	[49]
		SmelMYBL1	MN855525	51.61	[50]
VcMYB6	VaccDscaff31-processed-gene-75.6	MdMYB6	DQ074461	55.26	[51]
		IbMYB44	itf03g30290.t1	51.46	[52]
VcPL	VaccDscaff13-augustus-gene-110.26	OsPL	LOC_Os05g48010.1	75.19	[53]
VcMYB24L	VaccDscaff30-augustus-gene-338.25	MdMYB24L	XM_008343218.2	75.00	[54]
VcRVE8	VaccDscaff34-augustus-gene-308.23	PbRVE8	XP_009342285.1	68.22	[55]
VcCPC	VaccDscaff41-snap-gene-184.30	AtCPC	At2G46410	67.06	[56]
VcMYBC1	VaccDscaff32-augustus-gene-55.27	AaMYBC1	MN249175	57.41	[57]
VcMYB23	VaccDscaff46-processed-gene-168.9	MdMYB23	MDP0000230141	55.15	[58]
VcMYB340	VaccDscaff16-snap-gene-84.41	IbMYB340	itf12g05820.t1	55.02	[52]
VcPH4	VaccDscaff39-augustus-gene-189.28	CsPH4	Cs9g03070	54.14	[59]
VcMYBATV	VaccDscaff1069-augustus-gene-0.8	SlMYBATV	Solyc07g052490.4.1	53.57	[60]
VcMYB85	VaccDscaff36-augustus-gene-8.19	SiMYB85	Seita.4G086300	50.72	[61]

Table 3. Basic physicochemical properties of the identified blueberry ABRMs. CDS: coding sequence; PI: isoelectric point; MW: molecular weight; PI: isoelectric point.

Gene Name (ID)	CDS Length/bp	Protein Size/aa	MW/Da	PI	Instability Index	GRAVY	Subcellular Localization
VcMYB (VaccDscaff1486-snap-gene-0.3)	822	273	31,294.22	6.01	41.78	−0.749	Nucleus
VcMYB12 (VaccDscaff43-snap-gene-6.43)	1083	360	39,534.64	6.68	54.28	−0.603	Nucleus
VcMYB123 (VaccDscaff34-augustus-gene-10.31)	828	275	31,260.2	7.55	52.05	−0.705	Nucleus
VcMYBL2 (VaccDscaff28-augustus-gene-197.19)	702	233	26,419.91	8.40	51.94	−0.779	Nucleus

Table 3. *Cont.*

Gene Name (ID)	CDS Length/bp	Protein Size/aa	MW/Da	PI	Instability Index	GRAVY	Subcellular Localization
VcMYB6 (VaccDscaff31-processed-gene-75.6)	1080	359	39,018.97	6.26	61.47	−0.498	Nucleus
VcPL (VaccDscaff13-augustus-gene-110.26)	903	300	33,807.74	5.90	55.58	−0.702	Nucleus
VcMYB24L (VaccDscaff30-augustus-gene-338.25)	573	190	21,953.61	6.16	54.33	−0.854	Nucleus
VcRVE8 (VaccDscaff34-augustus-gene-308.23)	942	313	33,898.29	7.78	46.57	−0.462	Chloroplast; Nucleus
VcCPC (VaccDscaff41-snap-gene-184.30)	297	98	11,752.33	9.72	79.67	−1.014	Nucleus
VcMYBC1 (VaccDscaff32-augustus-gene-55.27)	798	265	30,057.3	8.57	58.02	−0.68	Nucleus
VcMYB23 (VaccDscaff46-processed-gene-168.9)	738	245	27,819.19	6.41	36.10	−0.668	Nucleus
VcMYB340 (VaccDscaff16-snap-gene-84.41)	744	247	28,372.85	7.66	54.91	−0.855	Nucleus
VcPH4 (VaccDscaff39-augustus-gene-189.28)	1086	361	40,095.85	8.72	48.94	−0.719	Nucleus
VcMYBATV (VaccDscaff1069-augustus-gene-0.8)	222	73	8407.08	4.31	61.99	−0.922	Nucleus
VcMYB85 (VaccDscaff36-augustus-gene-8.19)	912	303	33,738.76	8.37	60.02	−0.775	Nucleus

The gene structure analysis results showed that the exon numbers of the blueberry *ARBM* genes ranged between one and eight (Figure 2B). *VcRVE8* had the largest number of exons with eight, *VcMYB24L* had four exons, *VcPH4* and *VcMYB123* contained two exons, and *VcMYB6* had only one exon. All the other blueberry *ABRMs* contained three exons.

3.4. *Cis-Acting Elements in Promoters of Blueberry ABRM Genes*

We further analyzed the *cis*-acting elements in the promoters of the blueberry *ABRM* genes. The results showed that there were many light-responsive, phytohormone-responsive, stress-responsive, and growth and development related elements in their promoters (Figure 3). In total, we identified ten types of light-responsive elements in their promoters. All of the blueberry *ABRMs*, with the exception of *VcMYB*, contained a light-responsive Box4 element in their promoter regions, and the promoters of 12 blueberry *ABRM* genes (except *VcMYB24L*, *VcMYB6* and *VcMYBATV*) contained the light responsive G-box elements (Figure 3).

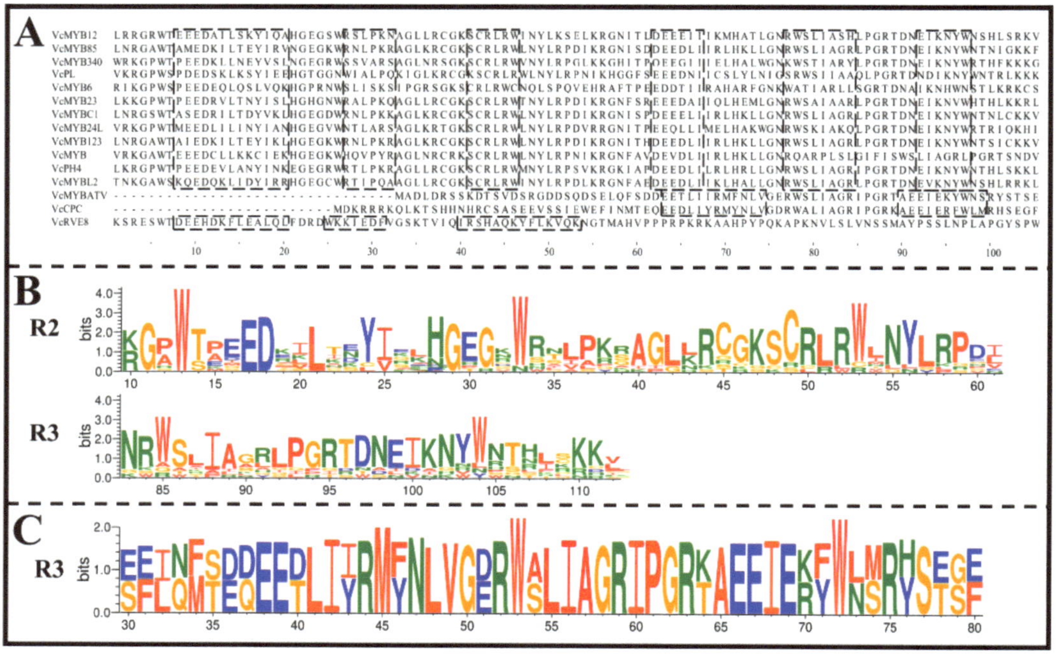

Figure 1. Repeat (R) domains sequences in blueberry ABRM proteins. (**A**): The R domain sequences of blueberry ABRM proteins. Sequences in dashed boxes represent the helix sequences in R domains; (**B**): Sequence logos for the R2 and R3 domain of R2R3-MYB type blueberry ABRMs; (**C**): Sequence logo for the R3 domain of 1R-MYB type blueberry ARBMs.

Nine types of phytohormone-responsive elements, including three gibberellin (GA)-responsive (P-box, TATC-box and GARE-motif), two methyl jasmonate (MeJA)-responsive (TGACG-motif and CGTCA-motif), one auxin-responsive (TAG-element), one abscisic acid (ABA)-responsive (ABRE), one salicylic acid (SA)-responsive (TCA-Element), and one ethylene-responsive elements (ERE), were found in the promoters of blueberry *ABRM* genes (Figure 3). With the exception of *VcMYB24L*, *VcMYB6* and *VcMYBATV*, all the other *ABRMs'* promoters contained ABA-responsive ABRE elements. With the exception of *VcCPC*, *VcMYB85*, *VcMYBC1* and *VcPH4*, the promoters of all the other *ABRM* genes contained the MeJA-responsive TGACG-motif and CGTCA-motif elements. With the exception of *VcPCP*, *VcMYB123*, *VcMYBC1*, *VcMYBL2* and *VcPL*, all the other *ABRMs'* promoters contained ethylene-responsive ERE elements. In addition, auxin-, SA- and GA-responsive elements were found in the promoters of six, five and four *ABRMs*, respectively.

In total, six kinds of stress-related elements, including low temperature-responsive element LTR, anaerobic-induction related element ARE, MYB drought-inducibility related element MBS, defense and stress related element TC-rich elements, anoxic specific inducibility related element GC-motif and wounding related element WUN-motif, were identified in the blueberry *ABRM* promoters (Figure 3). Moreover, there were eleven, nine, six, five, four and three *ABRMs* contained ARE, LTR, MBS, GC-motif, TC-rich repeats and WUN-motif in their promoters, respectively.

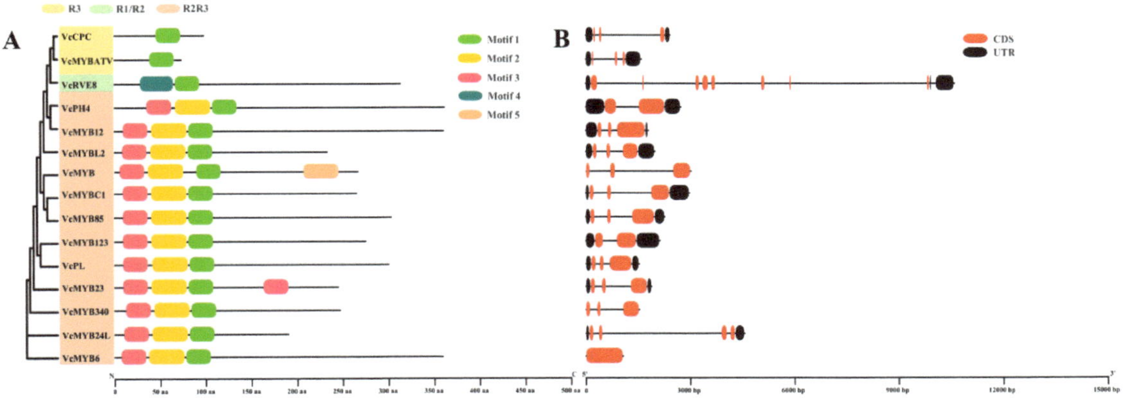

Figure 2. Conserved motifs distributions in blueberry ABRM proteins (**A**) and gene structures of their corresponding genes (**B**).

Additionally, we also identified many growth and development related *cis*-acting elements in the promoters of the blueberry *ABRM* genes (Figure 3). Eight *ABRMs* contained the regulatory A-box element in their promoters; seven *ABRMs*' promoters contained the zein metabolism regulation related O2-site element; seven *ABRMs*' promoters contained the MYBHV1-binding site element CCAAT-box; four *ABRMs*' promoters contained the meristem expression related element CAT-box; the promoters of *VcMYB24L* and *VcMYB340* contained the endosperm expression related GCN4_motif element; the promoter of *VcMYBL2* contained the circadian control related element. Moreover, the *VcMYB24L* promoter specifically contained the flavonoid biosynthetic related MBSI element.

3.5. Protein and Protein Interaction Analysis of Blueberry ABRM Proteins

Based on the Arabidopsis protein database, the STRING software was used to predict the interacting proteins of blueberry ABRMs (Figure 4). The results showed that VcCPC, VcMYB, VcMYB6, VcMYBATV, VcMYB23 and VcRVE8 were homologous protein of AtCPC (At2G46410), AtMYB114 (At1G66380), AtMYBR1 (At5G67300), AtTT2 (At5G35550), AtMYB15 (At3G23250) and AtRVE8 (At3G09600), respectively. AtPCP interacts with AtGL3 (At5G41315), AtEGL3 (At1G63650), AtGL2 (At1G79840) and AtTTG1 (At5G24520). AtMYB114 interacts with AtEGL3 (At1G63650), AtTT1 (At1G34790) and AtTTG1 (At5G24520). AtTT2 interacts with AtTT1 (At1G34790), AtTT8 (At4G09820), AtEGL3 (At1G63650), AtTTG1 (At5G24520) and AtGL3 (At5G41315). AtMYBR1 (At5G67300) interacts with AtRCAR3 (At5G53160). AtMYB15 (At3G23250) interacts with AtICE1 (At3G26744). And AtRVE8 interacts with AtLNK1 (At5G64170) and AtLNK2 (At3G54500).

3.6. Expression Analysis of Blueberry ABRM Genes

According to our transcriptome data, the expression patterns of the blueberry *ABRM* genes in fruits at different ripening stages were studied. It was found that *VcMYB6*, *VcMYB23*, *VcMYB*, *VcMYBL2* and *VcPH4* are expressed highly in blueberry fruits, but other *ABRMs*, such as *VcMYB12*, *VcMYB123*, *VcPL*, *VcMYB24L* and *VcMYB340*, showed either low expression (FPKM < 4) or are not expressed in the fruits (Figure 5). The expression level of *VcMYB* was low in GF (FPKM < 2), gradually increased in PiF (FPKM > 10), and then maintained at an abundant level in RF, PF and BF (FPKM > 40). The expression levels of *VcMYB* in RF and PF were both higher than that in BF. Moreover, its expression in the late three stages was found to be the highest among all the blueberry *ABRMs* and very significantly higher than that in GF and PiF, indicating that it might play an important role in regulating anthocyanin biosynthesis, particularly at the late fruit ripening stages. The expression level of *VcMYB6* in fruits at all stages was high (FPKM > 10), and its expression level changed slightly during blueberry fruit ripening. The expression of *VcMYB23* was the highest in RF and PF. *VcMYBL2* showed a 'fall-rise' expression pattern, and its expression level in PiF was the lowest, but there was no significant expression difference at the late three ripening stages. *VcPH4* also showed a 'fall-rise' expression profile during fruit ripening, but no significant difference was found among the fruits at different ripening stages.

	Function	Motif	VcCPC	VcMYB	VcMYB12	VcMYB123	VcMYB23	VcMYB24L	VcMYB340	VcMYB6	VcMYB85	VcMYBATV	VcMYBC1	VcMYBL2	VcPH4	VcPL	VcRVE8
Light responsive	Light	GT1-motif	2	2	1		2	3				2	2	1		1	2
		GATA-motif	2	1			2				1	1			1	2	
		Box 4	3		1	1	3	7	7	3	4	1	5	2	3	1	1
		G-Box	8	2	4	6	8		6		6		9	6	1	5	5
		Sp1	1		1						1						1
		I-box	1	1				1			2			2		1	2
		TCT-motif	1	1	1			1		1	1		3	2	1		1
		TCCC-motif		1								1		1	1		
		AE-box		1				4				2			3		
		MRE								2		2	1			1	
Phytohormone responsive	Gibberellin	P-box					1			2				1			
		TATC-box				2	1			1							
		GARE-motif				2											
	Auxin	TGA-element			4	4	1						1		1	1	
	Abscisic acid	ABRE	3	2	1	4	5		7		6		7	5	1	5	4
	Salicylic acid	TCA-element	1		1						1				1		1
	MeJA	TGACG-motif		1	1	1	1	1	2	1		1		1		3	3
		CGTCA-motif		1	1	1	1	1	2	1		1		1		3	3
	Ethylene	ERE		1	2		2	1	2	1	1	1			2		2
Stress responsive	Low-temperature	LTR	3	1		1			3			2	3	1		2	1
	Anaerobic induction	ARE	2	3			6	4	2	1	2	3	1		1	2	
	MYB drought-inducibility	MBS		3		3			2			1			1		1
	Defense and stress	TC-rich repeats					1				1		1				1
	Anoxic specific inducibility	GC-motif					1	1			1		1		1		
	Wound	WUN-motif					1			1				1			
Plant growth and development	Regulatory	A-box	1		2	1			1		1		1		1		1
	Circadian control	circadian												1			
	Flavonoid biosynthetic	MBSI						1									
	Endosperm expression	GCN4_motif						1	2								
	Meristem expression	CAT-box			3	1	1										1
	Zein metabolism regulation	O2-site				1		1	1	1	2		1		3		
	MYBHv1 binding site	CCAAT-box		1		2	1	1		2				1			1

Figure 3. The identified *cis*-acting elements in the promoters of blueberry anthocyanin biosynthesis related *MYB* genes.

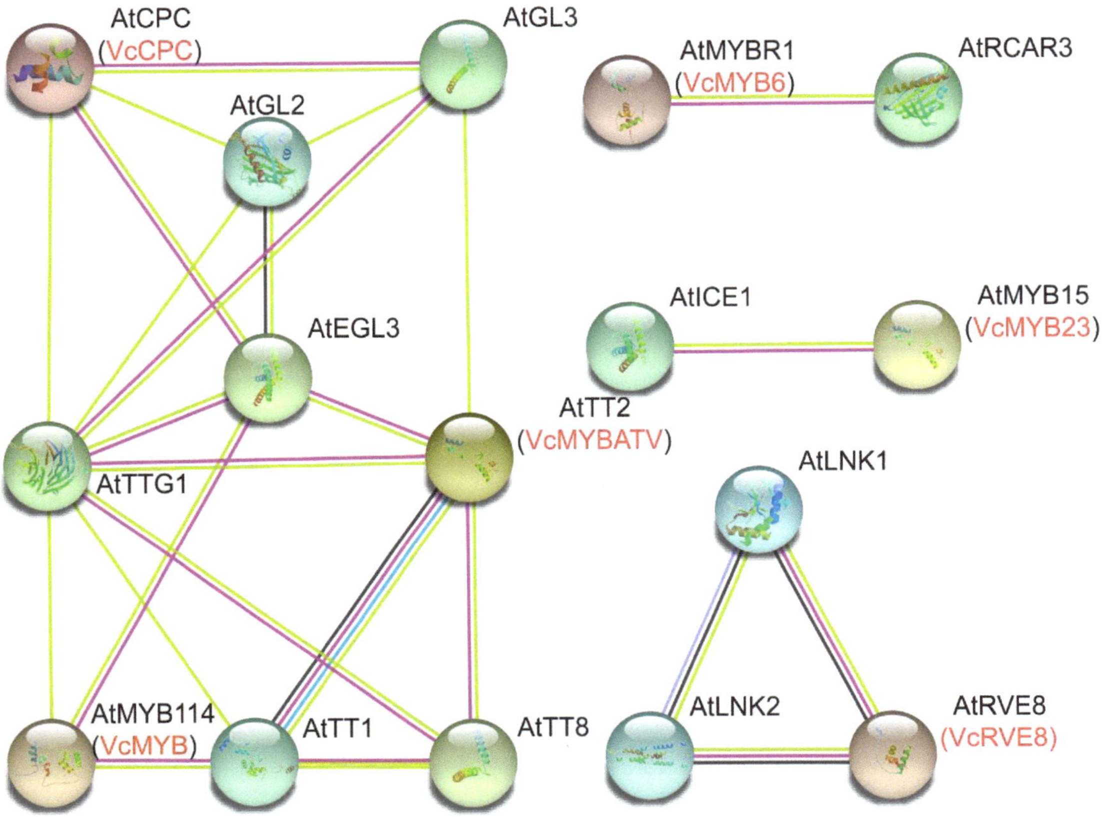

Figure 4. The predicted protein-protein interaction network for blueberry anthocyanin biosynthesis related MYBs based on the Arabidopsis protein database using STRING. At: *Arabidopsis thaliana*; Vc: *Vaccinium corymbosum*.

For the validation of the expression changes of the blueberry *ABRM* genes in fruits during ripening, quantitative real time PCR (qRT-PCR) analysis of four selected genes, including *VcMYB6*, *VcMYB23*, *VcMYBL2* and *VcMYB*, was performed. The results showed that the expression change patterns of these blueberry *ABRMs* during fruit ripening were mostly consistent with our transcriptome data (Figure 6), indicating that our transcriptome data is believable. Among them, the expression of *VcMYB* was found to be the lowest in GF, and its expression in PiF, RF, PF and BF was approximately 1.44-fold, 1.79-fold, 2.1-fold and 1.75-fold of GF, respectively.

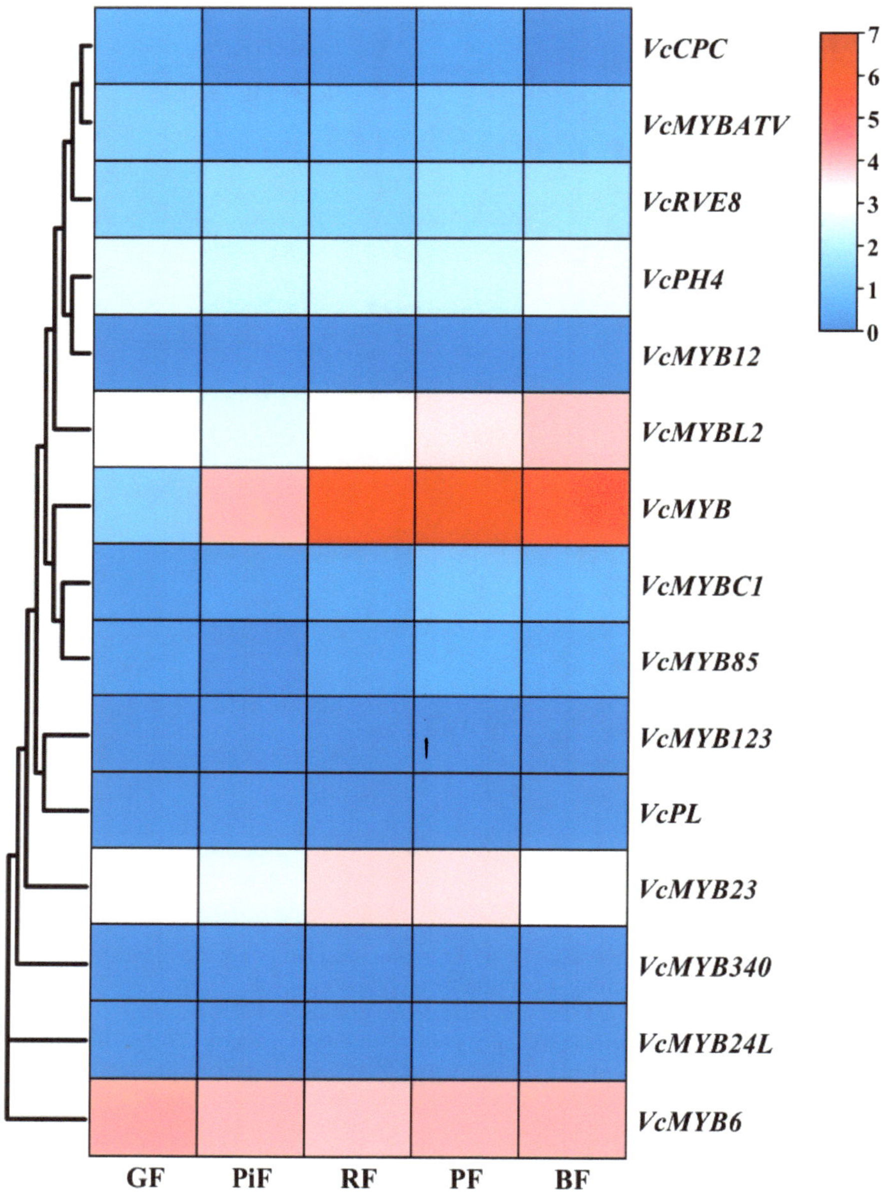

Figure 5. Expression analysis of blueberry *ABRM* genes in fruits at five different ripening stages based on transcriptome data. GF: green fruit; PiF: pink fruit; RF: red fruit; PF: purple fruit; BF: blue fruit. Log_2(FPKM + 1) values were used for heatmap drawing.

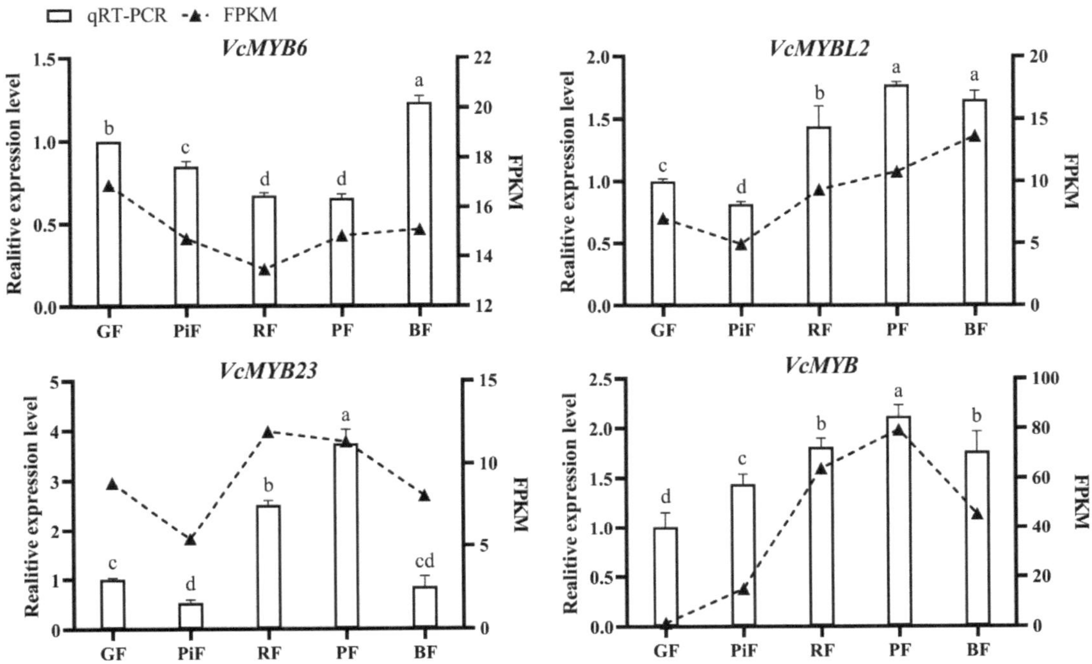

Figure 6. Quantitative real time PCR results of four selected blueberry *ABRM* genes. GF: green fruit; PiF: pink fruit; RF: red fruit; PF: purple fruit; BF: blue fruit. Different letters above columns represent significant difference at $p < 0.05$ level.

3.7. Transient Overexpression Analysis of Three VcMYB Variable Transcripts

Given the high similarity with more than ten reported *ABRM* genes and its high expression in fruits, particularly at the late ripening stages, *VcMYB* was proposed to play a key role in regulating blueberry anthocyanin biosynthesis. To confirm its function, the gene was successfully cloned by reverse transcription PCR (RT-PCR). The electrophoresis detection results showed that this gene had three variable transcripts, which were all shorter than the reference cDNA sequence (VaccDscaff1486-snap-gene-0.3) (Figure 7A). The sequencing results showed that the lengths of these three transcripts were 786 bp, 704 bp and 568 bp, respectively, and they were termed as *VcMYB-1*, *VcMYB-2* and *VcMYB-3*, in descending order of sequence length.

The three transcripts were separately inserted into the pBI123 vector, transformed into *A. tumefaciens* GV3101, and then transiently overexpressed in the young blueberry fruits. The results showed that only the transient overexpression of *VcMYB-1* triggered anthocyanin accumulation in the young blueberry fruits (Figure 7B). For the green blueberry fruits, at approximately 15 days post-flowering (dpf), at three days post-treatment, the areas injected with *A. tumefaciens* GV3101 carrying *VcMYB-1* became purple-red, and the L^*, a^* and b^* values of the fruit pericarps overexpressing *VcMYB-1* were all significantly lower than the control check group (CK) (Figure 7C). For the green fruits, at about 20 dpf, at three days post-treatment, the injected areas became much redder than the fruits at 15 dpf. By measuring the anthocyanin content, we found that the anthocyanin content in the *VcMYB-1* overexpressing fruit pericarps was approximately 1.56-fold of the CK (Figure 7D–F). However, no obvious pigmentation was found in the areas injected with *A. tumefaciens* GV3101 carrying *VcMYB-2* and *VcMYB-3*, indicating that their transient overexpression could not trigger anthocyanin accumulation in the blueberry fruit pericarps.

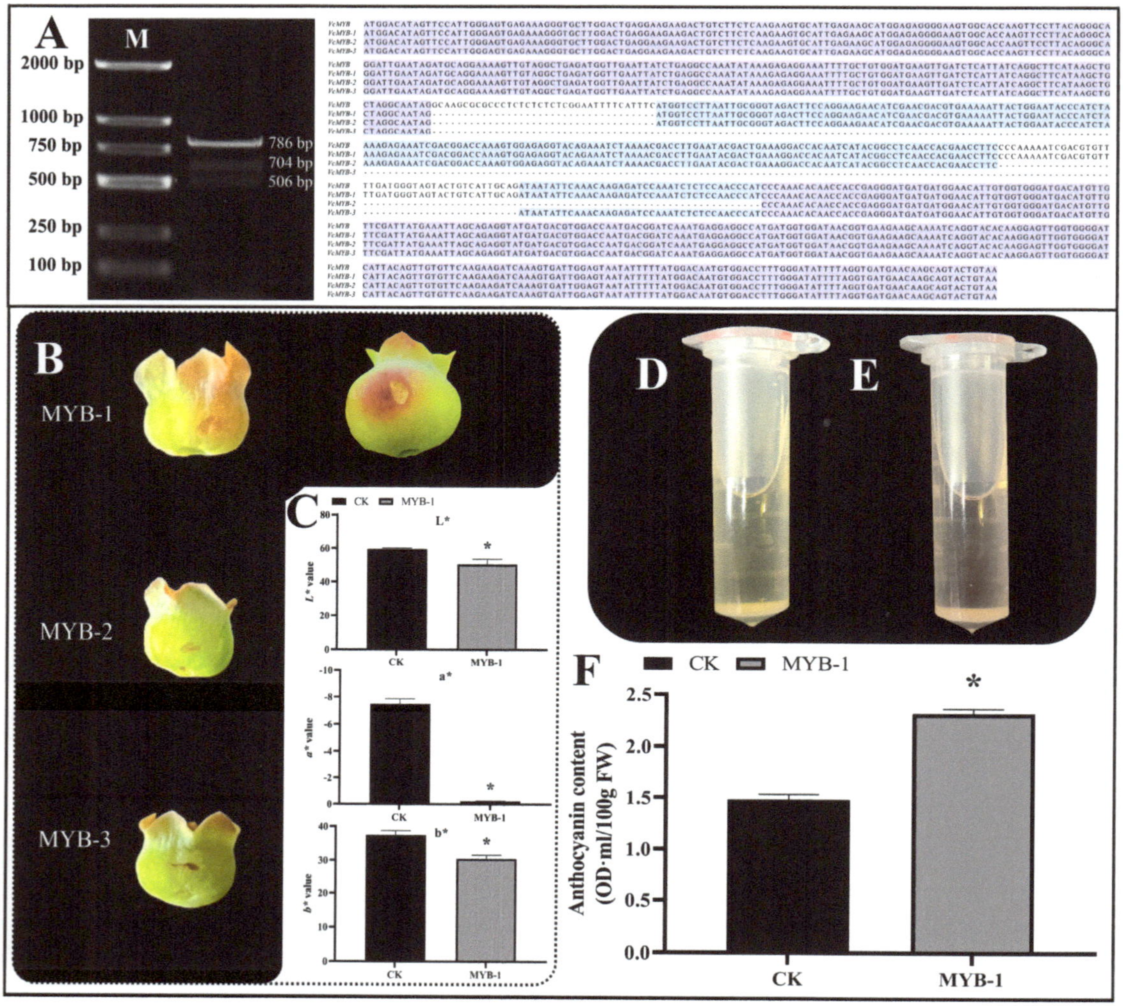

Figure 7. *VcMYB* gene has three variable transcripts and only the transient overexpression of *VcMYB-1* promotes the anthocyanin accumulation in blueberry fruits. (**A**): The three amplified transcripts of *VcMYB*, and sequence alignment results of their nucleotide sequences together with their reference *VcMYB* (VaccDscaff1486-snap-gene-0.3). (**B**): Transient overexpression of *VcMYB* variable transcripts in green blueberry fruits. For fruits of the MYB-1 group, the left one is green fruit at 15 days post flowering, and the right one is green fruit at 20 days post flowering (**C**): Colour parameters (L*, a* and b* values) of fruit pericarps at three days post treatment. (**D**): Anthocyanin extract solution of CK group. (**E**): Anthocyanin extract solution of MYB-1 group. (**F**): Anthocyanin content in blueberry fruit pericarps. The '*' above columns in C and F represents significant difference at $p < 0.05$ level.

4. Discussion

According to their R repeat numbers and sequences, MYBs could be classified into four major types, i.e., 1R-MYB/MYB-related, R2R3-MYB, 3R-MYB and 4R-MYB [25]. In plants, R2R3-MYBs and 1R-MYBs are the major types of MYBs. It was frequently discovered that most of the ABRMs belong to the R2R3 type, followed by the 1R type [62]. In this study, we identified fifteen ABRMs from the blueberry genome data through the homologous protein sequence alignment method. Among these ABRMs, twelve were R2R3-MYBs and three were 1R-MYBs. The subcellular localization prediction results showed that almost all of these ABRMs were nucleus-localized, which was consistent with the localization characteristics of the transcription factors. The gene structure analysis revealed that the exon number of blueberry *ABRMs* varied greatly, ranging between one and eight. The gene expression analysis showed that only *VcMYB6*, *VcMYB23*, *VcMYB*, *VcMYBL2* and *VcPH4* were expressed highly in fruits at different ripening stages, suggesting that they may play more important roles in regulating anthocyanin biosynthesis in blueberry.

MYBs function in regulating anthocyanin biosynthesis by associating with bHLH, WD40 and anthocyanin biosynthesis structural genes. Wang et al. [49] reported that exogenous cytokinin treatment could induce anthocyanin accumulation in red-fleshed apple callus by promoting the expression of *MdDFR*, *MdUFGT*, *MdMYB10* and *MdbHLH3* genes, and by suppressing the expression of *MYBL2*. Moreover, they found that MdMYBL2 could interact with MdbHLH3. In the callus overexpressing *MdMYBL2*, anthocyanin accumulation decreased, and the expression levels of *MdDFR*, *MdUFGT*, *MdMYB10* and *MdbHLH3* were all significantly inhibited. *Actinidia arguta* AaMYBC1 could interact with AabHLH42. The virus-induced gene silencing of *AaMYBC1* resulted in decreased anthocyanin accumulation and the reduced expression of anthocyanin biosynthesis structural genes in 'HB' kiwifruit [57]. Eggplant SmMYB86, a negative regulator of anthocyanin biosynthesis, can inhibit the expression of *SmCHS*, *SmF3H* and *SmANS* by binding to their promoters [63]. The down-regulation of *SmMYB86* resulted in anthocyanin content reduction and *SmCHS*, *SmF3H* and *SmANS* gene expression up-regulation, and its overexpression resulted in a significant decrease in the anthocyanin content in eggplant.

In this study, we predicted the protein-protein interaction network of blueberry ARBMs. According to their interacting proteins, it was predicted that the functions of these MYBs varied. VcCPC, VcMYB6 and VcMYBATV were found to be highly homologous to AtCPC, AtMYBR1 and AtTT2, respectively. AtCPC is a MYB that negatively regulates anthocyanin accumulation [64], suggesting that VcCPC might play a negative role in regulating the anthocyanin biosynthesis in blueberry. AtMYBR1 (also called AtMYB44) is involved in the abiotic stress responses of *A. thaliana* and exhibited a negative regulatory role in anthocyanin biosynthesis. Sweet potato IbMYB44 could interact with IbMYB340 and IbNAC56a/b, thereby inhibiting the formation of the MYB340-BHLH2-NAC56 complex and negatively regulating the accumulation of anthocyanins [52]. Apple MdMYB6 can bind to the promoters of *MdANS* and *MdGSTF12*, and the overexpression of *MdMYB6* in red apple callus would reduce anthocyanin accumulation and inhibit the expression of *MdANS* and *MdGSTF12* [51]. RCAR3, the interacting protein of AtMYBR1, is an ABA receptor regulator and plays a pivotal role in activating ABA signaling [65], suggesting that VcMYB6 might play a role in the ABA-regulated anthocyanin biosynthesis in blueberry. AtTT2 (AtMYB123) is mainly responsible for the biosynthesis regulation of tannins, such as proanthocyanidins [66]. VcMYB23 and VcRVE8 are homologous to AtMYB15 and AtRVE8, respectively. In Arabidopsis, the AtMYB15 interacts with AtICE1, and AtRVE interacts with AtLNK2 and AtLNK2. ICE1 plays an important role in plants' responses to low temperatures [67], while LNK1 and LNK2 play roles in the integrated regulation of light signal responses and circadian regulation [68,69], and they can also act as corepressors of phenylpropanoid metabolism by interacting with MYB [70], suggesting that VcMYB23 and VcREV8 might be involved in the low temperature and light triggered anthocyanin biosynthesis in blueberry, respectively.

It is worth noting that VcMYB shared the highest similarity with one reported blueberry MYB (VcMYBA) [32] and was identified as a homologous protein of many positive anthocyanin regulatory MYB proteins, such as Arabidopsis AtMYB114, AtMYB90 (PAP2), AtMYB75 (PAP1) and AtMYB113, *Helianthus tuberosus* HtMYB2, monkeyflower PELAN, *Eutrema salsugineum* EsMYB90, *Freesia hybrida* FhPAP1, grape hyacinth MaAN2 and apple MdMYB10. In Arabidopsis, the AtMYB75/90/113/114 members of the SG6 subfamily of the MYB family have been proven to be involved in the regulation of anthocyanin biosynthesis; the SG7 subfamily members AtMYB11/12/111 are involved in the flavonol biosynthesis regulation; the AtMYB5/123 are involved in the regulation of tannin biosynthesis, and AtMYB3/4/7/32 encode the transcriptional repressors [36]. The expression levels of *HtMYB2* in the root, stem, leaf and tuber epidermis of the red-skinned tubers variety 'QY1' are higher than those of the white-skinned tubers variety 'QY3'. The heterologous overexpression of *HtMYB2* in tobacco activates the anthocyanin biosynthesis pathway and accumulated pigments in leaves [37]. PyMYB114 and PyMYB10 can activate anthocyanin biosynthesis by interacting with PybHLH3; PyWRKY26 and PybHLH3 can synergistically target the promoter of *PyMYB114* and participate in the regulation of anthocyanin biosynthesis and transportation [71]. The co-expression of the *PalbHLH1* and *PalMYB90* genes in *Populus alba* could improve its disease resistance by promoting flavonoid synthesis [72]. The overexpression of *MdMYB90-like* upregulated the expression of the anthocyanin biosynthesis related structural genes and regulatory genes (including *MdCHS, MdCHI, MdANS, MdUFGT, MdbHLH3* and *MdMYB1*) and induced the accumulation of pigments in transgenic callus and pericarps [73]. The overexpression of onion *MYB1* can restore anthocyanin accumulation and the cyanic petal phenotype of the *A. majus R2R3-MYB* mutant, and heterologous co-expression of *MYB1* and *bHLH* can promote ectopic red pigmentation in garlic plants [74]. VcMYB is highly homologous to AtMYB114, which is a SG6 subfamily member of the Arabidopsis MYB family that positively regulates anthocyanin biosynthesis. Moreover, based on our transcriptome data, we found that the expression levels of *VcMYB* in RF, PF and BF were the highest among all members. Our qRT-PCR results also revealed that *VcMYB* was expressed significantly higher in the fruits at the red, purple and blue stages, accounting for 1.79-fold, 2.1-fold and 1.75-fold of that in GF, respectively. These results suggested that *VcMYB* may be a pivotal positive regulator of anthocyanin biosynthesis in blueberry. To verify the function of *VcMYB* in anthocyanin biosynthesis in blueberry, it was further cloned and functionally analyzed. Similar to sweet potato *IbMYB1* [75], the *VcMYB* also has three transcripts. The transient transformation results showed that only the longest *VcMYB* transcript, *VcMYB-1*, could promote anthocyanin accumulation in young blueberry fruit pericarps.

These results have demonstrated that the expression of anthocyanin regulatory *MYBs* was greatly influenced by light, ABA and MeJA [76–79]. Consistently, in this study, we identified many light-, ABA- and MeJA-elements in the promoters of blueberry *ABRMs*. Light quality and quantity greatly and widely affect the biosynthesis of anthocyanins or flavonoids in various plants [79,80]. UV-B treatment could upregulate the expression of the *HY5* gene that encodes the UV receptor at the green fruit stage, and HY5 promoted anthocyanin accumulation by upregulating the expression of *VcMYBA1*, while inhibiting the expression of *VcMYBC2* [30]. However, the expression of *VcMYBC2* was induced by UV-B treatment in the mature fruits, which could inhibit the excessive accumulation of anthocyanins. JA is an important signal for the biosynthesis of plants' secondary metabolites, and its induction role in anthocyanin accumulation has also been found in many plants [77]. Exogenous JA treatment can activate the expression of anthocyanin biosynthesis related structural genes and regulatory transcription factor genes, and enhance anthocyanin accumulation in Arabidopsis under far-red light [81]. In apple, JA can induce anthocyanin and proanthocyanin accumulations by regulating the JAZ-TRB1-MyB9 complex [78]. In addition, MeJA treatment could induce the expression of *MdMYB9* and *MdMYB11*, whose overexpression in apple callus could improve anthocyanin and proanthocyanin accumulation, and this promotion effect could be further enhanced by MeJA [82]. ABA is considered to be one of the most important positive regulators of the ripening of non-climacteric fruits and anthocyanin biosynthesis [32,83,84]. Karppinen et al. [85] found that exogenous ABA treatment could promote the accumulation of anthocyanins in *V. myrtillus*. Han et al. [32] found that the expression levels of *CHS, CHI, DRF, LDOX/ANS* and some other anthocyanin biosynthesis related genes were significantly up-regulated in blueberries when treated with ABA at the late ripening stages, and were highly positively correlated with the anthocyanin content. Consistently, in our present study, we found that 16 of the 19 identified blueberry *ABRM* genes contained the ABA-responsive ABRE element in their promoters, indicating that ABA have a significant function in influencing the expression of blueberry *ABRMs* and in regulating anthocyanin biosynthesis.

5. Conclusions

In this study, we identified fifteen candidate ABRM proteins from blueberry. Among them, twelve were R2R3-MYBs and three were 1R-MYBs, which could be well supported by their conserved motif types and numbers. With the exception of VcRVE8, which was localized in the chloroplast and nucleus, all of the blueberry ABRMs were predicted to be nucleus-localized. The exon numbers of the blueberry *ABRM* genes varied significantly. The promoters of *ABRMs* contain a large number of light-, ABA- and MeJA-responsive elements, indicating that the influences of these factors on anthocyanin accumulation were achieved, at least partially, by regulating the expression of the *ABRM* genes. *VcMYB* was highly expressed in blueberry fruits at the late ripening stages. This gene had three transcripts; however, only the transient overexpression of its longest transcripts (*VcMYB-1*) could promote anthocyanin accumulation in young blueberry fruits. Our study will provide a basis for the applications of the blueberry anthocyanin biosynthesis related *MYB* genes, and will lay the foundation for the high-anthocyanin aimed blueberry selection and breeding in the future.

Supplementary Materials: The following supporting information can be downloaded at: https://www.mdpi.com/article/10.3390/cimb45010027/s1. Figure S1: Phylogenetic analysis result of anthocyanin biosynthesis related MYB proteins from blueberry and some other plant species.

Author Contributions: Conceptualization, Y.Z., C.C. and D.Q.; methodology, Y.Z., B.W. and D.H.; software, Y.Z. and C.C.; validation, B.W. and Y.Z.; formal analysis, Y.Z., D.H. and C.C.; investigation, Y.Z., H.W., P.Q. and L.Y.; resources, D.Q., X.Y. and T.L.; data curation, C.C. and B.W.; writing—original draft preparation, Y.Z. and C.C.; writing—review and editing, C.C. and D.Q.; visualization, Y.Z. and D.H.; supervision, D.Q.; project administration, D.Q.; funding acquisition, D.Q., C.C., T.L. and X.Y. All authors have read and agreed to the published version of the manuscript.

Funding: The work supported by the Science and Technology Development Fund of Fujian Agriculture and Forestry University (Agriculture and Forestry University Section [2015] No. 36), the Construction of Plateau Discipline of Fujian Province (102/71201801101), the National Natural Science Foundation of China (31860225), the Natural Science Basic Research Program of Shanxi Province (202203021211267), the Fund for High-level Talents of Shanxi Agricultural University (2021XG010), the Reward Fund for PhDs and Postdoctors of Shanxi Province (SXBYKY2022004), Qian Kehe Foundation ([2019]1408), Qian Kehe platform Personnel ([2018]5781) and Special Fund for Scientific Research Institutes in the Public Interest of Fujian Province (2019R1028-1).

Institutional Review Board Statement: Not applicable.

Informed Consent Statement: Not applicable.

Data Availability Statement: All data are available in this article.

Conflicts of Interest: The authors declare no conflict of interest.

References

1. Joseph, J.A.; Shukitt-Hale, B.; Denisova, N.A.; Bielinski, D.; Martin, A.; Mcewen, J.J.; Bickford, P.C. Reversals of age-related declines in neuronal signal transduction, cognitive, motor behavioral deficits with blueberry, spinach, or strawberry dietary supplementation. *J. Neurosci.* **1999**, *19*, 8114–8121. [CrossRef] [PubMed]
2. Hou, D.X.; Fujii, M.; Terahara, N.; Yoshimoto, M. Molecular mechanisms behind the chemopreventive effects of anthocyanidins. *J. Biomed. Biotechnol.* **2004**, *5*, 321–325. [CrossRef] [PubMed]
3. Miguel, M. Anthocyanins: Antioxidant and/or anti-inflammatory activities. *J. Appl. Pharm. Sci.* **2011**, *1*, 7–15.
4. Mazza, G.; Kay, C.D.; Cottrell, T.; Holub, B.J. Absorption of anthocyanins from blueberries and serum antioxidant status in human subjects. *J. Agric. Food Chem.* **2002**, *50*, 7731–7737. [CrossRef] [PubMed]
5. Chun, O.K.; Kim, D.O.; Lee, C.Y. Superoxide radical scavenging activity of the major polyphenols in fresh plums. *J. Agric. Food Chem.* **2003**, *51*, 8067–8072. [CrossRef]
6. Williams, R.J.; Spencer, J.P.; Rice-Evans, C. Flavonoids: Antioxidants or signalling molecules? *Free Radic. Biol. Med.* **2004**, *36*, 838–849. [CrossRef]

7. Ma, Y.; Li, Y.; Zhang, H.; Wang, Y.; Wu, C.; Huang, W. Malvidin induces hepatic stellate cell apoptosis via the endoplasmic reticulum stress pathway and mitochondrial pathway. *Food Sci. Nutr.* **2020**, *8*, 5095–5106. [CrossRef]
8. Liu, Y.; Zhang, D.; Hu, J.; Liu, G.; Chen, J.; Sun, L.; Jiang, Z.; Zhang, X.; Chen, Q.; Ji, B. Visible light-induced lipid peroxidation of unsaturated fatty acids in the retina and the inhibitory effects of blueberry polyphenols. *J Agric. Food Chem.* **2015**, *63*, 9295–9305. [CrossRef]
9. Gapski, A.; Gomes, T.M.; Bredun, M.A.; Ferreira-Lima, N.E.; Ludka, F.K.; Bordignon-Luiz, M.T.; Burin, V.M. Digestion behavior and antidepressant-like effect promoted by acute administration of blueberry extract on mice. *Food Res. Int.* **2019**, *125*, 108618. [CrossRef]
10. Kalt, W.; Cassidy, A.; Howard, L.R.; Krikorian, R.; Stull, A.J.; Tremblay, F.; Zamora-Ros, R. Recent research on the health benefits of blueberries and their anthocyanins. *Adv. Nutr.* **2020**, *11*, 224–236. [CrossRef]
11. Yang, S.; Wang, C.; Li, X.; Wu, C.; Liu, C.; Xue, Z.; Kou, X. Investigation on the biological activity of anthocyanins and polyphenols in blueberry. *J. Food Sci.* **2021**, *86*, 614–627. [CrossRef]
12. Herrera-Balandrano, D.D.; Chai, Z.; Beta, T.; Feng, J.; Huang, W. Blueberry anthocyanins: An updated review on approaches to enhancing their bioavailability. *Trends Food Sci. Technol.* **2021**, *118*, 808–821. [CrossRef]
13. Li, X.; Pei, J.; Zhang, Z.; Wu, L.; Liu, H.; Li, Y.; Li, H. Molecular cloning and expression of chalcone synthase gene in blueberry. *J. Northeast For. Univ.* **2012**, *40*, 60–65. (In Chinese)
14. Zhang, C.; Guo, Q.; Liu, Y.; Liu, H.; Wang, F.; Jia, C. Molecular cloning and functional analysis of a flavanone 3-hydroxylase gene from blueberry. *J. Hortic. Sci. Biotechnol.* **2016**, *92*, 57–64. [CrossRef]
15. Song, Y.; Dou, L.; Zhang, H.; Zhang, Z.; Li, Y. Cloning and expression of *VcDFR* gene from 'Jersey' blueberry. *J. Fruit Sci.* **2014**, *31*, 784–792. (In Chinese)
16. Li, X.; Pei, J.; Zhang, Z.; Wu, L.; Liu, H.; Li, H.; Li, Y. Molecular cloning and expression analysis of *VcANS* gene in blueberry. *J. Northwest AF Univ. (Nat. Sci. Ed.)* **2012**, *40*, 201–209. (In Chinese)
17. Zhao, M.; Li, J.; Zhu, L.; Chang, P.; Li, L.; Zhang, L. Identification and characterization of MYB-bHLH-WD40 regulatory complex members controlling anthocyanidin biosynthesis in blueberry fruits development. *Genes* **2019**, *10*, 496. [CrossRef]
18. Yang, J.; Li, B.; Shi, W.; Gong, Z.; Chen, L.; Hou, Z. Transcriptional activation of anthocyanin biosynthesis in developing fruit of blueberries (*Vaccinium corymbosum* L.) by preharvest and postharvest UV irradiation. *J. Agric. Food Chem.* **2018**, *66*, 10931–10942. [CrossRef] [PubMed]
19. Lin, T.; Walworth, A.; Zong, X.; Danial, G.H.; Song, G.Q. VcRR2 regulates chilling-mediated flowering through expression of hormone genes in a transgenic blueberry mutant. *Hortic. Res.-Engl.* **2019**, *6*, 96. [CrossRef]
20. Die, J.V.; Jones, R.W.; Ogden, E.L.; Ehlenfeldt, M.K.; Rowland, L.J. Characterization and analysis of anthocyanin-related genes in wild-type blueberry and the pink-fruited mutant cultivar 'Pink Lemonade': New insights into anthocyanin biosynthesis. *Agronomy* **2020**, *10*, 1296. [CrossRef]
21. Günther, C.S.; Dare, A.P.; Mcghie, T.K.; Deng, C.; Lafferty, D.J.; Plunkett, B.J.; Grierson, E.R.P.; Turner, J.L.; Jaakola, L.; Albert, N.W.; et al. Spatiotemporal modulation of flavonoid metabolism in blueberries. *Front. Plant Sci.* **2020**, *11*, 545. [CrossRef] [PubMed]
22. Zhang, Y.; Liu, F.; Wang, B.; Wu, H.; Wu, J.; Liu, J.; Sun, Y.; Cheng, C.; Qiu, D. Identification, characterization and expression analysis of anthocyanin biosynthesis-related *bHLH* genes in blueberry (*Vaccinium corymbosum* L.). *Int. J. Mol. Sci.* **2021**, *22*, 13274. [CrossRef] [PubMed]
23. Paz-Ares, J.; Ghosal, D.; Wienand, U.; Peterson, P.A.; Saedler, H. The regulatory c1 locus of *Zea mays* encodes a protein with homology to *myb* proto-oncogene products and with structural similarities to transcriptional activators. *EMBO J.* **1987**, *6*, 3553–3558. [CrossRef] [PubMed]
24. Lloyd, A.M.; Walbot, V.; Davis, R.W. Arabidopsis and Nicotiana anthocyanin production activated by maize regulator R and C1. *Science* **1992**, *258*, 1773–1775. [CrossRef] [PubMed]
25. Dubos, C.; Stracke, R.; Grotewold, E.; Weisshaar, B.; Martin, C.; Lepiniec, L. MYB transcription factors in Arabidopsis. *Trends Plant Sci.* **2010**, *15*, 573–581. [CrossRef] [PubMed]
26. Lafferty, D.J.; Espley, R.V.; Deng, C.H.; Gunther, C.S.; Plunkett, B.; Turner, J.L.; Jaakola, L.; Karppinen, K.; Allan, A.C.; Albert, N.W. Hierarchical regulation of MYBPA1 by anthocyanin- and proanthocyanidin-related MYB proteins is conserved in Vaccinium species. *J. Exp. Bot.* **2021**, *73*, 1344–1356. [CrossRef]
27. Nguyen, C.T.T.; Lim, S.; Lee, J.G.; Lee, E.J. VcBBX, VcMYB21, and VcR2R3MYB transcription factors are involved in UV-B-Induced anthocyanin biosynthesis in the peel of harvested blueberry fruit. *J. Agric. Food Chem.* **2017**, *65*, 2066–2073. [CrossRef]
28. Liu, Z.S.; Yuan, Y.H.; Zhang, T.; Zhang, L.Y. Expression characteristics of the transcription factor *VcMYB21* in blueberry fruit coloration and response to UV in seedling. *Plant Physiol. J.* **2017**, *53*, 115–125. (In Chinese)
29. Plunkett, B.J.; Espley, R.V.; Dare, A.P.; Warren, B.A.W.; Grierson, E.R.P.; Cordiner, S.; Turner, J.L.; Allan, A.C.; Albert, N.W.; Davies, K.M.; et al. MYBA from blueberry (*Vaccinium* Section Cyanococcus) is a subgroup 6 type R2R3MYB transcription factor that activates anthocyanin production. *Front. Plant Sci.* **2018**, *9*, 1300. [CrossRef]
30. Li, T.; Yamane, H.; Tao, R. Preharvest long-term exposure to UV-B radiation promotes fruit ripening and modifies stage-specific anthocyanin metabolism in highbush blueberry. *Hortic. Res.* **2021**, *8*, 67. [CrossRef]
31. Wang, A.; Liang, K.; Yang, S.; Cao, Y.; Wang, L.; Zhang, M.; Zhou, J.; Zhang, L. Genome-wide analysis of MYB transcription factors of *Vaccinium corymbosum* and their positive responses to drought stress. *BMC Genom.* **2021**, *22*, 565. [CrossRef] [PubMed]

32. Han, T.; Wu, W.; Li, W. Transcriptome analysis revealed the mechanism by which exogenous ABA increases anthocyanins in blueberry fruit during veraison. *Front. Plant Sci.* **2021**, *12*, 758215. [CrossRef] [PubMed]
33. Gupta, V.; Estrada, A.D.; Blakley, I.; Reid, R.; Patel, K.; Meyer, M.D.; Andersen, S.U.; Brown, A.F.; Lila, M.A.; Loraine, A.E. RNA-Seq analysis and annotation of a draft blueberry genome assembly identifies candidate genes involved in fruit ripening, biosynthesis of bioactive compounds, and stage-specific alternative splicing. *Gigascience* **2015**, *4*, 5. [CrossRef] [PubMed]
34. Wang, J.; Wisecaver, J.H.; Yocca, A.E.; Alger, E.I.; Tang, H.; Xiong, Z.; Callow, P.; Ben-Zvi, G.; Brodt, A.; Baruch, K.; et al. Haplotype-phased genome and evolution of phytonutrient pathways of tetraploid blueberry. *Gigascience* **2019**, *8*, giz012.
35. Rowland, L.J.; Alkharouf, N.; Darwish, O.; Ogden, E.L.; Polashock, J.J.; Bassil, N.V.; Main, D. Generation and analysis of blueberry transcriptome sequences from leaves, developing fruit, and flower buds from cold acclimation through deacclimation. *BMC Plant Biol.* **2012**, *12*, 46. [CrossRef]
36. Stracke, R.; Werber, M.; Weisshaar, B. The R2R3-MYB gene family in *Arabidopsis thaliana*. *Curr. Opin. Plant Biol.* **2001**, *4*, 447–456. [CrossRef]
37. Gao, J.; Sun, X.; Zong, Y.; Yang, S.; Wang, L.; Liu, B. Functional MYB transcription factor gene HtMYB2 is associated with anthocyanin biosynthesis in *Helianthus tuberosus* L. *BMC Plant Biol.* **2020**, *20*, 247. [CrossRef]
38. Yuan, Y.W.; Sagawa, J.M.; Frost, L.; Vela, J.P.; Bradshaw, H.D., Jr. Transcriptional control of floral anthocyanin pigmentation in monkeyflowers (Mimulus). *New Phytol* **2014**, *204*, 1013–1027. [CrossRef]
39. Qi, Y.; Gu, C.; Wang, X.; Gao, S.; Li, C.; Zhao, C.; Li, C.; Ma, C.; Zhang, Q. Identification of the *Eutrema salsugineum EsMYB90* gene important for anthocyanin biosynthesis. *BMC Plant Biol.* **2020**, *20*, 186. [CrossRef]
40. Li, Y.; Shan, X.; Tong, L.; Wei, C.; Lu, K.; Li, S.; Kimani, S.; Wang, S.; Wang, L.; Gao, X. The conserved and particular roles of the R2R3-MYB regulator FhPAP1 from *Freesia hybrida* in flower anthocyanin biosynthesis. *Plant Cell Physiol* **2020**, *61*, 1365–1380. [CrossRef]
41. Chen, K.; Du, L.; Liu, H.; Liu, Y. A novel R2R3-MYB from grape hyacinth, MaMybA, which is different from MaAN2, confers intense and magenta anthocyanin pigmentation in tobacco. *BMC Plant Biol.* **2019**, *19*, 390. [CrossRef]
42. Jia, D.; Li, Z.; Dang, Q.; Shang, L.; Shen, J.; Leng, X.; Wang, Y.; Yuan, Y. Anthocyanin biosynthesis and methylation of the MdMYB10 promoter are associated with the red blushed-skin mutant in the red striped-skin "Changfu 2" apple. *J. Agric. Food Chem.* **2020**, *68*, 4292–4304. [CrossRef]
43. Chen, C.; Chen, H.; Zhang, Y.; Thormas, H.; Frank, M.; He, Y.; Xia, R. TBtools: An integrative toolkit developed for interactive analyses of big biological data. *Mol. Plant* **2020**, *13*, 1194–1202. [CrossRef]
44. Fuleki, T.; Francis, F. Quantitative methods for anthocyanins. *J. Food Sci.* **1968**, *33*, 266–274. [CrossRef]
45. Zhuang, L.; Huang, G.; Li, X.; Xiao, J.; Guo, L. Effect of different LED lights on aliphatic glucosinolates metabolism and biochemical characteristics in broccoli sprouts. *Food Res. Int.* **2022**, *154*, 111015. [CrossRef]
46. Stracke, R.; Jahns, O.; Keck, M.; Tohge, T.; Niehaus, K.; Fernie, A.R.; Weisshaar, B. Analysis of PRODUCTION OF FLAVONOL GLYCOSIDES-dependent flavonol glycoside accumulation in *Arabidopsis thaliana* plants reveals MYB11-, MYB12- and MYB111-independent flavonol glycoside accumulation. *New Phytol.* **2010**, *188*, 985–1000. [CrossRef]
47. Wang, L.; Tang, W.; Hu, Y.; Zhang, Y.; Sun, J.; Guo, X.; Lu, H.; Yang, Y.; Fang, C.; Niu, X.; et al. A MYB/bHLH complex regulates tissue-specific anthocyanin biosynthesis in the inner pericarp of red-centered kiwifruit *Actinidia chinensis* cv. Hongyang. *Plant J.* **2019**, *99*, 359–378. [CrossRef]
48. An, J.P.; An, X.H.; Yao, J.F.; Wang, X.N.; You, C.X.; Wang, X.F.; Hao, Y.J. BTB protein MdBT2 inhibits anthocyanin and proanthocyanidin biosynthesis by triggering MdMYB9 degradation in apple. *Tree Physiol* **2018**, *38*, 1578–1587. [CrossRef]
49. Wang, Y.; Sun, J.; Wang, N.; Xu, H.; Qu, C.; Jiang, S.; Fang, H.; Su, M.; Zhang, Z.; Chen, X. MdMYBL2 helps regulate cytokinin-induced anthocyanin biosynthesis in red-fleshed apple (*Malus sieversii* f. *niedzwetzkyana*) callus. *Funct. Plant Biol* **2019**, *6*, 187–196. [CrossRef]
50. Moglia, A.; Florio, F.E.; Iacopino, S.; Guerrieri, A.; Milani, A.M.; Comino, C.; Barchi, L.; Marengo, A.; Cagliero, C.; Rubiolo, P.; et al. Identification of a new R3 MYB type repressor and functional characterization of the members of the MBW transcriptional complex involved in anthocyanin biosynthesis in eggplant (*S. melongena* L.). *PLoS ONE* **2020**, *15*, e0232986.
51. Xu, H.; Zou, Q.; Yang, G.; Jiang, S.; Fang, H.; Wang, Y.; Zhang, J.; Zhang, Z.; Wang, N.; Chen, X. MdMYB6 regulates anthocyanin formation in apple both through direct inhibition of the biosynthesis pathway and through substrate removal. *Hortic. Res.* **2020**, *7*, 72. [CrossRef] [PubMed]
52. Wei, Z.Z.; Hu, K.D.; Zhao, D.L.; Tang, J.; Huang, Z.Q.; Jin, P.; Li, Y.H.; Han, Z.; Hu, L.Y.; Yao, G.F.; et al. MYB44 competitively inhibits the formation of the MYB340-bHLH2-NAC56 complex to regulate anthocyanin biosynthesis in purple-fleshed sweet potato. *BMC Plant Biol.* **2020**, *20*, 258. [CrossRef] [PubMed]
53. Akhter, D.; Qin, R.; Nath, U.K.; Eshag, J.; Jin, X.; Shi, C. A rice gene, *OsPL*, encoding a MYB family transcription factor confers anthocyanin synthesis, heat stress response and hormonal signaling. *Gene* **2019**, *699*, 62–72. [CrossRef] [PubMed]
54. Wang, Y.; Liu, W.; Jiang, H.; Mao, Z.; Wang, N.; Jiang, S.; Xu, H.; Yang, G.; Zhang, Z.; Chen, X. The R2R3-MYB transcription factor MdMYB24-like is involved in methyl jasmonate-induced anthocyanin biosynthesis in apple. *Plant Physiol. Biochem.* **2019**, *139*, 273–282. [CrossRef] [PubMed]
55. Li, X.; Wu, T.; Liu, H.; Zhai, R.; Wen, Y.; Shi, Q.; Yang, C.; Wang, Z.; Ma, F.; Xu, L. REVEILLE transcription factors contribute to the nighttime accumulation of anthocyanins in 'Red Zaosu' (*Pyrus bretschneideri* Rehd.) pear fruit skin. *Int. J. Mol. Sci.* **2020**, *21*, 1634. [CrossRef]

56. Zhu, H.F.; Fitzsimmons, K.; Khandelwal, A.; Kranz, R.G. CPC, a single-repeat R3 MYB, is a negative regulator of anthocyanin biosynthesis in Arabidopsis. *Mol Plant* **2009**, *2*, 790–802. [CrossRef]
57. Li, Y.; Cui, W.; Qi, X.; Lin, M.; Qiao, C.; Zhong, Y.; Hu, C.; Fang, J. MicroRNA858 negatively regulates anthocyanin biosynthesis by repressing AaMYBC1 expression in kiwifruit (*Actinidia arguta*). *Plant Sci.* **2020**, *296*, 110476. [CrossRef]
58. An, J.P.; Li, R.; Qu, F.J.; You, C.X.; Wang, X.F.; Hao, Y.J. R2R3-MYB transcription factor MdMYB23 is involved in the cold tolerance and proanthocyanidin accumulation in apple. *Plant J.* **2018**, *96*, 562–577. [CrossRef]
59. Zhang, Y.; Ye, J.; Liu, C.; Xu, Q.; Long, L.; Deng, X. Citrus PH4-Noemi regulatory complex is involved in proanthocyanidin biosynthesis via a positive feedback loop. *J. Exp. Bot.* **2020**, *71*, 1306–1321. [CrossRef]
60. Qiu, Z.; Wang, H.; Li, D.; Yu, B.; Hui, Q.; Yan, S.; Huang, Z.; Cui, X.; Cao, B. Identification of candidate HY5-dependent and -independent regulators of anthocyanin biosynthesis in tomato. *Plant Cell Physiol.* **2019**, *60*, 643–656. [CrossRef]
61. Bai, H.; Song, Z.; Zhang, Y.; Li, Z.; Wang, Y.; Liu, X.; Ma, J.; Quan, J.; Wu, X.; Liu, M.; et al. The bHLH transcription factor PPLS1 regulates the color of pulvinus and leaf sheath in foxtail millet (*Setaria italica*). *Appl. Genet.* **2020**, *133*, 1911–1926. [CrossRef]
62. Li, Y.; Liang, J.; Zeng, X.; Guo, H.; Luo, Y.; Kear, P.; Zhang, S.; Zhu, G. Genome-wide analysis of MYB gene family in potato provides insights into tissue-specific regulation of anthocyanin biosynthesis. *Hortic. Plant J.* **2021**, *7*, 129–141.
63. Li, L.; He, Y.; Ge, H.; Liu, Y.; Chen, H. Functional characterization of SmMYB86, a negative regulator of anthocyanin biosynthesis in eggplant (*Solanum melongena* L.). *Plant Sci.* **2021**, *302*, 110696. [CrossRef]
64. Zhang, W.; Ning, G.; Lv, H.; Liao, L.; Bao, M. Single MYB-type transcription factor AtCAPRICE: A new efficient tool to engineer the production of anthocyanin in tobacco. *Biochem. Biophys. Res. Commun.* **2009**, *388*, 742–747. [CrossRef]
65. Jaradat, M.R.; Feurtado, J.A.; Huang, D.; Lu, Y.; Cutler, A.J. Multiple roles of the transcription factor AtMYBR1/AtMYB44 in ABA signaling, stress responses, and leaf senescence. *BMC Plant Biol.* **2013**, *13*, 192. [CrossRef]
66. Nesi, N.; Jond, C.; Debeaujon, I.; Caboche, M.; Lepiniec, L. The Arabidopsis TT2 gene encodes an R2R3 MYB domain protein that acts as a key determinant for proanthocyanidin accumulation in developing seed. *Plant Cell* **2001**, *13*, 2099–2114. [CrossRef]
67. Xie, H.; Sun, Y.; Cheng, B.; Xue, S.; Cheng, D.; Liu, L.; Meng, L.; Qiang, S. Variation in ICE1 methylation primarily determines phenotypic variation in freezing tolerance in *Arabidopsis thaliana*. *Plant Cell Physiol.* **2019**, *60*, 152–165. [CrossRef]
68. Xie, Q.; Wang, P.; Liu, X.; Yuan, L.; Wang, L.; Zhang, C.; Li, Y.; Xing, H.; Zhi, L.; Yue, Z.; et al. LNK1 and LNK2 are transcriptional coactivators in the Arabidopsis circadian oscillator. *Plant Cell* **2014**, *26*, 2843–2857. [CrossRef]
69. Xing, H.; Wang, P.; Cui, X.; Zhang, C.; Wang, L.; Liu, X.; Yuan, L.; Li, Y.; Xie, Q.; Xu, X. LNK1 and LNK2 recruitment to the evening element require morning expressed circadian related MYB-like transcription factors. *Plant Signal Behav.* **2015**, *10*, e1010888. [CrossRef]
70. Zhou, M.; Zhang, K.; Sun, Z.; Yan, M.; Chen, C.; Zhang, X.; Tang, Y.; Wu, Y. LNK1 and LNK2 corepressors interact with the MYB3 transcription factor in phenylpropanoid biosynthesis. *Plant Physiol.* **2017**, *174*, 1348–1358. [CrossRef]
71. Li, C.; Wu, J.; Hu, K.D.; Wei, S.W.; Sun, H.Y.; Hu, L.Y.; Han, Z.; Yao, G.F.; Zhang, H. PyWRKY26 and PybHLH3 cotargeted the PyMYB114 promoter to regulate anthocyanin biosynthesis and transport in red-skinned pears. *Hortic Res.* **2020**, *7*, 37. [CrossRef] [PubMed]
72. Bai, Q.; Duan, B.; Ma, J.; Fen, Y.; Sun, S.; Long, Q.; Lv, J.; Wan, D. Coexpression of *PalbHLH1* and *PalMYB90* genes from *Populus alba* enhances pathogen resistance in poplar by increasing the flavonoid content. *Front. Plant Sci.* **2020**, *10*, 1772. [CrossRef] [PubMed]
73. Sun, C.; Wang, C.; Zhang, W.; Liu, S.; Wang, W.; Yu, X.; Song, T.; Yu, M.; Yu, W.; Qu, S. The R2R3-type MYB transcription factor MdMYB90-like is responsible for the enhanced skin color of an apple bud sport mutant. *Hortic. Res.* **2021**, *8*, 156. [CrossRef] [PubMed]
74. Schwinn, K.E.; Ngo, H.; Kenel, F.; Brummell, D.A.; Albert, N.W.; McCallum, J.A.; Pither-Joyce, M.; Crowhurst, R.N.; Eady, C.; Davies, K.M. The onion (*Allium cepa* L.) R2R3-MYB gene MYB1 regulates anthocyanin biosynthesis. *Front. Plant Sci.* **2016**, *7*, 1865. [CrossRef] [PubMed]
75. Tanaka, M.; Takahata, Y.; Kurata, R.; Nakayama, H.; Yoshinaga, M. Structural and functional characterization of *IbMYB1* genes in recent Japanese purple-fleshed sweetpotato cultivars. *Mol. Breed.* **2012**, *29*, 565–574. [CrossRef]
76. An, X.H.; Tian, Y.; Chen, K.Q.; Liu, X.J.; Liu, D.D.; Xie, X.B.; Cheng, C.G.; Cong, P.H.; Hao, Y.J. MdMYB9 and MdMYB11 are involved in the regulation of the JA-induced biosynthesis of anthocyanin and proanthocyanidin in apples. *Plant Cell Physiol.* **2015**, *56*, 650–662. [CrossRef]
77. Wasternack, C.; Strnad, M. Jasmonates are signals in the biosynthesis of secondary metabolites—Pathways, transcription factors and applied aspects—A brief review. *N Biotechnol* **2019**, *48*, 1–11. [CrossRef]
78. An, J.P.; Xu, R.R.; Liu, X.; Zhang, J.C.; Wang, X.F.; You, C.X.; Hao, Y.J. Jasmonate induces biosynthesis of anthocyanin and proanthocyanidin in apple by mediating the JAZ1-TRB1-MYB9 complex. *Plant J* **2021**, *106*, 1414–1430. [CrossRef]
79. Guan, L.; Dai, Z.; Wu, B.H.; Wu, J.; Merlin, I.; Hilbert, G.; Renaud, C.; Gomès, E.; Edwards, E.; Li, S.H.; et al. Anthocyanin biosynthesis is differentially regulated by light in the skin and flesh of white-fleshed and teinturier grape berries. *Planta* **2016**, *243*, 23–41. [CrossRef]
80. Ma, Y.; Ma, X.; Gao, X.; Wu, W.; Zhou, B. Light induced regulation pathway of anthocyanin biosynthesis in plants. *Int. J. Mol. Sci.* **2021**, *22*, 11116. [CrossRef]

81. Li, T.; Jia, K.P.; Lian, H.L.; Yang, X.; Li, L.; Yang, H.Q. Jasmonic acid enhancement of anthocyanin accumulation is dependent on phytochrome A signaling pathway under far-red light in Arabidopsis. *Biochem. Biophys. Res. Commun.* **2014**, *454*, 78–83. [CrossRef]
82. Dubos, C.; Le, G.J.; Baudry, A.; Huep, G.; Lanet, E.; Debeaujon, I.; Routaboul, J.M.; Alboresi, A.; Weisshaar, B.; Lepiniec, L. MYBL2 is a new regulator of flavonoid biosynthesis in *Arabidopsis thaliana*. *Plant J.* **2008**, *55*, 940–953. [CrossRef]
83. Li, D.D.; Luo, Z.S.; Mou, W.S.; Wang, Y.S.; Ying, T.J.; Mao, L.C. ABA and UV-C effects on quality, antioxidant capacity and anthocyanin contents of strawberry fruit (*Fragaria ananassa* Duch.). *Postharvest Biol. Technol.* **2014**, *90*, 56–62. [CrossRef]
84. Gupta, K.; Wani, S.H.; Razzaq, A.; Skalicky, M.; Samantara, K.; Gupta, S.; Pandita, D.; Goel, S.; Grewal, S.; Hejnak, V.; et al. Abscisic acid: Role in fruit development and ripening. *Front. Plant Sci.* **2022**, *13*, 817500. [CrossRef]
85. Karppinen, K.; Lafferty, D.J.; Albert, N.W.; Mikkola, N.; McGhie, T.; Allan, A.C.; Afzal, B.M.; Häggman, H.; Espley, R.V.; Jaakola, L. MYBA and MYBPA transcription factors co-regulate anthocyanin biosynthesis in blue-coloured berries. *New Phytol.* **2021**, *232*, 1350–1367. [CrossRef]

current issues in molecular biology

MDPI

Article

Genetic Mapping and QTL Analysis of Fruit Traits in Melon (*Cucumis melo* L.)

Haiyong Zhao [1,2], Taifeng Zhang [1,2], Xiaobing Meng [1,2], Jiayan Song [1,2], Chen Zhang [1,2] and Peng Gao [1,2,*]

[1] College of Horticulture and Landscape Architecture, Northeast Agricultural University, No. 600, Changjiang Road, Harbin 150030, China

[2] Key Laboratory of Biology and Genetic Improvement of Horticulture Crops (Northeast Region), Ministry of Agriculture and Rural Affairs, Harbin 150030, China

* Correspondence: gaopeng1982@neau.edu.cn or gaopeng_neau@163.com

Abstract: Melon (*Cucumis melo* L.) is an important horticultural cash crop and its quality traits directly affect consumer choice and market price. These traits are controlled by genetic as well as environmental factors. In this study, a quantitative trait locus (QTL) mapping strategy was used to identify the potential genetic loci controlling quality traits of melons (i.e., exocarp and pericarp firmness and soluble solid content) based on newly derived whole-genome single nucleotide polymorphism-based cleaved amplified polymorphic sequence (SNP-CAPS) markers. Specifically, SNPs of two melon varieties, M4-5 and M1-15, as revealed by whole-genome sequencing, were converted to the CAPS markers, which were used to construct a genetic linkage map comprising 12 chromosomes with a total length of 1414.88 cM, in the F_2 population of M4-5 and M1-15. The six identified QTLs included: *SSC6.1* and *SSC11.1* related to soluble solid content; *EF12.1* associated with exocarp firmness; and *EPF3.1*, *EPF3.2* and *EPF7.1* related to edible pericarp firmness. These genes were located on five chromosomes (3, 6, 7, 11, and 12) in the flanking regions of the CAPS markers. Moreover, the newly developed CAPS markers will be useful in guiding genetic engineering and molecular breeding in melon.

Keywords: QTL mapping; linkage map; exocarp firmness; edible pericarp firmness; soluble solid content

Citation: Zhao, H.; Zhang, T.; Meng, X.; Song, J.; Zhang, C.; Gao, P. Genetic Mapping and QTL Analysis of Fruit Traits in Melon (*Cucumis melo* L.). *Curr. Issues Mol. Biol.* **2023**, *45*, 3419–3433. https://doi.org/10.3390/cimb45040224

Academic Editor: Vijai Bhadauria

Received: 22 March 2023
Revised: 7 April 2023
Accepted: 11 April 2023
Published: 14 April 2023

1. Introduction

Melon (*Cucumis melo* L.) is a Cucurbitaceae vegetable crop widely cultivated in many countries, including China. In fact, melons constitute an important component of fruit and vegetable production in China [1]. Melon quality is determined by various characteristics [2]; for instance, melon is popular for its sweet taste, pleasant flavor, and high nutritional value [3]. The sensory quality of melon fruit depends largely on its soluble solid content (SSC) and its volatile aromatic components. In addition, melon fruit has a high SSC, including organic acids, multiple vitamins, and soluble proteins [4–6], which not only determines melon flavor but also is commonly used for the quality evaluation of melon. One of the most crucial characteristics related to SSC is the sugar concentration [6] as sugar is the main component that affects the quality and flavor of melon fruit, and is also the basic raw material for the synthesis of vitamins, pigments and aromatic substances. In addition, sugar provides the osmotic catalyst for fruit cell enlargement [7].

The firmness of melon fruit refers to the resistance of the flesh to external pressure. It is essential for sensory attributes, and can suit different needs of consumers, and thus has been used as one of the indicators of melon quality [8]. As a typical quantitative trait affected by genetic and environmental factors, the firmness trait of melon is quite complex. It can be affected by germplasm genetic differences and is typically manifested as the fruit harvest and ripening index [9,10]. The firmness of melon can be affected by cellular contents, such as intercellular space, cellulose, and starch contents. During the melon fruit ripening and softening process, the respiration rate increases, the starch and other substances are degraded, and pectin, which plays an essential role in the cell wall support, is decomposed into fructose, glucose, and other polysaccharide substances, resulting in reduced fruit firmness [11].

These traits have shown great variation since domestication, and their genetic bases have been extensively studied [12,13]. Multiple molecular tools have been developed to help researchers map traits in many crop varieties and link them to certain genomic loci or genes [14,15], thus accelerating plant breeding. The use of molecular markers has facilitated the unraveling of molecular mechanisms underlying different traits [16,17]. The identified markers have accelerated genetic crop breeding based on marker assisted selection (MAS). Several genetic maps of melon have been established since Pitrat [18] first used morphological markers to create a linkage map of melon [19–21]. QTL analysis has been used to link genomic regions to fruit traits, such as form, dimension, firmness, weight, and SSC. For instance, Argyris et al. identified 78 QTLs related to the traits of sugar and organic acid contents using near-isogenic and hybridized melon lines [6]. Paris et al. [22] identified 57 QTLs in 81 melon recombinant inbred lines (RILs) that are related to multiple quality traits, including 10 related to SSCs. In another study, 27 QTLs were linked to traits including shape, size, and pulp content in the F_2 population of melon [23].

The next–generation sequencing (NGS) method has been widely used to understand the genetic mechanisms regulating melon fruit diversity [24,25]. The Spanish Institute of Agricultural Sciences successfully constructed the first melon reference genome (the double-haploid line DHL92 of *C. melo* ssp. *Melo*) in 2012 [26]. In 2020, the latest version of the melon reference genome (Melon (DHL92) v4) was released [27]. This improved melon genome assembly has greatly facilitated the designing of high-density genetic markers [28].

In this study, the F_2 populations of the thick-skinned melon (M4-5, low SSC) and thin-skinned melon (M1-15, high SSC) were used to construct the CAPS markers, which were further used to identify QTLs related to melon quality traits. Based on previous studies that deployed CAPS markers to elucidate melon fruit quality traits [29], we further assessed the effectiveness of SNP-CAPS markers and using this method, we successfully identified a QTL (*SSC6.1*) that is associated with the trait of soluble solid content in melon. Thus, this study provides further knowledge on marker-assisted breeding of specific fruit varieties.

2. Materials and Methods

2.1. Plant Materials

P_1 (M4-5) has a round shape, low SSC, and slow growth rate; P_2 (M1-15) has an oval shape, high SSC, and fast plant growth rate. Usage of the P_1 (M4-5, female) and P_2 (M1-15, male) seeds was approved by the Key Laboratory of Biology and Genetic Improvement of Horticultural Crops (Northeast Region), Ministry of Agriculture, Northeast Agricultural University, China. P_1 (M4-5, female) and P_2 (M1-15, male) were crossed to generate the F_1 generation. F_1 was further selfed to obtain the F_2 generation. In a greenhouse at Northeast Agricultural University's Xiangyang Experiment Agricultural Station in Harbin, China (44°049′ N, 125°429′ E), 271 F_2 progenies, M4-5 (n = 10), M1-15 (n = 10), and their F_1 (n = 10) hybrids were planted in 2021. In 2022, F_2 (n = 393) and M4-5 (n = 10), and M1-15 (n = 10) and F_1 (n = 10) were planted in the greenhouse of the Facility Horticulture Engineering Center, Northeast Agricultural University, Harbin (45°774′ N, 126°727′ E), China. For Harbin's normal climatic circumstances, watering, extirpating weed, and vermin control were performed in accordance with industry standards. According to the climacteric or non-climacteric behavior of the fruit, different combinations of harvesting indicators are used for fruit harvesting. In the fruits of our F_2 population, there are both climactic and non-climactic fruits. Our harvesting index is based on the method of Obando-Ulloa et al. [30], wherein the melon fruit is picked when the fruit pedicel begin to split and an aromatic scent can be smelled.

2.2. Determination of Fruit Traits

A GY-4 digital fruit firmness tester with an 11 mm diameter probe (Aipli, China) was used to determine the firmness of the inner and outer peels of the same fruit three times each, and the average value calculated was represented as the hardness index (kg/cm^2) The juice in the middle of the fruit was taken and measured with a BM-02 digital display refraction instrument (Dongmei, China); light was avoided during the measurement. All measurements were repeated three times, and the average value was recorded.

2.3. Genomic DNA Sequencing

Fresh, 2-week-old leaves of the M4-5, M1-15, F_1 and F_2 populations were quick-frozen in liquid nitrogen, and gDNA was extracted using a modified cetyltrimethylammonium bromide (CTAB) method [31]. DNA concentration and quality were then determined using SMA3000 spectrophotometer (Plextech, Shenzhen, China) and 1% agarose gel electrophoresis, respectively. On a high-throughput Illumina sequencing platform, libraries prepared from the genomic DNA of the two parents, M4-5 and M1-15, was sequenced.

2.4. SNP Development, Annotation and CAPS Marker Development

The high-quality resequencing data of the parental were obtained using the preprocessing software FASTX-Toolkit (v0.0.13). The resequencing data were aligned with the Melon (DHL92) v3.6.1 Genome [32] using the Burrows–Wheeler Aligner (BWA). The parental resequencing data were aligned with the melon reference genome to the identified SNP sites using SAMtools (v1.12)-mpileup software. SNPs in the resequencing data were detected using VarScan (v2.0) to generate VCF files. SnpEff (v4.3) was used to export SNPs to web-based generated HTML, and according to their genetic variation effects introns, exons, start-stop codons, upstream-downstream regions, splice regions, and 5′ to 3′ end UTR regions for annotation.

For the development of CAPS markers, a total of 10–15 random SNP sites before and after 500 base pair sequences with suitable restriction endonucleases were mined across each chromosome using the SNP2CAPS [33]. The potential SNP sequences were transformed into CAPS markers following the manual settings of molecular parameters of the Primer Premier (v6.0) program.

2.5. CAPS Genotype Analysis

The CAPS primers were designed using Primer Premier (v6.0) program (The CAPS marker primer sequences used to construct the genetic linkage map are listed in Supplementary Table S1) and PCR amplification was performed for the parents and F_1 population. The obtained PCR products were cut to detect polymorphisms in the parents and F_1 population. CAPS markers with uniform location distribution and clear bands were selected for genotyping of the F_2 population. The components and concentrations of the PCR reaction system are shown in Supplementary Table S2. Three restriction enzymes were used to cleave the PCR products (*Eco*R I, *Hind* III, and *Bam*H I, 10 U/μL, TAKARA). According to the manufacturer's instructions, a mixture of 5 μL of PCR product, 0.2 μL of restriction enzyme (10 U/μL), 1.5 μL of enzyme-specific buffer, and 8.3 μL of sterile double distilled water was used for reaction in a 37 °C incubator for 4–5 h. All CAPS markers and products of both parents and F_1 generation were verified and examined using electrophoresis with 1% agarose gel. Images from gel electrophoresis were obtained using an image analysis system (Champ Gel 6000, Saizhi Entrepreneurship, Beijing, China).

2.6. Genetic Linkage Map Construction

QTL IciMapping (v4.0) and R/qtl (v1.5) [34] were used to construct linkage maps for the F_2 population. The genome-wide LOD threshold at $\alpha = 0.05$ was estimated using 1000 repeat replacement tests. LOD > 2.5 is used as the threshold for detecting the presence of QTLs. Composite interval mapping was used to scan the whole genome at a walking speed of 1.0 cM. The identified QTLs were coded using abbreviations for the traits, followed by the linkage group number and QTL number.

2.7. Statistical Analysis

All data were presented as mean ± SD and ranges. Analyses of distributions and correlations were performed using R (v4.2.0), for the correlation analysis of SSC, EF, and EPF. SPSS (v23.0) was used for the analysis of fruit quality traits in the parental lines and the F_1 and F_2 generations. Frequency distributions of SSC, EF, and EPF in the M4-5 and M1-15 derived F_2 populations were calculated based on counts. Microsoft Excel (v2021) was used to record band information and trait data (SSC, EF, and EPF) after enzyme digestion. GraphPad Prism (v8.0) software was used for *SSC6.1* CAPS marker genotype and SSC phenotype analysis.

3. Results

3.1. Phenotyping of Melon Morphological Traits

Two melon parental lines, M4-5 (P_1, female) and M1-15 (P_2, male), were selected as the experimental materials. The exocarp and pericarp firmness of M4-5 was 14.51 ± 0.37 kg/cm^2 and 4.17 ± 0.18 kg/cm^2, respectively; the SSC for M4-5 was 8.64 ± 0.30%. The exocarp and pericarp firmness of M1-15 was 14.11 ± 0.13 kg/cm^2 and 6.11 ± 0.14 kg/cm^2, respectively; the SSC for M1-15 was 10.21 ± 0.30%. The outer and longitudinal sections of both parents and F_1 melons are shown in Figure 1.

Figure 1. A picture of the melon parents (P_1 M4-5, P_2 M1-15) and F_1 progeny, showcasing their corresponding fruit longitudinal and lateral profiles.

3.1.1. Exocarp Firmness (EF) and Edible Pericarp Firmness (EPF)

EF of the two parental lines were quite similar, 14.51 ± 0.37 kg/cm^2 and 14.11 ± 0.13 kg/cm^2 for M4-5 and M1-15, respectively. The EF of the F_1 generation (15.46 ± 0.17 kg/cm^2) was higher than that of M4-5 and M1-15 (Table 1). The EF of the F_2 population was diverse, ranging from 6.20 kg/cm^2 to 25.20 kg/cm^2 (Figure 2A). M4-5 and M1-15 had an EPF of 4.17 ± 0.18 kg/cm^2 and 6.11 ± 0.14 kg/cm^2, respectively. The F_1 generation had an EPF of 5.14 ± 0.10 kg/cm^2, which was comparable to the parental lines (i.e., M4-5 and M1-15). The EPF of the F_2 population was diverse, ranging from 2.60 kg/cm^2 to 14.60 kg/cm^2 (Figure 2B).

Table 1. Traits data analysis of EF (kg/cm^2) and EPF (kg/cm^2) in P_1, P_2, F_1 and F_2 generations (2018 [29], 2021, 2022).

Traits	Years	P_1	P_2	F_1	F_2			
					Mean ± SD	Range	Kurtosis	Skewness
EF	2018	14.29 ± 0.34	14.12 ± 027	15.92 ± 0.76	14.64 ± 2.18	8.50–26.00	4.23	1.06
	2021	14.51 ± 0.37	14.11 ± 0.13	15.46 ± 0.17	14.35 ± 2.10	6.20–25.50	0.43	0.28
	2022	14.40 ± 0.27	13.56 ± 0.20	15.72 ± 0.45	17.87 ± 0.24	8.30–27.50	−0.11	−0.12
EPF	2018	4.44 ± 0.37	6.06 ± 0.17	4.93 ± 0.23	5.15 ± 1.64	1.00–10.60	0.27	0.63
	2021	4.17 ± 0.18	6.11 ± 0.14	5.14 ± 0.10	8.23 ± 2.37	2.60–14.60	−0.25	0.17
	2022	4.37 ± 0.21	6.28 ± 0.20	5.03 ± 0.33	4.25 ± 0.09	1.20–8.40	0.32	−0.98

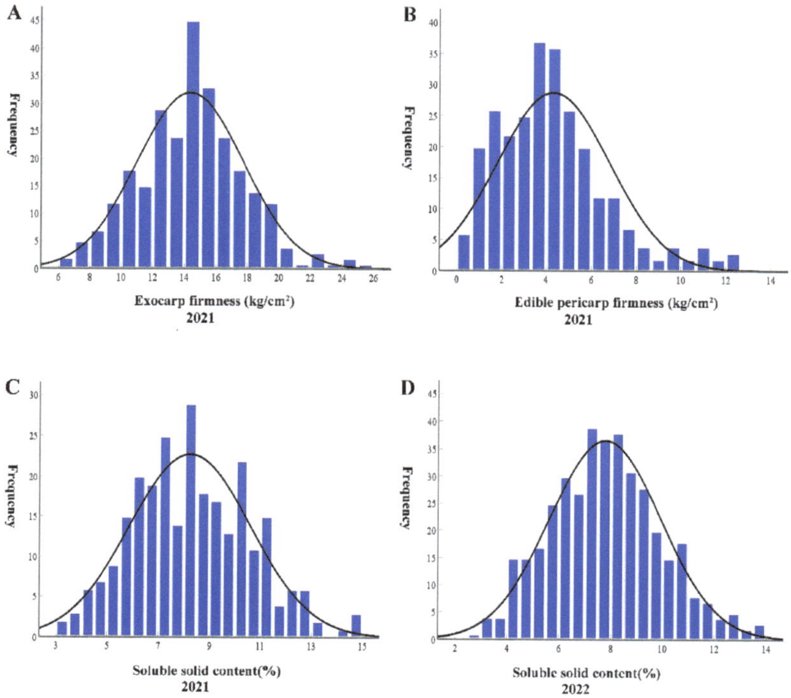

Figure 2. Distribution of EF (**A**), EPF (**B**), SSC 2021 (**C**), and SSC 2022 (**D**) frequencies in the F_2 populations.

3.1.2. Soluble Solid Content (SSC)

In a previous study [29], SSC of the parental lines, i.e., M4-5 and M1-15 lines, and the F_1 population were 9.24 ± 0.43%, 10.69 ± 0.26%, and 7.74 ± 0.58%, respectively (Table 2). In the F_2 population, SSC ranged from 4.00% to 14.00% (Supplementary Figure S1).

Table 2. Mean ± standard deviation (SD) and range of SSC (%) in the P_1, P_2, F_1, and F_2 generations over three years.

Year	P_1	P_2	F_1	F_2			
				Mean ± SD	Range	Kurtosis	Skewness
2018 [29]	9.24 ± 0.43	10.69 ± 0.26	7.74 ± 0.58	8.30 ± 2.20	4.00–14.00	−0.28	0.44
2021	8.64 ± 0.30	10.21 ± 0.30	7.93 ± 0.27	8.23 ± 2.37	2.60–14.60	−0.25	0.17
2022	9.73 ± 0.17	10.85 ± 0.07	7.77 ± 0.11	7.77 ± 2.15	2.80–13.60	−0.21	0.23

In 2021, we measured the SSC of the parents and their F_1 and F_2 populations. SSC of M1-15 (10.21 ± 0.30%) was higher than that of M4-5 (8.64 ± 0.30%). SSC of F_1 (7.93 ± 0.27%) was lower than that of M4-5 and M1-15 (Table 2). SSC of the F_2 population was diverse, ranging from 2.60% to 14.60% (Figure 2C).

In 2022, we further measured the SSC of the parents and their F_1 and F_2 populations. The SSC value of the two parents and F_1 population in 2022 was consistent with that from the previous two years (i.e., 2018 and 2021). However, the SSC value of the F_2 population in 2022 was lower than that of the previous two years. SSC of the F_2 generation showed continuous normal distribution (Figure 2D, Table 2).

3.1.3. Correlation Analysis

The correlation analysis showed that these fruit traits have a good relationship with each other (Table 3). Specifically, SSC was significantly correlated with EF, whereas EPF and EF, and SSC and EPF, were negatively correlated with each other.

Table 3. Correlation analysis of fruit traits in the F_2 plant population. *, $p < 0.05$.

Traits	SSC	EPF	EF
SSC	1	–	–
EPF	−0.02	1	–
EF	0.12 *	−0.06	1

3.2. *Single Nucleotide Polymorphisms (SNPs) and CAPS Marker Analysis*

The genome sequencing revealed 53.85% non-synonymous SNPs and 46.15% polymorphic SNPs, of which 44.68%, 21.39%, 22.27%, 0.01%, 0.01%, 0.17%, 0.74%, 0.51%, 8.01%, and 2.21% were located in the intergenic, downstream, upstream, splice acceptor, splice donor, splice region, 3′ UTR, 5′ UTR, intron, and exon regions, respectively (Figure 3).

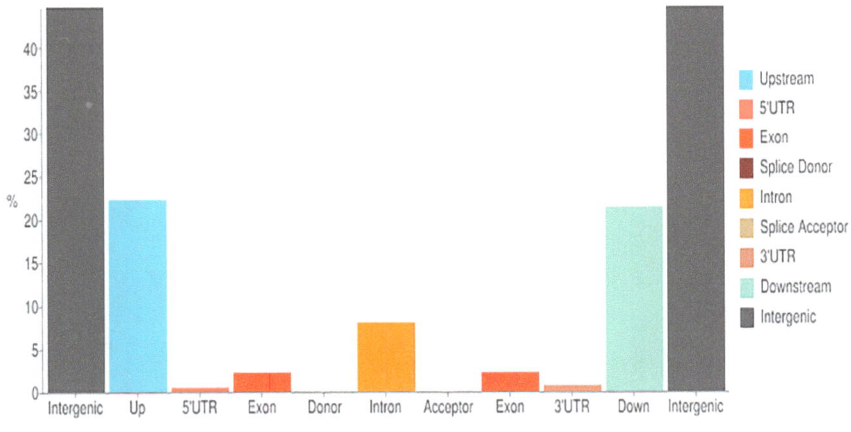

Figure 3. Location of various SNP variations found in newly sequenced areas of parental lines of melon. The *x*-axis represents the regions of distribution, and *y*-axis represents the proportion of SNPS in each location.

The total length of the whole genome was 375,360,399 bp. The SNP variant density was calculated in the chromosome region with a 1-Mb size window. A total of 2,388,036 SNP variants were detected, of which the highest number of SNP mutations were found on chromosome 1 (270,720), and the lowest number of SNP mutations on chromosome 9 (150,871). The SNP variant density in the whole genome was calculated using the CMplot R package, and the sequences of the two parental lines were compared with that of the melon reference genome (v3.6.1) (Supplementary Table S3, Figure 4).

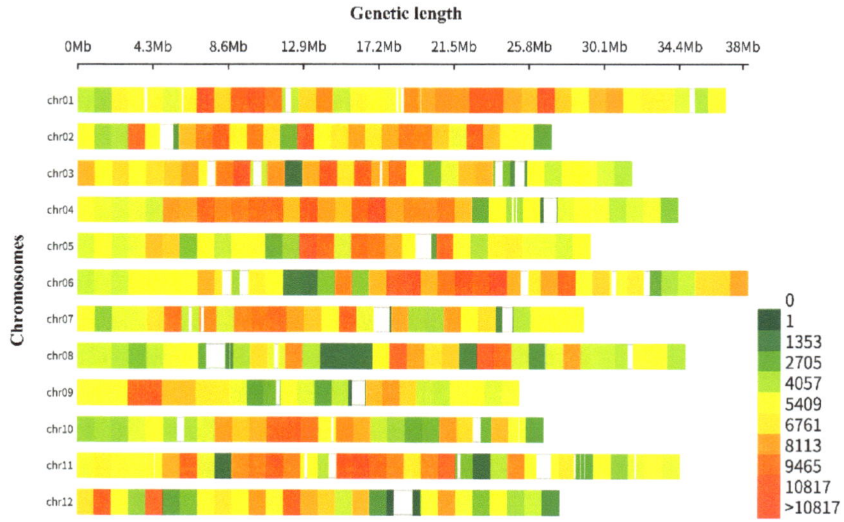

Figure 4. Total SNP variants detected within a 1-Mb window-sized chromosomal region of resequenced parental lines.

A total of 290 CAPS markers were designed, verified by P_1, P_2 and F_1, of which 116 CAPS markers were polymorphic, and the polymorphic rate was 40%. The restriction endonucleases (*Eco*R I, *Hind* III, and *Bam*H I, 10 U/μL, TAKARA) were used to cut the PCR products, which were examined using electrophoresis with 1% agarose gel.

3.3. Construction of Genetic Linkage Map

The genetic linkage maps of the 2021 F_2 populations of M4-5 and M1-15 were constructed with the 116 SNP-CAPS markers (Figure 5). Table 3 shows the marker distribution on each chromosome. The linkage map consisted of 12 linkage groups, with the largest number of CAPS markers located on chromosome 6, which has a total length of 118.34 cM and an average CAPS marker distance of 9.10 cM. The average genetic distance of the 116 CAPS markers was 12.20 cM, whereas the overall genetic distance of the 12 linkage maps (whole genome) was 1414.88 cM (Table 4).

Table 4. Details of the genetic linkage map (2021).

Chr No.	Marker No.	Linkage Group Distance (cM)	
		Genetic Distance	Average Distance
01	10	143.56	14.35
02	10	126.42	12.64
03	10	104.20	10.42
04	10	122.73	12.27
05	10	116.76	11.68
06	13	118.34	11.83
07	9	102.38	11.38
08	8	134.00	16.75
09	9	111.30	12.37
10	8	98.82	12.35
11	9	93.73	10.41
12	10	142.64	14.26
Total	116	1414.88	12.20

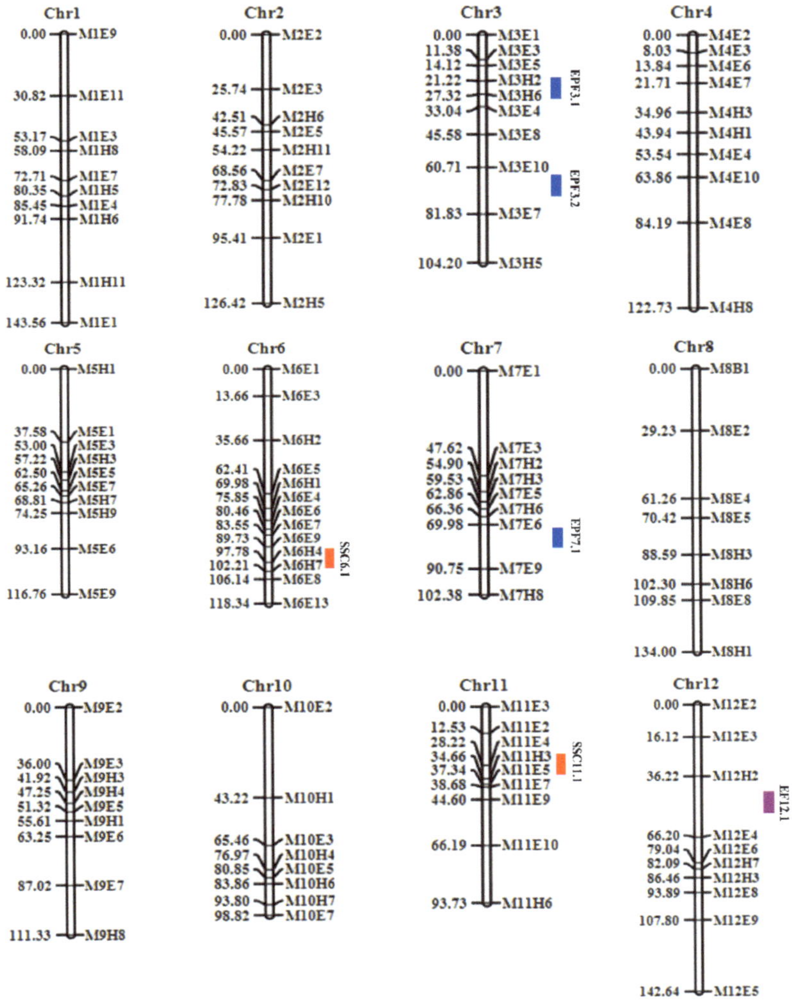

Figure 5. Based on the genotyping of 116 SNP-CAPS markers in the F_2 population, the genetic map was established. Purple (EF), blue (EPF), and red (SSC) represent mapped QTL areas. The genetic location of each chromosome is aligned on the left side, and the marker position on each chromosome is aligned on the right side.

3.4. QTLs Related to the Quality Traits of Melon

Using the melon genotype and phenotype data, six putative QTLs potentially related to melon fruit traits were identified, which are distributed on multiple chromosomes (Table 5; Figure 5). Specifically, two QTLs (*SSC6.1* and *SSC11.1*) were mapped on chromosomes 6 and 11 (Table 5 and Figure 5). Among all the six QTLs, *SSC6.1* had the highest logarithm of odds (LOD) score of 9.81, with an additive effect of −1.3466 and percentage phenotype variance (PVE%) of 16.32. It was positioned at 18 cM from the starting site of the flanking CAPS markers, *M6H4-M6E8* (21,063,516 bp–31,537,693 bp) (genetic confidence interval of 8.36 cM and physical interval of 10.47 Mb).

Table 5. QTLs information of EF, EPF, and SSC.

Traits	QTLs	Chr No.	Position (cM)	Marker Interval	Interval Mapping		Phenotypic Effect	
					LOD	PVE%	Additive	Dominance
EF	*EF12.1*	12	56	*M12H2-M12E4*	2.77	6.45	0.6332	1.8113
EPF	*EPF3.1*	3	26	*M3H2-M3E4*	3.56	6.14	0.0041	1.24
	EPF3.2	3	61	*M3E10-M3E7*	3.58	5.64	0.7995	0.42
	EPF7.1	7	75	*M7H6-M7E9*	2.90	4.89	0.9399	−0.3579
SSC	*SSC6.1*	6	102	*M6H4-M6E8*	9.81	16.32	−1.3466	−0.0064
	SSC11.1	11	37	*M11H3-M11E7*	2.72	6.27	−0.7174	0.0751

Two different QTL software (QTL IciMapping (v4.0) and R/qtl v1.5), identified a major effective QTL between CAPS markers *M6H4* and *M6E8* that is closely linked to the SSC trait. The intervals of CAPS markers in both software were roughly the same (Figure 6A,B).

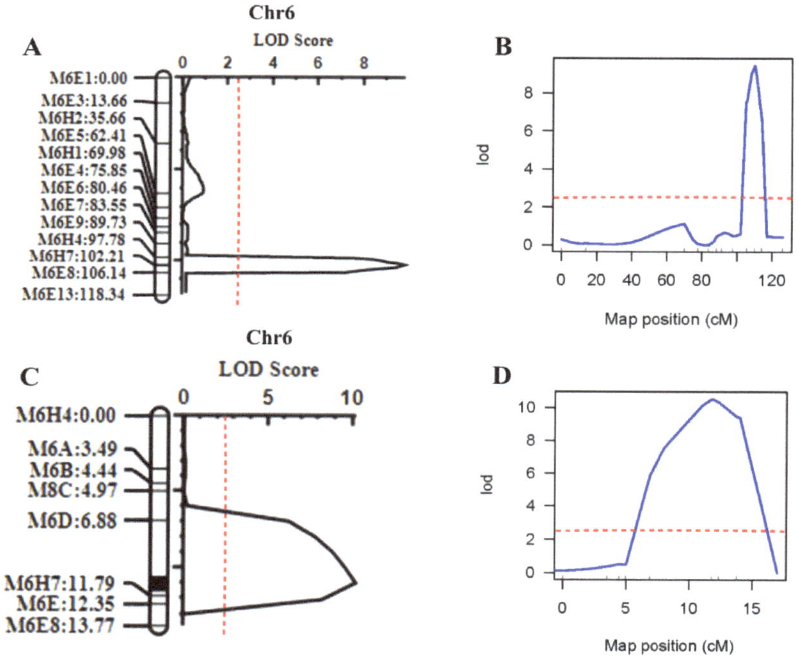

Figure 6. Positioning intervals for 2021 and 2022 SSC. (**A**) QTL IciMapping software 2021 QTL mapping results. (**B**) R/QTL software 2021 QTL mapping results. (**C**) QTL IciMapping software 2022 QTL mapping results. (**D**) R/QTL software 2022 QTL mapping results. The red dotted line indicates an LOD value of 2.5.

To locate the *SSC6.1* more precisely, based on QTL analysis in 2021, a larger F_2 population (n = 393) was planted in 2022 and eight new CAPS markers were identified in the flanking regions of the *M6H4-M6E8* markers. As a result, the range was narrowed to *M6D-M6E* (25,232,480–28,148,360 bp), and the physical distance was narrowed to 2.92 Mb (Figure 6C). The intervals (i.e., *M6H4-M6E8*) obtained using both software, i.e., R/qtl and QTL IciMapping, were the same (Figure 6D). The consistent interval calculated using the two software was similar, with LOD scores of 10.19 and 10.5, respectively.

On chromosome 12, a QTL (*EF12.1*) linked to EF was found at 56 cM (Table 5; Figure 5). The genetic distance between markers *M12H2* and *M12E4* was 29.98 cM, and a total of 6.45% PVE was detected at an LOD score of 2.77.

Three QTLs (*EPF3.1*, *EPF3.2*, and *EPF7.1*) were identified on chromosomes 3 and 7 (Table 5; Figure 5), with LOD scores of 3.56, 3.58, and 2.90; the additive effect values of 0.0041, 0.7995, and 0.9399; and the PVE% values of individual effect of 6.14, 5.64, and 4.89, respectively. On chromosome 3, the two QTLs for *EPF3.1* and *EPF3.2* were identified, with locations at 26 cM and 61 cM, and the genetic distance between the two markers at 11.82 cM (*M3H2-M3E4*) and 36.25 cM (*M3E8-M3E7*), respectively. One QTL (*EPF7.1*) is located at 75 cM on chromosome 7, and the genetic interval of the flanking markers *M7H6-M7E9* is 24.39 cM.

3.5. Marker Verification

Three markers flanking mapped QTL region and linked to the fruit SSC (Figure 7) were used to analyze the genotype–phenotype correlation of 2022 F_2 population to verify the accuracy of the QTL mapping results. The P_1, P_2, and F_1 populations had genotype A (low SSC), genotype B (high SSC), and genotype H (SSC between P_1 and P_2), respectively. In the F_2 population, the significance of *M6H7* and *M6E* among the three markers was most prominent, showing a greater probability of QTL between these two loci. These results further demonstrated the reliability of the QTL findings.

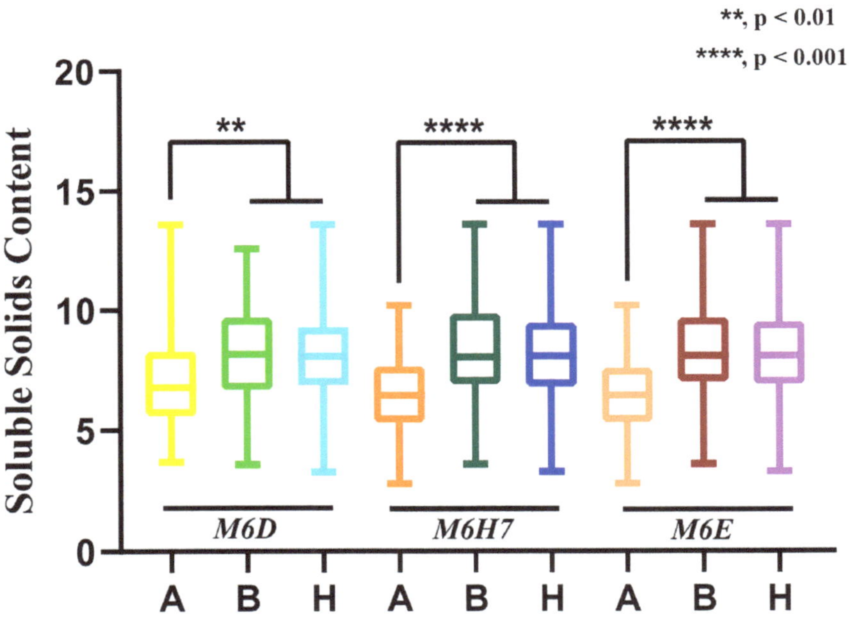

Figure 7. Three markers were used to conduct a correlation analysis between phenotypes and genotypes. The *X*-axis represents the three genotypes (A, B and H) of CAPS markers, and the *Y*-axis represents the value of SSC. Asterisk symbols (**, ****) represents the statistical significance; **, $p < 0.01$, ****, $p < 0.001$.

4. Discussion

Melon, a popular, horticultural crop, exhibits significant phenotypic variations. DNA-based genetic markers in the last few decades has been used to elucidate fruit quality traits in watermelon, melon and cucumber [35–39]. In this present study, we constructed a genetic linkage map using F_2 mapping populations over a three-year period at three distinct locations using genome-wide SNP-CAPS markers, and genetically mapped putative loci contributing to melon traits, such as exocarp and pericarp firmness and SSC (Figure 5).

In addition to the influence of external factors, such as cultivation environment and agronomic measures, the complex quantitative feature of sugar buildup in melon fruit is reportedly regulated by several genes [40]. In one study aimed to identify the genetic loci responsible for the sugar content, many QTLs and genes related to sugar content in melon fruit were identified [41]. However, at present, the reported QTLs related to the sugar content trait in melon can only explain a low level of phenotypic variation, which is unstable in different generations and under different environments and can interact with environmental factors, severely limiting the in-depth study of genetic mechanisms regulating this trait [42,43]. In previous studies, QTLs for SSC were found on chromosomes 2, 6, 7, and 9 [29]. However, in our 2021 and 2022 experiments, QTLs for SSC were found only on chromosome 6, in contrast to previous trials [29], and no QTLs for SSC were found on chromosomes 2, 7, and 9. Thus, we hypothesize that environmental factors may affect the phenotypes and interfere with the identification of the genetic loci contributing to the SSC trait.

Fruit SSC serves as a useful indicator for melon maturity. Chace et al. [44] proposed SSC as an objective quality indicator, and Mutton et al. [45] showed that both SSC and pulp firmness are key indicators for evaluating the quality of melons. Soluble solid content is usually expressed in % [45]. According to UNECE [46], SSC in the middle of the flesh of Charentais melons is $\geq$10%, whereas that of other types is $\geq$8%. In this experiment, the SSC ranges of the P_1, P_2, F_1, and F_2 populations were 8.64–9.73%, 10.21–10.85%, 7.74–7.93%, and 2.8–14.6%, respectively. According to several cultivar experiments, the reported SSC range of Galia melons was 10.2–16.1%, of melons was 11.0–14.6%, of Oriental crisp melons was 9.7–17.2%, of Canary melons was 10.4–15.2%, and of oriental melons was 10.5–14.2% [47–49].

The type, content, and composition of sugar can significantly affect fruit quality [5,50]. Thus, it is of practical significance to understand the genetic mechanism regulating sugar content in melon, which may be used in the breeding of high-quality melon varieties. Thus far, researchers have identified many QTLs in the sugar accumulation trait in melons, suggesting that this trait is highly complex and can be regulated by multiple genes [51,52]. Using the F_2 and DHL populations from a hybrid between the Piel de Sapo (PS) variety and the Korean germplasm PI161375 and using composite interval mapping, five QTLs related to SSC were identified [43]. By generating RILs from the parents of the melon varieties, TopMark and USDA-846-1, ten SSC-related QTLs were identified [22]. In the present study, two SSC related QTLs were identified and located on chromosomes 6 and 11 from the experiment performed in the first year (2021), with LOD and PVE% of 9.81 and 16.32% for *SSC6.1*, respectively. From the experiment performed in the second year, we have developed eight CAPS markers around SSC6.1, and expanded the F_2 population, which yielded significant results, with LOD and PVE% of 10.19 and 12.30%, respectively, exerting a strong potential influence on SSC.

Several mapping populations were established to investigate SSC traits in melon, and many QTLs related to the fruit sugar content trait were discovered. Currently, reported QTLs related to melon sugar content traits can only explain the low levels of phenotypic variation and exhibit inconsistency across different generations and environments; strong genotype–environment interaction further makes it challenging to investigate the genetic mechanism regulating this trait [42]. In this study, using SNP–CAPS markers established from the F_2 populations, and a QTL mapping strategy at three sites over a period of three years, we identified a stable QTL (*SSC6.1*) for SSC (The research results of 2018 are shown in Supplementary Figure and Table).

The exocarp and pericarp firmness of melons is particularly important for consumer choice as it is closely related to transportation and shelving. In particular, EF, which relates to shelf life and market price, is a significant index. Fruit firmness is an illustration of a complex quantitative attribute impacted by genes that regulate cell wall, cell dilatation, and stratum trait [53–55]. In a study by Moreno et al. [11], five QTLs (*FF2.2*, *FF3.5*, *FF8.2*, *FF8.4*, and *FF10.2*) related to flesh firmness were detected using PI 161375 (SC) and Piel de Sapo (PS). In another study by Harel-Beja et al. [50], using the RIL population developed by crossing two melon subspecies, PI414723 (*agrestis* subspecies) and Dulce (*melo* subspecies), they identified QTLs related to the fruit firmness on chromosomes 1 and 5. Herein, three QTLs related to fruit firmness were identified on chromosomes 3 and 7, including two QTLs present on chromosome 3. The LOD score of the two QTLs on chromosome 3 was 3.58 and 3.56, respectively, higher than that of the one on chromosome 7 (2.90). Among these, the position of *EPF3.2* is similar to that of *FF3.5* in the previous study by Moreno et al. [11].

Numerous studies have thoroughly examined the influences of genetic factors on traits, such as melon fruit firmness and other qualities. According to Beaulieu and Lea [56], melons get softer by 51.9% in 13 days between blooming and harvesting, with nearly a third of the softening process occurring between 35 and 38 days after flowering. The ethylene (ETH)-dependent regulation of cell wall-modifying proteins involved in melon ripening is linked to the maturation of ETH production. This is demonstrated by the observation that inhibition of 1-aminocyclopropane-1-carboxylate oxidase (ACO) gene expression can inhibit the hardness of transgenic fruits. However, after exogenous ETH administration, ACO expression was recovered and phenotype was restored [57]. Moreover, treatment with the ETH inhibitor 1-methylcyclopropene can delay the softening process of physiologically ripe melon fruit [58]. In previous studies [11], *FF3.5* and ETH (*ETH3.5*) were both identified on chromosome 3; *FF3.5* was a QTL related to fruit firmness, and *ETH3.5* was a QTL related to ETH content of fruits at maturity, leading to softening and reduction in fruit firmness. The QTL (*EPF3.2*) identified in our study was similar to *FF3.5* [11] (i.e., two QTLs positions are similar), indicating the reliability of our results.

Recent studies on watermelons have shown that tissue lignification promotes the formation of pericarp stone cells, thus increasing the firmness of the pericarp [59]. Different studies have been conducted exploring the molecular mechanisms underlying fruit firmness, mainly focusing on the ripening of fruit and the softening of cell walls caused by increased enzyme activity [53,60]. For example, β-D-xylosidase is involved in the breakdown of xylans and regulates fruit development and ripening in tomatoes [61] and Japanese pears [62]. This gene might also contribute to melon fruit firmness by controlling skin thickness and external pressure. Potential genes that can provide further molecular insights for upcoming CAPS marker-assisted selection breeding may be identified among the genetic loci we have identified.

5. Conclusions

NGS data can significantly improve the validity of genetic linkage maps and CAPS markers. Our constructed genetic linkage map covered 12 chromosomes, with a total genetic distance of 1414.88 cM. We mapped six QTLs related to fruit quality traits (EF, EPF and SSC). The presence of the QTL, *SSC6.1*, was verified in the F_2 populations in three independent experiments and at three different locations, which can facilitate MAS of potential genes in the breeding of different melon varieties.

Supplementary Materials: The following supporting information can be downloaded at: https://www.mdpi.com/article/10.3390/cimb45040224/s1, Supplementary Figure and Table: The position of the fruit traits QTL in the genetic linkage map in the previous research and QTL information for fruit traits. Supplementary Figure S1: Frequency distribution of BRC in F2 population derived from M4-5 and M1-15 (2018). Supplementary Table S1: CAPS makers primer sequence. Supplementary Table S2: The PCR reaction system. Supplementary Table S3: Summary of detected SNP variants distribution in re-sequenced parental lines of melon.

Author Contributions: Conceptualization, data curation, formal analysis, investigation, and writing—original draft, H.Z.; investigation, X.M. and J.S.; formal analysis, T.Z.; software, C.Z.; supervision, writing—review and editing, P.G. All authors have read and agreed to the published version of the manuscript.

Funding: This study was supported by the National Nature Science Foundation of China (No. U21A20229).

Institutional Review Board Statement: Not applicable.

Informed Consent Statement: Not applicable.

Data Availability Statement: Data will be made available upon request from the corresponding author.

Acknowledgments: We appreciate the important support from the laboratory's research group very much.

Conflicts of Interest: The authors declare no conflict of interest.

References

1. National Watermelon and Melon Industry Development Plan (2015–2020). Available online: http://www.moa.gov.cn/nybgb/2015/san/201711/t20171129_5923446.htm (accessed on 3 January 2023).
2. Mayobre, C.; Pereira, L.; Eltahiri, A.; Bar, E.; Lewinsohn, E.; Garcia-Mas, J.; Pujol, M. Genetic dissection of aroma biosynthesis in melon and its relationship with climacteric ripening. *Food Chem.* **2021**, *353*, 129484. [CrossRef] [PubMed]
3. Adıgüzel, P.; Nyirahabimana, F.; Shimira, F.; Solmaz, İ.; Taşkın, H. Applied Biotechnological Approaches for Reducing Yield Gap in Melon Grown Under Saline and Drought Stresses: An Overview. *J. Soil Sci. Plant Nutr.* **2022**, *23*, 139–151. [CrossRef]
4. Can, H.; TÜRkmen, Ö. Collection of local Kyrgyzstan Melon genotypes and determination of morphologicalrelationships between some Anatolian Melons. *Turk. J. Agric. For.* **2022**, *46*, 257–270. [CrossRef]
5. Zeb, A.; Qureshi, W.S.; Ghafoor, A.; Malik, A.; Imran, M.; Iqbal, J.; Alanazi, E. Is this melon sweet? A quantitative classification for near-infrared spectroscopy. *Infrared Phys. Technol.* **2021**, *114*, 103645. [CrossRef]
6. Argyris, J.M.; Diaz, A.; Ruggieri, V.; Fernandez, M.; Jahrmann, T.; Gibon, Y.; Pico, B.; Martin-Hernandez, A.M.; Monforte, A.J.; Garcia-Mas, J. QTL Analyses in Multiple Populations Employed for the Fine Mapping and Identification of Candidate Genes at a Locus Affecting Sugar Accumulation in Melon (*Cucumis melo* L.). *Front. Plant Sci.* **2017**, *8*, 1679. [CrossRef]
7. Berüter, J.; Feusi, M.E.S.; Rüedi, P. Sorbitol and sucrose partitioning in the growing apple fruit. *J. Plant Physiol.* **1997**, *151*, 269–276. [CrossRef]
8. Kyriacou, M.C.; Leskovar, D.I.; Colla, G.; Rouphael, Y. Watermelon and melon fruit quality: The genotypic and agro-environmental factors implicated. *Sci. Hortic.* **2018**, *234*, 393–408. [CrossRef]
9. Pereira, L.; Santo Domingo, M.; Ruggieri, V.; Argyris, J.; Phillips, M.A.; Zhao, G.; Lian, Q.; Xu, Y.; He, Y.; Huang, S.; et al. Genetic dissection of climacteric fruit ripening in a melon population segregating for ripening behavior. *Hortic. Res.* **2020**, *7*, 187. [CrossRef]
10. Dai, D.; Zeng, S.; Wang, L.; Li, J.; Ji, P.; Liu, H.; Sheng, Y. Identification of Fruit Firmness QTL *ff2.1* by SLAF-BSA and QTL Mapping in melon. *Euphytica* **2021**, *218*. [CrossRef]
11. Moreno, E.; Obando, J.M.; Dos-Santos, N.; Fernandez-Trujillo, J.P.; Monforte, A.J.; Garcia-Mas, J. Candidate genes and QTLs for fruit ripening and softening in melon. *Appl. Genet.* **2008**, *116*, 589–602. [CrossRef]
12. Monforte, A.J.; Diaz, A.; Cano-Delgado, A.; van der Knaap, E. The genetic basis of fruit morphology in horticultural crops: Lessons from tomato and melon. *J. Exp. Bot.* **2014**, *65*, 4625–4637. [CrossRef] [PubMed]
13. Chikh-Rouhou, H.; Mezghani, N.; Mnasri, S.; Mezghani, N.; Garcés-Claver, A. Assessing the Genetic Diversity and Population Structure of a Tunisian Melon (*Cucumis melo* L.) Collection Using Phenotypic Traits and SSR Molecular Markers. *Agronomy* **2021**, *11*, 1121. [CrossRef]
14. Wang, C.; Li, H.; Long, Y.; Dong, Z.; Wang, J.; Liu, C.; Wei, X.; Wan, X. A Systemic Investigation of Genetic Architecture and Gene Resources Controlling Kernel Size-Related Traits in Maize. *Int. J. Mol. Sci.* **2023**, *24*, 1025. [CrossRef] [PubMed]
15. Cui, H.; Fan, C.; Ding, Z.; Wang, X.; Tang, L.; Bi, Y.; Luan, F.; Gao, P. CmPMRl and CmPMrs are responsible for resistance to powdery mildew caused by Podosphaera xanthii race 1 in Melon. *Appl. Genet.* **2022**, *135*, 1209–1222. [CrossRef]
16. Santo Domingo, M.; Mayobre, C.; Pereira, L.; Argyris, J.; Valverde, L.; Martin-Hernandez, A.M.; Garcia-Mas, J.; Pujol, M. Fruit Morphology and Ripening-Related QTLs in a Newly Developed Introgression Line Collection of the Elite Varieties 'Vedrantais' and 'Piel de Sapo'. *Plants* **2022**, *11*, 3120. [CrossRef] [PubMed]
17. Zhang, T.; Cui, H.; Luan, F.; Liu, H.; Ding, Z.; Amanullah, S.; Zhang, M.; Ma, T.; Gao, P. A recessive gene Cmpmr2F confers powdery mildew resistance in melon (*Cucumis melo* L.). *Appl. Genet.* **2023**, *136*, 1–16. [CrossRef]
18. Pitrat, M. Linkage groups in *Cucumis melo* L. *J. Hered.* **1991**, *82*, 406–411. [CrossRef]
19. Zalapa, J.E.; Staub, J.E.; McCreight, J.D.; Chung, S.M.; Cuevas, H. Detection of QTL for yield-related traits using recombinant inbred lines derived from exotic and elite US Western Shipping melon germplasm. *Appl. Genet.* **2007**, *114*, 1185–1201. [CrossRef]
20. Fukino, N.; Ohara, T.; Monforte, A.J.; Sugiyama, M.; Sakata, Y.; Kunihisa, M.; Matsumoto, S. Identification of QTLs for resistance to powdery mildew and SSR markers diagnostic for powdery mildew resistance genes in melon (*Cucumis melo* L.). *Appl. Genet.* **2008**, *118*, 165–175. [CrossRef]
21. Amanullah, S.; Osae, B.A.; Yang, T.; Abbas, F.; Liu, S.; Liu, H.; Wang, X.; Gao, P.; Luan, F. Mapping of genetic loci controlling fruit linked morphological traits of melon using developed CAPS markers. *Mol. Biol. Rep.* **2022**, *49*, 5459–5472. [CrossRef]
22. Paris, M.K.; Zalapa, J.E.; McCreight, J.D.; Staub, J.E. Genetic dissection of fruit quality components in melon (*Cucumis melo* L.) using a RIL population derived from exotic × elite US Western Shipping germplasm. *Mol. Breed.* **2008**, *22*, 405–419. [CrossRef]
23. Diaz, A.; Martin-Hernandez, A.M.; Dolcet-Sanjuan, R.; Garces-Claver, A.; Alvarez, J.M.; Garcia-Mas, J.; Pico, B.; Monforte, A.J. Quantitative trait loci analysis of melon (*Cucumis melo* L.) domestication-related traits. *Appl. Genet.* **2017**, *130*, 1837–1856. [CrossRef] [PubMed]

24. Natarajan, S.; Kim, H.T.; Thamilarasan, S.K.; Veerappan, K.; Park, J.I.; Nou, I.S. Whole Genome Re-Sequencing and Characterization of Powdery Mildew Disease-Associated Allelic Variation in Melon. *PLoS ONE* **2016**, *11*, e0157524. [CrossRef]
25. Oren, E.; Tzuri, G.; Dafna, A.; Meir, A.; Kumar, R.; Katzir, N.; Elkind, Y.; Freilich, S.; Schaffer, A.A.; Tadmor, Y.; et al. High-density NGS-based map construction and genetic dissection of fruit shape and rind netting in *Cucumis melo*. *Appl. Genet.* **2020**, *133*, 1927–1945. [CrossRef] [PubMed]
26. Garcia-Mas, J.; Benjak, A.; Sanseverino, W.; Bourgeois, M.; Mir, G.; Gonzalez, V.M.; Henaff, E.; Camara, F.; Cozzuto, L.; Lowy, E.; et al. The genome of melon (*Cucumis melo* L.). *Proc. Natl. Acad. Sci. USA* **2012**, *109*, 11872–11877. [CrossRef]
27. Ruggieri, V.; Alexiou, K.G.; Morata, J.; Argyris, J.; Pujol, M.; Yano, R.; Nonaka, S.; Ezura, H.; Latrasse, D.; Boualem, A.; et al. An improved assembly and annotation of the melon (*Cucumis melo* L.) reference genome. *Sci. Rep.* **2018**, *8*, 8088. [CrossRef]
28. Amanullah, S.; Liu, S.; Gao, P.; Zhu, Z.; Zhu, Q.; Fan, C.; Luan, F. QTL mapping for melon (*Cucumis melo* L.) fruit traits by assembling and utilization of novel SNPs based CAPS markers. *Sci. Hortic.* **2018**, *236*, 18–29. [CrossRef]
29. Zhang, T.; Ding, Z.; Liu, J.; Qiu, B.; Gao, P. QTL mapping of pericarp and fruit-related traits in melon (*Cucumis melo* L.) using SNP-derived CAPS markers. *Sci. Hortic.* **2020**, *265*, 109243. [CrossRef]
30. Obando-Ulloa, J.M.; Moreno, E.; García-Mas, J.; Nicolai, B.; Lammertyn, J.; Monforte, A.J.; Fernández-Trujillo, J.P. Climacteric or non-climacteric behavior in melon fruit: 1. Aroma volatiles. *Postharvest Biol. Technol.* **2008**, *49*, 27–37. [CrossRef]
31. Allen, G.C.; Flores-Vergara, M.A.; Krasynanski, S.; Kumar, S.; Thompson, W.F. A modified protocol for rapid DNA isolation from plant tissues using cetyltrimethylammonium bromide. *Nat. Protoc.* **2006**, *1*, 2320–2325. [CrossRef]
32. Melon (DHL92) v3.6.1 Genome. Available online: http://cucurbitgenomics.org/organism/18 (accessed on 17 January 2023).
33. Thiel, T.; Kota, R.; Grosse, I.; Stein, N.; Graner, A. SNP2CAPS: A SNP and INDEL analysis tool for CAPS marker development. *Nucleic Acids Res.* **2004**, *32*, e5. [CrossRef] [PubMed]
34. Broman, K.W.; Wu, H.; Sen, S.; Churchill, G.A. R/qtl: QTL mapping in experimental crosses. *Bioinformatics* **2003**, *19*, 889–890. [CrossRef] [PubMed]
35. Osae, B.A.; Amanullah, S.; Liu, H.; Liu, S.; Saroj, A.; Zhang, C.; Liu, T.; Gao, P.; Luan, F. CAPS marker-base genetic linkage mapping and QTL analysis for watermelon ovary, fruit and seed-related traits. *Euphytica* **2022**, *218*, 39. [CrossRef]
36. Maragal, S.; Rao, E.S.; Reddy, D.C.L. Genetic mapping and meta-analysis identifies several candidate genes for watermelon (Citrullus lanatus) fruit quality traits. *Sci. Hortic.* **2023**, *308*, 111545. [CrossRef]
37. Riahi, C.; Reig-Valiente, J.L.; Picó, B.; Díaz, A.; Gonzalo, M.J.; Monforte, A.J. Evidence of the Role of QTL Epistatic Interactions in the Increase of Melon Fruit Flesh Content during Domestication. *Agronomy* **2020**, *10*, 1064. [CrossRef]
38. Fernandez-Silva, I.; Moreno, E.; Essafi, A.; Fergany, M.; Garcia-Mas, J.; Martín-Hernandez, A.M.; Álvarez, J.M.; Monforte, A.J. Shaping melons: Agronomic and genetic characterization of QTLs that modify melon fruit morphology. *Theor. Appl. Genet.* **2010**, *121*, 931–940. [CrossRef]
39. Weng, Y.; Colle, M.; Wang, Y.; Yang, L.; Rubinstein, M.; Sherman, A.; Ophir, R.; Grumet, R. QTL mapping in multiple populations and development stages reveals dynamic quantitative trait loci for fruit size in cucumbers of different market classes. *Theor. Appl. Genet.* **2015**, *128*, 1747–1763. [CrossRef] [PubMed]
40. Burger, Y.; Sa'ar, U.; Distelfeld, A.; Katzir, N.; Yeselson, Y.; Shen, S.; Schaffer, A.A. Development of Sweet Melon (*Cucumis melo*) Genotypes Combining High Sucrose and Organic Acid Content. *J. Am. Soc. Hortic. Sci. Jashs* **2003**, *128*, 537–540. [CrossRef]
41. Pereira, L.; Santo Domingo, M.; Argyris, J.; Mayobre, C.; Valverde, L.; Martin-Hernandez, A.M.; Pujol, M.; Garcia-Mas, J. A novel introgression line collection to unravel the genetics of climacteric ripening and fruit quality in melon. *Sci. Rep.* **2021**, *11*, 11364. [CrossRef]
42. Perpina, G.; Esteras, C.; Gibon, Y.; Monforte, A.J.; Pico, B. A new genomic library of melon introgression lines in a cantaloupe genetic background for dissecting desirable agronomical traits. *BMC Plant Biol.* **2016**, *16*, 154. [CrossRef]
43. Monforte, A.J.; Oliver, M.; Gonzalo, M.J.; Alvarez, J.M.; Dolcet-Sanjuan, R.; Arus, P. Identification of quantitative trait loci involved in fruit quality traits in melon (*Cucumis melo* L.). *Appl. Genet.* **2004**, *108*, 750–758. [CrossRef] [PubMed]
44. Chace, E.M.E.M.; Church, C.G.C.G.; Denny, F.E.F.E. *Relation between the Composition of California Cantaloupes and Their Commercial Maturity*; U.S. Dept. of Agriculture: Washington, DC, USA, 1924.
45. Mutton, L.L.; Cullis, B.R.; Blakeney, A.B. The objective definition of eating quality in rockmelons (*Cucumis melo*). *J. Sci. Food Agric.* **1981**, *32*, 385–391. [CrossRef]
46. Cantaloups, Honeydew, Honey Ball and Other Similar Melon. Available online: https://www.ams.usda.gov/sites/default/files/media/Cantaloup_Inspection_Instructions%5B1%5D.pdf (accessed on 23 January 2023).
47. Guan, W.J.; Zhao, X.; Treadwell, D.D.; Alligood, M.R.; Huber, D.J.; Dufault, N.S. Specialty Melon Cultivar Evaluation under Organic and Conventional Production in Florida. *HortTechnology* **2013**, *23*, 905–912. [CrossRef]
48. Schultheis, J.; Thompson, W. Specialty Melon Yield and Quality Response to Grafting in Trials Conducted in the Southeastern United States. *Acta Hortic.* **2015**, *1086*, 269–278. [CrossRef]
49. Soteriou, G.A.; Papayiannis, L.C.; Kyriacou, M.C. Indexing melon physiological decline to fruit quality and vine morphometric parameters. *Sci. Hortic.* **2016**, *203*, 207–215. [CrossRef]
50. Harel-Beja, R.; Tzuri, G.; Portnoy, V.; Lotan-Pompan, M.; Lev, S.; Cohen, S.; Dai, N.; Yeselson, L.; Meir, A.; Libhaber, S.E.; et al. A genetic map of melon highly enriched with fruit quality QTLs and EST markers, including sugar and carotenoid metabolism genes. *Appl. Genet.* **2010**, *121*, 511–533. [CrossRef]

51. Burger, Y.; Saar, U.; Katzir, N.; Paris, H.S. A Single Recessive Gene for Sucrose Accumulation in Cucumis melo Fruit. *J. Am. Soc Hort. Sci.* **2002**, *127*, 938–943. [CrossRef]
52. Diaz, A.; Fergany, M.; Formisano, G.; Ziarsolo, P.; Blanca, J.; Fei, Z.; Staub, J.E.; Zalapa, J.E.; Cuevas, H.E.; Dace, G.; et al. A consensus linkage map for molecular markers and Quantitative Trait Loci associated with economically important traits in melon (*Cucumis melo* L.). *BMC Plant Biol.* **2011**, *11*, 111. [CrossRef]
53. Zarid, M.; Garcia-Carpintero, V.; Esteras, C.; Esteva, J.; Bueso, M.C.; Canizares, J.; Pico, M.B.; Monforte, A.J.; Fernandez-Trujillo, J.P. Transcriptomic analysis of a near-isogenic line of melon with high fruit flesh firmness during ripening. *J. Sci. Food Agric.* **2021**, *101*, 754–777. [CrossRef]
54. Naets, M.; Wang, Z.; Verboven, P.; Nicolaï, B.; Keulemans, W.; Geeraerd, A. Size does matter—Susceptibility of apple for grey mould is affected by cell size. *Plant Pathol.* **2019**, *69*, 60–67. [CrossRef]
55. Sun, L.; Zhang, Y.; Cui, H.; Zhang, L.; Sha, T.; Wang, C.; Fan, C.; Luan, F.; Wang, X. Linkage Mapping and Comparative Transcriptome Analysis of Firmness in Watermelon (*Citrullus lanatus*). *Front. Plant Sci.* **2020**, *11*, 831. [CrossRef]
56. Beaulieu, J.C.; Lea, J.M. Quality changes in cantaloupe during growth, maturation, and in stored fresh-cut cubes prepared from fruit harvested at various maturities. *J. Am. Soc. Hortic. Sci.* **2007**, *132*, 720–728. [CrossRef]
57. Nishiyama, K.; Guis, M.; Rose, J.K.; Kubo, Y.; Bennett, K.A.; Wangjin, L.; Kato, K.; Ushijima, K.; Nakano, R.; Inaba, A.; et al. Ethylene regulation of fruit softening and cell wall disassembly in Charentais melon. *J. Exp. Bot.* **2007**, *58*, 1281–1290. [CrossRef] [PubMed]
58. Pech, J.C.; Bouzayen, M.; Latché, A. Climacteric fruit ripening: Ethylene-dependent and independent regulation of ripening pathways in melon fruit. *Plant Sci.* **2008**, *175*, 114–120. [CrossRef]
59. Yang, T.; Zhang, P.; Pan, J.; Amanullah, S.; Luan, F.; Han, W.; Liu, H.; Wang, X. Genome-Wide Analysis of the Peroxidase Gene Family and Verification of Lignin Synthesis-Related Genes in Watermelon. *Int. J. Mol. Sci.* **2022**, *23*, 642. [CrossRef]
60. Brummell, D.A. Cell wall disassembly in ripening fruit. *Funct. Plant Biol.* **2006**, *33*, 103–119. [CrossRef]
61. Itai, A.; Ishihara, K.; Bewley, J.D. Characterization of expression, and cloning, of beta-D-xylosidase and alpha-L-arabinofuranosidas in developing and ripening tomato (*Lycopersicon esculentum* Mill.) fruit. *J. Exp. Bot.* **2003**, *54*, 2615–2622. [CrossRef]
62. Tateishi, A.; Mori, H.; Watari, J.; Nagashima, K.; Yamaki, S.; Inoue, H. Isolation, characterization, and cloning of alpha-L-Arabinofuranosidase expressed during fruit ripening of Japanese pear. *Plant Physiol.* **2005**, *138*, 1653–1664. [CrossRef]

current issues in molecular biology

MDPI

Article

Comparative Transcriptome Analysis of Softening and Ripening-Related Genes in Kiwifruit Cultivars Treated with Ethylene

Han Ryul Choi [1,2], Min Woo Baek [1,2], Cheon Soon Jeong [1,2,*] and Shimeles Tilahun [1,3,4,*]

1 Department of Horticulture, Kangwon National University, Chuncheon 24341, Korea; hanryul192@kangwon.ac.kr (H.R.C.); minwoo100@kangwon.ac.kr (M.W.B.)
2 Interdisciplinary Program in Smart Agriculture, Kangwon National Uinversity, Chuncheon 24341, Korea
3 Agriculture and Life Science Research Institute, Kangwon National University, Chuncheon 24341, Korea
4 Department of Horticulture and Plant Sciences, Jimma University, Jimma 378, Ethiopia
* Correspondence: jeongcs@kangwon.ac.kr (C.S.J.); shimeles@kangwon.ac.kr (S.T.); Tel.: +82-033-250-6409 (C.S.J.)

Abstract: This work presents the transcriptome analysis of green 'Hayward' (*Actinidia deliciosa*) and gold 'Haegeum' (*Actinidia chinensis*) kiwifruit cultivars after treatment with ethylene for three days at 25 °C. Illumina high-throughput sequencing platform was used to sequence total mRNAs and the transcriptome gene set was constructed by de novo assembly. A total of 1287 and 1724 unigenes were differentially expressed during the comparison of ethylene treatment with control in green 'Hayward' and gold 'Haegeum', respectively. From the differentially expressed unigenes, 594 and 906 were upregulated, and 693 and 818 were downregulated in the green and gold kiwifruit cultivars, respectively, when treated with ethylene. We also identified a list of genes that were expressed commonly and exclusively in the green and gold kiwifruit cultivars treated with ethylene. Several genes were expressed differentially during the ripening of kiwifruits, and their cumulative effect brought about the softening- and ripening-related changes. This work also identified and categorized genes related to softening and other changes during ripening. Furthermore, the transcript levels of 12 selected representative genes from the differentially expressed genes (DEGs) identified in the transcriptome analysis were confirmed via quantitative real-time PCR (qRT-PCR) to validate the reliability of the expression profiles obtained from RNA-Seq. The data obtained from the present study will add to the information available on the molecular mechanisms of the effects of ethylene during the ripening of kiwifruits. This study will also provide resources for further studies of the genes related to ripening, helping kiwifruit breeders and postharvest technologists to improve ripening quality.

Keywords: kiwifruit; transcriptome analysis; gene expression; ethylene treatment; softening

Citation: Choi, H.R.; Baek, M.W.; Jeong, C.S.; Tilahun, S. Comparative Transcriptome Analysis of Softening and Ripening-Related Genes in Kiwifruit Cultivars Treated with Ethylene. *Curr. Issues Mol. Biol.* **2022**, *44*, 2593–2613. https://doi.org/10.3390/cimb44060177

Academic Editor: Vijai Bhadauria

Received: 17 May 2022
Accepted: 1 June 2022
Published: 2 June 2022

1. Introduction

Kiwifruit (*Actinidia* spp.) is a perennial deciduous fruit belonging to the family Actinidiaceae, widely cultivated in 41 countries around the world [1]. In 2019, the global production of kiwifruits was 4.35 million tons, of which the Republic of Korea accounted for 5622 [2]. There are more than 70 different kiwifruit species in the world, and they are classified according to the color of both skin and flesh of the fruit [3]. However, only a few representative cultivars, including green-fleshed *Actinidia deliciosa* and yellow-fleshed *Actinidia chinensis*, have dominated the international commercial market [2].

Kiwifruit has been called 'the king of fruits' owing to its excellent flavor and outstanding functional substances including minerals, vitamins, and antioxidants [4,5]. Several studies have recommended kiwifruit for diet, anti-cancer effects, relieving stress, and strengthening immunity [6]. Levels of primary and secondary metabolites and the genes encoding their expressions vary depending on genotype, maturity stage, and storage period [2].

Kiwifruit is a climacteric fruit that generates autocatalytic ethylene through the respiration process during ripening [7]; it can be harvested at physiological maturity at an unripe stage [8]. To fulfill consumers' preference for "ready to eat" kiwifruit, it is imperative to use postharvest techniques such as exogenous ethylene treatment [3]. Exogenous ethylene treatment in kiwifruit has the advantage of rapidly and uniformly ripening the fruit up to the optimal edible stage [9]. The kiwifruit ripening process of involves changes in many gene expressions, biochemical and physiological processes [1].

In recent years, various studies related to ripening and storage of kiwifruit have been reported. Tilahun et al. [3] treated kiwifruit cultivars with exogenous ethylene to determine ripening quality based on sensory evaluation and physicochemical criteria, reporting that gold 'Haegeum' and red 'Hongyang' kiwifruits attained eating quality on the 2nd day of ripening, whereas green 'Hayward' attained eating quality on the 4th day, irrespective of harvest time. The effect of cold storage was also investigated on ripening quality, primary and secondary metabolites, antioxidant activities, and ripening and stress-related genes, to determine biological markers for indication of storability and ripening quality in kiwifruits [2].

Transcriptome profiling analyses related to ripening have been reported for other fruits, including persimmon [10], banana [11], orange [12], watermelon [13] and strawberry [14]. Similarly, the transcript profile of kiwifruit during ripening has been studied. Hydrogen sulfide (H_2S) treatment can delay the ripening of kiwifruit by regulating cell-wall- and ethylene-related genes [4]. Tilahun et al. [15] reported transcriptome analysis of gold 'Haeguem' kiwifruit treated with ethylene to improve postharvest ripening quality. However, comparative transcriptome analysis information has been lacking for the two cultivars in relation to kiwifruit softening and ripening-related changes following ethylene treatment.

In this study, we conducted comparative transcriptome analysis and classified by cultivar the candidate genes related to softening- and ripening-related changes. This enabled our assessment of the differences in postharvest fruit quality, sensory acceptance, and transcriptome profile between naturally ripe and exogenous ethylene-induced ripe kiwifruit during ripening. By analyzing the DEG of kiwifruit cultivars following exogenous ethylene treatment, candidate genes that could be engaged in softening- and ripening-related changes were identified. The expression profiles of 12 selected representative differentially expressed genes were confirmed by quantitative real-time PCR to validate the RNA sequencing results. The transcriptome profile provided by our study will provide useful information on the effects of ethylene treatment on softening- and ripening-related changes in kiwifruit at the genomic level. It could be helpful for further research into genes related to ripening for use in kiwifruit breeding and postharvest technology.

2. Materials and Methods

2.1. Plant Material and Ethylene Treatment

Two kiwifruit (*Actinidia* spp.) cultivars, green 'Hayward' and gold 'Haegeum', were used in this study. Fruits were harvested at commercial maturity (170 days after full bloom) [3] on 23 October 2020 in Jangheung, South Korea. Fruits were immediately transferred to the postharvest laboratory at Kangwon University. After careful selection, uniform-sized fruits free from physical defects were treated with exogenous ethylene at 100 $\mu L\ kg^{-1}$ [15], in a sealed 62 L container for 3 days at 25 °C. Air in the sealed container was ventilated and distributed by a fan (Coolertec CT8025L12RA-3P, Zhengzhou, China). Four containers were used and 60 fruits were placed in each container. Fruits were regularly inspected and data were collected at 0 d and on the third day; from ten biological replicates for firmness, soluble solids content (SSC), titratable acidity (TA), brix-acid ratio (BAR), and overall acceptability of the fresh fruit; and from five replicates for respiration rate and ethylene production rate. Samples of fruit flesh were also taken for pectin content, polygalacturonase (PG) activity, and transcriptome analysis. All samples were frozen in liquid nitrogen and stored in a deep freezer (−80 °C). Then, samples for analysis of secondary metabolites were freeze-dried with a vacuum freeze dryer (FDT-8650, Operon, Korea) and the dried samples were ground to powder.

2.2. Transcriptome Sequencing

Samples of each cultivar (green, gold) and each treatment (control, ethylene) were sent for sequencing on the third day. Three replicates were used for transcriptome sequencing. RNA sequencing was performed at DNACARE (Seoul, Korea) using HISAT v2.1.0. Total RNA was extracted and pooledin equal volumes from each sample of the control and ethylene-treated groups [16]. Total RNA was isolated from frozen pulp samples with the Robospin Plant TM Kit (GeneAll, Korea) following the manufacturer's protocol, and genomic DNA was removed with RNA-free DNase I (Sigma, St. Louis, MO, USA). The quality and content of the extracted RNA were measured using a Nano-drop and cDNA was synthesized with oligo d (T)18 primer and SuperScript® III Reverse Transcriptase (Life Technologies, Carlsbad, CA, USA) from 5 µg of total RNA [10]. RNA sequencing was performed using the Illumina high-throughput sequencing platform according to the manufacturer's protocol (Illumina Inc., San Diego, CA, USA) in the National Instrumentation Center for Environmental Management (NICEM), Seoul National University in Korea [10]. Raw reads were processed using Trimmomatic(v0.38). Then, reads for each sample were mapped to the reference genome (Hongyang v3.0) by HISAT2(v2.0.13) (http://ccb.jhu.edu/software/hisat2/index.shtml accessed on 10 April 2021). The read count of the transcript expression level was calculated using StringTie(v1.3.4d). Differentially expressed genes (DEGs) were putatively identified using DESeq by comparing the transcript levels in control and ethylene treated kiwifruit using two criteria; false discovery rate (FDR) at $p < 0.05$ and $|\log_2 \text{fold change}| \geq 1$. The flow chart of mRNA processing is indicated in Figure S1.

2.3. Identification of DEGs and Functional Enrichment Analysis

To classify all genes into annotated functional subcategories, blast (e-value 1×10^{-4}) analysis was performed using the Refseq plant protein sequence of UniProt, TAIR, and NCBI as a database. Gene ontology and pathway analysis were performed using BLAST2GO (version 5.2.4) and the InterProScan program targeting Pfam and KEGG databases.

2.4. Verification of DEGs by qRT-PCR

Transcript accumulation of *EXPA8*, *EXPA11*, *QRT1*, *ACO1*, *ACO3*, *ACS3*, *ERF061*, *ERF062*, *TLP1*, *LOX1.5*, *LOX6*, *CYP75B1* was evaluated via quantitative real-time RCR (qRT-PCR), as described by [16], using gene-specific primers (Table S1).

2.5. Measurement of Firmness, Pectin Content, EIS and Polygalacturonase(PG) Activity

The flesh firmness of kiwifruit was measured by a Rheometer (Sun Scientific Co., Ltd., Tokyo, Japan) from 10 fruits (mean of two measurements per fruit) by a puncture at the equator with a 3 mm diameter round stainless-steel probe with a flat end and a maximum force of 10 kg [17]. The pectin content, EIS, and polygalacturonase (PG) activity of kiwifruit were measured and expressed as described by [3].

2.6. Measurement of Soluble Solids Content (SSC), Titratable Acidity (TA), Brix-Acid Ratio (BAR) and Overall Acceptability

Soluble solids content (SSC) from the juice of each fruit was measured using a digital refractometer (Atago Co., Ltd., Tokyo, Japan) at 20 °C. The unit of measurement was expressed in percent [1]. Titratable acidity (TA) was measured using a DL22 food and beverage analyzer (Mettler Toledo Ltd., Zurich, Switzerland). Diluted kiwifruit juice (1 mL juice to 19 mL distilled water) was used for titration by 0.1 N NaOH up to pH 8.1 to obtain

TA, expressed as mg of citric acid per kg of fresh kiwifruit weight. BAR was found by dividing the SSC by the titratable acidity [17]. Overall acceptability was identified as the mean value of the subjective scale for flavor, sweetness, chewiness, and appearance during the ripening period [17]. It was evaluated by 10 trained panels of graduate students, and successive digits were assigned to each rating from 1 = bad to 5 = excellent [17].

2.7. Weight Loss, Ethylene Production and Respiration Rates

Fresh weight loss was measured as described by [3]. Kiwifruits were weighed before treatment and weighed again after three days to calculate the percentage (%) weight loss during ripening. The ethylene production rate and respiration rate of kiwifruit were measured and expressed as described by [3].

2.8. Total Phenolics, Total Flavonoids, and Vitamin C

Total phenolics and total flavonoids contents were measured from freeze-dried kiwifruit samples, according to the methodology implemented previously in our laboratory and described by [18]. Extraction of ascorbic acid was performed according to [19] with some modification. Kiwifruit powder sample (1 g) was extracted by 10 mL 3% (w/v) meta-phosphoric acid. The sonicated sample was centrifuged (12,578$\times$ g for 10 min), the liquid layer of extracts was membrane-filtered (0.22 µm) (Advantec, Tokyo, Japan), and analyzed as described by [20]. Meta-phosphoric acid (0.1%) was used as the mobile phase.

2.9. Statistical Analysis

The results of the collected quality parameters were analyzed using SPSS 20.0 and are expressed as mean $\pm$ SE. The data were subjected to analysis of variance (ANOVA) to determine the significance of differences between cultivars and treatments ($p < 0.05$). Tests for significance between cultivars and treatments were done using a t-test.

3. Results and Discussion

3.1. Assembly and Annotation

The sequencing and mapping results are summarized in parts A and B in Table 1. A total of 72.14 and 64.09 million reads were generated from the green 'Hayward' kiwifruit in the control and ethylene libraries, respectively. A total of 51.44 and 52.72 million reads were generated from gold 'Haegeum' kiwifruit in the control and ethylene libraries, respectively. From the total reads, 65.60, 59.06, 48.45, and 49.89 million were mapped with a high mapping rate (>84%) to the reference genome (*A. Chinensis* 'Hong yang' v3 Genome) from the green control, green ethylene, gold control, and gold ethylene, respectively.

Table 1. Summary of raw and trimmed data (A) and mapping rate (B) and differentially expressed genes (DEGs) during the comparison of control vs. ethylene-treated green 'Hayward' and gold 'Haegeum' kiwifruit cultivars (C).

(A)

Sample	Raw Data		Trimmed Data		
	Total reads	Total read bases (bp)	Total reads	Total read bases (bp)	%
Green_Con	85,610,346	13,520,193,606	85,610,346	12,683,777,637	93.8%
Green_Eth	75,655,556	11,974,790,146	75,655,556	11,152,550,066	93.1%
Gold_Con	61,430,214	6,204,451,614	58,940,256	5,834,133,639	94.0%
Gold_Eth	61,777,836	6,239,561,436	59,494,040	5,873,712,144	94.1%

Table 1. *Cont.*

(B)				
Sample	**Progressed read**	**Mapped reads**	**Properly paired reads**	**Mapping rate**
Green_Con	85,610,346	72,135,718	65,600,622	84.3%
Green_Eth	75,665,556	64,094,382	59,061,836	84.7%
Gold_Con	58,940,256	51,437,355	48,447,084	87.3%
Gold_Eth	59,494,040	52,721,368	49,888,316	88.6%

(C)				
	Total transcripts		**$p < 0.05$, $\lvert \log_2$ fold change$\rvert \geq 1$**	
		UP	**DOWN**	**TOTAL**
Green Con vs. Eth	26,130	594	693	1287
Gold Con vs. Eth	28,605	906	818	1724

Figure 1 shows the percentage and number of genes in different gene ontology (GO) classifications of the DEGs of control vs. ethylene in green 'Hayward' and gold 'Haegeum' kiwifruit. The identified unigenes were classified into three functional categories. The genes in the cellular component were mainly categorized as "membrane" in both cultivars treated with ethylene. Moreover, a higher number of genes in cellular components were mainly involved in "binding" and "catalytic activity" in both cultivars treated with ethylene. The genes in the biological process were enriched "cellular process" and "metabolic process" types in both cultivars treated with ethylene.

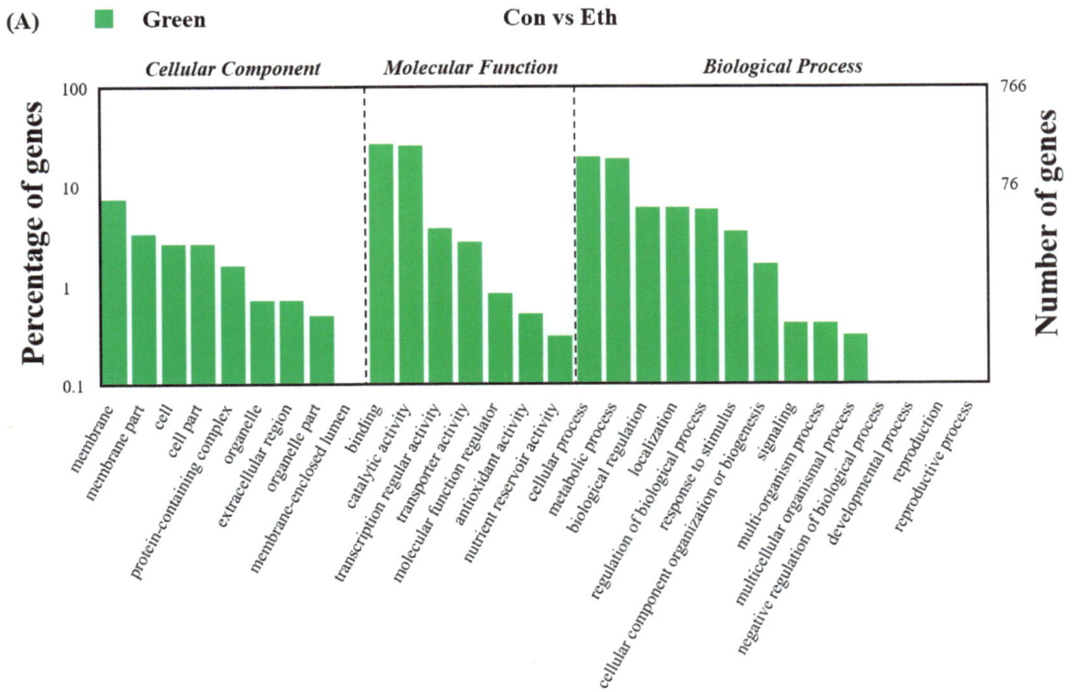

Figure 1. *Cont.*

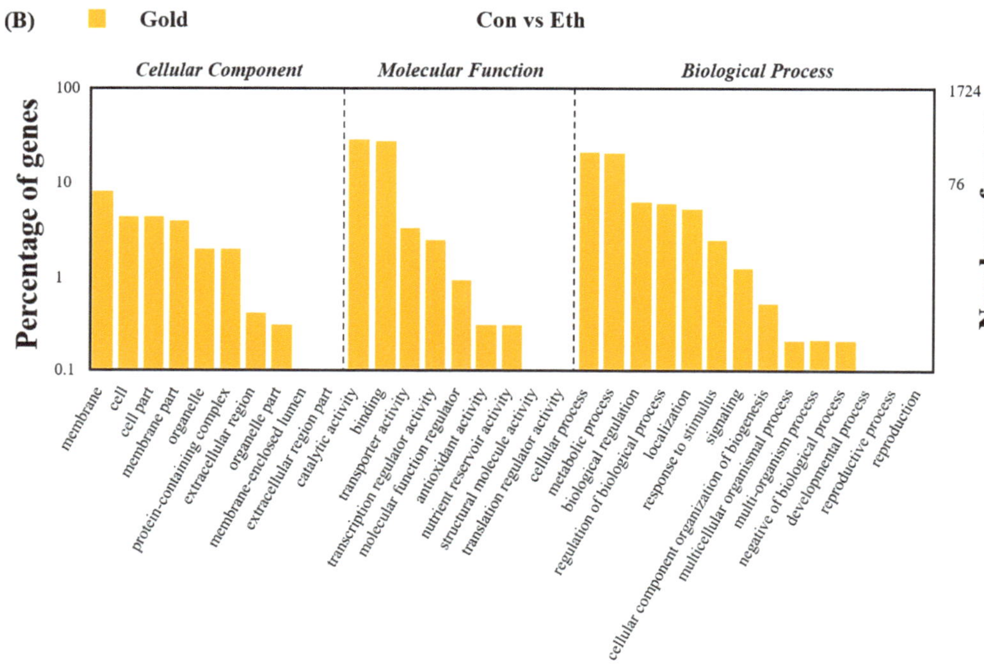

Figure 1. GO classification of the DEGs of control vs. ethylene in (**A**) green 'Hayward' and (**B**) gold 'Haegeum' kiwifruit.

We identified a total of 26,130 and 28,605 transcripts during the comparison of control vs. ethylene-treated green 'Hayward' and gold 'Haegeum' kiwifruit, respectively (part C in Table 1). Differentially expressed genes (DEGs) were compared based on $\log_2$ fold change and $p < 0.05$ during comparison of control to ethylene. A total of 1287 and 1724 unigenes were differentially expressed during the comparison of control vs. ethylene in green 'Hayward' and gold 'Haegeum' kiwifruit, respectively.

The number of DEGs in the control vs. ethylene treatment groups was compared (Figure 2). Higher DEG numbers were shown in the gold 'Haegeum' kiwifruit than in green 'Hayward' kiwifruit, as indicated by the heat map, MA plot, and volcano plot.

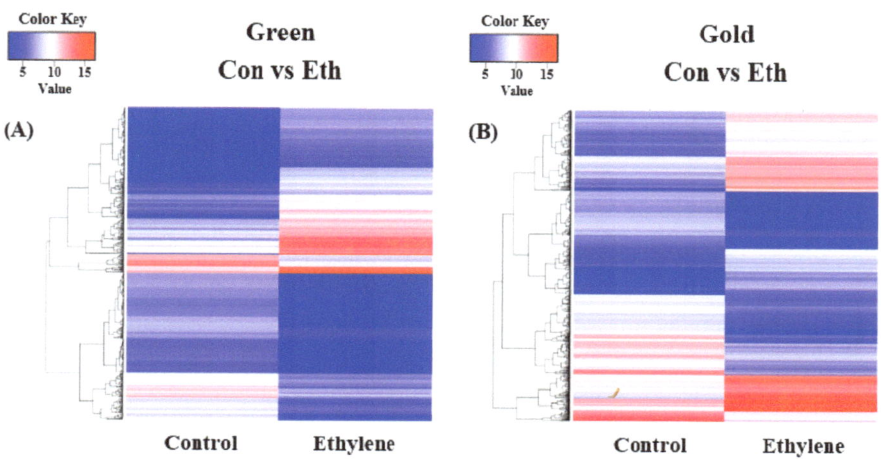

Figure 2. *Cont.*

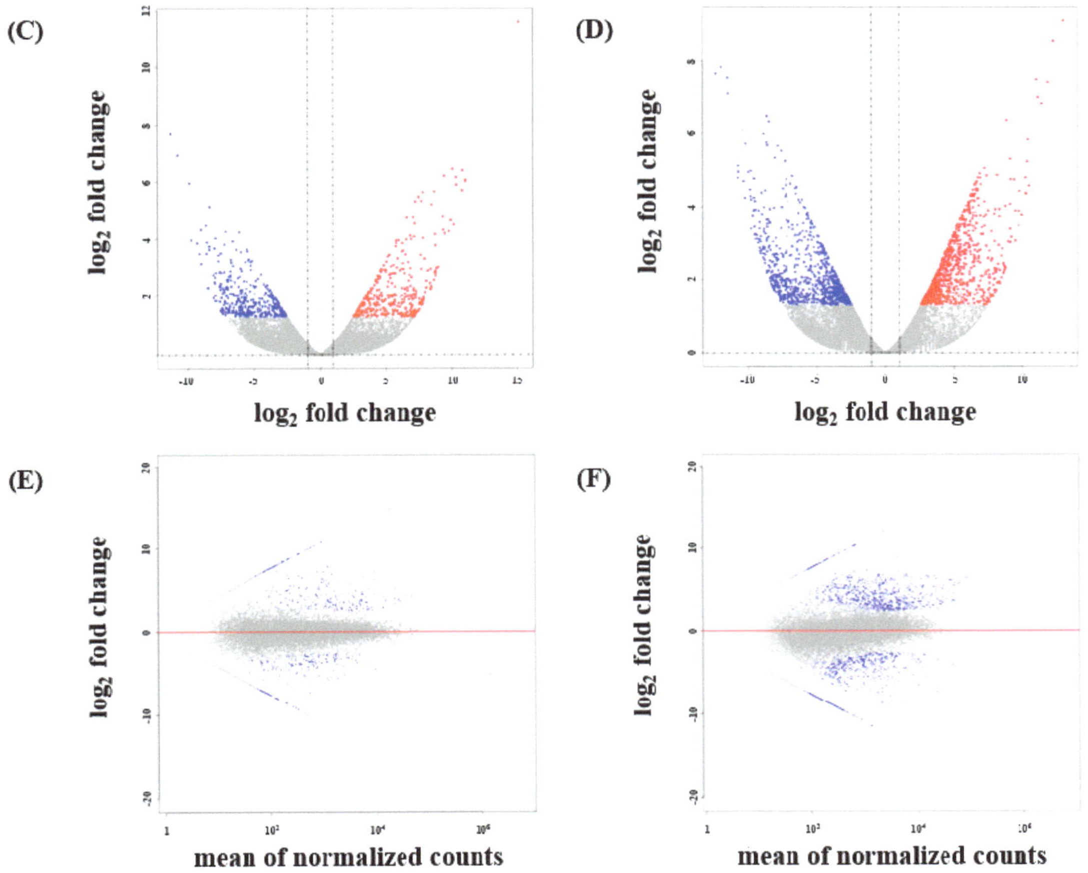

Figure 2. The transcriptional profiles of control vs. ethylene in green 'Hayward' and gold 'Haegeum' kiwifruit. (**A**,**B**) Heat map visualization. Each gene is represented by a single row. Red indicates relatively high levels, genes with relatively low levels are shown in blue. (**C**,**D**) Volcano plot of the identified genes. The DEGs are shown in red and blue, black indicates genes that were not differentially expressed. (**E**,**F**) MA plot of the identified genes. The DEGs are shown in gray ($p > 0.05$) and blue ($p < 0.05$).

Go enrichment analysis was applied to investigate the major biological processes affected by ethylene treatment (Figure 3A,B). We identified 20 Go terms that were represented (p-adj < 0.01) in response to ethylene treatment. The lowest p values for activated biological processes were for "organic acid metabolic process", "carboxylic acid metabolic process", "carboxylic acid metabolic process", "acyltransferase activity", "acyltransferase activity transferring groups other than amino-acyl groups", "phenylpropaniod biosynthetic process", "fatty acid synthase activity", and "oligosaccharide metabolic process" in green 'Hayward' kiwifruit; and "heme binding", "cellular developmental process", "oxidoreductase activity, acting on peroxide as acceptor", "antioxidant activity", and "peroxidase activity" in gold 'Haegeum' kiwifruit. On the other hand, "envelope", "organelle envelope", "plastid envelope", "chloroplast envelope", "thylakoid", "thylakoid membrane", "photosynthetic membrane", "plastid membrane", "chloroplast thylakoid", and "plastid thylakoid" were suppressed in green 'Hayward' kiwifruit; "regulation of developmental process", "meristem development", "regulation of cellular component organization", "regulation of developmental growth", "regulation of post-embryonic development", "regulation of chromosome organization", "thylakoid", "plastid thylakoid", and "chloroplast thylakoid" were suppressed in gold 'Haegeum' kiwifruit. KEGG pathway enrichment analysis of DEGs identified significantly enriched "metabolic pathways" and "biosynthesis of secondary metabolites" in green 'Hayward', and it identified "metabolic pathways" and "biosynthesis of amino acids" in gold 'Haegeum' kiwifruit (Figure 3C,D).

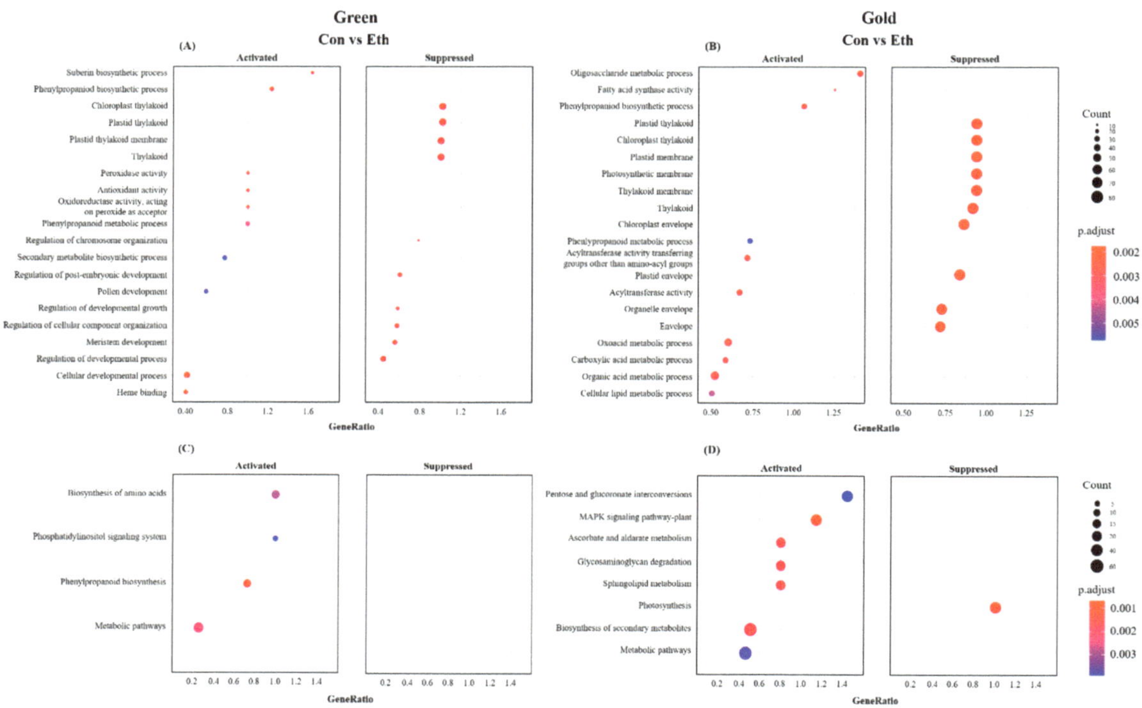

Figure 3. (**A**,**C**) Go enrichment analysis of the differentially expressed genes in green 'Hayward' and gold 'Haegeum' kiwifruit cultivars. The horizontal axis represents the gene ratio corresponding to the pathway, and the vertical axis represents the pathway name. *p*-adjust values are represented by the color of the points. The gene count in each pathway is indicated by size of point and (**B**,**D**) KEGG pathway enrichment scatter analysis of the differentially expressed genes. The horizontal axis represents the gene ratio corresponding to the pathway, and the vertical axis represents the pathway name. *p*-adjust values are represented by the color of the points. The gene count in each pathway is indicated by size of point.

The Venn diagram shows 397 commonly expressed genes and 890 and 1327 exclusively expressed genes during the comparison of control vs. ethylene in green 'Hayward' and gold 'Haegeum' kiwifruit (Figure 4). From the differentially expressed unigenes, 594 (46%) and 906 (53%) were upregulated, and 693 (54%) and 818 (47%) were downregulated in green 'Hayward' and gold 'Haegeum' kiwifruit, respectively.

Figure 4. Venn diagram showing the number of differentially expressed genes ($\log_2$ fold change equal or greater than 1 and $p < 0.05$) total (**A**), upregulated (**B**), and downregulated (**C**), between green 'Hayward' and gold 'Haegeum' kiwifruit cultivars after 3 days of ripening with ethylene treatment or without ethylene treatment (control) at 25 °C.

Figure 5 shows commonly and exclusively expressed genes of identified DEGs related to softening- and ripening-related changes in the comparison of control vs. ethylene green 'Hayward' and gold 'Haegeum' kiwifruit by heat map. To identify genes related to softening and ripening, a list of commonly and exclusively expressed genes was aligned (Tables 2–4). In addition, some of the identified genes were briefly discussed in comparison with ripening-related parameters.

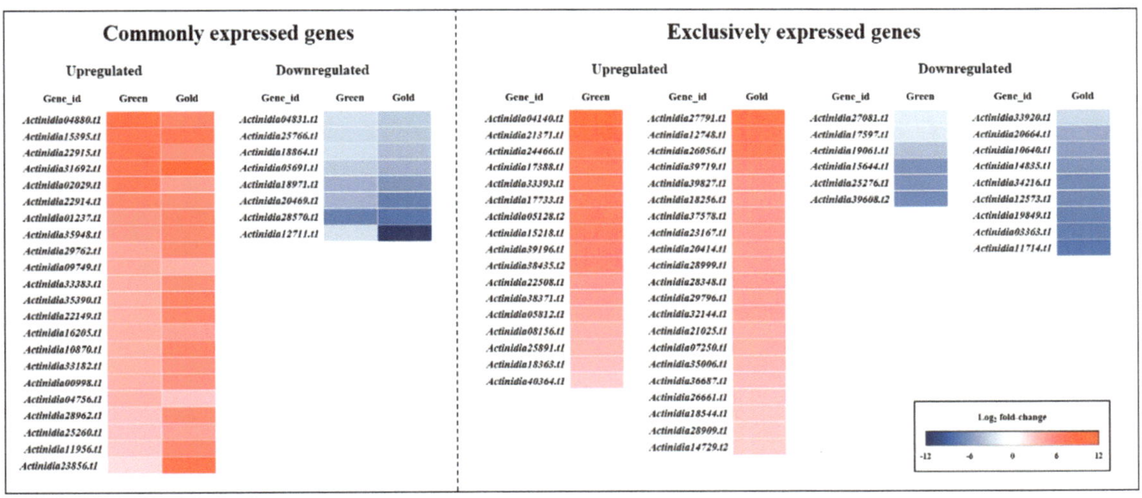

Figure 5. Heat map of commonly and exclusively expressed and identified DEGs related to softening- and ripening-related changes in the comparison of control vs. ethylene-treated green 'Hayward' and gold 'Haegeum' kiwifruit on the third day of ripening.

Table 2. List of DEGs common to green 'Hayward' and gold 'Haegeum' kiwifruit in the comparison of ethylene treated vs. control.

Gene_id	Gene Descriptions	Log_2 Fold Change		*p* Value	
		Green	Gold	Green	Gold
Upregulated					
Actinidia04880.t1	*expansin-A1*	8.41	6.65	0.000	0.024
Actinidia15395.t1	*aldehyde oxidase GLOX-like*	7.59	7.52	0.003	0.016
Actinidia22915.t1	*chitinase 1-like*	7.50	5.58	0.000	0.000
Actinidia31692.t1	*pathogenesis-related protein 4*	7.44	8.77	0.005	0.000
Actinidia02029.t1	*thaumatin-like protein*	7.24	4.74	0.000	0.002
Actinidia22914.t1	*basic endochitinase*	6.30	5.83	0.000	0.002
Actinidia01237.t1	*methanol O-anthraniloyltransferase-like*	5.23	6.43	0.000	0.002
Actinidia35948.t1	*probable pectate lyase 8*	5.02	6.13	0.000	0.002
Actinidia29762.t1	*expansin-A11*	4.33	6.18	0.000	0.002
Actinidia09749.t1	*1-aminocyclopropane-1-carboxylate oxidase 3*	4.31	3.93	0.000	0.011
Actinidia33383.t1	*laccase-15-like*	3.98	5.91	0.039	0.020
Actinidia35390.t1	*pectinesterase QRT1-like*	3.97	6.68	0.000	0.000
Actinidia22149.t1	*ethylene-responsive transcription factor ERF061-like*	3.95	6.41	0.000	0.000
Actinidia16205.t1	*probable pectate lyase 18*	3.89	4.61	0.008	0.022
Actinidia10870.t1	*beta-amylase 3, chloroplastic*	3.89	6.42	0.000	0.003
Actinidia33182.t1	*ethylene-responsive transcription factor ERF062*	3.83	5.51	0.014	0.005
Actinidia00998.t1	*expansin-A8-like*	3.78	5.50	0.000	0.002
Actinidia04756.t1	*cellulose synthase A catalytic subunit 2*	3.71	2.80	0.000	0.034
Actinidia28962.t1	*polygalacturonase-like*	2.84	6.10	0.027	0.006
Actinidia25260.t1	*1-aminocyclopropane-1-carboxylate oxidase 1*	2.84	4.52	0.029	0.004
Actinidia11956.t1	*1-aminocyclopropane-1-carboxylate synthase 3*	2.65	6.17	0.009	0.000
Actinidia23856.t1	*GDSL esterase/lipase*	1.93	8.06	0.045	0.000
Downregulated					
Actinidia04831.t1	*flavonoid 3′-monooxygenase-like isoform X2*	−2.66	−3.54	0.035	0.013
Actinidia25766.t1	*lipoxygenase 6, chloroplastic*	−2.69	−3.63	0.016	0.012
Actinidia18864.t1	*putative UDP-glucose flavonoid 3-O-glucosyltransferase 3*	−2.46	−3.90	0.034	0.004
Actinidia05691.t1	*probable linoleate 9S-lipoxygenase 5 isoform X2*	−3.29	−4.35	0.001	0.002
Actinidia18971.t1	*LOW QUALITY PROTEIN: flavonol sulfotransferase-like*	−4.29	−5.46	0.003	0.000
Actinidia20469.t1	*endoglucanase 24*	−4.05	−6.74	0.022	0.022
Actinidia28570.t1	*probable WRKY transcription factor 70*	−6.74	−7.44	0.023	0.031
Actinidia12711.t1	*transcription factor MYB41*	−2.42	−11.42	0.024	0.000

Table 3. List of DEGs exclusive to green 'Hayward' kiwifruit in the comparison of ethylene treated vs. control.

Gene_id	Gene Descriptions	Log_2 Fold Change	*p* Value
Upregulated			
Actinidia04140.t1	*xyloglucan endotransglucosylase/hydrolase protein 22-like*	8.68	0.000
Actinidia21371.t1	*probable pectinesterase 53*	8.16	0.001
Actinidia24466.t1	*flavonoid 3′,5′-hydroxylase 2*	7.40	0.006
Actinidia17388.t1	*endochitinase-like*	6.97	0.011
Actinidia33393.t1	*endochitinase EP3*	6.96	0.000
Actinidia17733.t1	*pectin acetylesterase 12*	6.77	0.021
Actinidia05128.t2	*thioredoxin-like protein CXXS1*	6.64	0.009
Actinidia15218.t1	*1-aminocyclopropane-1-carboxylate oxidase 5*	6.52	0.000
Actinidia39196.t1	*glucan endo-1,3-beta-glucosidase, acidic-like*	6.22	0.038
Actinidia38435.t2	*probable S-adenosylmethionine-dependent methyltransferase At5g38100*	6.03	0.032
Actinidia22508.t1	*beta-galactosidase BG1-like precursor*	4.69	0.034
Actinidia38371.t1	*alpha-xylosidase 1-like*	4.64	0.009
Actinidia05812.t1	*probable xyloglucan endotransglucosylase/hydrolase protein 30*	4.22	0.000

Table 3. *Cont.*

Gene_id	Gene Descriptions	Log2 Fold Change	*p* Value
Actinidia08156.t1	*polygalacturonase At1g48100-like isoform X1*	3.65	0.030
Actinidia25891.t1	*ethylene-responsive transcription factor 2-like*	3.42	0.006
Actinidia18363.t1	*pathogenesis-related protein PR-4-like*	3.18	0.002
Actinidia40364.t1	*glucan endo-1,3-beta-glucosidase, basic vacuolar isoform-like*	2.73	0.028
Downregulated			
Actinidia27081.t1	*electron transfer flavoprotein-ubiquinone oxidoreductase, mitochondrial isoform X1*	−2.03	0.036
Actinidia17597.t1	*protein COBRA-like*	−2.20	0.023
Actinidia19061.t1	*flavonol synthase/flavanone 3-hydroxylase-like*	−3.78	0.016
Actinidia15644.t1	*sucrose synthase-like*	−5.89	0.034
Actinidia25276.t1	*probable caffeoyl-CoA O-methyltransferase At4g26220 isoform X3*	−5.95	0.015
Actinidia39608.t2	*probable pectinesterase/pectinesterase inhibitor 12*	−6.39	0.000

Table 4. List of DEGs exclusive to gold 'Haeguem' kiwifruit in the comparison of ethylene treated vs. control.

Gene_id	Gene Descriptions	Log2 Fold Change	*p* Value
Upregulated			
Actinidia27791.t1	*polygalacturonase At1g48100*	7.99	0.019
Actinidia12748.t1	*thioredoxin-like protein CITRX1, chloroplastic*	7.29	0.041
Actinidia26056.t1	*beta-galactosidase 17-like*	7.08	0.021
Actinidia39719.t1	*beta-galactosidase-like isoform X1*	5.64	0.000
Actinidia39827.t1	*ethylene-responsive transcription factor ERF118*	5.05	0.010
Actinidia18256.t1	*shikimate O-hydroxycinnamoyltransferase-like*	4.81	0.001
Actinidia37578.t1	*probable pectate lyase 8 isoform X1*	4.77	0.007
Actinidia23167.t1	*expansin-A4-like*	4.68	0.004
Actinidia20414.t1	*ethylene-responsive transcription factor TINY-like*	4.63	0.032
Actinidia28999.t1	*probable glucan endo-1,3-beta-glucosidase A6*	4.56	0.019
Actinidia28348.t1	*1-aminocyclopropane-1-carboxylate synthase-like*	4.55	0.016
Actinidia29796.t1	*1-aminocyclopropane-1-carboxylate oxidase 1 isoform X1*	4.44	0.005
Actinidia32144.t1	*beta-galactosidase-like*	4.39	0.002
Actinidia21025.t1	*ethylene-responsive transcription factor ERF073*	3.79	0.006
Actinidia07250.t1	*glucan endo-1,3-beta-glucosidase 14*	3.78	0.010
Actinidia35006.t1	*thioredoxin-like 1–2, chloroplastic*	3.47	0.010
Actinidia36687.t1	*probable ribose-5-phosphate isomerase 2*	3.27	0.019
Actinidia26661.t1	*S-adenosylmethionine synthase 1*	3.05	0.022
Actinidia18544.t1	*probable pectate lyase 18 isoform X1*	2.91	0.041
Actinidia28909.t1	*S-adenosylmethionine synthase 3*	2.82	0.035
Actinidia14729.t2	*pectin acetylesterase 8-like*	2.73	0.042
Downregulated			
Actinidia33920.t1	*leucoanthocyanidin dioxygenase-like*	−3.07	0.044
Actinidia20664.t1	*cellulose synthase A catalytic subunit 3*	−4.71	0.001
Actinidia10640.t1	*cellulose synthase A catalytic subunit 2*	−5.30	0.017
Actinidia14835.t1	*phosphoenolpyruvate carboxylase 4 isoform X2*	−5.75	0.000
Actinidia34216.t1	*probable pectinesterase/pectinesterase inhibitor 34*	−6.32	0.000
Actinidia12573.t1	*phosphoenolpyruvate carboxylase, housekeeping isozyme*	−6.32	0.001
Actinidia19849.t1	*glutathione S-transferase U17*	−6.79	0.028
Actinidia03363.t1	*cinnamoyl-CoA reductase-like SNL6*	−7.03	0.031
Actinidia11714.t1	*endoglucanase 25-like*	−7.75	0.019

3.2. Firmness, Total Pectin, EIS, PG Activity, and Related Genes

Firmness and firmness-related parameters are the most important indicators that can determine the quality of fruit [3]. Firmness, total pectin, and EIS were shown at 0 d and on the third day of ripening at 25 °C with or without ethylene treatment (Figure 6A). Firmness, total pectin, and EIS tended to decrease as the ripening period proceeded, and in particular, significant differences were observed in the ethylene treatment group on the third day. Tilahun et al. [3] reported that 5–10 N firmness values for kiwifruit cultivars can fulfill customers' preferences. In this study, eating quality (5–10 N) after ethylene treatment was attained in green 'Hayward' and gold 'Haegeum' kiwifruit on the third day, irrespective of cultivars. Concurrently, a significant reduction in EIS and total pectin and a significant increase in PG activity were confirmed in ethylene-treated fruit of both cultivars on the third day. The solubilization of pectin can explain the decrease in total pectin and EIS due to the increase in PG activity as the ripening proceeded in kiwifruit cultivars [17].

The solubilization of pectin polysaccharides in fruit is achieved by the interactive activities of several enzymes such as polygalacturonase (PG), pectate lyase (PL), pectinesterase (PE), pectin acetylesterase (PAE) and beta-galactosidase (b-gal) [21,22]. In this study, the genes that encode *probable pectate lyase 8*, *probable pectate lyase 18*, and *polygalacturonase-like* were commonly upregulated in both cultivars (Table 2). In addition, the genes that encode *probable pectinesterase 53*, *pectin acetylesterase 12*, *beta-galactosidase 13-like*, *beta-galactosidase BG1-like precursor*, and *polygalacturonase At1g48100-like isoform X1* were exclusively upregulated in green 'Hayward' kiwifruit (Table 3). In contrast, the genes that encode *pectin acetylesterase 8*, *beta-galactosidase-like*, *beta-galactosidase 17-like*, *beta-galactosidase-like isoform X1*, *probable pectate lyase 8 isoform X1*, *probable pectate lyase 18 isoform X1*, and *polygalacturonase At1g48100* were exclusively upregulated in gold 'Haegeum' kiwifruit (Table 3). Consistent with this study, Tilahun et al. [15] reported that the genes that encode *polygalacturonase*, *pectate lyase*, *pectin acetlyesterase*, and *beta-galacturonase* were upregulated due to ethylene treatment in gold 'Haegeum' kiwifruit (*Actinidia chinensis*). In this study, the gene encoding *pectinesterase QRT1-like*, involved in the progress of pectin degradation, was commonly upregulated in both cultivars. Consistent with this study, Guo et al. [23] reported that the gene expression of pectinesterase QRT was cultivar-dependent and much higher in 'Fengzao' than 'Kyoho' grape cultivars. In addition, Zhu et al. [13] reported that protein inhibitors can synchronize the action of pectinesterase, and the gene encoding *pectinesterase inhibitor* was downregulated in two watermelon cultivars during ripening. In this study, the genes that encode *probable pectinesterase/pectinesterase inhibitors 12* and *probable pectinesterase/pectinesterase inhibitor 34* were exclusively downregulated in green 'Hayward' and gold 'Haegeum' kiwifruit, respectively. *Alpha-xylosidase 1-like*, which plays an essential role in cell wall modification [24], was exclusively upregulated in green 'Hayward' kiwifruit. Endoglucanase is an enzyme that loosens the network of xyloglucan-cellulose and hydrolyzes the internal sites of the cellulose chain within the cell wall [25]. In this study, the gene encoding *endoglucanase 24* was commonly downregulated in both cultivars. Furthermore, the gene encoding *endoglucanse 25-like* was exclusively downregulated in gold 'Haegeum' kiwifruit. Xyloglucan endotransglucosylase/hydrolase protein participates in cell wall construction by cleaving and linking the xyloglucan polymer that stabilizes the cellulose-hemicellulose framework [15]. In this study, the genes encoding *xyloglucan endotransglucosylase/hydrolase protein 22-like* and *probable xyloglucan endotransglucosylase/hydrolase protein 30* were exclusively upregulated in green 'Hayward' kiwifruit. Then, the upregulation of *xyloglucan endotransglucosylase/hydrolase protein 22-like* and *probable xyloglucan endotransglucosylase/hydrolase protein 30* genes could lead to the softening and ripening of kiwifruit. Expansins may play essential roles in cell wall disassembly and precede the action of various cell wall hydrolases [26]. In this study, the genes that encode *expansin-A1*, *expansin-A8-like*, and *expansin-A11* were commonly upregulated in both cultivars. The gene encoding *expansin-A4-like* was exclusively upregulated in gold 'Haegeum' kiwifruit. Consistent with this study, Palapol et al. [27] reported that the expression of expansin genes in durian fruit pulp was hastened by ethylene, and ethylene promoted the expression of genes such as DzEXP1 and DzEXP2. The cellulose synthase A catalytic subunit is involved in secondary cell wall formation [28]. In this study, the genes that encode *cellulose synthase A catalytic subunit 2* and *cellulose synthase A catalytic subunit 3* were exclusively downregulated in gold 'Haegeum' kiwifruit. Tilahun et al. [15] reported that the gene encoding *cellulose synthase A catalytic subunit 8-like* was downregulated in gold kiwifruit. Malladi et al. [29] and Cao et al. [30] reported that a gene encoding *COBRA* plays an essential role in the regulation of cell wall architecture in fruit. The gene encoding *protein COBRA-like* was exclusively downregulated in green 'Hayward' kiwifruit, consistent with Cao et al. [30], who reported the expression of the COBRA gene declined rapidly during ripening in tomatoes. The gene that encodes *shikimate hyroxycinnamoyl transferase* is involved in forming lignin, a significant component of stone cells in pear fruit [31]. In this study, the gene encoding *shikimate O-hydroxycinnamoyltransferase-like* was exclusively upregulated in gold 'Haegeum' kiwifruit. The results could be explained by the response of the enzyme to faster ripening after ethylene treatment.

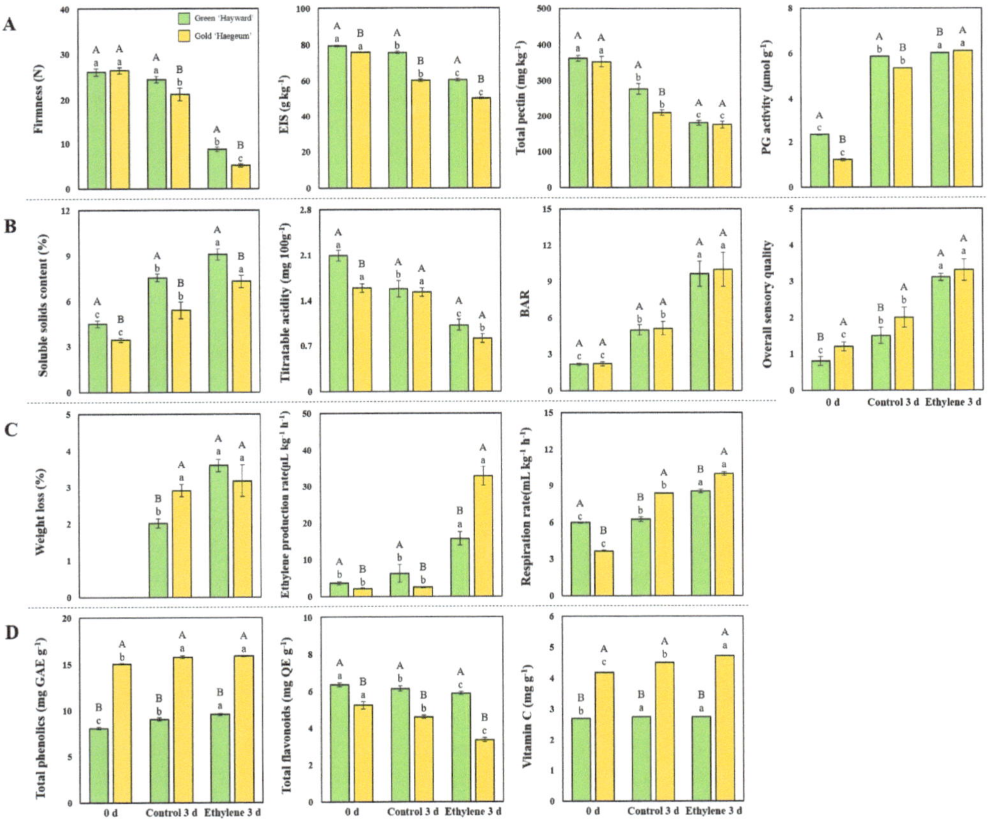

Figure 6. Firmness, EIS, total pectin and PG activity (**A**); soluble solids content (SSC), titratable acidity (TA), Brix/acid ratio (BAR) and overall sensory quality (**B**); weight loss, ethylene production and respiration rates (**C**); total phenolics, total flavonoids and vitamin C (**D**) of green 'Hayward' and gold 'Haegeum' kiwifruit on day 0 and third day of ripening at 25 °C with ethylene treatment or without treatment (control). Data are presented as a mean ± standard errors in 10 replicates for firmness, SSC, TA, BAR and overall sensory quality; and in 5 replicates for the other parameters. The bars with different upper-case letters indicate a significant difference ($p < 0.05$) between cultivars, the bars with different lower-case letters indicate a significant difference ($p < 0.05$) between treatments.

3.3. SSC, TA, BAR and Sensory Evaluation and the Related Genes

In this study, both cultivars showed a tendency to increase SSC, BAR, and sensory evaluation scores during the ripening process compared to immediately after harvest. Significant differences were observed, especially on the third day after exogenous ethylene treatment (Figures 6B and S2). Conversely, in both cultivars, titratable acidity during the ripening process tended to decrease compared to immediately after harvest, and a significant difference was observed, especially on the third day after exogenous ethylene treatment.

Beta-amylase is a very important enzyme in the starch degradation process because it produces maltose by cleaving the starch chain. As a result, it affects the sweetness of ripe fruit [15,32]. In this study, the genes encoding *beta-amylase 3* and *chloroplastic* were commonly upregulated in both cultivars. Chen et al. [33] reported that *beta-amylase 3* was upregulated in ethylene-treated fruit in African Pride atemoya, suggesting that ethylene treatment might accelerate starch degradation. Chen et al. [34] suggested that sucrose synthase is the key enzyme catalyzing the process of sucrose degradation of kiwifruit. In this study, the gene encoding *sucrose synthase-like* was exclusively downregulated in green 'Hayward' kiwifruit. *Phosphoenolpyruvate carboxylase* is the critical enzyme in organic acid biosynthesis in plants, and might be linked to the conversion of organic acids into sugars [15,35]. In this study, the genes encoding *phosphoenolpyruvate carboxylase 4 isoform X2* and *phosphoenolpyruvate carboxylase, housekeeping isozyme* were exclusively downregulated after ethylene treatment due to the reduction of acidity in gold 'Haegeum' kiwifruit. In general, *lipoxygenase* is known to be related to fruit ripening quality characteristics such as aroma development in kiwifruit [36]. In this study, the genes encoding *lipoxygenase 6, chloroplastic* and *probable linoleate 9S-lipoxygenase 5 isoform X2* were commonly downregulated in both cultivars. Consistent with this study, Zhang et al. [36] reported the gene expression levels of *LOX 2*, *LOX 3*, *LOX 4*, and *LOX 6* showed a tendency to decrease during kiwifruit ripening and may contribute to producing aroma after ethylene treatment. *Methanol O-anthraniloyltransferase* is an enzyme solely responsible for the production of O-methyl anthranilate, a compound with aroma and flavor in the grapefruit [37]. In this study, the gene encoding *methanol O-anthraniloyltransferase-like* was commonly upregulated in both cultivars. This implies that O-methyl anthranilate can directly influence the aroma and flavor of kiwifruit after ethylene treatment. Zhang et al. [38] and Tilahun et al. [15] reported that ribose-5-phosphate isomerase is essential in the glycolysis and TCA cycle of plants. In this study, the gene encoding *probable ribose-5-phosphate isomerase 2* was exclusively upregulated in gold 'Haeguem' kiwifruit. The results imply that the enzyme reaction could lead to faster ripening after ethylene treatment compared to the control.

3.4. Ethylene Production and Respiration Rates and Related Genes

Figure 6C shows that the weight loss, ethylene production, and respiration rates were the highest in ethylene-treated green 'Hayward' and gold 'Haegeum' kiwifruit compared to the values at 0 d and on the third day of the control. Consistent with our results, Tilahun et al. [3] reported a tendency to increase SSC and respiration rate, but a tendency to decrease titratable acidity was observed after exogenous ethylene treatment of kiwifruit. The results may imply that the reduction in TA could be due to the change of organic acid into sugars during ripening.

S-adenosyl methionine (SAM) synthase is an enzyme that converts methionine to SAM in the ethylene pathway [39]. Aminocyclopropane-1-carboxylic acid (ACC) synthase is involved in the conversion of SAM to ACC [40]. ACC oxidase is an enzyme that catalyzes the conversion of ACC to ethylene [41]. In this study, the genes that encode *1-aminocyclopropane-1-carboxylate synthase 3*, *1-aminocyclopropane-1-carboxylate oxidase*, and *1-aminocyclopropane-1-carboxylate oxidase* were commonly upregulated in both cultivars. The genes that encode *1-aminocyclopropane-1-carboxylate oxidase 5*, and *probable S-adenosylmethionine-dependent methyltransferase At5g38100* were exclusively upregulated in green 'Hayward' kiwifruit. The genes that encode *1-aminocyclopropane-1-carboxylate synthase-like, 1-aminocyclopropane-1-carboxylate oxidase 1 isoform X1, S-adenosylmethionine synthase 1,* and *S-adenosylmethionine synthase 3* were also exclusively upregulated in gold 'Haegeum' kiwifruit. This suggests that exogenous ethylene application induces more ethylene biosynthesis and thus may increase respiration rate, biochemical changes, color changes, and softening (Figure 6C). *Ethylene response factors (ERFs),* which play a pivotal role in plant responses to biotic or abiotic stresses, are involved in the ethylene signaling and response pathway [42]. In this study, *ethylene-responsive transcription factor ERF 061-like* and *ethylene-responsive transcription factor ERF 062* were commonly upregulated in both cultivars. The gene encoding *ethylene-responsive transcription factor 2-like* was exclusively upregulated in green 'Hayward' kiwifruit. The genes that encode *ethylene-responsive transcription factor ERF073, ethylene-responsive transcription factor ERF118,* and *ethylene-responsive transcription factor TINY-like* were exclusively upregulated in gold 'Haegeum' kiwifruit. Consistent with this study, Tilahun et al. [10] reported that increased levels in expression of ethylene-related gene families such as *ERF3, ERF10, ERF18, ERF 23,* and *ERF 24, ERS1 ethylene receptor,* and *ethylene-responsive transcription factor RAP2-11,* correlated with ripening and softening in astringent persimmon. Ethylene action is achieved by regulating ethylene receptors and triggering signal transduction reactions, and ultimately by controlling relevant gene expression in the fruit [16]. It has been shown that upregulation of ethylene-related gene families after exogenous ethylene treatment has led to higher respiration rates and faster ripening.

3.5. Stress-Related Genes Due to Ethylene Treatment

Secondary metabolites are involved in plant defense and responses against biotic and abiotic stresses [43]. These compounds, such as polyphenols and terpenoids, are also responsible for the fruits' organoleptic, color, and nutritional characteristics [1]. In this study, total phenolics, flavonoids, and vitamin C were significantly different after exogenous ethylene treatment in green 'Hayward' and gold 'Haegeum', respectively (Figure 6D). Generally, total phenolics and vitamin C contents were higher in gold 'Haegeum' than in green 'Hayward' kiwifruit. Conversely, total flavonoid content was higher in green 'Hayward' than in gold 'Haegeum' kiwifruit.

The gene that encodes *Aldehyde oxidase GLOX-like,* which might play a defense mechanism role in attacking pathogens in grapevines, was commonly upregulated in both cultivars [44]. The gene encoding *chitinase 1-like,* the enzyme that degrades fungal cell wall components, resulting in inhibited fungal growth [45], was commonly upregulated in both cultivars. *Endochitinase,* one of the pathogenesis-related proteins, plays a vital role in plant defense mechanisms against fungal pathogens [46]. In this study, the gene encoding *basic endochitinase* was commonly upregulated in both cultivars. Moreover, the genes encoding *endochitinase-like* and *endochitinase EP3* were exclusively upregulated in green 'Hayward' kiwifruit. Glucan endo-1,3-beta-glucosidase may play an essential role in degrading fungal cell wall polysaccharides. In this study, the genes encoding *glucan endo-1,3-beta-glucosidase, acidic-like,* and *glucan endo-1,3-beta-glucosidase, basic vacuolar isoform-like* were exclusively upregulated in green 'Hayward', and *probable glucan endo-1,3-beta-glucosidase A6* and *glucan endo-1,3-beta-glucosidase 14* were exclusively upregulated in gold 'Haegeum' kiwifruit. Consistent with this study, Wurms et al. [47] reported that the expression levels of the gene encoding *glucan endo-1,3-beta-glucosidase* increased in postharvest ripe rot disease in gold 'Hort 16A' kiwifruit. Pathogenesis-related protein 4 has powerful antifungal activity against plant pathogens such as *Trichoderma harzianum, Fusarium culmorum, F. graminearum,* and *B. cinerea* [46]. In this study, the gene encoding *pathogenesis-related protein 4* was commonly upregulated in both cultivars. The gene encoding *pathogenesis-related protein PR-4-like* was exclusively upregulated in gold 'Haegeum' kiwifruit. Bai et al. [48] reported that pathogenesis-related protein 4 is involved in defense responses against *B.dothidea* in *Malus domestica* apple. *Thaumatin-like protein (TLP)* is one of the protein families and plays a vital role in plant defense mechanisms against various biotic and abiotic stresses [49]. In this study, the gene encoding *thaumatin-like protein* was commonly upregulated in both cultivars. Choi et al. [42] reported that *thaumatin-like protein 1* protects tissues from pathogen infection in peach. The GDSL esterase/lipase protein, which plays an important role in plant defense and growth, is a multifunctional hydrolase and has many functions in secondary metabolism, abiotic stress, morphogenesis, seed development, and pathogen defense [50,51]. In this study, the gene encoding *GDSL esterase/lipase* was commonly upregulated in both cultivars, implying that the exogenous application of ethylene could stimulate the defense mechanism due to stresses. The gene encoding *cinnamoyl-CoA reductase-like SNL6*, an enzyme that resists pathogenic infection, was also exclusively downregulated in gold 'Haegeum' treated with ethylene [52]. Tilahun et al. [15] reported that Cinnamoyl-CoA reductase could stimulate lignin biosynthesis in kiwifruit due to stresses such as exogenous ethylene treatment and low temperature. And Giordano et al. [53] reported that caffeoyl-CoA O-methyltransferase could be responsible for anthocyanin methylation activity under drought stress conditions in grape berries. In this study, the gene encoding *caffeoyl-CoA O-methyltransferase* was commonly downregulated in both cultivars, and the gene encoding *probable caffeoyl-CoA O-methyltransferase At4g26220 isoform X3* was exclusively downregulated in green 'Hayward' kiwifruit. The *transcription factor MYB1R1* enhances drought tolerance by regulating water loss in potatoes [54]. In this study, the gene encoding *transcription factor MYB41* was commonly downregulated in both cultivars. This implies that MYB1R1 functions as a transcription factor involved in stress-related genes. In addition, Chen et al. [55] reported that WRKY transcription factors play vital roles in regulating stress responses in plants. In this study, the genes encoding *probable WRKY transcription factor 70* were commonly downregulated in both cultivars. This implies that the downregulation of the genes encoding the above enzymes could be the response of the kiwifruit to stress caused by the exogenous ethylene treatment. We identified the genes encoding the enzymes involved in the flavonoid biosynthesis pathway, such as *flavonol synthase/flavanone 3-hydroxylase-like, flavonoid 3′-monooxygenase, flavonoid 3-O-glucosyltransferase, flavonol sulfotransferase, flavonoid 3′,5′-hydroxylase, flavanone 3-hydroxylase, flavoprotein-ubiquinone oxidoreductase, putative UDP-glucose flavonoid 3-O-glucosyltransferase 3, leucoanthocyanidin dioxygenase,* and *glutathione-S-transferase,* in the transcriptome data [56]. In this study, the genes encoding *flavonoid 3′-monooxygenase-like isoform X2, putative UDP-glucose flavonoid 3-O-glucosyltransferase 3*, and *flavonol sulfotransferase-like* were commonly downregulated in both cultivars. Moreover, the gene encoding *flavonoid 3′5′-hydroxylase 2* was exclusively upregulated, and the genes encoding *electron transfer flavoprotein-ubiquinone oxidoreductase, mitochondrial isoform X1* and *flavonol synthase/flavanone 3-hydroxylase-like* were exclusively downregulated in green 'Hayward' kiwifruit. In addition, the gene encoding *glutathione S-transferase 3-like* was exclusively upregulated in gold 'Haegeum' kiwifruit, and the genes encoding *leucoanthocyanidin dioxygenase-like* and *glutathione S-transferase U17* were exclusively downregulated in gold 'Haegeum' kiwifruit. Glutathione-S-transferase is essential in transporting flavonoids synthesized in the cytosol to vacuoles and other locations in plants [56]. Kobayashi et al. [57] reported that UDP-glucose flavonoid 3-O-glucosyltransferase plays an essential role in anthocyanin biosynthesis in the grape berry. The present study shows that the genes related to the flavonoid biosynthesis pathway interacted with the stress resistance mechanism after exogenous ethylene treatment. *Laccase* is a well-known redox enzyme that oxidizes various phenols and aroma substances based on oxidation-reduction reactions [58]. In this study, the gene encoding *laccase-15-like* was commonly upregulated in both cultivars. Yihui et al. [59] reported that laccase-related genes such as *LAC7* and *LAC9* in 'Red Delicious' and *LAC7, LAC9, LAC14, LAC15,* and *LAC16* in 'Cortland' were upregulated after 4 and 7 months of storage in apples. Thioredoxin plays an important role in plants' tolerance of oxidative stress [60]. The genes encoding *thioredoxin-like protein CXXS1* in green 'Hayward', and those for *thioredoxin-like protein CITRX1, chloroplastic* and *thioredoxin-like 1–2, chloroplastic* in gold 'Haegeum' kiwifruit were exclusively upregulated. Consistent with this study, [60] reported that expression of *thioredoxin* such as *MaTrx6* and *MaTrx12* was upregulated in banana fruit after ethylene treatment.

3.6. Verification of DEGs by qRT-PCR

To validate the RNA-Seq results, we selected 12 representative genes related to softening and ripening-related changes from the differentially expressed genes (Figure 7). Transcript accumulation of the selected representative genes (EXPA8, EXPA11, QRT1, ACO1, ACO3, ACS3, ERF061, ERF062, TLP1, LOX1.5, LOX6, CYP75B1) from the differentially expressed genes was evaluated by quantitative qRT-PCR [42]. The gene expression results by qRT-PCR were consistent with those obtained from RNA-seq for the 12 observed genes (Figure 7, Table 2). Based on these results, the coincided expression patterns of these representative genes in the RNA-seq assay and qRT-PCR suggest the reliability of the RNA-seq data.

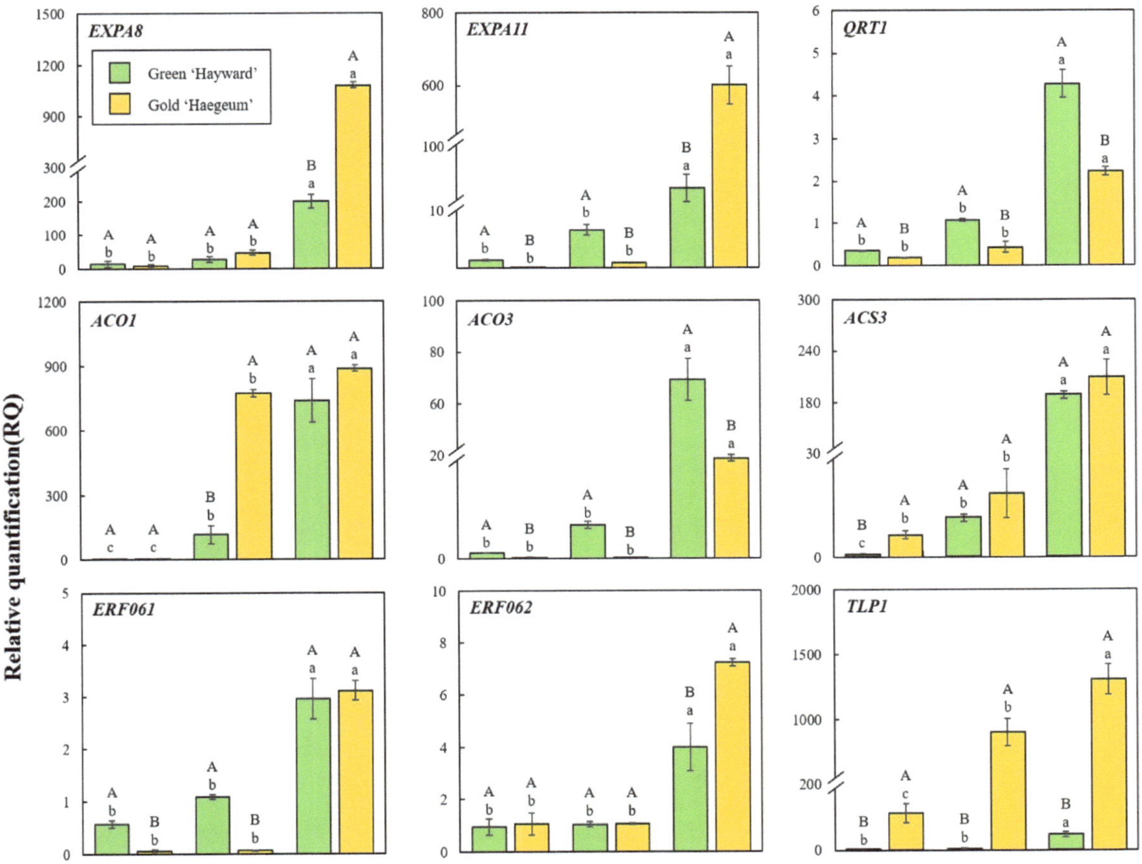

Figure 7. *Cont.*

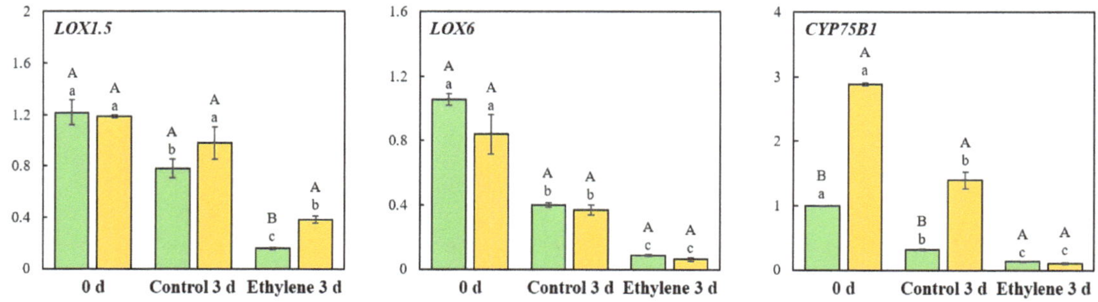

Figure 7. qRT-PCR transcript accumulation of the selected DEGs of green 'Hayward' and gold 'Haegeum' kiwifruit for the comparison of fruit on day 0 and third day of ripening at 25 °C with ethylene treatment or without treatment (control). Vertical bars represent standard errors of the means ($n = 3$). The bars with different upper-case letters indicate a significant difference ($p < 0.05$) between cultivars, whereas the bars with different lower-case letters indicate a significant difference ($p < 0.05$) between treatments. The names of genes are indicated: *EXPA8 (expansin-A8-like), EXPA11 (expansin-A11), QRT1 (pectinesterase QRT1), ACO1 (1-aminocyclopropane-1-carboxylate oxidase 1), ACO3 (1-aminocyclopropane-1-carboxylate oxidase 3), ACS3 (1-aminocyclopropane-1-carboxylate synthase 3), ERF061 (ethylene-responsive transcription factor ERF061), ERF062 (ethylene-responsive transcription factor ERF062), TLP1 (thaumatin-like protein), LOX1.5 (probable linoleate 9S-lipoxygenase 5 isoform X2), LOX6 (lipoxygenase 6, chloroplastic)*, and *CYP75B1 (flavonoid 3′-monooxygenase-like isoform X2).*

4. Conclusions

This study reported the comparative transcriptome analysis of kiwifruit cultivars after treatment with ethylene for three days at 25 °C. We compared ethylene-treated fruit with the control, to study the candidate genes related to softening and other changes during ripening. Several genes were expressed commonly and exclusively in the green and gold kiwifruit cultivars treated with ethylene. The findings also showed that various genes were expressed differentially during the ripening of kiwifruits with exogenous ethylene application, and the cumulative effect brought softening- and ripening-related changes. In addition, this work identified and categorized genes related to softening and other changes during ripening. The data obtained from the present study will add to the information available on the molecular mechanisms of the effects of ethylene during kiwifruit ripening. This study will provide resources for further study of the genes related to ripening, for kiwifruit breeding and postharvest technologists to improve ripening quality.

Supplementary Materials: The following supporting information can be downloaded at: https://www.mdpi.com/article/10.3390/cimb44060177/s1, Figure S1: Flow chart of mRNA processing; Table S1: Selected genes and primers used for validation of gene expression during qRT-PCR; Figure S2: Green 'Hayward' and Gold 'Haegeum' kiwifruit cultivars at 0 d and after 3 days of ripening with ethylene treatment or without ethylene treatment (control) at 25 °C.

Author Contributions: Conceptualization and methodology, H.R.C.; software and formal analysis, M.W.B. and H.R.C.; resources investigation, M.W.B. and H.R.C.; resources, C.S.J.; data curation, H.R.C. and S.T.; writing—original draft preparation, H.R.C.; writing—review and editing, S.T.; visualization, H.R.C.; supervision, C.S.J. and S.T.; project administration, C.S.J.; funding acquisition, C.S.J. All authors have read and agreed to the published version of the manuscript.

Funding: This research was funded by the Korea Institute of Planning and Evaluation for Technology in Food, Agriculture, Forestry and Fisheries (Project no. 120030-1) and the National Research Foundation (NRF), Korea, under BK21 FOUR project.

Institutional Review Board Statement: Not applicable.

Informed Consent Statement: Not applicable.

Data Availability Statement: All data sets are available upon reasonable request from the corresponding author.

Conflicts of Interest: The authors declare no conflict of interest.

References

1. Choi, H.R.; Baek, M.W.; Cheol, L.H.; Jeong, C.S.; Tilahun, S. Changes in metabolites and antioxidant activities of green 'Hayward'and gold 'Haegeum' kiwifruits during ripening with ethylene treatment. *Food Chem.* **2022**, *384*, 132490. [CrossRef] [PubMed]
2. Choi, H.R.; Baek, M.W.; Tilahun, S.; Jeong, C.S. Long-term cold storage affects metabolites, antioxidant activities, and ripening and stress-related genes of kiwifruit cultivars. *Postharvest Biol. Technol.* **2022**, *189*, 111912. [CrossRef]
3. Tilahun, S.; Choi, H.R.; Lee, Y.M.; Choi, J.H.; Baek, M.W.; Hyok, K.; Park, S.M.; Jeong, C.S. Ripening quality of kiwifruit cultivars is affected by harvest time. *Sci. Hortic.* **2020**, *261*, 108936. [CrossRef]
4. Lin, X.; Yang, R.; Dou, Y.; Zhang, W.; Du, H.; Zhu, L.; Chen, J. Transcriptome analysis reveals delaying of the ripening and cell-wall degradation of kiwifruit by hydrogen sulfide. *J. Sci. Food Agric.* **2020**, *100*, 2280–2287. [CrossRef]
5. Hunter, D.C.; Skinner, M.A.; Ferguson, A.R. Kiwifruit and health. In *Fruits, Vegetables, and Herbs*; Academic Press: Cambridge, MA, USA, 2016; pp. 239–269. [CrossRef]
6. Meena, N.K.; Baghel, M.; Jain, S.K.; Asrey, R. Postharvest biology and technology of kiwifruit. In *Postharvest Biology and Technology of Temperate Fruits*; Springer: Cham, Switzerland, 2018; pp. 299–329. [CrossRef]
7. Chiaramonti, N.; Barboni, T. Relationship between the physicochemical parameters and the ethylene emission during cold storage of kiwifruits. *Int. J. Food Sci. Technol.* **2010**, *45*, 1513–1516. [CrossRef]
8. Schroder, R.; Atkinson, R.G. Kiwifruit cell walls: Towards an understanding of softening? *N. Z. J. For. Sci.* **2006**, *36*, 112.
9. Lim, S.; Lee, J.G.; Lee, E.J. Comparison of fruit quality and GC–MS-based metabolite profiling of kiwifruit 'Jecy green': Natural and exogenous ethylene-induced ripening. *Food Chem.* **2017**, *234*, 81–92. [CrossRef]
10. Tilahun, S.; Park, K.C.; Choi, I.Y.; Jeong, C.S. Transcriptome analysis of astringent 'Cheongdo-Bansi'persimmon fruit treated with ethylene for removal of astringency. *Postharvest Biol. Technol.* **2019**, *150*, 52–59. [CrossRef]
11. Asif, M.H.; Lakhwani, D.; Pathak, S.; Gupta, P.; Bag, S.K.; Nath, P.; Trivedi, P.K. Transcriptome analysis of ripe and unripe fruit tissue of banana identifies major metabolic networks involved in fruit ripening process. *BMC Plant Biol.* **2014**, *14*, 316. [CrossRef]
12. Yu, K.; Xu, Q.; Da, X.; Guo, F.; Ding, Y.; Deng, X. Transcriptome changes during fruit development and ripening of sweet orange (*Citrus sinensis*). *BMC Genom.* **2012**, *13*, 10. [CrossRef]
13. Zhu, Q.; Gao, P.; Liu, S.; Zhu, Z.; Amanullah, S.; Davis, A.R.; Luan, F. Comparative transcriptome analysis of two contrasting watermelon genotypes during fruit development and ripening. *BMC Genom.* **2017**, *18*, 3. [CrossRef] [PubMed]
14. Wang, Q.H.; Zhao, C.; Zhang, M.; Li, Y.Z.; Shen, Y.Y.; Guo, J.X. Transcriptome analysis around the onset of strawberry fruit ripening uncovers an important role of oxidative phosphorylation in ripening. *Sci. Rep.* **2017**, *7*, 41477. [CrossRef]
15. Tilahun, S.; Choi, H.R.; Kwon, H.; Park, S.M.; Park, D.S.; Jeong, C.S. Transcriptome analysis of 'Haegeum' gold kiwifruit following ethylene treatment to improve postharvest ripening quality. *Agronomy* **2020**, *10*, 487. [CrossRef]
16. Tilahun, S.; Heo, J.Y.; Jeong, C.S. Quality and expression of ethylene response genes of 'Daebong' persimmon fruit during ripening at different temperatures. *Postharvest Biol. Technol.* **2017**, *133*, 57–63. [CrossRef]
17. Choi, H.R.; Tilahun, S.; Lee, Y.M.; Choi, J.H.; Baek, M.W.; Jeong, C.S. Harvest time affects quality and storability of kiwifruit (*Actinidia* spp.): Cultivars during long-term cool storage. *Sci. Hortic.* **2019**, *256*, 108523. [CrossRef]
18. Baek, M.W.; Choi, H.R.; Solomon, T.; Jeong, C.S.; Lee, O.H.; Tilahun, S. Preharvest methyl jasmonate treatment increased the antioxidant activity and glucosinolate contents of hydroponically grown pak choi. *Antioxidants* **2021**, *10*, 131. [CrossRef]
19. Krupa, T.; Latocha, P.; Liwińska, A. Changes of physicochemical quality, phenolics and vitamin C content in hardy kiwifruit (*Actinidia arguta* and its hybrid) during storage. *Sci. Hortic.* **2011**, *130*, 410–417. [CrossRef]
20. Tilahun, S.; Choi, H.R.; Baek, M.W.; Cheol, L.H.; Kwak, K.W.; Park, D.S.; Solomon, T.; Jeong, C.S. Antioxidant Properties, γ-Aminobutyric Acid (GABA) Content, and Physicochemical Characteristics of Tomato Cultivars. *Agronomy* **2021**, *11*, 1204. [CrossRef]
21. Fabi, J.P.; Broetto, S.G.; Silva, S.L.G.L.D.; Zhong, S.; Lajolo, F.M.; do Nascimento, J.R.O. Analysis of papaya cell wall-related genes during fruit ripening indicates a central role of polygalacturonases during pulp softening. *PLoS ONE* **2014**, *9*, e105685. [CrossRef]
22. Xiao, L.; Li, T.; Jiang, G.; Jiang, Y.; Duan, X. Cell wall proteome analysis of banana fruit softening using iTRAQ technology. *J. Proteom.* **2019**, *209*, 103506. [CrossRef]
23. Guo, D.L.; Xi, F.F.; Yu, Y.H.; Zhang, X.Y.; Zhang, G.H.; Zhong, G.Y. Comparative RNA-Seq profiling of berry development between table grape 'Kyoho'and its early-ripening mutant 'Fengzao'. *BMC Genom.* **2016**, *17*, 795. [CrossRef] [PubMed]
24. Wu, T.; Cao, J. Differential gene expression of tropical pumpkin (*Cucurbita moschata* Duchesne) bush mutant during internode development. *Sci. Hortic.* **2008**, *117*, 219–224. [CrossRef]
25. Jara, K.; Castro, R.I.; Ramos, P.; Parra-Palma, C.; Valenzuela-Riffo, F.; Morales-Quintana, L. Molecular insights into FaEG1, a strawberry endoglucanase enzyme expressed during strawberry fruit ripening. *Plants* **2019**, *8*, 140. [CrossRef] [PubMed]
26. Sane, V.A.; Chourasia, A.; Nath, P. Softening in mango (*Mangifera indica* cv. Dashehari) is correlated with the expression of an early ethylene responsive, ripening related expansin gene, MiExpA1. *Postharvest Biol. Technol.* **2005**, *38*, 223–230. [CrossRef]

27. Palapol, Y.; Kunyamee, S.; Thongkhum, M.; Ketsa, S.; Ferguson, I.B.; van Doorn, W.G. Expression of expansin genes in the pulp and the dehiscence zone of ripening durian (*Durio zibethinus*) fruit. *J. Plant Physiol.* **2015**, *182*, 33–39. [CrossRef]
28. Kotake, T.; Aohara, T.; Hirano, K.; Sato, A.; Kaneko, Y.; Tsumuraya, Y.; Takatsuji, H.; Kawasaki, S. Rice *Brittle culm* 6 encodes a dominant-negative form of CesA protein that perturbs cellulose synthesis in secondary cell walls. *J. Exp. Bot.* **2011**, *62*, 2053–2062 [CrossRef]
29. Malladi, A. Molecular physiology of fruit growth in apple. *Hortic. Rev.* **2020**, *47*, 1–42. [CrossRef]
30. Cao, Y.; Tang, X.; Giovannoni, J.; Xiao, F.; Liu, Y. Functional characterization of a tomato COBRA-likegene functioning in fruit development and ripening. *BMC Plant Biol.* **2012**, *12*, 211. [CrossRef]
31. Ma, C.; Zhang, H.; Li, J.; Tao, S.; Qiao, X.; Korban, S.S.; Zhang, S.; Wu, J. Genome-wide analysis and characterization of molecular evolution of the HCT gene family in pear (*Pyrus bretschneideri*). *Plant Syst. Evol.* **2017**, *303*, 71–90. [CrossRef]
32. Setiabudi, E.; Meitha, K.; Dwivany, F.M. In silico characterization and comparison of the fruit ripening related beta-amylase (BAM) gene family in banana genome A and B. *Indones. J. Biotechnol.* **2021**, *26*, 175–182. [CrossRef]
33. Chen, J.; Duan, Y.; Hu, Y.; Li, W.; Sun, D.; Hu, H.; Xie, J. Transcriptome analysis of atemoya pericarp elucidates the role of polysaccharide metabolism in fruit ripening and cracking after harvest. *BMC Plant Biol.* **2019**, *19*, 219. [CrossRef] [PubMed]
34. Chen, C.; Yuan, Y.; Zhang, C.; Li, H.; Ma, F.; Li, M. Sucrose phloem unloading follows an apoplastic pathway with high sucrose synthase in Actinidia fruit. *Plant Sci.* **2017**, *255*, 40–50. [CrossRef] [PubMed]
35. Moing, A.; Rothan, C.; Svanella, L.; Just, D.; Diakou, P.; Raymond, P.; Gaudillère, J.P.; Monet, R. Role of phosphoenolpyruvate carboxylase in organic acid accumulation during peach fruit development. *Physiol. Plant.* **2000**, *108*, 1–10. [CrossRef]
36. Zhang, B.; Yin, X.R.; Li, X.; Yang, S.L.; Ferguson, I.B.; Chen, K.S. Lipoxygenase gene expression in ripening kiwifruit in relation to ethylene and aroma production. *J. Agric. Food Chem.* **2009**, *57*, 2875–2881. [CrossRef] [PubMed]
37. Wang, J.; Luca, V.D. The biosynthesis and regulation of biosynthesis of Concord grape fruit esters, including 'foxy' methylanthranilate. *Plant J.* **2005**, *44*, 606–619. [CrossRef]
38. Zhang, R.G.; Andersson, C.E.; Savchenko, A.; Skarina, T.; Evdokimova, E.; Beasley, S.; Arrowsmith, C.H.; Edwards, A.M.; Joachimiak, A.; Mowbray, S.L. Structure of *Escherichia coli* ribose-5-phosphate isomerase: A ubiquitous enzyme of the pentose phosphate pathway and the Calvin cycle. *Structure* **2003**, *11*, 31–42. [CrossRef]
39. Pattyn, J.; Vaughan-Hirsch, J.; Van de Poel, B. The regulation of ethylene biosynthesis: A complex multilevel control circuitry. *New Phytol.* **2021**, *229*, 770–782. [CrossRef]
40. Kende, H. Enzymes of ethylene biosynthesis. *Plant Physiol.* **1989**, *91*, 1–4. [CrossRef]
41. John, P. How plant molecular biologists revealed a surprising relationship between two enzymes, which took an enzyme out of a membrane where it was not located, and put it into the soluble phase where it could be studied. *Plant Mol. Biol. Report.* **1991**, *9*, 192–194. [CrossRef]
42. Choi, H.R.; Jeong, M.J.; Baek, M.W.; Choi, J.H.; Lee, H.C.; Jeong, C.S.; Tilahun, S. Transcriptome Analysis of Pre-Storage 1-MCP and High CO_2-Treated 'Madoka' Peach Fruit Explains the Reduction in Chilling Injury and Improvement of Storage Period by Delaying Ripening. *Int. J. Mol. Sci.* **2021**, *22*, 4437. [CrossRef]
43. Leitzmann, C. Characteristics and health benefits of phytochemicals. *Complementary Med. Res.* **2016**, *23*, 69–74. [CrossRef] [PubMed]
44. Balestrini, R.; Ghignone, S.; Quiroga, G.; Fiorilli, V.; Romano, I.; Gambino, G. Long-term impact of chemical and alternative fungicides applied to Grapevine *cv* Nebbiolo on Berry Transcriptome. *Int. J. Mol. Sci.* **2020**, *21*, 6067. [CrossRef] [PubMed]
45. Kitajima, S.; Sato, F. Plant pathogenesis-related proteins: Molecular mechanisms of gene expression and protein function. *J. Biochem.* **1999**, *125*, 1–8. [CrossRef] [PubMed]
46. Selitrennikoff, C.P. Antifungal proteins. *Appl. Environ. Microbiol.* **2001**, *67*, 2883–2894. [CrossRef]
47. Wurms, K.; Cui, W.; Ah-Chee, A.; Rees-George, J.; Bublin, M.; Breiteneder, H. Down Regulation of Putative Defence-associated Transcripts Correlates with Ripe Rot Symptoms on Kiwifruit (*Actinidia chinensis*). *J. Phytopathol.* **2011**, *159*, 435–442. [CrossRef]
48. Bai, S.; Dong, C.; Li, B.; Dai, H. A PR-4 gene identified from *Malus domestica* is involved in the defense responses against *Botryosphaeria dothidea*. *Plant Physiol. Biochem.* **2013**, *62*, 23–32. [CrossRef]
49. Cao, J.; Lv, Y.; Hou, Z.; Li, X.; Ding, L. Expansion and evolution of thaumatin-like protein (TLP) gene family in six plants. *Plant Growth Regul.* **2016**, *79*, 299–307. [CrossRef]
50. Li, H.; Han, X.; Qiu, W.; Xu, D.; Wang, Y.; Yu, M.; Wang, Y.; Zhuo, R. Identification and expression analysis of the GDSL esterase/lipase family genes, and the characterization of SaGLIP8 in *Sedum alfredii* Hance under cadmium stress. *PeerJ* **2019**, *7*, e6741. [CrossRef]
51. Ni, P.Y.; Ji, X.R.; Guo, D.L. Genome-wide identification, characterization, and expression analysis of GDSL-type esterases/lipases gene family in relation to grape berry ripening. *Sci. Hortic.* **2020**, *264*, 109162. [CrossRef]
52. Singh, A.; Singh, S.; Singh, R.; Kumar, S.; Singh, S.K.; Singh, I.K. Dynamics of Zea mays transcriptome in response to a polyphagous herbivore, *Spodoptera litura*. *Funct. Integr. Genom.* **2021**, *21*, 571–592. [CrossRef]
53. Giordano, D.; Provenzano, S.; Ferrandino, A.; Vitali, M.; Pagliarani, C.; Roman, F.; Cardinale, F.; Castellarin, S.D.; Schubert, A. Characterization of a multifunctional caffeoyl-CoA O-methyltransferase activated in grape berries upon drought stress. *Plant Physiol. Biochem.* **2016**, *101*, 23–32. [CrossRef] [PubMed]

54. Shin, D.; Moon, S.J.; Han, S.; Kim, B.G.; Park, S.R.; Lee, S.K.; Yoon, H.J.; Lee, H.E.; Kwon, H.B.; Baek, D.W.; et al. Expression of StMYB1R-1, a novel potato single MYB-like domain transcription factor, increases drought tolerance. *Plant Physiol.* **2011**, *155*, 421–432. [CrossRef] [PubMed]
55. Chen, L.; Song, Y.; Li, S.; Zhang, L.; Zou, C.; Yu, D. The role of WRKY transcription factors in plant abiotic stresses. *Biochim. Biophys. Acta (BBA)-Gene Regul. Mech.* **2012**, *1819*, 120–128. [CrossRef] [PubMed]
56. Liang, W.; Ni, L.; Carballar-Lejarazú, R.; Zou, X.; Sun, W.; Wu, L.; Yuan, X.; Mao, Y.; Huang, W.; Zou, S. Comparative transcriptome among *Euscaphis konishii* Hayata tissues and analysis of genes involved in flavonoid biosynthesis and accumulation. *BMC Genom.* **2019**, *20*, 24. [CrossRef] [PubMed]
57. Kobayashi, S.; Ishimaru, M.; Ding, C.K.; Yakushiji, H.; Goto, N. Comparison of UDP-glucose: Flavonoid 3-O-glucosyltransferase (UFGT) gene sequences between white grapes (*Vitis vinifera*) and their sports with red skin. *Plant Sci.* **2001**, *160*, 543–550. [CrossRef]
58. Oh, T.G.; Jo, J.A.; Lee, S.J. Evaluation of time–temperature integrator for indicating the ripeness of kiwifruit in plastic container at home. *J. Food Sci.* **2021**, *86*, 2872–2885. [CrossRef]
59. Yihui, G.; Song, J.; Du, L.; Vinqvist, M.; Palmer, L.C.; Fillmore, S.; Pang, X.; Zhang, Z. Characterization of laccase from apple fruit during postharvest storage and its response to diphenylamine and 1-methylcyclopropene treatments. *Food Chem.* **2018**, *253*, 314–321. [CrossRef]
60. Wu, F.; Li, Q.; Yan, H.; Zhang, D.; Jiang, G.; Jiang, Y.; Duan, X. Characteristics of three thioredoxin genes and their role in chilling tolerance of harvested banana fruit. *Int. J. Mol. Sci.* **2016**, *17*, 1526. [CrossRef]

current issues in molecular biology

MDPI

Communication

Hidden Pitfalls of Using Onion Pollen in Molecular Research

Majd Mardini [1,†], Aleksey Ermolaev [1,2,†] and Ludmila Khrustaleva [1,3,4,*]

1. Center of Molecular Biotechnology, Russian State Agrarian University—Moscow Timiryazev Agricultural Academy (RSAU-MTAA), 49, Timiryazevskaya Str., 127550 Moscow, Russia
2. Laboratory of Applied Genomics and Crop Breeding, All-Russia Research Institute of Agricultural Biotechnology, Timiryazevskaya Str. 42, 127550 Moscow, Russia
3. Department of Botany, Breeding and Seed Production of Garden Plants, Russian State Agrarian University—Moscow Timiryazev Agricultural Academy (RSAU-MTAA), 49, Timiryazevskaya Str., 127550 Moscow, Russia
4. Plant Cell Engineering Laboratory, All-Russian Research Institute of Agricultural Biotechnology, Timiryazevskay 42 Str., 127550 Moscow, Russia

* Correspondence: ludmila.khrustaleva19@gmail.com or khrustaleva@rgau-msha.ru

† These authors contributed equally to this work.

Abstract: There is little information on the use of pollen in molecular research, despite the increased interest in genome editing by pollen-mediated transformation. This paper presents an essential toolbox of technical procedures and observations for molecular studies on onion (*Allium cepa* L.) pollen. PCR is a useful tool as an express method to evaluate editing results before pollination. A direct PCR protocol for pollen suspension has been adapted without needing DNA pre-extraction. We showed that the outer layer of lipids known as pollenkitt is a limiting factor for successful PCR on pollen. A simple pre-washing step of pollen suspension was able to eliminate the pollenkitt and enormously affect the PCR results. Additionally, our pollenkitt study helped us develop a simple and effective pollination method using wetted onion pollen grains. Classical manual pollination usually is conducted by intact pollen without wetting. Most existing methods of the editing system delivery into pollen are carried out in a wet medium with consequent drying before pollination, which adversely affects the viability of pollen. The optimal medium for wet pollination was 12% sucrose water solution. Our method of using wetted pollen grains for pollination might be very beneficial for pollen genetic manipulation.

Keywords: onion; PCR inhibition; pollination; pollenkitt; wetted pollen

Citation: Mardini, M.; Ermolaev, A.; Khrustaleva, L. Hidden Pitfalls of Using Onion Pollen in Molecular Research. *Curr. Issues Mol. Biol.* **2023**, *45*, 1065–1072. https://doi.org/10.3390/cimb45020070

Academic Editor: Vijai Bhadauria

Received: 10 January 2023
Revised: 24 January 2023
Accepted: 26 January 2023
Published: 29 January 2023

1. Introduction

For many years, pollen has been an appealing subject in a wide spectrum of studies, such as paleontology [1–3], forensics [4], vaccine development [5,6], allergology [7] and many others. Furthermore, pollen research is also considered a major player in the field of plant breeding and crop improvement. The fact that a pollen grain consists of 2–3 haploid cells gives a unique chance to conduct fundamental studies like DNA genotyping and population genetics [8–14]. Another promising line of research is the use of pollen as a natural carrier of exogenous DNA for genome editing [15–23]. Although many articles have been published on the study of pollen, there is scarce information on the use of pollen in molecular research.

PCR is the most used instrument in molecular biology for the analysis of DNA variations. In the case of genetic manipulation of pollen, PCR might be used as an express method to evaluate results before pollination. Since the genetic material inside pollen grains is well protected by the rigid wall, DNA extraction is considered a critical step for successful PCR. This step is usually conducted either by mechanical crushing or enzymatic lysis or a combination of the two [10,12,18]. However, some observations suggest that DNA might be passively released from pollen grains due to the high temperature during PCR amplification [24], or in some certain species, pollen grains simply burst upon contact with liquid [25]. Another issue is the possible inhibition of PCR due to the natural presence of certain substances in pollen grains. A pre-washing step before PCR is obviously required in many protocols [3]; however, several studies claim that constituents of pollen grains have no effect on PCR activity [12]. Nevertheless, PCR on pollen might be a species-specific procedure. In addition, there is a marked lack of research on commercially important crops, highlighting the need to adapt protocols for new species.

Pollen-mediated genetic manipulation is a revolutionary approach in crop improvement. This promising technique gives the ability to readily obtain transformed plants in the form of seeds, which means bypassing the laborious steps of tissue culture and regeneration. Since the mid-1970s, scientists have been suggesting a wide range of methods to solve this problem, including free uptake of DNA [16], particle bombardment [19,22], electroporation of pollen tubes [17,18], pollen magnetofection [21], and uptake of nanoassemblies [23]. Regardless of the many distinctive variations among these methods, all of them require a *wet medium* to facilitate the main procedure. However, in most available protocols, little attention has been paid to the impact of wet mediums on pollen viability and subsequent pollination.

This paper presents an essential toolbox of technical procedures and observations for molecular studies on onion (*Allium cepa* L.) pollen. A direct PCR protocol for pollen suspension has been adapted without needing DNA pre-extraction. We showed that a limiting factor for successful PCR on pollen suspension is a material known as pollenkitt. Pollenkitt is characteristic of many entomophilous species but lacking in those which are wind pollinated. Pollenkit was first named by Knoll in 1930 and described as all substances which connect single pollen grains to pollen clumps [26]. Pollenkitt is a hydrophobic mixture, composed mainly of saturated and unsaturated lipids and also carotenoids, flavonoids, proteins, and carbohydrates [27]. The physico-chemical properties of pollenkitt depend on the substances in the mixture and their proportions [28]. Pollenkitt is present in large quantities in exine cavities of pollen [26]. A simple pre-washing step of pollen suspension was able to eliminate the pollenkitt and enormously affect the PCR results.

Additionally, phase-contrast and dark-field microscopy were used to determine the dynamics of pollenkitt release from pollen grains upon contact with a liquid medium. It was clear that pollenkitt plays a major role in increasing the surface area of the pollen grain, hence affecting floating and adhesion behavior. Our investigation on pollenkitt helped us to develop a simple and efficient method for pollination of onion flowers using wetted pollen grains. All the observations and methods in this paper will be useful for future research on gene editing and manipulation of onion pollen.

2. Materials and Methods

2.1. Plant Materials and Pollen Collection

Bulbs of onion (*Allium cepa* L.) cultivar "Myachkovsky 300" were kindly provided by the Federal Scientific Vegetable Center. By the end of April 2022, vernalized onion bulbs were planted within a 40 cm distance in an open field at The Center of Molecular Biotechnology—Russian State Agrarian University, Moscow. Flowering starts by the end of July 2022. Since flowering within one umbel is not synchronized, pollen was collected once or twice a day for 3–4 days. The pollen was used immediately or stored in a small Petri dish at −20 °C for further use. Mature seeds were collected by the end of August 2022.

2.2. PCR on Pollen Suspension

Using a sensitive analytical balance (Ohaus-PA214C), 0.5 mg of pollen was weighed inside a 1.5 mL microtube. A simple washing technique was performed by vortexing in 500 μL of distilled water for 10 s and centrifuged for 5 s using minispin centrifuge (750× *g*). After discarding the supernatant, fresh distilled water was added and the same procedure was repeated for 10 rounds. To ensure that no precipitated pollen grains are lost during washing, only 450 μL of the supernatant was carefully discarded between rounds. As a result, after discarding the last supernatant, 50 μL of a pollen grains suspension with a concentration of 10 μg/μL was obtained.

PCR was performed using 10 μg of mature intact pollen (1 μL of the pre-washed suspension), or 50 ng of *A. cepa* genomic DNA as a positive control. Genomic DNA was isolated from young leaves by a standard CTAB-method [29]. 25 μL of 1× PCR mix contained 2.5 μL of 10× Taq Turbo buffer (25 mM $MgCl_2$, pH = 8.3) (Evrogen, Moscow, Russia), 0.2 mM of each dNTP (Evrogen, Moscow), 0.2 μM both forward and reverse primers and 5 U of Taq-polymerase (Evrogen, Moscow). Primers (F: TCAAGCACTGCAAACCTCTTC; R: ATGGAGCTAAATCCTGGTGCA) were designed for amplification of ORF (Open Reading Frame) of the LFS gene (GenBank: AB089203.1). Touchdown PCR was performed to amplify desirable amplicons without a non-specific PCR product. Cycling was performed as follows: 10 min of initial denaturation at 95 °C, 10 cycles (touchdown stage)—30 s of denaturation at 95 °C, 30 s of annealing (from 67 °C to 62 °C, decrement at 0.5 °C/cycle), 45 s of elongation at 72 °C, 30 cycles (standard PCR)—30 s of denaturation at 95 °C, 30 s of annealing at 62 °C, 45 s of elongation at 72 °C, 10 min of final elongation at 72 °C. PCR product was visualized in agarose gel (2% of agarose in 0.5× TBE buffer, 0.5 μg/mL of ethidium bromide, 3 V/cm).

2.3. Evaluating the Effect of Pollenkitt on PCR

To test our hypothesis of pollenkitt inhibition of PCR, we conducted a series of PCR experiments with different dilutions of pollenkitt, using genomic DNA of *A. cepa* as a template and the LFS gene as an amplification target. PCR mix, cycling conditions, and gel electrophoresis were performed as in the previous section (positive control). Pollenkitt was extracted from pollen grains as the following: in a 1.5 mL tube, 1 mg pollen was thoroughly vortexed with 100 μL ddH_2O for 30 s then centrifuged for 5 s using minispin centrifuge (750× *g*). Since pollen grains are much heavier than the pollenkitt debris, they are precipitated first, and the latter remains in the supernatant. 1 μL of the supernatant was added to the PCR reaction either immediately (concentrated) or as an aqueous dilution (1:10 and 1:100).

2.4. Microscopy of the Onion Pollenkitt

For microscope preparation, a large droplet of distilled water (100 μL) was placed on a glass slide. Using a microspatula or dissecting needle, clumps of pollen (~1 mg) were carefully placed on the droplet. Microscopy was performed using Zeiss SterREO LUMAR.V12 and Zeiss Axiolab 5.

2.5. Wet Manual Pollination of Onion Flowers

Two different liquid mediums were tested in wet pollination experiments: (1) 12% (*w*/*v*) sucrose water solution and (2) pollen growing medium according to Brewbaker and Kwack [30] as following: 0.1 g/L H_3BO_3, 0.3 g/L $Ca(NO_3)_2{\cdot}4H_2O$, 0.3 g/L $MgSO_4{\cdot}7H_2O$, 0.1 g/L KNO_3 dissolved in 12% sucrose. The pollen clump was added to 100 μL of liquid medium placed on a glass slide. The floating cover of pollen was easily transferred to the stigma using a small bacterial inoculation loop.

About 70–75% of the flowers in selected umbels have been removed to make subsequent steps more convenient. Umbels were proven against flying insects using isolation cages made of 4-layered medical gauze. Castration was conducted on a daily basis. Due to the non-synchronous flowering, the umbels were constantly checked to immediately remove the immature anthers from the newly opened flowers. The flowering period for a single umbel (the time between the opening of the first flower until the drying of the last flower) lasted for 8–10 days on average.

3. Results

During experimental studies on the genetic manipulation of onion pollen, we encountered a number of hidden pitfalls due to the lack of a basic methodology for evaluating results and post-experimental application. To address the lack of practical knowledge, we investigated the factors that influence the polymerase chain reaction (PCR) on pollen grains, the viability of pollen, and its further use for pollination and seed setting.

3.1. Direct PCR on Pollen Suspension of Onion

A gene encoding Lachrymatory Factor Synthase (LFS) was used as a target for PCR on pollen. This is a single-locus gene encoding enzyme involved in the synthesis of the volatile sulfur compounds released during wounding [31,32]. In the first attempt of PCR using fresh intact pollen as a template we faced a complete inhibition of amplification. During the early stages of our experiment, we believed that the inaccessibility of pollen DNA could be the reason for unsuccessful PCR. Most of the available protocols for PCR on pollen require DNA extraction [10,12,18,33]. Accordingly, we tried different techniques of DNA extraction through mechanical damage of the pollen wall: vortexing, grinding with a pestle, steel beads, or glass powder. None of these techniques showed any positive PCR results (data not shown). Finally, a simple washing procedure using distilled water was able to restore normal amplification (Figure 1). We concluded that an unknown factor/substance coating the surface of the pollen grain might be responsible for PCR inhibition.

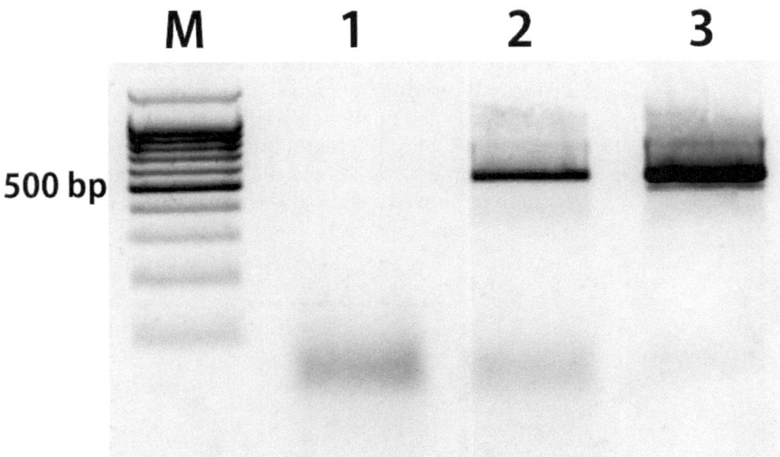

Figure 1. PCR amplification of the gene encoding LFS in onion: M—DNA Ladder 100bp+; lane 1—mature pollen without washing; lane 2—mature pollen after 10-times washing in ddH_2O; lane 3—genomic DNA of *A. cepa*.

In all angiosperms, the common material for coating pollen grains is the sticky tapetal tissue secretion, also known as pollenkitt [34]. In addition to a wide range of functions [26], pollen serves as a binding agent in entomophilous plants [35]. According to Dobson [36], pollenkitt is a hydrophobic mixture of materials composed mainly of saturated and unsaturated lipids. Lipids are known to be major inhibitors of PCR [37].

During the washing procedure of the pollen suspension, we noticed that the discarded supernatant always had a cloudy appearance. Since pollenkitt is a hydrophobic substance, we assumed that the cloudiness is probably due to the insoluble debris of the pollenkitt, which is released into the water during washing (vortexing and centrifugation). This was confirmed using dark field microscopy (Figure 2a,b).

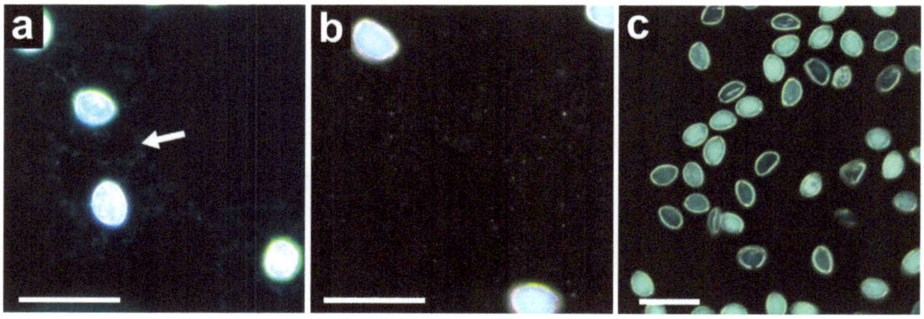

Figure 2. The dark field microscopy of pollen suspension. (**a**) pollen before washing—pollenkitt appears as a bright material around each pollen grain. White arrow pointing to the released pollenkitt from pollen grains; (**b**) supernatant after vortexing and centrifugation—pollenkitt is no longer accumulated around each pollen grain, but rather scattered into fine granular particles (appears here as bright white dots); (**c**) pollen suspension after washing—the pollenkitt is removed. Bars represent 50 μm.

Upon repeated washing of the suspension, pollenkitt residues were removed together with the supernatant until their concentration became negligible for PCR inhibition (Figure 2c). For onion pollen, 10-times washing by vortexing in 500 μL of distilled water for 10 s and centrifuging for 5 s using minispin centrifuge (750× *g*) was optimal.

To verify the relationship between PCR inhibition and pollenkitt, we conducted a series of PCR with different dilutions of pollenkitt added to genomic DNA of *A. cepa* as an amplification template and the LFS gene as a target (see Materials and Methods). Electrophoresis showed a clear correlation between pollenkitt concentration and the intensity of the PCR band (Figure 3).

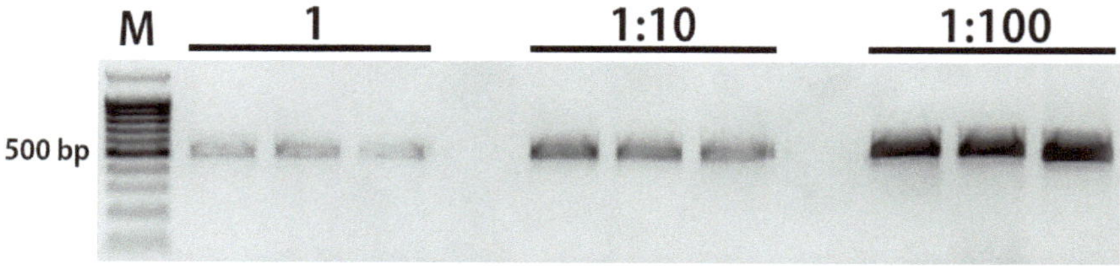

Figure 3. Inhibition of PCR upon addition of aqueous dilutions of pollenkitt to *A. cepa* genomic DNA as an amplification template and the LFS gene as a target. The PCR experiment was repeated three times with each dilution. Dilution 1 represents the equivalent amount of pollenkitt eluated from 10 μg of pollen.

3.2. Hand Pollination with Wetted Pollen Grains

Most existing methods of delivery of the editing systems into pollen are carried out in a wet medium. One of the main objectives of our study was to verify whether it is possible to pollinate onion flowers using wetted pollen grains. Since hydration is an inevitable step in every technique of pollen genetic manipulation, it is crucial to verify such matters. In vitro experiments of pollen tube growing we found that onion pollen grains cannot grow after dehydration. Thus, drying the suspension was excluded. The necessity of the use of wet pollen led us to conduct detailed studies of the dynamics of pollen wetting using microscopy.

Imaging under a stereo microscope showed a rapid movement of the pollen clump immediately upon contact with liquid. This rapid movement was mainly caused by the momentary swelling of the pollen grains (Video S1). Despite the vigorous movement, it was noticeable that the neighboring pollen grains could not come into direct contact with each other, as if they were surrounded by an invisible structure (Video S2). In less than a minute, the rapid movement stopped, and all the pollen grains that were previously in one clump floated steadily on the surface of the droplet. Using phase-contrast microscopy a light-dense structure surrounding the floating pollen grains was observed (Figure 4a). We suggest that this structure is in fact the pollenkitt released from the surface of pollen grains upon contact with water. Within 4–5 min after contact with the liquid surface, adjacent fragments of pollenkitt began to join together, forming an interwoven "blanket" of pollenkitt components (Figure 4b).

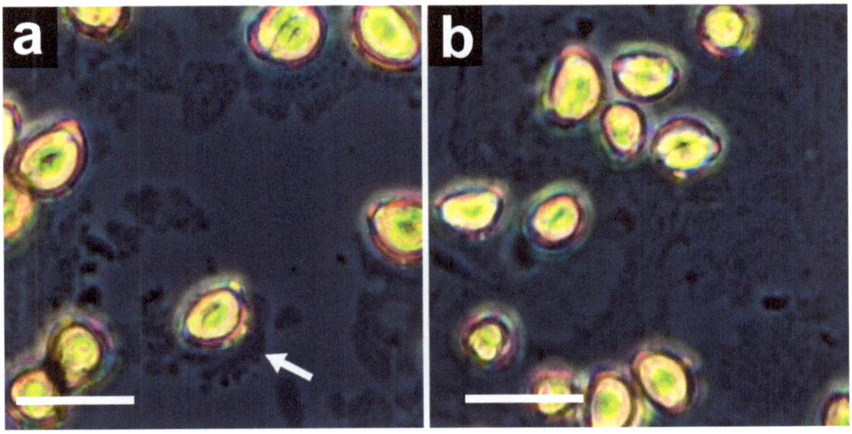

Figure 4. The phase-contrast microscopy of the pollenkitt in *A. cepa*: (**a**) pollen grains immediately after placing on a droplet of distilled water; pollenkitt appears as fragments released from the surface of pollen grains (white arrow). (**b**) the same pollen grains after 5 min; pollenkitt appears as a "blanket" that keeps pollen grains on the surface of the droplet. Bars represent 50 μm.

This "blanket" obviously affects the floating and adhesion properties of pollen grains. Based on this observation the optimal technique of transferring the wet pollen grains onto the stigma was developed. Using an inoculation loop, a lump of the blanket was easily collected from a liquid surface by the hollow center of the loop which enabled us to easily apply an even layer of wetted pollen grains to the stigmatic surface (Video S3).

Two different liquid mediums were tested: (1) 12% (w/v) sucrose water solution and (2) pollen growing medium according to Brewbaker and Kwack [30] as following: 0.1 g/L H_3BO_3, 0.3 g/L $Ca(NO_3)_2{\cdot}4H_2O$, 0.3 g/L $MgSO_4{\cdot}7H_2O$, 0.1 g/L KNO_3 dissolved in 12% sucrose. There were two positive controls: (3) open pollination with insects and (4) hand pollination with fresh intact pollen grains using a brush. A negative control (5) was also used to test castration and inflorescence isolation.

The seed set percentage using pollen growing medium was the lowest—35.7%, while open pollination by insects was the highest—93.5% (Table 1). The percentage of set seeds in the experiment with 12% sucrose water solution was almost the same as in the positive control of intact pollen grains (63.5% and 65.5%, respectively). This result means that 12% sucrose water solution and the method of applying pollen using a loop are the most optimal for further work related to pollen genetic manipulations.

Table 1. The effect of different pollination methods on seed setting in the *A. cepa* umbels.

Treatment	Number of Flowers in Umbel	Number Flowers that Set Seeds, (%)
Wet pollination with pollen growing medium	56	20 (35.7)
Wet pollination with 12% sucrose solution	52	33 (63.5)
Manual pollination with un-wetted pollen grains	61	40 (65.5)
Open pollination by insects	62	58 (93.5)
Negative control	0	0 (0)

In summary, our results demonstrate a feasible technique of using wetted pollen grains to pollinate onion flowers without affecting viability and seed setting. Minimizing the influence of technical procedures is always favorable, especially in experiments such as genetic manipulation. Therefore, engaging our wet pollination technique, and PCR using pollen DNA as a template might be very beneficial in future experiments based on genetic manipulation of onion pollen aimed at producing seeds with an edited genome.

Supplementary Materials: The following supporting information can be downloaded at: https://www.mdpi.com/article/10.3390/cimb45020070/s1, Video S1: The wetting behavior of onion pollen after contact with liquid. The movie was recorded in real-time using Zeiss SterREO LUMAR.V12 (objective: NeoLumer S 0.8x FWD 80mm); Video S2: Pollenkitt release from onion pollen grains after contact with liquid. The movie was recorded in real-time using Zeiss SterREO LUMAR.V12 (objective: NeoLumer S 0.8x FWD 80 mm). The fixed image is the same area shown under phase-contrast microscope Zeiss Axiolab 5; Video S3: The technique we used for manual pollination of onion flower shown.

Author Contributions: M.M.—investigation, validation, visualization, writing—original draft, methodology; A.E.—investigation, validation, writing—original draft, methodology, L.K. — conceptualization, funding acquisition, project administration, supervision, writing—original draft, writing—review & editing. All authors have read and agreed to the published version of the manuscript.

Funding: This research was funded by the Russian Foundation for Basic Research, grant number 20-016-00065.

Institutional Review Board Statement: Not applicable.

Informed Consent Statement: Not applicable.

Data Availability Statement: Data available in a publicly accessible repository.

Acknowledgments: We thank Alexander Voronkov, K.A. Timiryazev Institute of Plant Physiology RAS, for practical advice on pollen collection.

Conflicts of Interest: The authors declare no conflict of interest.

References

1. Paffetti, D.; Vettori, C.; Caramelli, D.; Vernesi, C.; Lari, M.; Paganelli, A.; Paule, L.; Giannini, R. Unexpected presence of Fagus orientalis complex in Italy as inferred from 45,000-year-old DNA pollen samples from Venice lagoon. *BMC Evol. Biol.* **2007**, *7*, S6. [CrossRef] [PubMed]
2. Parducci, L.; Suyama, Y.; Lascoux, M.; Bennett, K.D. Ancient DNA from pollen: a genetic record of population history in Scots pine. *Mol. Ecol.* **2005**, *14*, 2873–2882. [CrossRef] [PubMed]
3. Suyama, Y.; Kawamuro, K.; Kinoshita, I.; Yoshimura, K.; Tsumura, Y.; Takahara, H. DNA sequence from a fossil pollen of Abies spp. from Pleistocene peat. *Genes Genet. Syst.* **1996**, *71*, 145–149. [CrossRef] [PubMed]
4. Alotaibi, S.S.; Sayed, S.M.; Alosaimi, M.; Alharthi, R.; Banjar, A.; Abdulqader, N.; Alhamed, R. Pollen molecular biology: Applications in the forensic palynology and future prospects: A review. *Saudi J. Biol. Sci.* **2020**, *27*, 1185–1190. [CrossRef] [PubMed]
5. Lale, S.V.; Gill, H.S. Pollen grains as a novel microcarrier for oral delivery of proteins. *Int. J. Pharm.* **2018**, *552*, 352–359. [CrossRef]
6. Atwe, S.U.; Ma, Y.; Gill, H.S. Pollen grains for oral vaccination. *J. Control. Release* **2014**, *194*, 45–52. [CrossRef]
7. Ravindra, K.; Goyal, A.; Mor, S. Pollen allergy: Developing multi-sectorial strategies for its prevention and control in lower and middle-income countries. *Int. J. Hyg. Environ. Health* **2022**, *242*, 113951. [CrossRef]
8. Ziegenhagen, B.; Schauerte, M.; Kormutak, A.; Scholz, F. Plastid DNA polymorphism of megagametophytes and pollen in two Abies species. *Silvae Genet.* **1996**, *45*, 355–358.
9. Aziz, A.N.; Seabrook, J.E.; Tai, G.C. Amplification of RAPD markers from single pollen grains of diploid (2N= 2X= 24) potato. *Am. J. Potato Res.* **1999**, *76*, 179–182. [CrossRef]
10. Matsuki, Y.; Isagi, Y.; Suyama, Y. The determination of multiple microsatellite genotypes and DNA sequences from a single pollen grain. *Mol. Ecol. Notes* **2007**, *7*, 194–198. [CrossRef]
11. Aziz, A.; Sauve, R. Genetic mapping of Echinacea purpurea via individual pollen DNA fingerprinting. *Mol. Breed.* **2008**, *21*, 227–232. [CrossRef]
12. Chen, P.H.; Pan, Y.B.; Chen, R.K. High-throughput procedure for single pollen grain collection and polymerase chain reaction in plants. *J. Integr. Plant Biol.* **2008**, *50*, 375–383. [CrossRef] [PubMed]

13. Mase, N.; Sawamura, Y.; Yamamoto, T.; Takada, N.; Nishio, S.; Saito, T.; Iketani, H. Direct genotyping of single pollen grains of a self-compatible mutant of Japanese pear (Pyrus pyrifolia) revealed inheritance of a duplicated chromosomal segment containing a second S-haplotype. *Euphytica* **2014**, *200*, 297–304. [CrossRef]
14. Matsunaga, S.; Schütze, K.; Donnison, I.S.; Grant, S.R.; Kuroiwa, T.; Kawano, S. Single pollen typing combined with laser-mediated manipulation. *Plant J.* **1999**, *20*, 371–378. [CrossRef] [PubMed]
15. Hess, D.; Gresshoff, P.M.; Fielitz, U.; Gleiss, D. Uptake of protein and bacteriophage into swelling and germinating pollen of Petunia hybrida. *Z. Pflanzenphysiol.* **1975**, *74*, 371–376. [CrossRef]
16. Ohta, Y. High-efficiency genetic transformation of maize by a mixture of pollen and exogenous DNA. *Proc. Natl. Acad. Sci. USA* **1986**, *83*, 715–719. [CrossRef]
17. Booy, G.; Krens, F.; Huizing, H. Attempted pollen-mediated transformation of maize. *J. Plant Physiol.* **1989**, *135*, 319–324. [CrossRef]
18. Abdul-Baki, A.A.; Saunders, J.A.; Matthews, B.F.; Pittarelli, G.W. DNA uptake during electroporation of germinating pollen grains. *Plant Sci.* **1990**, *70*, 181–190. [CrossRef]
19. Twell, D.; Klein, T.M.; McCormick, S. Transformation of pollen by particle bombardment. In *Plant Tissue Culture Manual*; Springer: Berlin/Heidelberg, Germany, 1991; pp. 631–644.
20. Schreiber, D.N.; Dresselhaus, T. In vitro pollen germination and transient transformation ofZea mays and other plant species. *Plant Mol. Biol. Rep.* **2003**, *21*, 31–41. [CrossRef]
21. Zhao, X.; Meng, Z.; Wang, Y.; Chen, W.; Sun, C.; Cui, B.; Cui, J.; Yu, M.; Zeng, Z.; Guo, S.; et al. Pollen magnetofection for genetic modification with magnetic nanoparticles as gene carriers. *Nat. Plants* **2017**, *3*, 956–964. [CrossRef]
22. Nagahara, S.; Higashiyama, T.; Mizuta, Y. Detection of a biolistic delivery of fluorescent markers and CRISPR/Cas9 to the pollen tube. *Plant Reprod.* **2021**, *34*, 191–205. [CrossRef] [PubMed]
23. Gogoi, N.; Kanwal, M.; Norman, M.; Downs, J.; Ahmad, N.; Mago, R.; Bariana, H.; Muellner, M.; Bansal, U.; Jones, B.J. Wheat pollen uptake of CRISPR/Cas9 RNP-PDMAEMA nanoassemblies results in targeted loss of gene function in progeny. *bioRxiv* **2022**. [CrossRef]
24. Levin, I.; Gilboa, N. Direct PCR using tomato pollen grain suspensions. *BioTechniques* **1997**, *23*, 986–990. [CrossRef] [PubMed]
25. Petersen, G.; Johansen, B.; Seberg, O. PCR and sequencing from a single pollen grain. *Plant Mol. Biol.* **1996**, *31*, 189–191. [CrossRef] [PubMed]
26. Pacini, E.; Hesse, M. Pollenkitt–its composition, forms and functions. *Flora-Morphol. Distrib. Funct. Ecol. Plants* **2005**, *200*, 399–415. [CrossRef]
27. Prisle, N.L.; Lin, J.J.; Purdue, S.; Lin, H.; Meredith, J.C.; Nenes, A. Cloud condensation nuclei activity of six pollenkitts and the influence of their surface activity. *Atmos. Chem. Phys.* **2019**, *19*, 4741–4761. [CrossRef]
28. Chichiriccò, G.; Pacini, E.; Lanza, B. Pollenkitt of some monocotyledons: lipid composition and implications for pollen germination. *Plant Biol.* **2019**, *21*, 920–926. [CrossRef]
29. Rogers, S.O.; Bendich, A.J. Extraction of DNA from milligram amounts of fresh, herbarium and mummified plant tissues. *Plant Mol. Biol.* **1985**, *5*, 69–76. [CrossRef]
30. Brewbaker, J.L.; Kwack, B.H. The essential role of calcium ion in pollen germination and pollen tube growth. *Am. J. Bot.* **1963**, *50*, 859–865. [CrossRef]
31. Imai, S.; Tsuge, N.; Tomotake, M.; Nagatome, Y.; Sawada, H.; Nagata, T.; Kumagai, H. An onion enzyme that makes the eyes water. *Nature* **2002**, *419*, 685–685. [CrossRef]
32. Masamura, N.; McCallum, J.; Khrustaleva, L.; Kenel, F.; Pither-Joyce, M.; Shono, J.; Suzuki, G.; Mukai, Y.; Yamauchi, N.; Shigyo, M. Chromosomal organization and sequence diversity of genes encoding lachrymatory factor synthase in Allium cepa L. *G3 Genes Genomes Genet.* **2012**, *2*, 643–651. [CrossRef] [PubMed]
33. Suyama, Y. Procedure for single-pollen genotyping. In *Single-Pollen Genotyping*; Springer: Berlin/Heidelberg, Germany, 2011; pp. 7–15.
34. Heslop-Harrison, J. Tapetal origin of pollen-coat substances in Lilium. *New Phytol.* **1968**, *67*, 779–786. [CrossRef]
35. Hesse, M. Pollenkitt development and composition in Tilia platyphyllos (Tiliaceae) analysed by conventional. *Tapetum Cytol. Funct. Biochem. Evol.* **2012**, *7*, 39.
36. Dobson, H.E. Survey of pollen and pollenkitt lipids–chemical cues to flower visitors? *Am. J. Bot.* **1988**, *75*, 170–182. [CrossRef]
37. Wilson, I.G. Inhibition and facilitation of nucleic acid amplification. *Appl. Environ. Microbiol.* **1997**, *63*, 3741–3751. [CrossRef]

current issues in molecular biology

MDPI

Article

Characterizations of a Class-I BASIC PENTACYSTEINE Gene Reveal Conserved Roles in the Transcriptional Repression of Genes Involved in Seed Development

Xianjin Ma [1,2,3,†], Yifan Yu [1,2,3,†], Zhikang Hu [1,2,3], Hu Huang [1,2,3], Sijia Li [1,2,3] and Hengfu Yin [1,3,*]

1 State Key Laboratory of Tree Genetics and Breeding, Research Institute of Subtropical Forestry, Chinese Academy of Forestry, Hangzhou 311400, China
2 College of Information Science and Technology, Nanjing Forestry University, Nanjing 210037, China
3 Key Laboratory of Forest Genetics and Breeding, Research Institute of Subtropical Forestry, Chinese Academy of Forestry, Hangzhou 311400, China
* Correspondence: hfyin@caf.ac.cn; Tel.: +86-571-63105-09320
† These authors are equally contributed to this work.

Abstract: The developmental regulation of flower organs involves the spatio-temporal regulation of floral homeotic genes. *BASIC PENTACYSTEINE* genes are plant-specific transcription factors that is involved in many aspects of plant development through gene transcriptional regulation. Although studies have shown that the *BPC* genes are involved in the developmental regulation of flower organs, little is known about their role in the formation of double-flower due. Here we characterized a Class I BPC gene (*CjBPC1*) from an ornamental flower—*Camellia japonica*. We showed that *CjBPC1* is highly expressed in the central whorls of flowers in both single and doubled varieties. Overexpression of *CjBPC1* in *Arabidopsis thaliana* caused severe defects in siliques and seeds. We found that genes involved in ovule and seed development, including *SEEDSTICK, LEAFY COTYLEDON2, ABSCISIC ACID INSENSITIVE 3* and *FUSCA3*, were significantly down-regulated in transgenic lines. We showed that the histone 3 lysine 27 methylation levels of these downstream genes were enhanced in the transgenic plants, indicating conserved roles of *CjBPC1* in recruiting the Polycomb Repression Complex for gene suppression.

Keywords: BPC transcription factor; flower development; histone methylation; ovule; seed

Citation: Ma, X.; Yu, Y.; Hu, Z.; Huang, H.; Li, S.; Yin, H. Characterizations of a Class-I BASIC PENTACYSTEINE Gene Reveal Conserved Roles in the Transcriptional Repression of Genes Involved in Seed Development. *Curr. Issues Mol. Biol.* **2022**, *44*, 4059–4069. https://doi.org/10.3390/cimb44090278

Academic Editor: Vijai Bhadauria

Received: 15 August 2022
Accepted: 2 September 2022
Published: 7 September 2022

1. Introduction

Floral development requires concerted expression of genes involved in determining organ identity. With the establishment and continuous improvement of the ABC model, the floral homeotic genes, classified into A-, B-, C-, D-, E- categories, have been identified, which forms a complex regulatory network in the control of spatiotemporal gene expression [1,2]. The establishment and maintenance of expression domain of floral homeotic genes are determined by multi-layered regulators [3]; and subtle modification of expression can lead to alterations of floral forms [4].

Transcription factors are often defined as proteins that bind to DNA in a sequence-specific manner and regulate the transcription process [5,6]. In plants, different families of transcription factors are classified based on their specific DNA-binding domains, which play an important role in regulating gene expression in plant growth, development, and stress response [7,8]. The BASIC PENTACYSTEINE (BPC) transcription factors belong to a small plant-specific gene family that contains a highly conserved DNA-binding domain, including five conserved cysteine residues at the C-terminus of proteins [9,10]. BPC genes are found to preferentially bind the "RGARAGRRAA" (GA-rich box or C-box) DNA elements to regulate the expression of downstream genes [9]. In *Arabidopsis thaliana*, *BPC* gene family has seven members; and based on the N-terminus sequences, *BPC* genes can be divided into three subclasses: Class I contains BPC1/2/3; Class II contains BPC4/6; Class III contains BPC7; BPC5 is likely a pseudogene [10,11]. All *Arabidopsis BPC* genes are ubiquitously expressed in various tissues [12]. It is found that simultaneous mutation of four or five *BPC* genes in *Arabidopsis* cause severe developmental defects in multiple organs, but a weaker phenotype of a single mutation implies functional redundancy among members of the *BPC* family [11,12]. BPC proteins can form homologous and heterologous complexes, indicating that there are complex regulatory relationships between *BPC* genes in the regulation of specific downstream genes [10,13].

Several studies have reported that BPC proteins recruit a high-order protein complex of Polycomb Repressive Complex (PRC) for the suppression of genes through the trimethylation of histone 3 proteins at the lysine 27 position (H3K27me3) [14,15]. Considering that the PRC protein complex does not have the binding ability of specific DNA sites, BPC protein-mediated DNA binding site is crucial in determining the gene expression of floral organ specificity. Indeed, for the refinement of expression domain of C function gene *AGAMAOUS* (*AG*), the GATA motif inside a Polycomb Response Element (PRE)—located in the second intron of *AG*—is recognized by the BPC proteins to recruit the PRC for gene suppression [16,17]. In addition, the *BPC* genes can regulate the expression of HOMEOBOX transcription factors, including *SHOOTMERISTEMLESS* and *BREVIPEDICELLUS*, to maintain the activity of shoot apical meristem. [18]. In the cytokinin signaling pathway, *BPCs* are found to regulate a subset of transcription factors involved in cytokinin actions [19]. In *A. thaliana*, *BPC* genes are initially found to limit the expression of the D function gene (*STK*, *SEEDSTICK*) to regulate ovule development through transcriptional inhibition [9,13]. During the seed development, *BPC* genes are required to determinate the expression of *INNER NO OUTER* (*INO*) [20], *LEAFY COTYLEDON2* (*LEC2*) [21], *ABSCISIC ACID INSENSITIVE 3* (*ABI3*) [22], and *FUSCA3* (*FUS3*) [23].

Camellia japonica is an important ornamental flower and possesses different kinds of cultivars with diverse floral variations [24]. The changes of floral pattern and organ shapes in ornamental varieties of *C. japonica* provide a unique resource to interrogate the molecular regulation of floral gene expression [25]. Although the studies have indicated that the *BPC* genes are involved in regulating the expression of floral homeotic genes, it remains unknown how the protein complex mediated by BPCs is regulated to determine the organ-specific expression of floral homeotic genes, particularly in the regulation of double flower formation. In this study, we characterized a Class-I BPC homolog, *CjBPC1*, from *C. japonica*. We showed that the expression of *CjBPC1* is highly expressed in the inner flower tissues in both single and doubled varieties. We demonstrated that *CjBPC1* possessed conserved functions of regulating genes involved in ovule and seed development through the H3K27me3 modifications of downstream genes. Our results indicate *BPC* genes are potential regulators of double flower formation in *C. japonica*.

2. Materials and Methods

2.1. Plant Materials and Growth Conditions

The *Camellia japonica* L. materials used in this experiment were planted in the Research Institute of Subtropical Forestry, China Academy of Forestry (119°57′ N, 30°04′ E, Fuyang, China). The flower bud materials of *C. japonica* were collected and classified and immediately frozen in liquid nitrogen and stored in a refrigerator at −80 °C. The *Arabidopsis* (Colombia ecotype) was used and maintained in a growth chamber (long-day conditions, 16 h light/8 h dark, set at 22 °C).

2.2. Identification and Phylogenetic Analysis of BPC Genes

Seven BPC protein sequences from *A. thaliana* were downloaded from TAIR (https://www.Arabidopsis.org/ (accessed on 10 May 2020)). The software BioEdit software was used to identify BPC sequences using local *C. japonica* datasets through the BLASTP program (E-value cutoff 10–15). The conserved domains of the *BPC* genes were analyzed on the SMART [26] with default settings. Protein sequences from *C. japonica* and *A. thaliana* are used to construct a phylogenetic tree by the MEGA 7.0 software with parameters described before [27]. The online tool MEME (meme-suite.org) was used to identify the conserved motifs of BPC proteins [28]. The parameters were as follows: the distribution of arbitrary number of repeat sites, the found six motifs, and the minimum and maximum motif widths were six and 50, respectively. The subcellular localization of CjBPC proteins was predicted by CELLO v.2.5 server (http://cello.life.nctu.edu.tw/cello2go/ (accessed on 10 May 2020)) [29,30].

2.3. RNA Extraction and qRT-PCR Analysis

Total RNA was extracted using a plant total RNA extraction kit (RNAprep Pure polysaccharide polyphenol, TIANGEN, Beijing, China), and then the first strand of cDNA was synthesized using PrimeScript™ II 1st Strand cDNA Synthesis Kit (Takara, Dalian, China). Real-time quantitative PCR (qRT-PCR) primers were designed by Primer Express 3.0.1 to amplify short fragment DNA (50–100 bp). SYBR® Premix Ex TaqTM II (Takara, Dalian, China) reagents and an ABI 7300 REAL-TIME PCR instrument were used for PCR analysis. Three replicates per test were used for obtaining the gene expression results, and data analysis was based on the $2^{-\Delta\Delta CT}$ method [31].

2.4. Vector Construction and Arabidopsis Transformation

The full-length sequences of *CjBPC1* gene were cloned using specific primers (Supplementary Table S1), and cloned into pEXT06/g vector (Baige, Suzhou, China) according to the user's manual. The recombinant plasmid was validated and transformed into *Agrobacterium tumefaciens* C58 (pGV3101) strain by heat shock method, and then used for the transformation of *A. thaliana* by the floral-dip method [32].

2.5. Phenotypic Analysis of Transgenic Lines

The transformed *Arabidopsis* T1 seeds were germinated on 1/2 Murashige and Skoog medium containing hygromycin (10 mg/L) for selection. The genomic DNA of transgenic plants was extracted from the young leaves (Two-week old), and PCR validation was performed using hygromycin and construct-specific primers to determine positive transgenic lines. In addition, qRT-PCR was performed to reveal the ectopic expression of *CjBPC1* in transgenic plants. The length and number of seeds of deformed and shortened pods of transgenic lines were measured.

2.6. Chromatin Immunoprecipitation PCR (ChIP-PCR)

The ChIP-PCR experiment was performed according to previous studies [33]. The H3K27me3 antibody was obtained from Sigma-Aldrich (Catalogue Number: 07-449). For chromatin isolation and immunoprecipitation, around 1.5 g inflorescence of transgenic *A. thaliana* were taken to extract DNA for PCR analysis. Two independent chromatin isolations were performed, and three PCR amplification replicates were used for quantitative analysis according to $2^{-\Delta\Delta CT}$ method [31].

2.7. Statistical Analysis

The one-way analysis of variance (ANOVA) test is used to perform significance test for groups of samples by SPSS software [34]. Different letters indicate the significant differences (p-value < 0.01). The unpaired two-tailed Student's tests are used for to determine the statistical significance. * $p < 0.05$; ** $p < 0.01$.

3. Results

3.1. Identification and Characterization of BPC Genes in C. japonica

To identify BPC family genes, we performed a homology search in the transcriptome sequence database of *C. japonica* [25,35] using sequences of *Arabidopsis* BPC proteins. Six candidates were identified based on the similarity of their coding sequences. Then we cloned the full-length coding sequences of the candidates using gene-specific primers

(Supplementary Table S1) and validated the transcript sequences. The length of deduced amino acids (AAs) from the transcripts ranged from 235 to 332 (Supplementary Table S1). The predicted isoelectric points of all proteins were relatively high ($p > 9$; Supplementary Table S2), largely due to the abundance of alkaline AAs.

To study the sequence characteristics, we performed sequence-alignment analysis by using the *BPC* genes from *C. japonica* and *A. thaliana*. We found all of the six candidates from *C. japonica* contain the highly conserved motif in which the conserved arrangement of the five cysteine residues was revealed (Figure 1A). To gain a deep understanding of the evolution of the *BPC* genes in *C. japonica*, we constructed a phylogenetic tree. We showed that BPC proteins from *C. japonica* were categorized into three groups: two members in the Class I clade, two members in the Class II clade and one member in the Class III clade (Figure 1B). Through sequence alignment and motif analyses, we found that, the subclades of BPCs possessed several conserved motifs in the N-terminal regions (Figure 1A,C). The classification of sub-clades of *BPC* genes is consistent with previous studies [11], suggesting functional conservations of *BPC* genes in different plant lineages.

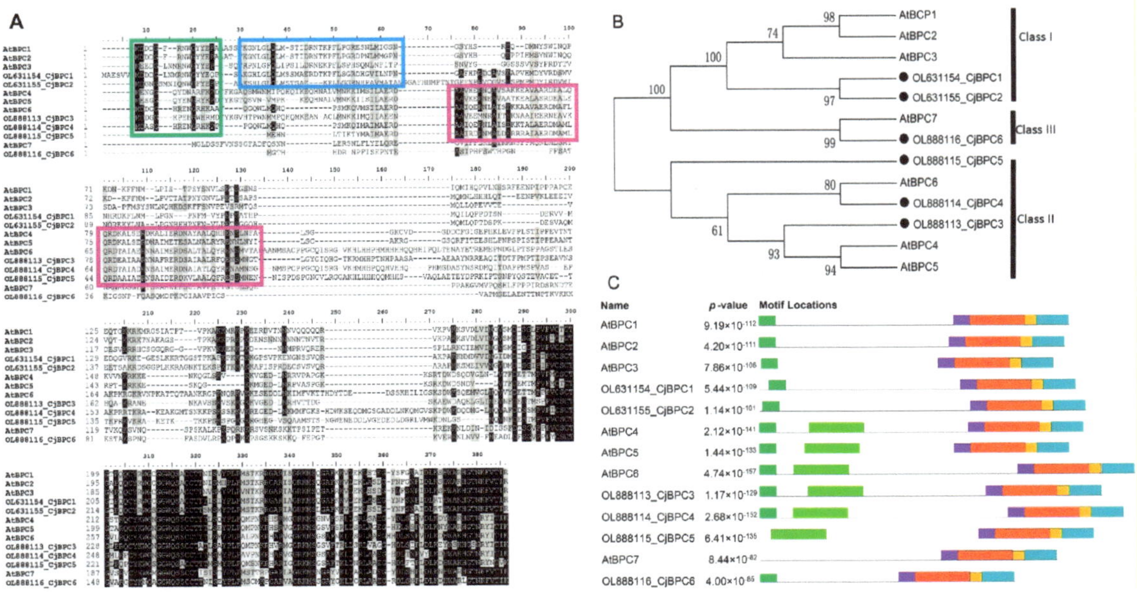

Figure 1. Sequence alignment and phylogenetic analysis of BPC family genes from *C. japonica*. (**A**), protein sequence alignment of BPCs from *C. japonica* and *A. thaliana*. Green rectangle indicates the conserved N-terminal sequences in Class I and Class II *BPC* genes; blue rectangle indicates a conserved domain in the Class I *BPC* genes; magenta rectangles indicate the conserved domains in the Class II *BPC* genes. (**B**), a phylogenetic tree containing the *BPC* genes from *C. japonica* and *A. thaliana*. Three classes are indicated by black bars. (**C**), the distribution of motifs that are predicted by SMART analysis. Details of domains are presented in Supplementary Figure S1. Different color bars indicate the consensus sequences of motifs.

3.2. Expression Analysis of CjBPC1 in Floral Tissues of Single and Double Flowers of C. japonica

To reveal potential roles of *CjBPC* genes in floral development, we analyzed the expression profiles of *CjBPCs* in different tissue types. We found that the members of *CjBPCs* displayed different overall expression levels: *CjBPC1* and *CjBPC5* were abundantly detected in all tissue types; *CjBPC2*, *CjBPC3*, and *CjBPC6* were relatively lowly expressed; *CjBPC4* was barely detected (Supplementary Figure S2) [35]. *CjBPC1* was highly expressed in floral bud, indicating a potential function of regulating floral development. We further evaluated the *CjBPC1* expression in floral organs of single and doubled flowers. The four floral tissues types (sepal, petal, stamen and carpel) from wild *C. japonica*, together with floral tissues from anemone double and formal double varieties, were studied (Figure 2A). We found that the expression of *CjBPC1* was detectable in all examined tissues (Figure 2B), and *CjBPC1* displayed higher expression levels in carpels in wild *C. japonica*. In doubled cultivars, *CjBPC1* was markedly up-regulated in the central tissues of doubled flowers (carpel in anemone double and inner petals of formal double) (Figure 2B).

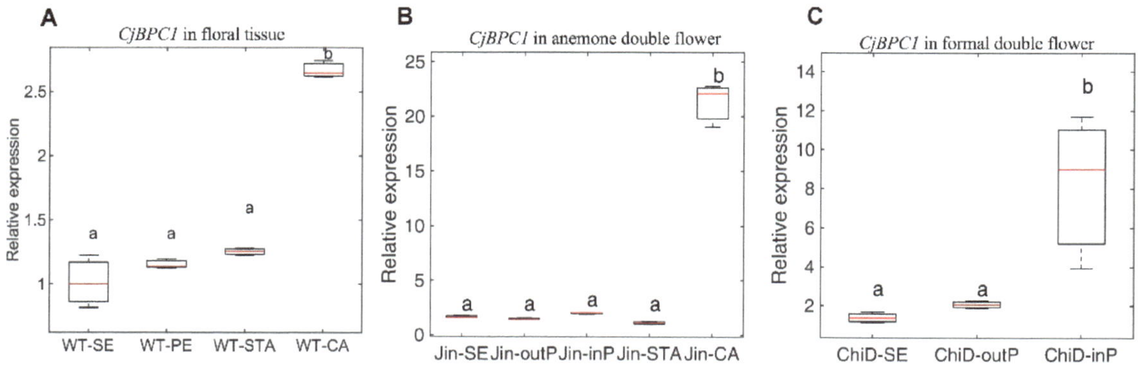

Figure 2. Expression analysis of *CjBPC1* in wild and doubled *C. japonica* plants. (**A**), the expression of *CjBPC1* in four floral organs in wild single *C. japonica*. WT-SE, sepal; WT-PE, petal; WT-STA, stamen; WT-CA, carpel. (**B**), the expression of *CjBPC1* in anemone double variety 'Jinpanlizhi'. Jin-SE, sepal; Jin-outP, outer whorl of petal; JIN-inP, inner whorl of petal; Jin-STA, stamen; Jin-CA, carpel. (**C**), the expression of *CjBPC1* in formal double variety 'Chidan'. ChiD-SE, sepal; ChiD-outP, outer whorl of petal; ChiD-inP, inner whorl of petal. Three biological replicates are used for each tissue and two technical replicates of real-time qPCR analysis are performed. The GAPDH gene was used as the internal reference for gene expression analysis (Supplementary Table S1). For each boxplot, the red bar indicates the mean value. Box limits represent the upper and lower quartiles. Whiskers represent minimum to the lower quartiles and maximum to the upper quartiles. One-way ANOVA and multiple comparisons are performed for significance tests (p-value < 0.01), and different letters indicate significant changes.

3.3. Ectopic Expression of CjBPC1 in Arabidopsis Causes Ovule Abortion

To understand the function of *CjBPC1*, we generated the transgenic *Arabidopsis* lines through ectopic expression of *CjBPC1*. The transgenic lines were verified by PCR amplification and gene expression analysis (Supplementary Figure S3), and we selected five lines with ectopic expression of *CjBPC1* for the phenotypical evaluations (Supplementary Figure S3B,C). Moreover, we showed that the CjBPC1 predominantly localized in the nuclei (Figure 3), suggesting its functions in the regulation of gene expression.

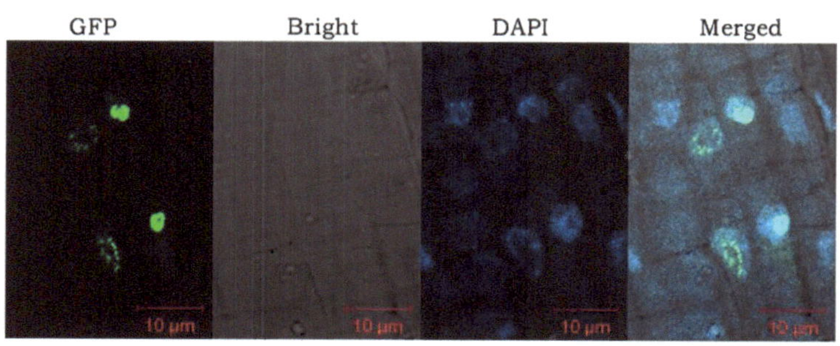

Figure 3. Subcellular localization of CjBPC1 in transgenic *A. thaliana* lines. From left to right: the light field, GFP signal, (DAPI 4′,6-diamidino-2-phenylindole) staining of nuclei, merged image. The root tips of transgenic lines are used for confocal microscopy analysis. Bars = 10 μm.

We analyzed the phenotypical alterations using transgenic lines with the high-level expression of *CjBPC1* (line 3 and line 6) (Supplementary Figure S3C). We found no visible morphological changes during the vegetative growth; during the reproductive development, the siliques in transgenic lines had conspicuous defects: the shortened and deformed silique were frequently produced in the inflorescences of transgenic plants (Figure 4A–C). To compare the phenotypical effects, we classified the abnormal siliques into "strong" and "weak" categories by the length of the siliques: strong, less than 6 mm; weak, between 6–10 mm (Figure 4D,E). We found that there were varying degrees of aborted ovules throughout the development of ovules in the five assessed transgenic lines (Figure 4).

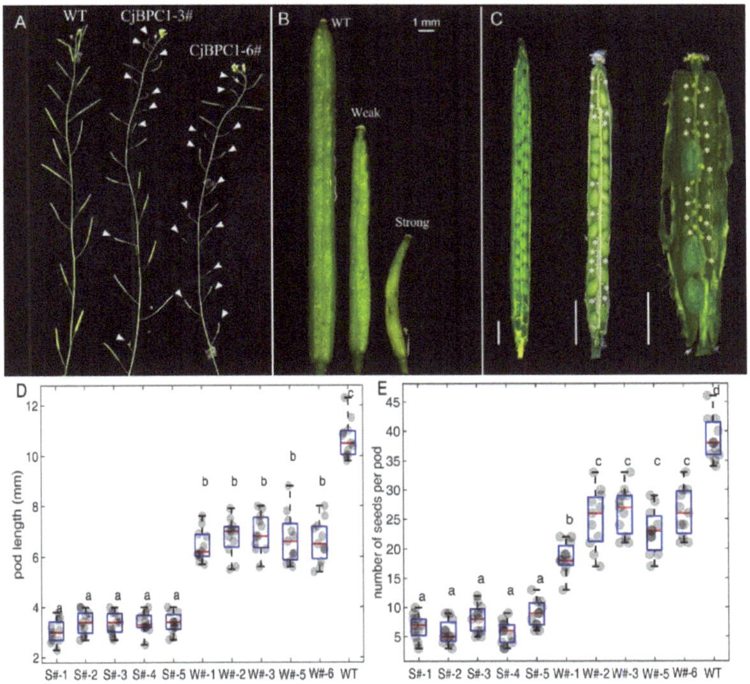

Figure 4. Overexpression of *CjBPC1* causes seed abortion in *A. thaliana*. (**A**), the comparison of inflorescences in WT and *CjBPC1* transgenic lines (line CjBPC1-3 and line CjBPC1-6). Arrowheads indicate the shortened or deformed siliques. (**B**), close-up images of siliques, including strong and week phenotypical changes comparing to the WT. Bar = 1 mm. (**C**), the dissected siliques of transgenic lines. From left to right: mature siliques with normal, weakly deformed and strongly deformed phenotypes. The stars indicate the aborted seeds. Bars = 1 mm. (**D**), Measurements of the length of mature pods in different transgenic lines (Line 1, 2, 3, 5, 6) and WT. (**E**), Measurements of the number of seeds per pod. (**D**,**E**), The pods from transgenic lines are categorized into strong (S#) and weak (W#). The one-way ANOVA test is used to perform significance test. Each group contains 11 measurements. Different letters on top of boxes indicate the significant differences (p-value < 0.01). For each boxplot, the red bar indicates the mean value. Box limits represent the upper and lower quartiles. Whiskers represent minimum to the lower quartiles and maximum to the upper quartiles. The grey dots indicate the raw values of each measurement.

3.4. Overexpression of CjBPC1 Leads to Down-Regulation of Genes Involved in the Seed Development

The *BPC* genes have been found to be involved in several aspects of ovule and seed development through regulating the expression of downstream genes [23]. We investigated the expression of genes that are involved in the seed development using the transgenic lines of *CjBPC1*. The significant down-regulation of *STK*, *LEC2*, *ABI3* and *FUS3* was revealed in both strong and weak transgenic lines (Figure 5). Consistently, strong lines displayed severer reduction of gene expression than that of weak lines (Figure 5). These results indicate that these genes are probably downstream genes of *CjBPC1* that are responsible for the defects of siliques; and *CjBPC1* might share conserved functions of regulating the expression of genes involved in seed development.

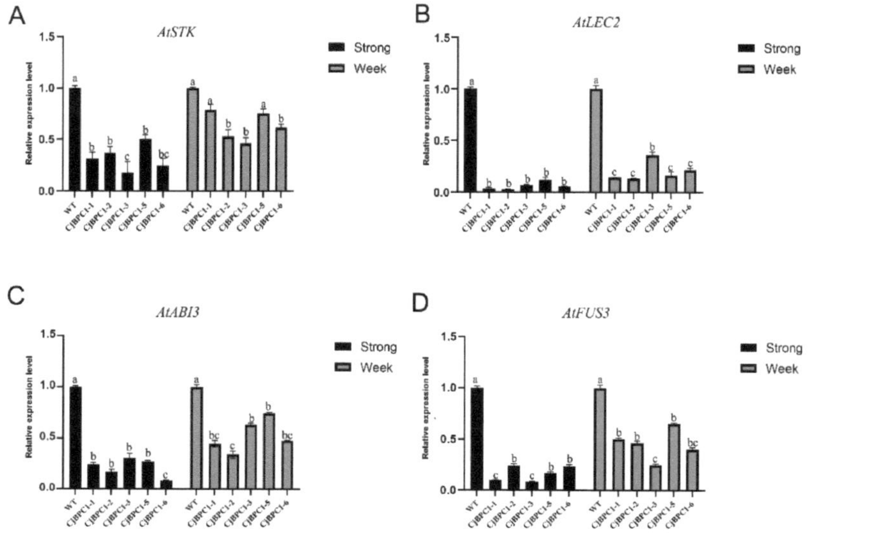

Figure 5. Expression analysis of genes involved in seed development from *A. thaliana*. The expression of *STK* (**A**), *LEC2* (**B**), *ABI3* (**C**), *FUS3* (**D**) in different transgenic lines. **A-D**: strong and weak siliques are used for gene expression separately. Three biological replicates of each sample are used, and two technical replicates of real-time PCR experiments are performed to obtain the gene expression levels. *ACTIN* gene was used as the internal reference (Supplementary Table S1). Error bars indicate standard deviations of all the replicates in each genetic background. The one-way ANOVA test is used to perform significance test. Different letters on top of columns indicate the significant differences (p-value < 0.01).

3.5. Transgenic Lines of CjBPC1 Displays Enhanced H3K27me3 Levels of Downstream Genes

The mechanism of BPC mediated gene repression is involved in the recruitment of PRC complex that leads to the enhanced levels of H3K27me3 [13]. To further investigate if overexpression of *CjBPC1* caused direct changes of histone status of the downstream genes, we investigated the H3K27me3 levels in transgenic *Arabidopsis* lines. We performed the Chromatin Immunoprecipitation PCR (ChIP-PCR) analysis of H3K27me3 at the regions of the downstream genes [21,23,36,37] (Figure 6). We showed that the increased levels of H3K27me3 were identified in all the tested downstream genes at different levels (Figure 6), in which *STK* and *ABI3* had over three folds of methylation levels. These results indicate that *CjBPC1* might be directly involved in the regulation of H3K27me3 levels through recruiting the PRC complex in *A. thaliana*.

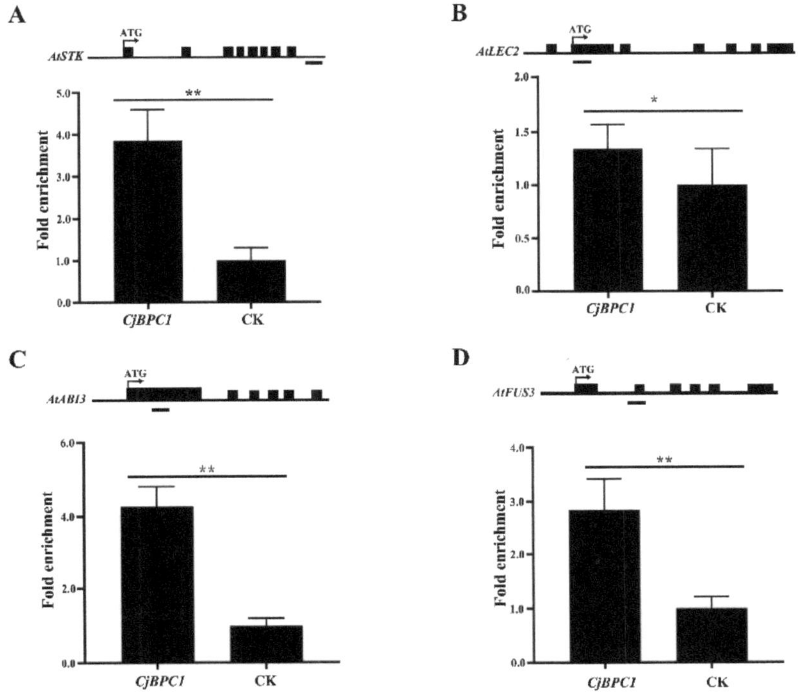

Figure 6. Analysis of H3K27me3 levels in WT and transgenic lines of *CjBPC1*. The H3K27me3 level of *STK* (**A**), *LEC2* (**B**), *ABI3* (**C**), *FUS3* (**D**) loci were measured by ChIP-PCR using the fragment indicated on top of each panel (indicated by the black bars under the gene models). The immunoprecipitation experiment was performed twice and three replicates were performed to obtain quantification values. Error bars indicate standard deviations of all the replicates. The student's t tests are used for the comparisons between control group (CK) and transgenic lines. * $p < 0.05$; ** $p < 0.01$.

4. Discussion

The *BPC* gene family is unique among plants and usually has a small number of members. By recruiting PRC complexes for their role in gene regulation, *BPCs* may affect far more genes in regulation of many aspects of plant development than has been revealed so far [38]. A recent study based on the ChIP sequencing analysis of *BPC6* uncovered thousands of potential direct binding sites in *Arabidopsis* genome [19]. It is not known, however, whether and how different classes of *BPC* genes acquire sequence-preferences in regulating their downstream genes.

Although, in *A. thaliana*, all members of *BPCs* exhibit ubiquitous expression in different tissue types, the transcriptional regulation of *BPCs* remains an important aspect of functional divergence [11]. We found that, in *C. japonica*, the expression patterns of different *BPCs* differed greatly in terms of the expression abundance, based on transcriptomics analysis using relatively mature organs (Supplementary Figure S1). Particularly, *CjBPC1* (a Class I member) and *CjBPC5* (a Class II member) were highly expressed in all tissues, indicating the major roles of maintaining the gene expression profiles (Figure 1; Supplementary Figure S1). Thus, different members of *BPCs* might play diverse functions at different stages of plant development or in certain specified tissues. The expression level of *CjBPC1* was detected in all floral organs of wild *Camellia* flower and markedly accumulated in the carpel tissues (Figure 2A), and this result was in good agreement with the phenotypic changes observed in transgenic *Arabidopsis* (Figure 3), which indicated that *CjBPC1* is mainly involved in ovule development. However, the expression of *CjBPC1* did not appear to be entirely tissue-specific: in the formal double variety "Chidan", which is completely devoid of ovule development, *CjBPC1* remained highly expressed in the central regions of floral buds, where ovules in wild *Camellia* differentiate (Figure 2C). Therefore *CjBPC1* is potentially involved not only in specifying ovule tissues but also in maintaining the floral meristem activity.

Overexpression of different *BPC* members can reveal functional specificity in the regulation of gene expression. For example, ectopic expression of *BPC2* caused down-regulation of *LATE EMBRYOGENESIS ABUNDANT* (*LEA*) genes, which led to enhanced susceptibility of osmotic stress in *A. thaliana* [39]. In rice, overexpression of a Class-II BPC gene revealed divergent functions in regulation flowering time [40]. We showed, in the *Arabidopsis* system the regulation of ovule/seed development was affected by the heterogeneous expression of *CjBPC1* (Figure 3). This result indicated *CjBPC1* might have specific functions in ovule or seed development. In *Arabidopsis*, the Class-I *BPC* genes were also found to regulate *HOMEOBOX* genes for the meristem maintenance [18]. In *CjBPC1* transgenic lines, because no visible phenotypes at the vegetative stage (e.g., leaf development and flowering time) were observed, it is likely that expression of *HOMEOBOX* genes is not altered. Gene expression analysis of potential downstream genes supported that *CjBPC1* is mainly involved in regulating expression of ovule- and seed-related genes (Figures 5 and 7). We also found that the H3K27me3 levels of downstream genes are enhanced in transgenic lines (Figure 6). These results are consistent with molecular functions of *BPC* genes [13,41]. We postulated that *CjBPC1* share conserved functions of regulating ovule and seed development through regulating the subset of downstream genes in *C. japonica*. We thereby proposed a model of the functions in the transgenic lines of *Arabidopsis* which involved the recruitment of PRC and suppression of genes involved in seed development (Figure 7).

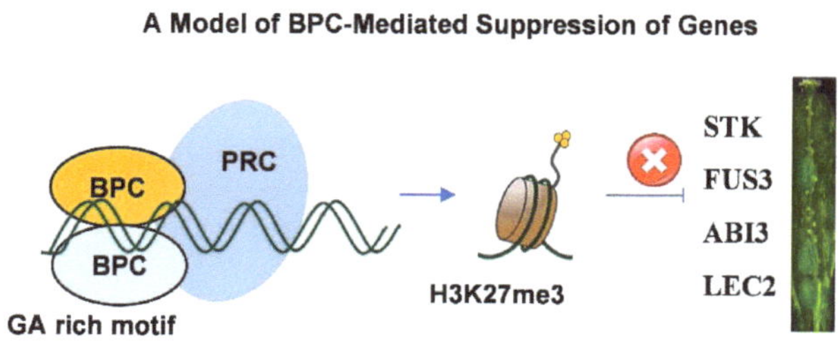

Figure 7. A molecular model of *BPC* genes and their roles in the regulation of seed development. The BPC proteins can form hetero- or homo- dimers during the binding of the GA-rich motifs [6,7]; this leads to the tri-methylation of histone 3 lysine 27 and further suppress the expression of genes including *STK*, *FUS3*, *ABI3* and *LEC2*.

The spatiotemporal expression of floral homeotic genes is critical for floral development and is controlled by multi-layered regulators [42,43]. For the regulation of floral homeotic genes, it was found that the *BPC* genes are necessary to limit the expression domain of *AG* and *STK* [9,13,16]. The functional action of *BPCs* mainly involves two aspects: First, the binding to the GA-rich box to regulate downstream genes; Second, interact with proteins to recruit the PRC complexes. Based on a molecular model of PRE mediated gene suppression [17,44], BPCs and GA-rich motif are components of the regulatory mechanism [44]. It is attempting to hypothesize that the BPC-mediated gene suppression is an upstream mechanism for ensuring the correct expression of floral homeotic genes. Double flowers in *C. japonica* exhibits extensive changes in floral organ identity, number of flower organs and arrangement of flower organs; and the alterations in the expression of floral homeotic genes (e.g., B- and C- class genes) have been revealed to be associated with the formation of double flowers [24,45]. Further studies of BPCs and GA-rich motifs could lead to the discovery of potential regulatory mechanisms underlying the modifications of gene expression in double flowers of *C. japonica*.

5. Conclusions

In conclusion, this work revealed that *CjBPC1*—the Class-I homolog of BPC family—has conserved functions of gene suppression in regulating flower and seed development; and BPC members from *C. japonica* are potential regulators for establishing and maintaining the expression pattern of floral homeotic genes during double flower formation.

Supplementary Materials: The following supporting information can be downloaded at: https://www.mdpi.com/article/10.3390/cimb44090278/s1.

Author Contributions: H.Y. conceived the research objectives. X.M., Y.Y. and Z.H. performed gene expression and functional analyses of BPC genes. H.H. and S.L. participated in the sample collection. X.M. and H.Y. drafted the manuscript and all authors contributed and approved the paper. All authors have read and agreed to the published version of the manuscript.

Funding: This research is supported by Nonprofit Research Projects (CAFYBB2021QD001-1) of Chinese Academy of Forestry and National Science Foundation of China (31870578).

Institutional Review Board Statement: Not applicable.

Informed Consent Statement: Not applicable.

Data Availability Statement: The data presented in this study are available from the web link of the manuscript.

Conflicts of Interest: The authors declare no conflict of interest.

References

1. Bowman, J.L.; Smyth, D.R.; Meyerowitz, E.M. The ABC model of flower development: Then and now. *Development* **2012**, *139*, 4095–4098. [CrossRef] [PubMed]
2. Kramer, E.M.; Hall, J.C. Evolutionary dynamics of genes controlling floral development. *Curr. Opin. Plant Biol.* **2005**, *8*, 13–18. [CrossRef]
3. Liu, Z.; Mara, C. Regulatory mechanisms for floral homeotic gene expression. In *Seminars in Cell & Developmental Biology*; Elsevier: Amsterdam, The Netherlands, 2010; pp. 80–86.
4. Specht, C.D.; Howarth, D.G. Adaptation in flower form: A comparative evodevo approach. *New Phytol.* **2015**, *206*, 74–90. [CrossRef]
5. Latchman, D.S. Transcription factors: An overview. *Int. J. Biochem. Cell Biol.* **1997**, *29*, 1305–1312. [CrossRef]
6. Suter, D.M. Transcription factors and DNA play hide and seek. *Trends Cell Biol.* **2020**, *30*, 491–500. [CrossRef] [PubMed]
7. Strader, L.; Weijers, D.; Wagner, D. Plant transcription factors—Being in the right place with the right company. *Curr. Opin. Plant Biol.* **2022**, *65*, 102136. [CrossRef] [PubMed]
8. Yamasaki, K.; Kigawa, T.; Seki, M.; Shinozaki, K.; Yokoyama, S. DNA-binding domains of plant-specific transcription factors: Structure, function, and evolution. *Trends Plant Sci.* **2013**, *18*, 267–276. [CrossRef]
9. Kooiker, M.; Airoldi, C.A.; Losa, A.; Manzotti, P.S.; Finzi, L.; Kater, M.M.; Colombo, L. BASIC PENTACYSTEINE1, a GA binding protein that induces conformational changes in the regulatory region of the homeotic Arabidopsis gene SEEDSTICK. *Plant Cell* **2005**, *17*, 722–729. [CrossRef]
10. Theune, M.L.; Bloss, U.; Brand, L.H.; Ladwig, F.; Wanke, D. Phylogenetic analyses and GAGA-motif binding studies of BBR/BPC proteins lend to clues in GAGA-motif recognition and a regulatory role in Brassinosteroid signaling. *Front. Plant Sci.* **2019**, *10*, 466. [CrossRef]
11. Monfared, M.M.; Simon, M.K.; Meister, R.J.; Roig-Villanova, I.; Kooiker, M.; Colombo, L.; Fletcher, J.C.; Gasser, C.S. Overlapping and antagonistic activities of BASIC PENTACYSTEINE genes affect a range of developmental processes in Arabidopsis. *Plant J.* **2011**, *66*, 1020–1031. [CrossRef]
12. Petrella, R.; Caselli, F.; Roig-Villanova, I.; Vignati, V.; Chiara, M.; Ezquer, I.; Tadini, L.; Kater, M.M.; Gregis, V. BPC transcription factors and a Polycomb Group protein confine the expression of the ovule identity gene SEEDSTICK in Arabidopsis. *Plant J.* **2020**, *102*, 582–599. [CrossRef] [PubMed]
13. Simonini, S.; Roig-Villanova, I.; Gregis, V.; Colombo, B.; Colombo, L.; Kater, M.M. Basic pentacysteine proteins mediate MADS domain complex binding to the DNA for tissue-specific expression of target genes in Arabidopsis. *Plant Cell* **2012**, *24*, 4163–4172. [CrossRef]
14. Hecker, A.; Brand, L.H.; Peter, S.; Simoncello, N.; Kilian, J.; Harter, K.; Gaudin, V.; Wanke, D. The Arabidopsis GAGA-binding factor basic pentacysteine6 recruits the polycomb-repressive complex1 component like heterochromatin protein1 to GAGA DNA motifs. *Plant Physiol.* **2015**, *168*, 1013–1024. [CrossRef]
15. Mozgova, I.; Hennig, L. The polycomb group protein regulatory network. *Annu. Rev. Plant Biol.* **2015**, *66*, 269–296. [CrossRef] [PubMed]

16. Pelayo, M.A.; Yamaguchi, N.; Ito, T. One factor, many systems: The floral homeotic protein AGAMOUS and its epigenetic regulatory mechanisms. *Curr. Opin. Plant Biol.* **2021**, *61*, 102009. [CrossRef] [PubMed]
17. Xiao, J.; Jin, R.; Yu, X.; Shen, M.; Wagner, J.D.; Pai, A.; Song, C.; Zhuang, M.; Klasfeld, S.; He, C. Cis and trans determinants of epigenetic silencing by Polycomb repressive complex 2 in Arabidopsis. *Nat. Genet.* **2017**, *49*, 1546–1552. [CrossRef]
18. Simonini, S.; Kater, M.M. Class I BASIC PENTACYSTEINE factors regulate HOMEOBOX genes involved in meristem size maintenance. *J. Exp. Bot.* **2014**, *65*, 1455–1465. [CrossRef]
19. Shanks, C.M.; Hecker, A.; Cheng, C.Y.; Brand, L.; Collani, S.; Schmid, M.; Schaller, G.E.; Wanke, D.; Harter, K.; Kieber, J.J. Role of BASIC PENTACYSTEINE transcription factors in a subset of cytokinin signaling responses. *Plant J.* **2018**, *95*, 458–473. [CrossRef]
20. Meister, R.J.; Williams, L.A.; Monfared, M.M.; Gallagher, T.L.; Kraft, E.A.; Nelson, C.G.; Gasser, C.S. Definition and interactions of a positive regulatory element of the Arabidopsis INNER NO OUTER promoter. *Plant J.* **2004**, *37*, 426–438. [CrossRef]
21. Berger, N.; Dubreucq, B.; Roudier, F.; Dubos, C.; Lepiniec, L. Transcriptional regulation of Arabidopsis LEAFY COTYLEDON2 involves RLE, a cis-element that regulates trimethylation of histone H3 at lysine-27. *Plant Cell* **2011**, *23*, 4065–4078. [CrossRef]
22. Mu, Y.; Zou, M.; Sun, X.; He, B.; Xu, X.; Liu, Y.; Zhang, L.; Chi, W. BASIC PENTACYSTEINE proteins repress ABSCISIC ACID INSENSITIVE4 expression via direct recruitment of the Polycomb-Repressive Complex 2 in Arabidopsis root development. *Plant Cell Physiol.* **2017**, *58*, 607–621. [CrossRef] [PubMed]
23. Wu, J.; Mohamed, D.; Dowhanik, S.; Petrella, R.; Gregis, V.; Li, J.; Wu, L.; Gazzarrini, S. Spatiotemporal restriction of FUSCA3 expression by class I BPCs promotes ovule development and coordinates embryo and endosperm growth. *Plant Cell* **2020**, *32*, 1886–1904. [CrossRef] [PubMed]
24. Sun, Y.; Fan, Z.; Li, X.; Liu, Z.; Li, J.; Yin, H. Distinct double flower varieties in *Camellia japonica* exhibit both expansion and contraction of C-class gene expression. *BMC Plant Biol.* **2014**, *14*, 288. [CrossRef] [PubMed]
25. Li, X.; Li, J.; Fan, Z.; Liu, Z.; Tanaka, T.; Yin, H. Global gene expression defines faded whorl specification of double flower domestication in Camellia. *Sci. Rep.* **2017**, *7*, 3197. [CrossRef]
26. Letunic, I.; Doerks, T.; Bork, P. SMART 6: Recent updates and new developments. *Nucleic Acids Res.* **2009**, *37* (Suppl. 1), D229–D232. [CrossRef]
27. Kumar, S.; Stecher, G.; Li, M.; Knyaz, C.; Tamura, K. MEGA X: Molecular evolutionary genetics analysis across computing platforms. *Mol. Biol. Evol.* **2018**, *35*, 1547. [CrossRef]
28. Bailey, T.L.; Johnson, J.; Grant, C.E.; Noble, W.S. The MEME suite. *Nucleic Acids Res.* **2015**, *43*, W39–W49. [CrossRef]
29. Yu, C.S.; Lin, C.J.; Hwang, J.K. Predicting subcellular localization of proteins for Gram-negative bacteria by support vector machines based on n-peptide compositions. *Protein Sci.* **2004**, *13*, 1402–1406. [CrossRef]
30. Yu, C.S.; Chen, Y.C.; Lu, C.H.; Hwang, J.K. Prediction of protein subcellular localization. *Proteins Struct. Funct. Bioinform.* **2006**, *64*, 643–651. [CrossRef]
31. Livak, K.J.; Schmittgen, T.D. Analysis of relative gene expression data using real-time quantitative PCR and the $2^{-\Delta\Delta CT}$ method. *Methods* **2001**, *25*, 402–408. [CrossRef]
32. Clough, S.J.; Bent, A.F. Floral dip: A simplified method for Agrobacterium-mediated transformation of *Arabidopsis thaliana*. *Plant J.* **1998**, *16*, 735–743. [CrossRef] [PubMed]
33. Yamaguchi, N.; Winter, C.M.; Wu, M.-F.; Kwon, C.S.; William, D.A.; Wagner, D. PROTOCOLS: Chromatin immunoprecipitation from Arabidopsis tissues. *Arab. Book/Am. Soc. Plant Biol.* **2014**, *12*, e0170. [CrossRef] [PubMed]
34. Park, H.M. *Comparing Group Means: T-Tests and One-Way ANOVA Using Stata, SAS, R, and SPSS*; Indiana University: Bloomington, IN, USA, 2009.
35. Hu, Z.; Lyu, T.; Yan, C.; Wang, Y.; Ye, N.; Fan, Z.; Li, X.; Li, J.; Yin, H. Identification of alternatively spliced gene isoforms and novel noncoding RNAs by single-molecule long-read sequencing in Camellia. *RNA Biol.* **2020**, *17*, 966–976. [CrossRef] [PubMed]
36. Yang, C.; Bratzel, F.; Hohmann, N.; Koch, M.; Turck, F.; Calonje, M. VAL-and AtBMI1-mediated H2Aub initiate the switch from embryonic to postgerminative growth in Arabidopsis. *Curr. Biol.* **2013**, *23*, 1324–1329. [CrossRef]
37. Molitor, A.M.; Bu, Z.; Yu, Y.; Shen, W.-H. Arabidopsis AL PHD-PRC1 complexes promote seed germination through H3K4me3-to-H3K27me3 chromatin state switch in repression of seed developmental genes. *PLoS Genet.* **2014**, *10*, e1004091. [CrossRef]
38. Berger, N.; Dubreucq, B. Evolution goes GAGA: GAGA binding proteins across kingdoms. *Biochim. Biophys. Acta-Gene Regul. Mech.* **2012**, *1819*, 863–868. [CrossRef]
39. Liu, Y.; Xu, Y.; Jia, X. The complete chloroplast genome of *Camellia osmantha*, an edible oil Camellia. *Mitochondrial DNA. Part B Resour.* **2021**, *6*, 3169–3170. [CrossRef]
40. Gong, R.; Cao, H.; Zhang, J.; Xie, K.; Wang, D.; Yu, S. Divergent functions of the GAGA-binding transcription factor family in rice. *Plant J.* **2018**, *94*, 32–47. [CrossRef]
41. Mozgova, I.; Köhler, C.; Hennig, L. Keeping the gate closed: Functions of the polycomb repressive complex PRC 2 in development. *Plant J.* **2015**, *83*, 121–132. [CrossRef]
42. Dennis, L.; Peacock, J. Genes directing flower development in Arabidopsis. *Plant Cell* **2019**, *31*, 1192–1193. [CrossRef]
43. Thomson, B.; Wellmer, F. Molecular regulation of flower development. *Curr. Top. Dev. Biol.* **2019**, *131*, 185–210. [PubMed]
44. Zhou, Y.; Wang, Y.; Krause, K.; Yang, T.; Dongus, J.A.; Zhang, Y.; Turck, F. Telobox motifs recruit CLF/SWN–PRC2 for H3K27me3 deposition via TRB factors in Arabidopsis. *Nat. Genet.* **2018**, *50*, 638–644. [CrossRef]
45. Sun, Y.; Fan, Z.; Li, X.; Li, J.; Yin, H. The APETALA1 and FRUITFUL homologs in *Camellia japonica* and their roles in double flower domestication. *Mol. Breed.* **2014**, *33*, 821–834. [CrossRef]

current issues in molecular biology

MDPI

Article

Altering Stomatal Density for Manipulating Transpiration and Photosynthetic Traits in Rice through CRISPR/Cas9 Mutagenesis

Sakthi Ambothi Rathnasamy [1], Rohit Kambale [1], Allimuthu Elangovan [1], Williams Mohanavel [1], Priyanka Shanmugavel [2], Gowtham Ramasamy [2], Senthil Alagarsamy [3], Rajavel Marimuthu [3], Veera Ranjani Rajagopalan [1], Sudha Manickam [1], Valarmathi Ramanathan [4], Raveendran Muthurajan [1,*] and Geethalakshmi Vellingiri [2,*]

[1] Department of Plant Biotechnology, Tamil Nadu Agricultural University, Coimbatore 641 003, Tamil Nadu, India
[2] Agro-Climatology Research Centre, Tamil Nadu Agricultural University, Coimbatore 641 003, Tamil Nadu, India
[3] Department of Crop Physiology, Tamil Nadu Agricultural University, Coimbatore 641 003, Tamil Nadu, India
[4] Sugarcane Breeding Institute, Coimbatore 641 007, Tamil Nadu, India
* Correspondence: raveendrantnau@gmail.com or raveendran.m@tnau.ac.in (R.M.); geetha@tnau.ac.in (G.V.)

Abstract: Stomata regulates conductance, transpiration and photosynthetic traits in plants. Increased stomatal density may contribute to enhanced water loss and thereby help improve the transpirational cooling process and mitigate the high temperature-induced yield losses. However, genetic manipulation of stomatal traits through conventional breeding still remains a challenge due to problems involved in phenotyping and the lack of suitable genetic materials. Recent advances in functional genomics in rice identified major effect genes determining stomatal traits, including its number and size. Widespread applications of CRISPR/Cas9 in creating targeted mutations paved the way for fine tuning the stomatal traits for enhancing climate resilience in crops. In the current study, attempts were made to create novel alleles of *OsEPF1* (Epidermal Patterning Factor), a negative regulator of stomatal frequency/density in a popular rice variety, ASD 16, using the CRISPR/Cas9 approach. Evaluation of 17 T_0 progenies identified varying mutations (seven multiallelic, seven biallelic and three monoallelic mutations). T_0 mutant lines showed a 3.7–44.3% increase in the stomatal density, and all the mutations were successfully inherited into the T_1 generation. Evaluation of T_1 progenies through sequencing identified three homozygous mutants for one bp insertion. Overall, T_1 plants showed 54–95% increased stomatal density. The homozygous T_1 lines (# E1-1-4, # E1-1-9 and # E1-1-11) showed significant increase in the stomatal conductance (60–65%), photosynthetic rate (14–31%) and the transpiration rate (58–62%) compared to the nontransgenic ASD 16. Results demonstrated that the genetic alterations in *OsEPF1* altered the stomatal density, stomatal conductance and photosynthetic efficiency in rice. Further experiments are needed to associate this technology with canopy cooling and high temperature tolerance.

Keywords: rice; *OsEPF1*; stomatal density; photosynthetic efficiency

Citation: Rathnasamy, S.A.; Kambale, R.; Elangovan, A.; Mohanavel, W.; Shanmugavel, P.; Ramasamy, G.; Alagarsamy, S.; Marimuthu, R.; Rajagopalan, V.R.; Manickam, S.; et al. Altering Stomatal Density for Manipulating Transpiration and Photosynthetic Traits in Rice through CRISPR/Cas9 Mutagenesis. *Curr. Issues Mol. Biol.* **2023**, *45*, 3801–3814. https://doi.org/10.3390/cimb45050245

Academic Editors: Angel Llamas and Sung-Kun (Sean) Kim

Received: 27 December 2022
Revised: 12 February 2023
Accepted: 28 February 2023
Published: 30 April 2023

1. Introduction

Global population is expected to cross 9.8 billion by 2050, which necessitates the doubling of food production. Being a staple cereal food crop, rice is consumed by more than 50% of the global population. Hence, increasing the production of rice plays a key role in achieving global food security [1]. Rice yield has undergone two major leaps in the past viz. first, during the 1960s, i.e., the green revolution when the semi-dwarfing gene was introduced, and secondly, through the introduction of hybrids during the 1980s [2,3]. Afterwards, no major advances have been reported in rice yield. Drought and high-temperature events are fast becoming major threats to increasing rice productivity under marginal environments. Any further increase in the rice yields must overcome major challenges viz yield plateau, declining land, labour, water resources and finally, the increasing effects of climate change. The climate projections indicate an increased frequency in the occurrence of extreme weather events, thus affecting the rice yields significantly [4–6]. Hence, it has become important to develop the 'climate-smart' rice varieties that can alleviate the risk of crop failures due to extreme weather events [1,7,8].

Drought remains at the top in affecting rice productivity under both irrigated as well as rainfed conditions [6,9,10], followed by salinity [11–13] and flooding [14,15]. During recent years, high temperature has become a major yield-limiting factor, especially during flowering and grain-filling stages [16]. Substantial progress has been made in unravelling the genetic basis of drought tolerance traits and yield under drought conditions. However, only limited progress has been made in dissecting the physiological and genetic basis of high temperature tolerance traits in rice. A survey of literature indicated that apart from tissue tolerance and avoidance traits, canopy cooling may serve as a potential option to mitigate heat-induced yield losses. Genes involved in stomatal development and patterning have been functionally validated through transgenic and genome editing approaches [17–21]. Altering the stomatal density was proved to alter both stomatal conductance as well as CO_2 assimilation rate, thus affecting the growth and metabolism of plants [17,22]. The genes belonging to the EPIDERMAL PATTERNING FACTOR-LIKE (EPFL) family (EPF1, EPF2 and EPFL9/STOMAGEN) encode for small secretory mobile peptides regulating stomatal development and its patterning [23]. As reported earlier, it is possible to alter the stomatal density by manipulating the expression patterns of EPF1, EPF2 and/or EPFL9 to attain the desired photosynthetic/transpiration traits [24–26]. In rice, the overexpression of the negative regulator *OsEPF1* reduces the stomatal density and the corresponding stomatal conductance [24,26]. Rice lines with reduced stomatal density exhibited enhanced water use efficiency and performed better under dry conditions [24,27]. On the other hand, the overexpression of the positive regulator *OsEPF9* increases the stomatal density [28], and knockout of *OsEPFL10* showed reduction in the stomatal density without any impact on stomatal conductance or carbon assimilation [27].

This may require development of a suitable genetic material harbouring increased/decreased stomatal distribution which can be used for analyzing the influence of stomatal traits on improving the photosynthetic efficiency and yield traits. Outcomes of evaluation revealed that the knockout mutants of *OsEPF1* showed increased stomatal density which, in turn, enhanced the stomatal conductance and photosynthetic efficiency.

2. Materials and Methods

2.1. Plant Materials

Seedlings of a popular rice variety ASD 16 were raised at Paddy Breeding Station, Tamil Nadu Agricultural University, Coimbatore, Tamil Nadu, India and transplanted after 21 days and maintained using prescribed procedures. Panicles were harvested at 12–15 days after flowering, and immature embryos were isolated and used for genetic transformation.

2.2. Construction of CRISPR/Cas9 Expression Vectors

Nucleotide sequence of *OsEPF1* (LOC_Os04g54490) was retrieved from the Ensembl Plants database (https://plants.ensembl.org/index.html (accessed on 27 December 2022)) The synthetic guide RNA that targets the region before the signal peptide cleavage site of the first exon (5′-GCTACTCTTGCTGACGCCCG-3′) was designed using the CRISPOR online tool (http://crispor.tefor.net/ (accessed on 27 December 2022)) [29]. The oligonucleotides were synthesized with adapter sequences 5′-GGCA-3′ attached to the forward primer and 5′-GGCA-3′ attached to the reverse primer so that the oligo-duplex can anneal and ligate with *Bsa*I digested pRGEB31 plasmid (Addgene plasmid # 51295) [30]. The pRGEB31 plasmid harbours guide RNA scaffold region and Cas9 gene, driven by Os U3 promoter and CaMV35S promoter, respectively. The oligonucleotides were annealed for duplex formation, and the sgRNA with sticky ends of *Bsa*I enzyme was cloned into *Bsa*I-digested pRGEB31 vector. The cloned sgRNA sequence was then confirmed by the Sanger sequencing method and mobilized into *Agrobacterium tumifaciens* strain LBA4404 using the freeze-thaw method [31]. *Agrobacterium* colonies were then screened using vector-specific primers (M13 reverse primer and Guide RNA reverse primer, Table 1) and used for transformation.

Table 1. List of the primers used in the study.

Name of the Construct/Organism Name of the Gene/	Selection Marker	Forward and Reverse Primers	Annealing Temperature	Size of the Amplicon
Agrobacterium tumefaciens harboring pRGEB31-*EPF1*-sgRNA	*CaMV35S* and *hpt*	hpt–F: 5′ TACACAGCCATCGGTCCA 3′ CaMV35S -R: 5′ ACCTCCTCGGATTCCATTGC 3′	59 °C	1.3 kb
E. coli harboring pRGEB31-*EPF1*-sgRNA	*M13* and *EPFs* gRNA	*M13* -R: 5′ TCACACAGGAAACAGCTATG 3′ *EPF* sgRNA *R*- AAACCGGGCGTCAGCAAGAGTAGCG	55 °C	445 bp
Oryza sativa (ASD 16)	*CaMV35S* and *hpt*	hpt–F: 5′ TACACAGCCATCGGTCCA 3′ 59 °C 1.3 kb CaMV35S -R: 5′ ACCTCCTCGGATTCCATTGC 3′	59 °C	1.3 kb
	Gene specific primer *EPF1*	EPF F- CAATGGCTGCACACACATATAC EPFR-CAAGCAAGCACATTCGGTAAG	58 °C	400 bp

2.3. *Agrobacterium*-Mediated Transformation

Immature seeds dissected from the panicles harvested at 12–15 days after flowering were surface sterilized with 1.5% sodium hypochlorite along with a drop of Tween 20 for 3 min, followed by 70% ethanol treatment for 1 min and washed 4–5 times using sterile water. Isolated immature embryos were used for co-cultivation by the *Agrobacterium*-mediated transformation method [32]. Briefly, the isolated immature embryos were pretreated by incubating in a water bath at 42 °C for 30 min, followed by centrifugation at 1100× *g* for 10 min, after cooling the tube in ice for 1 min. Then, the embryos were transferred to the co-cultivation medium and infected with 5 µL of 3-day-old *Agrobacterium* culture. After co-cultivation, the proliferated calli were separated from the shoots and transferred to the first resting medium for 5 days and then to the second resting medium for 10 days. The calli proliferated in the resting medium were subjected to selection (twice) using 50 mg/L hygromycin and 250 mg/L cefotoxime. The selected calli were then regenerated using 1 mg/L NAA and 3 mg/L BAP with glutamine (30 mg/L). Then, the regenerated shoots were rooted in $\frac{1}{2}$ MS medium containing 30 mg/L hygromycin. The putative transgenic plants were then hardened in green house facility available at the Centre for Plant Molecular Biology and Biotechnology, Tamil Nadu Agricultural University, Coimbatore.

2.4. Molecular Characterization of Transgenic Plants

The genomic DNA was isolated from both the transgenic and nontransgenic plants of ASD 16 using CTAB method [33] and used for PCR analysis. All the plants were screened using vector-specific primers (*hpt* forward primer and CaMV reverse primer), and the PCR-positive transgenic plants were analysed for the presence of mutation(s) through sequencing the target region. Table 1 shows a list of primers used for screening the transgenic plants and sequencing. The sequencing data of the target region was analysed using the ICE v3 CRISPR analysis tool (https://ice.synthego.com/ (accessed on 27 December 2022)) and DSDecode [34].

2.5. Measuring Stomatal Density and Stomatal Size

In order to measure the stomatal density, a fully expanded leaf was selected, and both abaxial and adaxial surfaces were cleaned using a tissue paper. Then, the stomatal impressions were made from the middle of the leaves using a transparent nail polish. Adequate care was taken to prepare impressions on both the sides of the midrib so as to avoid the bias in measuring the stomata number. After 10 min, the stomatal imprints were gently peeled off and immediately mounted on the glass slide. The number of stomata in both abaxial and adaxial surfaces of the leaf was counted using the images captured at 40× magnification, in a field of 0.2897 mm × 0.2184 mm = 0.06327 mm^{-2} using a fluorescence microscope (Radical RXLr-5, Ambala Cantt, India). The stomatal density (SD) was calculated using the formula: stomatal density = number of stomata/area in view. A total of ten fields of view were randomly observed in each impression. In each field view, a total of 10 stomata were randomly selected and used for recording the length and width of the stomata. Stomata size (SS) was calculated using the formula: $SS = \pi \times a \times b$, where a and b denote the semi-major and semi-minor axes, respectively.

2.6. Measurement of Gas Exchange Parameters

A portable photosynthesis analyser (LCpro SD, ADC BioScientific, Hoddesdon, UK) was used for the measurement of all the gas exchange parameters, such as the photosynthetic rate, stomatal conductance and the transpiration rate. All the measurements were recorded at the middle of the fully expended 3rd leaf. All the measurements were taken after the instrument reached the stability (2–10 min). The temperature maintained in the chamber was 28 °C, and the flow rate was 500 mL/min at a relative humidity of 60–65%. The CO_2 concentration was 400 µmol mol^{-1} and the PPFD (Photosynthetic Photon Flux Density) was 1500 µmol m^{-1} s^{-1}. The parameters measured are as follows: photosynthetic rate (A, µmol CO_2 m^{-2} s^{-1}), stomata conductance (gs mol H_2O m^{-2} s^{-1}) and transpiration rate (E, mmol H_2O m^{-2} s^{-1}). All the measurements were taken under light-saturated conditions. The water use efficiency was derived from the measured photosynthetic parameters. The ratio of photosynthetic CO_2 assimilation (A) to transpiration rate (E) was calculated as the intrinsic water use efficiency.

2.7. Statistical Analysis

All of the required statistical analysis (ANOVA and DMRT) was performed using the R-Studio. Correlation analysis was also performed by using the R package by adopting the Pearson correlation method.

3. Results

3.1. Cloning of gRNA Targeting OsEPF1 and the Generation of Gene-Edited Rice Lines

The nucleotide sequence, encoding *OsEPF1* (LOC_Os04g54490), was retrieved from the Ensembl Plants database. The *OsEPF1* gene contains three exons separated by an intron (Figure 1a). After the in silico evaluation of the target sequence, a small-guide RNA (sgRNA) targeting the exon1 of *OsEPF1* gene (LOC_Os04g54490) was designed using the CRISPOR online tool ((http://crispor.tefor.net/ (accessed on 27 December 2022)). The DNA oligos were synthesized along with forward (5′-GGCA-3′) and reverse (5′-AAAC-3′) adapter sequences (Figure 1a) and were cloned into pRGEB31 under the control of OsU3 promoter (Figure 1b,c). This plasmid was mobilized into *Agrobacterium* and used to transform the popular rice variety i.e., ASD 16.

About 175 immature embryos of the rice variety ASD 16 transformed using pRGEB31, harboring Cas9- *OsEPF1* gRNA, were subjected to two rounds of hygromycin selection. Putative transformed embryos survived after hygromycin selection were transferred to pre-regeneration followed by regeneration (refer to the Supplementary Figure S1). A total of 17 T_0 progenies were regenerated, and PCR analysis using vector-specific and gene-specific primers revealed that all the 17 T_0 lines were positive (Figure 1d).

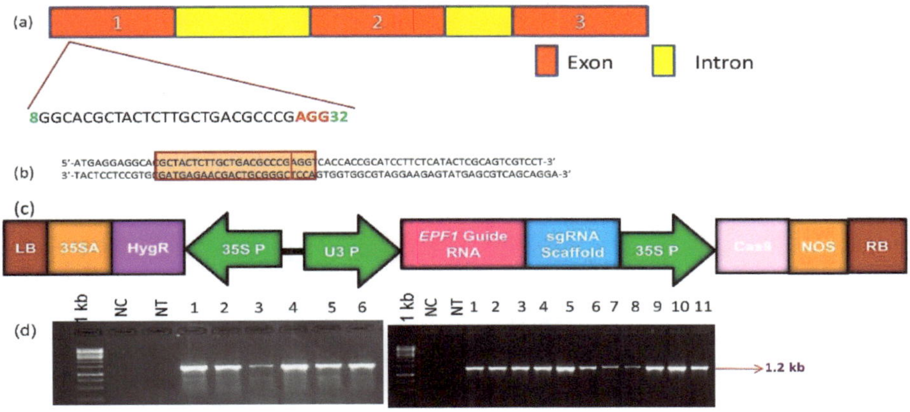

Figure 1. Development of the rice lines harboring mutations in *OsEPF1*: (**a**) Structure of *OsEPF1* (LOC_Os04g54490 on Chromosome 4) showing three exons and two introns; (**b**) The highlighted sequence shown in the red box represents the guide RNA, whereas the inner box represents the PAM; (**c**) T-DNA cassette of pRGEB31 harboring *OsEPF1* guide RNA. The expression of guide RNA scaffold is driven by OsU3 small nuclear RNA promoter; Cas9 gene is driven by CaMV35S promoter; hygromycin gene is driven by enhanced CaMV35S promoter; LB-left border RB-right border of T-DNA (**d**) PCR screening of transgenic plants (1–6 in the left panel and 1–11 in the right panel) using vector-specific primers (Table 1) 1 kb, DNA ladder, NC, negative control and \NT, Non transgenic ASD 16.

3.2. Characterization of T_0 Lines

The genomic DNA, isolated from the gene-edited transgenic lines, were subjected to PCR amplification using primers flanking the target site, and the amplicons were subjected to Sanger sequencing. Non-transgenic (NT) ASD 16 was used as a reference. Outcomes of sequence analysis confirmed the presence of targeted mutation(s) in the 1st exon of *OsEPF1* across all the T_0 lines (Figure 2). The ICE V2 CRISPR analysis identified seven multiallelic (E1-3, E 1-7, E 1-8, E1-9 and E2-4), seven biallelic (E1-1, E1-2, E1-4, E1-5, E1-6, E2-2 and E2-6) and three monoallelic mutations (E2-2, E2-3 and E2-5) (Table 2 and Supplementary Figure S2). T_0 lines were found to harbour five different types of deletion mutations in the range of −1 bp to −7 bp and two types of insertion (+1 and +2 allele) mutations.

Table 2. Mutations observed in the *OsEPF1* gene-edited transgenic lines (T_0) of ASD 16.

Line No	Mutation Type	Mutation Allele Size
E1-1	Biallelic	+1, −1
E1-2	Biallelic	+1, −1
E1-3	Multiallelic	+1, −7, −7
E1-4	Biallelic	+1, −1
E1-5	Biallelic	+1, −1
E1-6	Biallelic	+1, −1
E1-7	Multiallelic	−3, 0, −2, −3
E1-8	Multiallelic	−3, 0, −2, −3
E1-9	Triallelic	+1, −1, 0
E2-1	Monoallelic	+1
E2-2	Biallelic	+1, +2
E2-3	Monoallelic	+1
E2-4	Triallelic	+1, −1, 0
E2-5	Monoallelic	+1
E2-6	Biallelic	+1, −1
E2-7	Multiallelic	+1, −1, 0, −5
E2-8	Multiallelic	+1, −1, 0, −5

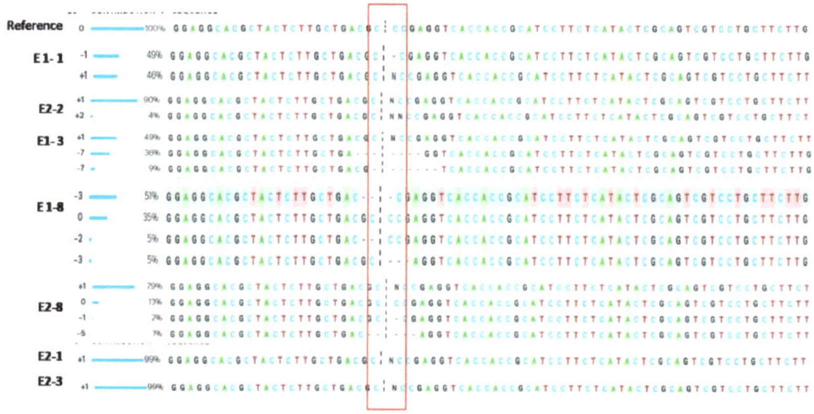

Figure 2. Sequence analysis of targeted region among the gene-edited lines (T_0) using ICE CRISPR tool. Putative mutations were detected upon comparison against the reference sequence i.e., non-transgenic ASD 16 on the top. Sequence analysis of events E2-1 and E2-3 showed monoallelic mutation; E1-1 and E2-2 showed biallelic mutation, and E1-3, E1-8 and E2-18 showed multiallelic mutation. The vertical, black-dotted line represents the cut site of cas9 whereas the horizontal black hyphen denotes deletions and N indicates the insertion.

3.3. Phenotypic Evaluation of T_0 Mutant Lines for Stomatal Density

All the 17 T_0 lines were evaluated for stomatal density (SD) along with the nontransgenic ASD 16. Six lines showed significantly higher SD in both abaxial and adaxial surface than the nontransgenic ASD 16 (Figure 3). The line #E1-1 showed maximum (44.3%) increase in the abaxial SD followed by the lines viz. E1-5, E2-2, E1-3, E2-6, E1-6, E1-2, E2-1 and E1-4 with a 3.7 to 33.7% increase. Overall, the stomatal density of the adaxial leaf surface was found to be lower than the abaxial surface. The mutant line #E1-1 showed significantly higher abaxial and adaxial stomatal density compared to the nontransgenic ASD 16. Mutant lines #E1-7, E1-8, and E2-5 have displayed reduced stomatal density accompanied by decreased row density and increased stomatal size.

3.4. Phenotypic Evaluation of T_1 Progenies

The inheritance pattern of the altered stomatal traits was studied in the T_1 generation. The mutant line #E1-1 harbouring a biallelic mutation (+1/−1) and significantly higher stomatal density in the T_0 generation was forwarded to the T_1 generation. Six progenies (# E1-1-1, # E1-1-3, # E1-1-4, # E1-1-6, # E1-1-9 and # E1-1-11) were analysed against the non-transgenic ASD 16. Sequence analysis identified three lines, namely, E1-1-4, #E1-1-9 and #E1-1-11 showing homozygous mutation (+1 bp insertion) and remaining three mutant lines (#E1-1-1, #E1-1-3 and #E1-1-6) showing the mutation(s) under heterozygous mutation (+1/−1) (Figure 4; Supplementary Figure S3).

All the three homozygous T_1 mutant lines i.e., #E1-1-4, #E1-1-9 and #E1-1-11 were found to possess significantly higher number of stomata (Figure 5). The line #E1-1-11 was found to possess 95% increased stomatal density in the abaxial surface than the nontransgenic ASD 16. This increase in the stomatal density was accompanied by a corresponding reduction in the stomatal size. Mutant lines showed a 50–64% reduction in the size of stomata in the abaxial surface and 35–48% reduction in the size of stomata in the adaxial surface (Table 3). Mutants had an increased number of stomatal files leading to increased stomatal density.

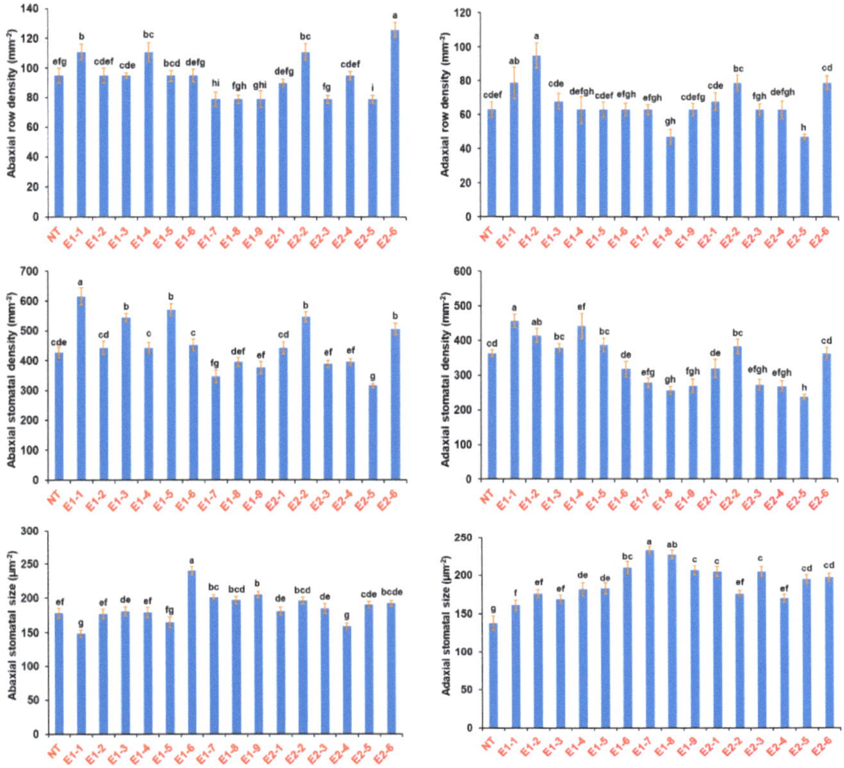

Figure 3. Adaxial and abaxial row density, stomatal density and stomatal size among the putative transgenic lines (T_0) compared to the non-transgenic ASD 16 (NT). Gene edited lines showed altered stomatal parameters on both adaxial and abaxial side of leaves. Each data point represents an average of ten replications and error bar represents ± standard error of the mean data. Statistical significance was tested using Duncan's multiple range test (DMRT). Different alphabets indicate the significance at $p < 0.05$.

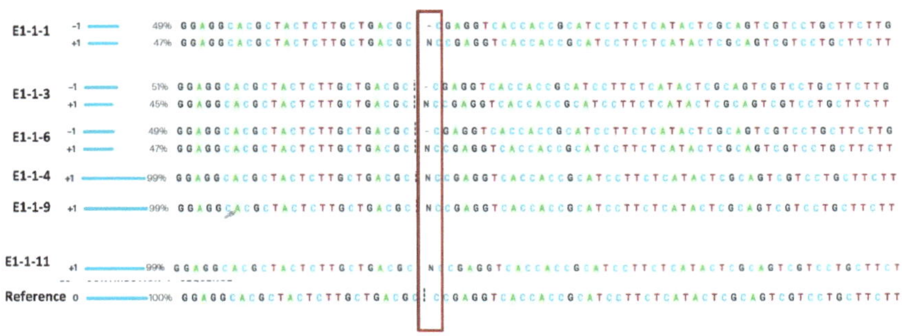

Figure 4. Sequence analysis of the target region among the T_1 progenies of # E1-1 using ICE CRISPR tool. Mutants #E1-1-4, E1-1-9, E1-1-11 showed homozygous monoallelic mutation and the progenies E1-1-1, E1-1-3, E1-1-6 showed biallelic mutation on sequence analysis. The vertical, black-dotted line represents the cut site of cas9 whereas the horizontal black hyphen denotes the deletions and N indicates the insertion.

Table 3. Stomatal density and percent increase in adaxial and abaxial stomatal densities of the homozygous T_1 progenies of *OsEPF1* gene-edited plants compared against the nontransgenic ASD 16. Data of stomatal density and stomatal size represented as ±standard error; each value represents the mean of 10 replications and different letters indicate the significance at $p < 0.05$.

Event	Abx_SD_mm	(%) Increase in Abaxial_SD (mm^{-2})	Adx_SD_mm	(%) Increase in Adaxial_SD (mm^{-2})	Abaxial Stomatal Size (μm^2)	Adaxial Stomatal Size (μm^2)
NTL	456 ± 10.7^{c}	-	391 ± 9.98^{b}	-	219.8 ± 5.37^{a}	197.2 ± 6.6^{a}
E1-1-4	706 ± 16.11^{b}	54	557 ± 19.14^{a}	42.5	109.9 ± 2.53^{b}	127.1 ± 2.8^{b}
E1-1-9	701 ± 19.82^{b}	53.7	391 ± 17.02^{b}	0	93.2 ± 2.18^{c}	102.6 ± 2.78^{c}
E1-1-11	891 ± 21.26^{a}	95	543 ± 14.29^{a}	38.8	79.5 ± 2.47^{c}	108.2 ± 2.98^{c}

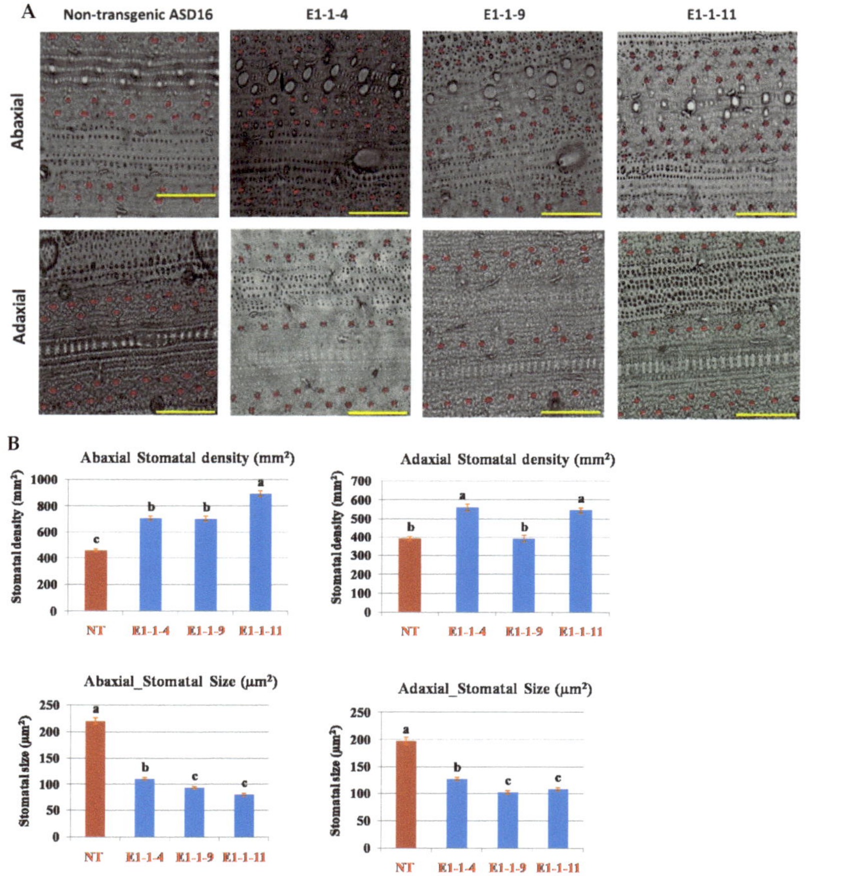

Figure 5. *OsEPF1* edited mutants showed altered stomatal density and stomatal size. (**A**). Stomatal distribution in the T_1 progenies of *OsEPF1* edited lines against its wild type ASD 16. Homozygous mutants showed increased stomatal density and altered stomatal file density compared to non-transgenic ASD 16. The clustered stomata showed reduction in stomatal size. The top row represents the stomatal density in abaxial side, whereas the bottom row represents the stomatal density in adaxial leaf surface. Scale bar = 100 µm. (**B**). Graphical representation of stomatal density and stomatal size in the T_1 transgenic lines compared against the non-transgenic ASD 16 (NT). Each data point represents an average of 10 replications and error bar represents ± standard error of the mean data. Statistical significance was tested using Duncan's multiple range test (DMRT). Different alphabets indicate the significance at $p < 0.05$.

3.5. Physiological Evaluation of the T_1 Mutant Lines

The T_1 mutant progenies (#E1-1-4, #E1-1-9 and #E1-1-11) harbouring homozygous mutation (+1 bp) were evaluated for stomatal conductance (g_s), photosynthetic rate (A) and the transpiration rate (E) (Figure 6). The mutant (T_1) lines showed a 60–65% increased stomatal conductance than the nontransgenic ASD 16. The line #E1-1-11 recorded the maximum increase in its stomatal conductance (65.2%) and photosynthetic rate (31%) over the nontransgenic ASD 16, followed by E1-1-4 and E1-1-9. Similarly, the transpiration rate was also found to be significantly higher in all three T_1 progenies i.e., 58–62% higher transpiration rate than NT ASD 16. The correlation analysis results showed a significant positive association between stomatal density and other parameters, such as g_s, A and E. Further, the stomatal density was found to be negatively correlated with stomatal size (Figure 7). The mutants showed 20–29% decrease in water use efficiency compared to nontransgenic ASD 16.

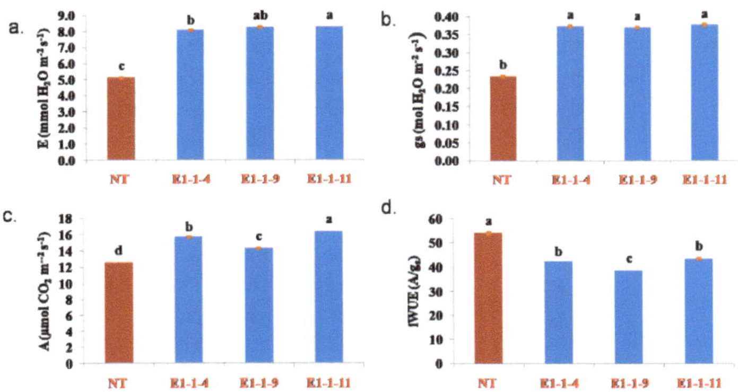

Figure 6. Evaluation of the homozygous mutant lines for stomatal conductance, photosynthetic rate and transpiration rate. NT, nontransgenic ASD 16; #E1-1-4, E1-1-9 and #E1-1-11 represented the *OsEPF1* gene edited lines. (**a**). transpiration rate; (**b**). stomatal conductance, (**c**). carbon dioxide assimilation rate (**d**). water use efficiency (A/gs). All values of stomatal density and stomatal size are mean ± standard error; each value represents the mean of ten replications and different letters indicates significance at $p < 0.05$ compared.

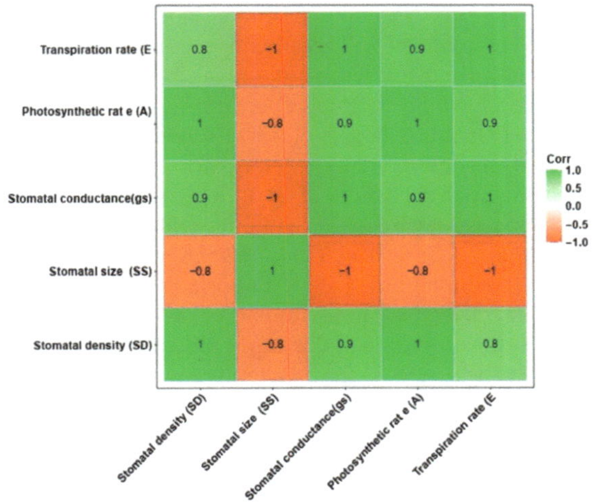

Figure 7. Correlation analysis between the stomatal traits and physiological parameters. Stomatal density showed positive correlation with stomatal conductance, photosynthetic rate and transpiration rate and negative correlation to stomatal size.

4. Discussion

Extreme weather events viz. drought vs. flood and high temperature vs. cold stresses are predicted to occur at a high frequency in the coming years due to rapidly changing climatic conditions. Genetic improvement of the staple cereal rice for enhanced tolerance to heat and drought stresses will help in sustaining rice production under future climatic conditions. Maintenance of water relations between the plant and soil is vital for the regulation of plants' responses to both drought and heat stresses. Stomata plays a vital role in regulating gas and water exchange in plants by facilitating the physiological processes, such as the transpiration, stomatal conductance and photosynthesis [35]. Precise genetic manipulation of the stomatal traits may help in altering the crops' responses to drought and high temperature. Any increase in the stomatal density enhances the stomatal conductance (gs) which, in turn, favours high evaporative cooling to overcome heat-induced damages [20,36,37]. This warrants a thorough understanding of the molecular events regulating stomatal traits, such as the stomatal files, density and size. At present, the knowledge generated from the species *Arabidopsis thaliana* is utilized for the genetic improvement of stomatal traits in other grass species [26,38–40]. Several investigations have been conducted to alter the stomatal traits (density and size) through functional genomics approaches. The EPIDERMAL PATTERNING FACTOR and EPIDERMAL PATTERNING FACTOR (LIKE) family of genes (*EPF1*, *EPF2* and *EPFL9*/STOMAGEN) were found to regulate the stomatal density [24–28,41,42]. The effect of altered stomatal density on the photosynthetic traits was shown to depend on plant species and the environmental conditions [43–45]. The transgenic *Arabidopsis thaliana* plants with reduced stomatal density exhibited a reduction in leaf transpiration. On the other hand, the *epf1*/*epf2* knock out mutants with increased stomatal density exhibited a high transpiration rate [22]. Plants with an increased transpiration rate recorded a reduction in leaf temperature than their controls. This revealed the role of stomatal density in regulating transpiration rate [22]. Generally, the plants with increased stomatal density exhibit enhanced stomatal conductance, carbon dioxide assimilation and photosynthetic rate [17,18]. Sakoda, et al. [42] attempted to manipulate the CO_2 diffusion by altering the stomatal density so as to enhance the photosynthetic capacity in plants, using *EPF1* knockout mutants. Several other studies have also revealed that the stomatal density is positively correlated with photosynthetic capacity and stomatal conductance [46,47]. The PATROL1 over-expression line with faster stomatal opening responses and the slac1 (slow anion channel-associated 1) and ost1 (open stomata 1) mutants with stay-open stomata all displayed greater photosynthetic rates and growth rate under variable light conditions than the wild type [48]. The improvement of stomatal response in fluctuating light conditions in a natural environment will greatly improve photosynthesis rate and yield.

In the current study, the authors used CRISPR/Cas9 approach to alter the stomatal density of a lowland-irrigated rice variety, ASD 16, through targeted suppression of *Os-EPF1*. The immature embryos of ASD 16 were genetically transformed using *OsEPF1* gRNA harbouring a genome editing vector which produced a total of 17 mutant progenies. The sequencing analysis results of 17 T_0 lines identified seven progenies with multi-allelic mutations, seven with biallelic and three with monoallelic mutations. The T_0 lines showed significant increase (3.7–44.3%) in their stomatal density compared to the nontransgenic ASD 16. The mutant line # E1-1, harboring a biallelic mutation (+1/−1), showed a maximum of 44.3% increase in its stomatal density. The mutants E1-7, E-8, and E2-5 showed increased stomatal size and decreased row density.

As per the previous reports, altering expression of *EPF1*, *EPF2* and *EPFL9* was reported to alter the stomatal density [24–26,42] and water use efficiency [27]. *EPF1* and *EPF2* were reported to function as negative regulators of the stomatal development, whereas *EPFL9* was shown to be a positive regulator [41,42]. The *Arabidopsis* lines that over-express the STOMAGEN/*EPFL9*- line (ST-OX) showed a 268.1% increase in the stomatal density [42]. The EPIDERMAL PATTERNING FACTOR 1 knockout line (*epf1*) of *Arabidopsis* showed a 46.5% increased stomatal density than the wild type [42].

The inheritance of mutation and its effect on stomatal characteristics was assessed in the T_1 generation. In the current study, evaluation of T1 progenies of the T_0 line # E1-1 identified three progenies, namely, # E1-1-4, # E1-1-9 and # E1-1-11 possessing +1 bp mutation under the homozygous condition. The mutant line E1-1-11 showed 95% increase in stomatal density. The homozygous mutants showed increased stomatal file density compared to nontransgenic ASD 16.

The homozygous progenies showed a greater increase in stomatal conductance (gs), photosynthetic rate (A) and transpiration rate (E) than the nontransgenic ASD 16. The mutant progenies showed a 60–65% increase in their stomatal conductance, 14–31% increase in their photosynthetic rate and a 58–62% increase in their transpiration rate. Among staple crops, rice is reported to possess small-sized stomata with high density [36]. The transgenic plants of IR 64, over-expressing *OsEPF1*, showed an increase in the canopy temperature (0.3 °C warmer) under adequately irrigated conditions [24]. The speed at which the stomata can respond to the changing environment is important to improve the short-term water use efficiency [36,49,50]. Pitaloka et al. [50] inferred that the small-sized stomata respond much faster than the larger ones.

An *Arabidopsis* mutant with increased stomatal density showed a 30% higher CO_2 assimilation rate than the wild type under high light intensities [18]. A similar condition is predicted under elevated CO_2 levels with decreased stomatal conductance since it limits the CO_2 fixation rate [51]. However, the crops with high density of stomata possess high gas exchange potential, an important trait to mitigate the effects of heat stress by transpiration-mediated cooling. Poor stomatal conductance may be a major cause of photosynthetic limitation. In order to overcome the insufficient photosynthesis as a result of low stomatal density, genome engineering may pave the way to increase the stomatal density by improving the stomatal conductance, photosynthetic efficiency as well as biomass production in field crops. In the present study, we have attempted to increase the stomatal conductance of rice by altering the stomatal density. The lines generated in the study will be further evaluated for their responses to high temperature stress conditions; hence, tolerant lines under diverse environmental conditions will be identified for future crop improvement programmes.

Supplementary Materials: The following supporting information can be downloaded at: https://www.mdpi.com/article/10.3390/cimb45050245/s1, Figure S1. *Agrobacterium*-mediated transformation of immature embryos of rice cultivar ASD16 with pRGEB31 vector harbouring *OsEPF1*-sgRNA. (a) Co-cultivated immature embryos in co-cultivation media; (b) Immature embryo derived calli after 7 days of co-cultivation; (c and d) Calli at first and second resting stage; (e) Calli at selection stage after 32 days of co-cultivation; (f) Calli at shoot regeneration satge; (g) Regenerated shoots under first rooting; (h) Second stage rooting; (i) Hardening of putative transgenic plants in portrays; Figure S2. Chromatograms of T0 *OsEPF1* gene edited lines: Sequencing results (chromatogram) showing the altered sequence at the target region in the genome edited lines compared to the non-transgenic control. For each sample the top chromatogram is of the edited line and the bottom one is of non-transgenic. The horizontal black underline represents the guide RNA sequence. The horizontal red underline indicates the PAM site. The vertical black dotted line represents the mutated site which is also highlighted using vertical red box; Figure S3. Sanger sequencing analysis of T_1 lines: Sequencing results (chromatogram) showing the altered sequence at the target region in the genome edited lines compared to the non-transgenic control. For each sample the top chromatogram is of the edited line and the bottom one is of non-transgenic. Homozygous mutant lines showed insertion (+1 bp) at 3 bp upstream of PAM site. The horizontal black underlined region represents the guide RNA sequence. The horizontal red underline is the PAM site. The vertical black dotted line represents the mutated site. Red box indicates the insertions (T). In each black rectangle, bottom lane represents the wild allele and the upper lane represents the corresponding edited line.

Author Contributions: Conceptualization, writing—review and editing and funding acquisition, G.V. and R.M. (Raveendran Muthurajan); methodology and writing—original draft, V.R., S.A.R. and R.K.; software, A.E. and V.R.R.; validation, A.E., W.M. and R.K.; formal analysis, S.A. and R.M. (Rajavel Marimuthu); investigation, W.M., A.E., P.S. and G.R.; resources, R.M. (Raveendran Muthurajan) and G.V.; data curation, V.R.R., A.E., S.A.R., P.S. and G.R.; visualization, S.A. and R.M. (Rajavel Marimuthu); supervision, S.M. and V.R., S.M.; project administration, V.R. and R.M. (Raveendran Muthurajan); All authors have read and agreed to the published version of the manuscript.

Funding: This research was funded by Department of Biotechnology, BT/PR25820.

Institutional Review Board Statement: Not applicable as this study does not involve any humans or animals.

Data Availability Statement: Not applicable.

Acknowledgments: Authors profusely thank www.addgene.org for providing the vector.

Conflicts of Interest: The authors declare no conflict of interest.

References

1. Gadal, N.; Shrestha, J.; Poudel, M.N.; Pokharel, B. A review on production status and growing environments of rice in Nepal and in the world. *Arch. Agric. Environ. Sci.* **2019**, *4*, 83–87. [CrossRef]
2. Yadav, S.; Anand, S. Green Revolution and Food Security in India: A Review. *Natl. Geogr. J. India* **2022**, *65*, 312–323.
3. Singh, R. Environmental consequences of agricultural development: A case study from the Green Revolution state of Haryana, India. *Agric. Ecosyst. Environ.* **2000**, *82*, 97–103. [CrossRef]
4. Challinor, A.J.; Watson, J.; Lobell, D.B.; Howden, S.; Smith, D.; Chhetri, N. A meta-analysis of crop yield under climate change and adaptation. *Nat. Clim. Chang.* **2014**, *4*, 287–291. [CrossRef]
5. Calanca, P.P. Effects of abiotic stress in crop production. In *Quantification of Climate Variability, Adaptation and Mitigation for Agricultural Sustainability*; Springer: Berlin/Heidelberg, Germany, 2017; pp. 165–180.
6. Serraj, R.; McNally, K.L.; Slamet-Loedin, I.; Kohli, A.; Haefele, S.M.; Atlin, G.; Kumar, A. Drought resistance improvement in rice: An integrated genetic and resource management strategy. *Plant Prod. Sci.* **2011**, *14*, 1–14. [CrossRef]
7. Reay, D. Climate-Smart Rice. In *Climate-Smart Food*; Springer: Berlin/Heidelberg, Germany, 2019; pp. 121–133.
8. Singh, A.; Choudhury, B.; Bouman, B. Effects of rice establishment methods on crop performance, water use, and mineral nitrogen. In *Water-Wise Rice Production*; International Rice Research Institute: Los Baños, Philippines, 2002; pp. 237–246.
9. Elert, E. Rice by the numbers: A good grain. *Nature* **2014**, *514*, S50. [CrossRef]
10. Swain, P.; Raman, A.; Singh, S.; Kumar, A. Breeding drought tolerant rice for shallow rainfed ecosystem of eastern India. *Field Crops Res.* **2017**, *209*, 168–178. [CrossRef] [PubMed]
11. Quan, R.; Wang, J.; Hui, J.; Bai, H.; Lyu, X.; Zhu, Y.; Zhang, H.; Zhang, Z.; Li, S.; Huang, R. Improvement of salt tolerance using wild rice genes. *Front. Plant Sci.* **2018**, *8*, 2269. [CrossRef] [PubMed]
12. Chinnusamy, V.; Jagendorf, A.; Zhu, J.K. Understanding and improving salt tolerance in plants. *Crop Sci.* **2005**, *45*, 437–448. [CrossRef]
13. Radanielson, A.; Gaydon, D.; Li, T.; Angeles, O.; Roth, C. Modeling salinity effect on rice growth and grain yield with ORYZA v3 and APSIM-Oryza. *Eur. J. Agron.* **2018**, *100*, 44–55. [CrossRef]
14. Niroula, R.K.; Pucciariello, C.; Ho, V.T.; Novi, G.; Fukao, T.; Perata, P. SUB1A-dependent and-independent mechanisms are involved in the flooding tolerance of wild rice species. *Plant J.* **2012**, *72*, 282–293. [CrossRef] [PubMed]
15. Hendrawan, V.S.A.; Komori, D. Developing flood vulnerability curve for rice crop using remote sensing and hydrodynamic modeling. *Int. J. Disaster Risk Reduct.* **2021**, *54*, 102058. [CrossRef]
16. Lin, C.-J.; Li, C.-Y.; Lin, S.-K.; Yang, F.-H.; Huang, J.-J.; Liu, Y.-H.; Lur, H.-S. Influence of high temperature during grain filling on the accumulation of storage proteins and grain quality in rice (*Oryza sativa* L.). *J. Agric. Food Chem.* **2010**, *58*, 10545–10552. [CrossRef] [PubMed]
17. Tanaka, Y.; Sugano, S.S.; Shimada, T.; Hara-Nishimura, I. Enhancement of leaf photosynthetic capacity through increased stomatal density in Arabidopsis. *New Phytol.* **2013**, *198*, 757–764. [CrossRef] [PubMed]
18. SchluÈter, U.; Muschak, M.; Berger, D.; Altmann, T. Photosynthetic performance of an Arabidopsis mutant with elevated stomatal density (sdd1-1) under different light regimes. *J. Exp. Bot.* **2003**, *54*, 867–874. [CrossRef] [PubMed]
19. Wang, Y.; Xue, X.; Zhu, J.-K.; Dong, J. Demethylation of ERECTA receptor genes by IBM1 histone demethylase affects stomatal development. *Development* **2016**, *143*, 4452–4461. [CrossRef]
20. Hara, K.; Kajita, R.; Torii, K.U.; Bergmann, D.C.; Kakimoto, T. The secretory peptide gene EPF1 enforces the stomatal one-cell-spacing rule. *Genes Dev.* **2007**, *21*, 1720–1725. [CrossRef]
21. Hunt, L.; Gray, J.E. The signaling peptide EPF2 controls asymmetric cell divisions during stomatal development. *Curr. Biol.* **2009**, *19*, 864–869. [CrossRef]

22. Doheny-Adams, T.; Hunt, L.; Franks, P.J.; Beerling, D.J.; Gray, J.E. Genetic manipulation of stomatal density influences stomatal size, plant growth and tolerance to restricted water supply across a growth carbon dioxide gradient. *Philos. Trans. R. Soc. B Biol. Sci.* **2012**, *367*, 547–555. [CrossRef]
23. Bergmann, D.C.; Sack, F.D. Stomatal development. *Annu. Rev. Plant Biol.* **2007**, *58*, 163–181. [CrossRef]
24. Caine, R.S.; Yin, X.; Sloan, J.; Harrison, E.L.; Mohammed, U.; Fulton, T.; Biswal, A.K.; Dionora, J.; Chater, C.C.; Coe, R.A. Rice with reduced stomatal density conserves water and has improved drought tolerance under future climate conditions. *New Phytol.* **2019**, *221*, 371–384. [CrossRef] [PubMed]
25. Yin, X.; Biswal, A.K.; Dionora, J.; Perdigon, K.M.; Balahadia, C.P.; Mazumdar, S.; Chater, C.; Lin, H.-C.; Coe, R.A.; Kretzschmar, T. CRISPR-Cas9 and CRISPR-Cpf1 mediated targeting of a stomatal developmental gene EPFL9 in rice. *Plant Cell Rep.* **2017**, *36*, 745–757. [CrossRef] [PubMed]
26. Mohammed, U.; Caine, R.; Atkinson, J.; Harrison, E.; Wells, D.; Chater, C.; Gray, J.; Swarup, R.; Murchie, E. Rice plants overexpressing OsEPF1 show reduced stomatal density and increased root cortical aerenchyma formation. *Sci. Rep.* **2019**, *9*, 5584.
27. Karavolias, N.; Patel, D.; Seong, K.; Tjahjadi, M.; Gueorguieva, G.-A.; Tanaka, J.; Dahlbeck, D.; Cho, M.-J.; Niyogi, K.K.; Staskawicz, B.J. CRISPR/Cas9 knockout of EPFL10 reduces stomatal density while maintaining photosynthesis and enhancing water conservation in rice. *bioRxiv* **2021**. [CrossRef]
28. Hunt, L.; Bailey, K.J.; Gray, J.E. The signalling peptide EPFL9 is a positive regulator of stomatal development. *New Phytol.* **2010**, *186*, 609–614. [CrossRef]
29. Concordet, J.-P.; Haeussler, M. CRISPOR: Intuitive guide selection for CRISPR/Cas9 genome editing experiments and screens. *Nucleic Acids Res.* **2018**, *46*, W242–W245. [CrossRef] [PubMed]
30. Xie, K.; Yang, Y. RNA-guided genome editing in plants using a CRISPR–Cas system. *Mol. Plant* **2013**, *6*, 1975–1983. [CrossRef]
31. Weigel, D.; Glazebrook, J. Vectors and Agrobacterium hosts for Arabidopsis transformation. *Cold Spring Harb. Protoc.* **2006**, *2006*, pdb-ip29. [CrossRef]
32. Hiei, Y.; Komari, T. Agrobacterium-mediated transformation of rice using immature embryos or calli induced from mature seed. *Nat. Protoc.* **2008**, *3*, 824–834. [CrossRef]
33. Porebski, S.; Bailey, L.G.; Baum, B.R. Modification of a CTAB DNA extraction protocol for plants containing high polysaccharide and polyphenol components. *Plant Mol. Biol. Report.* **1997**, *15*, 8–15. [CrossRef]
34. Liu, W.; Xie, X.; Ma, X.; Li, J.; Chen, J.; Liu, Y.-G. DSDecode: A web-based tool for decoding of sequencing chromatograms for genotyping of targeted mutations. *Mol. Plant* **2015**, *8*, 1431–1433. [CrossRef] [PubMed]
35. Brownlee, C. Intracellular signalling: Sphingosine-1-phosphate branches out. *Curr. Biol.* **2001**, *11*, R535–R538. [CrossRef] [PubMed]
36. McAusland, L.; Vialet-Chabrand, S.; Davey, P.; Baker, N.R.; Brendel, O.; Lawson, T. Effects of kinetics of light-induced stomatal responses on photosynthesis and water-use efficiency. *New Phytol.* **2016**, *211*, 1209–1220. [CrossRef] [PubMed]
37. Kostaki, K.-I.; Coupel-Ledru, A.; Bonnell, V.C.; Gustavsson, M.; Sun, P.; McLaughlin, F.J.; Fraser, D.P.; McLachlan, D.H.; Hetherington, A.M.; Dodd, A.N. Guard cells integrate light and temperature signals to control stomatal aperture. *Plant Physiol.* **2020**, *182*, 1404–1419. [CrossRef] [PubMed]
38. Abrash, E.; Anleu Gil, M.X.; Matos, J.L.; Bergmann, D.C. Conservation and divergence of YODA MAPKKK function in regulation of grass epidermal patterning. *Development* **2018**, *145*, dev165860. [CrossRef] [PubMed]
39. Lu, J.; He, J.; Zhou, X.; Zhong, J.; Li, J.; Liang, Y.-K. Homologous genes of epidermal patterning factor regulate stomatal development in rice. *J. Plant Physiol.* **2019**, *234*, 18–27. [CrossRef]
40. Wu, Z.; Chen, L.; Yu, Q.; Zhou, W.; Gou, X.; Li, J.; Hou, S. Multiple transcriptional factors control stomata development in rice. *New Phytol.* **2019**, *223*, 220–232. [CrossRef] [PubMed]
41. Sugano, S.S.; Shimada, T.; Imai, Y.; Okawa, K.; Tamai, A.; Mori, M.; Hara-Nishimura, I. Stomagen positively regulates stomatal density in Arabidopsis. *Nature* **2010**, *463*, 241–244. [CrossRef]
42. Sakoda, K.; Yamori, W.; Shimada, T.; Sugano, S.S.; Hara-Nishimura, I.; Tanaka, Y. Higher stomatal density improves photosynthetic induction and biomass production in Arabidopsis under fluctuating light. *Front. Plant Sci.* **2020**, *11*, 589603. [CrossRef]
43. Schuler, M.L.; Sedelnikova, O.V.; Walker, B.J.; Westhoff, P.; Langdale, J.A. SHORTROOT-mediated increase in stomatal density has no impact on photosynthetic efficiency. *Plant Physiol.* **2018**, *176*, 757–772. [CrossRef]
44. Drake, P.L.; Froend, R.H.; Franks, P.J. Smaller, faster stomata: Scaling of stomatal size, rate of response, and stomatal conductance. *J. Exp. Bot.* **2013**, *64*, 495–505. [CrossRef] [PubMed]
45. Papanatsiou, M.; Amtmann, A.; Blatt, M.R. Stomatal spacing safeguards stomatal dynamics by facilitating guard cell ion transport independent of the epidermal solute reservoir. *Plant Physiol.* **2016**, *172*, 254–263. [CrossRef]
46. Franks, P.J.; W. Doheny-Adams, T.; Britton-Harper, Z.J.; Gray, J.E. Increasing water-use efficiency directly through genetic manipulation of stomatal density. *New Phytol.* **2015**, *207*, 188–195. [CrossRef] [PubMed]
47. McElwain, J.C.; Yiotis, C.; Lawson, T. Using modern plant trait relationships between observed and theoretical maximum stomatal conductance and vein density to examine patterns of plant macroevolution. *New Phytol.* **2016**, *209*, 94–103. [CrossRef] [PubMed]
48. Kimura, H.; Hashimoto-Sugimoto, M.; Iba, K.; Terashima, I.; Yamori, W. Improved stomatal opening enhances photosynthetic rate and biomass production in fluctuating light. *J. Exp. Bot.* **2020**, *71*, 2339–2350. [CrossRef] [PubMed]
49. Lawson, T.; Vialet-Chabrand, S. Chlorophyll fluorescence imaging. In *Photosynthesis*; Springer: Berlin/Heidelberg, Germany, 2018; pp. 121–140.

50. Pitaloka, M.K.; Caine, R.S.; Hepworth, C.; Harrison, E.L.; Sloan, J.; Chutteang, C.; Phuntong, C.; Nongngok, R.; Toojinda, T.; Ruengpayak, S. Induced Genetic Variations in Stomatal Density and Size of Rice Strongly Affect Water-Use Efficiency, Drought Tolerance, and Responses to Abiotic Stresses. *Front. Plant Sci.* **2021**, *13*, 801706. [CrossRef]
51. Sreeharsha, R.V.; Sekhar, K.M.; Reddy, A.R. Delayed flowering is associated with lack of photosynthetic acclimation in Pigeon pea (*Cajanus cajan* L.) grown under elevated CO_2. *Plant Sci.* **2015**, *231*, 82–93. [CrossRef]

MDPI AG
Grosspeteranlage 5
4052 Basel
Switzerland
Tel.: +41 61 683 77 34

Current Issues in Molecular Biology Editorial Office
E-mail: cimb@mdpi.com
www.mdpi.com/journal/cimb